D1259107

Lecture Notes in Physics

Springer

Berlin
Heidelberg
New York
Hong Kong
London
Milan
Paris
Tokyo

Physics and Astronomy

springeronline.com

The Editorial Policy for Edited Volumes

The series *Lecture Notes in Physics* (LNP), founded in 1969, reports new developments in physics research and teaching - quickly, informally but with a high degree of quality. Manuscripts to be considered for publication are topical volumes consisting of a limited number of contributions, carefully edited and closely related to each other. Each contribution should contain at least partly original and previously unpublished material, be written in a clear, pedagogical style and aimed at a broader readership, especially graduate students and nonspecialist researchers wishing to familiarize themselves with the topic concerned. For this reason, traditional proceedings cannot be considered for this series though volumes to appear in this series are often based on material presented at conferences, workshops and schools.

Acceptance

A project can only be accepted tentatively for publication, by both the editorial board and the publisher, following thorough examination of the material submitted. The book proposal sent to the publisher should consist at least of a preliminary table of contents outlining the structure of the book together with abstracts of all contributions to be included. Final acceptance is issued by the series editor in charge, in consultation with the publisher, only after receiving the complete manuscript. Final acceptance, possibly requiring minor corrections, usually follows the tentative acceptance unless the final manuscript differs significantly from expectations (project outline). In particular, the series editors are entitled to reject individual contributions if they do not meet the high quality standards of this series. The final manuscript must be ready to print, and should include both an informative introduction and a sufficiently detailed subject index.

Contractual Aspects

Publication in LNP is free of charge. There is no formal contract, no royalties are paid, and no bulk orders are required, although special discounts are offered in this case. The volume editors receive jointly 30 free copies for their personal use and are entitled, as are the contributing authors, to purchase Springer books at a reduced rate. The publisher secures the copyright for each volume. As a rule, no reprints of individual contributions can be supplied.

Manuscript Submission

The manuscript in its final and approved version must be submitted in ready to print form. The corresponding electronic source files are also required for the production process, in particular the online version. Technical assistance in compiling the final manuscript can be provided by the publisher's production editor(s), especially with regard to the publisher's own LaTeX macro package which has been specially designed for this series.

LNP Homepage (springerlink.com)

On the LNP homepage you will find:
−The LNP online archive. It contains the full texts (PDF) of all volumes published since 2000. Abstracts, table of contents and prefaces are accessible free of charge to everyone. Information about the availability of printed volumes can be obtained.
−The subscription information. The online archive is free of charge to all subscribers of the printed volumes.
−The editorial contacts, with respect to both scientific and technical matters.
−The author's / editor's instructions.

H.-T. Elze (Ed.)

Decoherence and Entropy in Complex Systems

Selected Lectures from DICE 2002

Springer

Editor

Hans-Thomas Elze
Universidade Federal do Rio de Janeiro
Instituto de Fisica, FIN
Caixa Postal 68528
21941-972 Rio de Janeiro, RJ, Brasil

Cataloging-in-Publication Data applied for

A catalog record for this book is available from the Library of Congress.

Bibliographic information published by Die Deutsche Bibliothek
Die Deutsche Bibliothek lists this publication in the Deutsche Nationalbibliografie;
detailed bibliographic data is available in the Internet at http://dnb.ddb.de

ISSN 0075-8450
ISBN 3-540-20639-6 Springer-Verlag Berlin Heidelberg New York

Springer-Verlag is a part of Springer Science+Business Media

springeronline.com

© Springer-Verlag Berlin Heidelberg 2004 Printed in Germany

Typesetting: Camera-ready by the authors/editor
Data conversion: PTP-Berlin Protago-TeX-Production GmbH
Cover design: *design & production*, Heidelberg

Printed on acid-free paper
54/3141/du - 5 4 3 2 1 0

Preface

The contributions to this volume are based on lectures selected by participating advisory board members at the **First International Workshop on Decoherence, Information, Complexity and Entropy DICE 2002,1** which was held at Castello di Piombino (Tuscany) in September 2002.[1]

The aim of this collection of lectures is to reflect upon the stimulating exchange of ideas at this workshop, with its lively discussions of interdisciplinary aspects of quantum theory ranging from general relativity to statistical physics. Many of the reported results are original and published here for the first time. In particular results from rapidly developing interdisciplinary research areas are presented. The topics discussed have previously only been addressed partly, separately, or indeed not at all at recent specialized conferences, and they include

- the foundations of quantum theory and its generalization in applications to gravity and cosmology
- quantum decoherence and the emergence of classicality in open quantum systems, the properties and manipulation of their environments
- mesoscopic systems and quantum information processing
- entropy and information theory related to complex systems

About one hundred years after the first steps into "Quantum Land" were taken, the lectures presented here show the subject alive and well. They anticipate further profound experimental as well as theoretical advances, still undiscovered, yet likely to be seen in the near future.

The great success of the workshop must be attributed not only to all colleagues who have participated enthusiastically in the meeting or are contributing their work here, but also equally to the wonderful hospitality which the Comune di Piombino and its citizens extended to us and to the initiative shown there while embarking on this project.

A crowded public opening lecture by Gabriele Veneziano from CERN – 'Prima del Big Bang: una storia piú lunga del tempo'– communicated several of the physics topics dealt with here to the captivated audience, with a question and answer session running until late into the night, and prepared an exciting and productive atmosphere which stayed with us.

A workshop such as this and this collection of lectures could not have materialized without the feedback and encouragement from colleagues and friends.

---

[1] See: http://www.if.ufrj.br/thomas/DICEhp.html

We cordially thank the citizens of Piombino for their friendly and warm hospitality and for their interest and support, extended to us through the Mayor Luciano Guerrieri, the Vice-mayor Umberto Canovaro, and by Pablo Gorini, the Assessore ai Beni Culturali e Archeologici, who had the vision that this all could and should be done.

The workshop was generously sponsored by the Fondazione Cassa Di Risparmi Di Livorno represented by the President Luciano Barsotti. We owe him a great debt of gratitude for supporting a cultural initiative in the form of a scientific symposium.

The organizational matters have been dealt with very efficiently by the Servizio Promozione Culturale, Comune di Piombino. It is a pleasure to thank Carla Boggero, Aldo Falchi, Tiziana Ghini, Maria Gianfranchi, Luciano Giannoni, Lucia Grilli, and Piera Venturi.

I warmly thank Gustav Obermaier, co-organizer of the workshop, in particular, for guiding me from the initial idea towards its realization. Carolina Nemes, Tomaz Prosen, and Claus Kiefer are thanked for their special effort in writing introductions to parts of this volume. Last but not least, I gratefully acknowledge help with the editorial task and advice from Julian Barbour, Christian Caron, Diego Carvalho, and Claus Kiefer.

A special "Thank You" goes to Maria Gianfranchi, Hans W. Elze, and Melissa Pesce-Rollins, our conference secretary, who all helped in diverse and most essential ways, and especially to Laura Pesce for her continuous encouragement and persistent support.

Rio de Janeiro and Campiglia Marittima,   *Hans-Thomas Elze*
July 2003

Contents

Part V   Entropy, Chaos, and Complexity

List of Contributors

F. Tito Arecchi
Istituto Nazion. di Ottica Applicata
Largo Enrico Fermi 6
50125 Firenze, Italia
arecchi@ino.it

Julian Barbour
College Farm
South Newington, Banbury
Oxon, OX15 4JG, UK
julian@platonia.com

Orfeu Bertolami
Departamento de Física
Instituto Superior Técnico
Av. Rovisco Pais
1049-001 Lisboa, Portugal
orfeu@cosmos.ist.utl.pt

Iwo Bialynicki-Birula
Center for Theoretical Physics
Polish Academy of Sciences
Lotnikow 32/46
02-668 Warsaw, Polonia
birula@cft.edu.pl

Tamas S. Biró
MTA KFKI RMKI, P.O.Box 49
1525 Budapest, Hungary
tsbiro@sunserv.kfki.hu

Massimo Blasone
Blackett Laboratory
Imperial College, Prince Consort Rd.
London, SW7 2BW, UK
m.blasone@imperial.ac.uk

Todd A. Brun
Institute for Advanced Study
Princeton, NJ 08540, USA
tbrun@ias.edu

Giulio Casati
Center for Nonlinear and
Complex Systems
Università dell'Insubria
INFM, Unità di Como
Via Valleggio 11
22100 Como, Italia
giulio.casati@uninsubria.it

Ariel Caticha
Department of Physics
University at Albany, SUNY
Albany, NY 12222, USA
ariel@albany.edu

Luiz Davidovich
Instituto de Física
Univers. Federal do Rio de Janeiro
C.P. 68.528
21941-972 Rio de Janeiro (RJ), Brasil
ldavid@if.ufrj.br

Lajos Diósi
Research Institute
for Particle and Nuclear Physics
1525 Budapest 114, P.O.Box 49
Hungary
diosi@rmki.kfki.hu

Hans-Thomas Elze
Instituto de Física
Univers. Federal do Rio de Janeiro
C.P. 68.528
21941-972 Rio de Janeiro (RJ), Brasil
thomas@if.ufrj.br

Zbigniew Haba
Institute of Theoretical Physics
University of Wroclaw
Plac Maxa Borna 9
50-204 Wroclaw, Polonia
zhab@ift.uni.wroc.pl

Jonathan Halliwell
Blackett Laboratory
Imperial College
London, SW7 2BZ, UK
j.halliwell@imperial.ac.uk

Petr Jizba
Institute of Theoretical Physics
University of Tsukuba
Ibaraki 305-8571, Japan
petr@cm.ph.tsukuba.ac.jp

Viv Kendon
QOLS, Blackett Laboratory
Imperial College London
London, SW7 2BW, UK
v.kendon@imperial.ac.uk

Claus Kiefer
Institut für Theoretische Physik
Universität zu Köln
Zülpicher Str. 77
50937 Köln, Deutschland
kiefer@thp.uni-koeln.de

Hagen Kleinert
Institut für Theoretische Physik
Freie Universität Berlin
Arnimallee 14
14195 Berlin, Deutschland
kleinert@physik.fu-berlin.de

Seth Lloyd
d'Arbeloff Laboratory
Massachusetts Institute of Technol.
Department of Mechan. Engineering
Cambridge, MA 02139, USA
slloyd@mit.edu

Sergei G. Matinyan
Yerevan Physics Institute
Yerevan, Armenia
sgmatin@aol.com

Nicolaos E. Mavromatos
Department of Physics
King's College
London, WC2R 2LS, UK
nikolaos.mavromatos@kcl.ac.uk

Simone Montangero
Scuola Normale Superiore
NEST-INFM
Piazza dei Cavalieri, 7
56126 Pisa, Italia
monta@sns.it

Berndt Müller
Department of Physics
Duke University
Durham, NC 27708, USA
muller@phy.duke.edu

Maria Carolina Nemes
Departamento de Física
Univers. Federal de Minas Gerais
C.P. 702
30123-970 Belo Horizonte (MG)
Brasil
carolina@fisica.ufmg.br

A. Keith Powell
Department of Physics
King's College
London, WC2R 2LS, UK
keith.powell@kcl.ac.uk

Tomaz Prosen
Physics Department, FMF
University of Ljubljana
Ljubljana, Slovenia
prosen@dioniz.fiz.uni-lj.si

Johann Rafelski
Department of Physics
University of Arizona
Tucson, AZ 85721, USA
rafelski@physics.arizona.edu

Carlo Rovelli
Centre de Physique Théorique
de Luminy, CNRS, Case 907
13288 Marseille, France
crovelli@perimeterinstitute.ca

Travis Sherman
Department of Physics
University of Arizona
Tucson, AZ 85721, USA

Daniel L. Stein
Departments of Physics
and Mathematics
University of Arizona,
Tucson, AZ 85721, USA
dls@physics.arizona.edu

Ben Tregenna
QOLS, Blackett Laboratory
Imperial College London
London, SW7 2BW, UK

Constantino Tsallis
Centro Brasileiro de Pesqu. Fisicas
Xavier Sigaud 150
22290-180 Rio de Janeiro (RJ), Brasil
tsallis@cbpf.br

Giuseppe Vitiello
Dipartimento di Fisica
"E.R.Caianiello"
INFN/INFM
Università di Salerno
84100 Salerno, Italia
vitiello@sa.infn.it

Yaakov S. Weinstein
Massachusetts Institute of Technol.
Department of Nuclear Engineering
Cambridge, MA 02319, USA
yaakov@mit.edu

Christof Wetterich
Institut für Theoretische Physik
Universität Heidelberg
Philosophenweg 16
69120 Heidelberg, Deutschland
c.wetterich
@thphys.uni-heidelberg.de

Part I

Prologue

An Overview

Hans-Thomas Elze

Instituto de Física, Universidade Federal do Rio de Janeiro, C.P. 68.528,
21941-972 Rio de Janeiroe, RJ, Brazil

The material of this volume is organized in four parts, with separate introductions, which roughly follow the proceedings of the workshop. The overlap of these research areas is a recent feature which we would like to introduce here.

It is common wisdom by now that the pillars of modern physics, quantum theory and general relativity, coexist very successfully, having in their respective domains impressive observational support. However, the black hole entropy puzzle and conceptual problems in applying quantum theory to the Universe are stumbling blocks on the way to a deeper understanding. They are as hard as before, even though much experience has been gained during the last decade with various promising approaches to quantum gravity and quantum cosmology. These issues are addressed in Part II.

While these problems are particularly pressing for the attempts to accommodate quantum theory and general relativity in a common framework, they do reach beyond. Some of them have been around since the early days. It is most interesting to see how attention has gradually shifted away from rather general discussions of the quantum formalism, especially of the measurement problem, of the Copenhagen versus many-worlds interpretations, for example, and of epistemological or ontological questions [1,2].

With the blooming experimental progress, which has led to unforeseen capabilities of realizing the "Gedankenexperimente" discussed in textbooks before, such as the Einstein-Rosen-Podolsky experiment, the impetus is now towards probing further the predictions of quantum mechanics [3]. Or, rather, towards making fully use of them. This is particularly visible in the realm of quantum information, the motivation being largely the quest for quantum computing. A variety of related topics is discussed in Part IV.

From fundamental as well as from practical points of view, closely related is the issue of classical behavior in complex quantum systems, i.e. decoherence [4,5]. One of the central questions is, how one can possibly protect a quantum system against the disturbing influence of its environment, which is bound to destroy the quantum coherence (with its desirable properties for quantum information processing). It has recently been studied extensively in mesoscopic systems, leading to an understanding of the emergence of classicality, especially of the classical macroscopic world of our daily experience, from a quantum mechanical description.

H.-T. Elze, An Overview, Lect. Notes Phys. **633**, 3–5 (2004)
http://www.springerlink.com/   © Springer-Verlag Berlin Heidelberg 2004

Notions such as "environment induced superselection" and "pointer states", describing key effects of decoherence processes, are here to stay. They refer to quantum mechanical states which show particular stability features with respect to interaction with the environment of the quantum system under study. Thus, they qualify as correspondents of a pointer with its distinguishable positions in a classical measuring apparatus. Decoherence here, more recently also for classically chaotic systems, is well understood. The theory provides a comforting picture of the transition between quantum and classical physics, and, last not least, succeeded to replace the physically and often mathematically dubious limit $\hbar \to 0$, so freely invoked in the past.

However, the implications for the resolution of the fundamental measurement problem and the related questions about the role of the observer in quantum theory, even more so in quantum gravity and quantum cosmology – for example, "Does it make sense at all to speak of the Universe being in a pure or mixed quantum state?" – are much less clear [6]. In fact, this problem is been studied intensely and the debate is as active as thirty years ago, when Ted Bastin's volume summarized the state of the art [1, 7].

Furthermore, relevant decohering degrees of freedom in microscopic and general relativistic systems are still much less understood. Attempts to quantize gravity have stimulated studies of the classical dynamics and of the statistical properties of "timeless" reparametrization-invariant systems. There, the absence of a well-defined "time" observable is most accentuated in the Wheeler-DeWitt equation, $H|\Psi\rangle = 0$, supposedly describing the quantum evolution of the Universe. Via the decoherence bridge to the classical domain it is hoped to bring more light into the analysis of these difficulties.

A central problem in such systems with many degrees of freedom is to determine the observables that are consistent with the constraints. Various related problems are treated in Parts II, III, and V.

Although not always emphasized, the study of these topics includes an analysis of the kind of information that exists in such systems with typically long-range interactions and of the basic conditions under which it can be extracted. This has been investigated in much detail in statistical theories of spin glasses [8]. On the other hand, corresponding nonextensive entropies and measures of complexity have been elaborated [9]. These subjects are included in Part V.

Novel studies, which might seem close to heresy to some, are covered by lectures in Part III. Likely encouraged by considerations put forth by Gerard 't Hooft in recent years [10], several contributions in Part III look afresh at the possibility to find deterministic classical structures beneath quantum mechanics. If such an approach succeeds, of course, most of what has been said here before would appear in new light, raising new questions, but promising to solve some long-standing puzzles: quantum mechanics might turn out to be an emergent "effective" theory after all! As much as *decoherence* has contributed to understand the transition between quantum and classical physics, a new paradigm of *determinism* might be ahead of us, leading the opposite pathway from more fundamental classical (bits?) to quantized degrees of freedom.

In any case, the general impression we should like to take from here is that future advances in these matters will presumably depend on an intimate fusion and new offspring of ideas coming from quantum theory as much as from information theory, from general relativity as much as from statistical mechanics. Without doubt, things will continue to look strange "out there" and a timely step here is hoped to further such research.

References

1. *Quantum Theory and Beyond*, ed. by T. Bastin (Cambridge University Press, Cambridge 1971)
2. A reprint volume is: *Quantum Theory and Measurement*, ed. by J.A. Wheeler and W.H. Zurek (Princeton University Press, Princeton 1983)
3. For a recent overview, see: M. Tegmark and J.A. Wheeler: '100 Years of the Quantum', Scientific American **284**, 68 (2001)
4. W.H. Zurek: 'Decoherence and the transition from quantum to classical', Physics Today **44**, 36 (1991); an updated version is : quant-ph/0306072; 'Decoherence, einselection, and the quantum origins of the classical', Rev. Mod. Phys. 75, 715 (2003)
5. E. Joos, H.D. Zeh, C. Kiefer, D. Giulini, J. Kupsch, and I.-O. Stamatescu: *Decoherence and the appearance of a classical world in quantum theory*, 2nd edition (Springer-Verlag, Berlin 2003)
6. Several of these topics were already visible in the proceedings: *Complexity, Entropy, and the Physics of Information*, ed. by W.H. Zurek, Santa Fe Institute Studies in the Sciences of Complexity, Vol. VIII (Addison-Wesley, Redwood City 1990)
7. See, for example, the recent article and its valuable list of references: S. Adler: 'Why Decoherence has not Solved the Measurement Problem: A Response to P.W. Anderson', quant-ph/0112095
8. A reprint collection with a very useful and still thought provoking introduction is: *Spin Glass Theory and Beyond*, ed. by M. Mezard, G. Parisi and M.A. Virasoro, World Scientific Lecture Notes in Physics Vol. 9 (World Scientific, Singapore 1987)
9. A collection of reviews and research articles is: *Nonextensive Statistical Mechanics and Thermodynamics*, ed. by S.R.A. Salinas and C. Tsallis, Brazil. Journ. Phys. (Special Issue) **29**, No 1 (1999)
10. G. 't Hooft: 'Quantum Mechanics and Determinism'. In: *Particles, Strings and Cosmology*, ed. by P. Frampton and J. Ng (Rinton Press, Princeton, 2001), p. 275; hep-th/0105105; 'Determinism Beneath Quantum Mechanics', quant-ph/0212095

Part II

Gravity and Cosmology
  – Classical and Quantum Aspects

Introduction:
The Relevance of Gravity for DICE

Claus Kiefer

Institut für Theoretische Physik, Universität zu Köln, Zülpicher Str. 77,
50937 Köln, Germany.

Gravity is unique because it acts universally on all interactions in Nature. For this reason it is also of fundamental relevance for the issues discussed in this volume – **D**ecoherence, **I**nformation, **C**omplexity, **E**ntropy. Decoherence – the irreversible emergence of classical properties for a quantum system due to the interaction with its environment – can only be understood if a special initial condition holds [1]. This must be an initial condition of lacking (or small) entanglement with the environment. Since most subsystems in the universe cannot be treated as isolated, such a special initial condition must be of a *cosmological* nature. On cosmological scales, however, gravity is the decisive interaction, and the initial condition cannot be understood without a careful investigation into the nature of the gravitational field.

Information, on the other hand, is inextricably linked with decoherence, as is discussed extensively in this volume. The occurrence of quasiclassical domains leads to increasing information about robust quantities. Complexity is linked with the emergence of structure, and this is again a cosmological problem. As indicated by observations of the cosmological microwave background, the early universe was surprisingly smooth. Only small primordial fluctuations were present. The gravitational interaction then created the observed structure in the universe (galaxies, clusters of galaxies) out of these fluctuations.

The increase of entropy in the universe can be understood only if a special boundary condition holds at 'early' times [2, 3]. The Bekenstein-Hawking formula for the entropy of a black hole indicates that the maximal entropy of the universe would be reached if all matter were in a single black hole. Again, the question arises why the universe started with a low gravitational entropy, i.e. in an approximately homogeneous state.

The investigation of these questions leads to a problem. The beginning of the universe – and therefore the origin of the special boundary condition – takes place when the classical theory of gravity, Einstein's theory of general relativity, breaks down. The general expectation is that a *quantum* theory of gravity is needed. What are the main motivations for such a theory [4]?

- **Unification.** The history of science shows that a reductionist viewpoint has been very fruitful in physics. The standard model of particle physics is a *quantum* field theory which has united in a certain sense all non-gravitational interactions. The universal coupling of gravity to all forms of energy would

C. Kiefer, Introduction: The Relevance of Gravity for DICE, Lect. Notes Phys. **633**, 9–14 (2004)
http://www.springerlink.com/   © Springer-Verlag Berlin Heidelberg 2004

make it plausible that gravity has to be implemented in a quantum framework, too. Moreover, attempts to construct an exact semiclassical theory, where gravity stays classical but all other fields are quantum, have failed up to now. This demonstrates in particular that classical and quantum *concepts* (phase space versus Hilbert space, etc.) are most likely incompatible.

- **Cosmology and Black Holes.** As the *singularity theorems* and the ensuing breakdown of general relativity demonstrate, a fundamental understanding of the early universe – in particular its initial conditions near the 'big bang' – and of the final stages of black-hole evolution requires an encompassing theory. From the historical analogue of quantum mechanics (which due to its stationary states has 'rescued' the atoms from collapse) the general expectation is that this encompassing theory is a *quantum* theory. It must be emphasised that *if* gravity is quantized, the kinematical nonseparability of quantum theory demands that the whole universe must be described in quantum terms. This leads to the concepts of quantum cosmology and the wave function of the universe.

- **Problem of Time.** Quantum theory and general relativity (in fact, every general covariant theory) contain drastically different concepts of time (and spacetime). Strictly speaking, they are incompatible. In quantum theory, time is an external (absolute) element, *not* described by an operator (in special relativistic quantum field theory, the role of time is played by the external Minkowski spacetime). In contrast to this, spacetime is a dynamical object in general relativity. It is clear that a unification with quantum theory must lead to modifications of the concept of time. Related problems concern the role of background structures in quantum gravity, the role of the diffeomorphism group (Poincaré invariance, as used in ordinary quantum field theory, is no longer a symmetry group), and the notion of 'observables'.

What are the relevant scales on which effects of quantum gravity should be unavoidable? As has already been shown by Max Planck in 1899, the fundamental constants speed of light (c), gravitational constant (G), and quantum of action ($\hbar$) can be combined in a unique way (up to a dimensionless factor) to yield units of length, time, and mass. In Planck's honour they are called Planck length, l_P, Planck time, t_P, and Planck mass, m_P, respectively. They are given by the expressions

$$l_\mathrm{P} = \sqrt{\frac{\hbar G}{c^3}} \approx 1.62 \times 10^{-33} \ \mathrm{cm} \ , \tag{1}$$

$$t_\mathrm{P} = \frac{l_\mathrm{P}}{c} = \sqrt{\frac{\hbar G}{c^5}} \approx 5.40 \times 10^{-44} \ \mathrm{s} \ , \tag{2}$$

$$m_\mathrm{P} = \frac{\hbar}{l_\mathrm{P} c} = \sqrt{\frac{\hbar c}{G}} \approx 2.17 \times 10^{-5} \ \mathrm{g} \approx 1.22 \times 10^{19} \ \mathrm{GeV} \ . \tag{3}$$

The Planck mass seems to be a rather large quantity on microscopic standards. One has to keep in mind, however, that this mass (energy) must be concentrated in a region of linear dimension l_P in order to see direct quantum-gravity effects.

In fact, the Planck scales are attained for an elementary particle whose Compton wavelength is (apart from a factor of 2) equal to its Schwarzschild radius,

$$\frac{\hbar}{m_{\mathrm{P}}c} \approx R_{\mathrm{S}} \equiv \frac{2Gm_{\mathrm{P}}}{c^2} \ ,$$

which means that the spacetime curvature of an elementary particle is non-negligible. A truly unified theory may, of course, contain further parameters. An example is string theory where the fundamental 'string length' l_{s} appears.

A quantity expressing the ratio of atomic scales to the Planck scale is the 'fine structure constant of gravity' defined by

$$\alpha_{\mathrm{g}} = \frac{Gm_{\mathrm{pr}}^2}{\hbar c} \equiv \left(\frac{m_{\mathrm{pr}}}{m_{\mathrm{P}}}\right)^2 \approx 5.91 \times 10^{-39} \ , \tag{4}$$

where m_{pr} denotes the proton mass. Its smallness is responsible for the unimportance of quantum-gravitational effects on laboratory and astrophysical scales, and for the separation between micro- and macrophysics. It is interesting that structures in the universe occur for masses which contain simple powers of α_{g} in terms of m_{pr}, cf. [5]. For example, stellar masses are of the order $\alpha_{\mathrm{g}}^{-3/2}m_{\mathrm{pr}}$, while stellar lifetimes are of the order $\alpha_{\mathrm{g}}^{-3/2}t_{\mathrm{P}}$. It is also interesting to note that the size of human beings is roughly the geometric mean of Planck length and size of the observable universe. It is an open question whether a fundamental theory of quantum gravity can provide an explanation for such values, e.g. for the ratio $m_{\mathrm{pr}}/m_{\mathrm{P}}$, or not. If not, only an anthropic principle could yield a – not very satisfying – explanation.

How can one construct a consistent quantum theory of gravity? Unfortunately, experimental clues are still elusive [4]. A direct probe of the Planck scale (3) in high-energy experiments would be illusory. In fact, an accelerator of current technology would have to have the size of several thousand lightyears in order to probe the Planck energy $m_{\mathrm{P}}c^2 \approx 10^{19}$ GeV. However, it is imaginable that effects of quantum gravity can in principle occur at lower energy scales. Possibilities could be non-trivial applications of the superposition principle for the quantized gravitational field or the existence of discrete quantum states in black-hole physics or the early universe. But one might also be able to observe quantum-gravitational correction terms to established theories, such as correction terms to the functional Schrödinger equation in an external spacetime or effective terms violating the weak equivalence principle. Such effects could potentially be measured in the anisotropy spectrum of the cosmic microwave background radiation or in the forthcoming satellite tests of the equivalence principle, such as STEP.

A truly fundamental theory should have such a rigid structure that all phenomena in the low-energy regime, such as particle masses or coupling constants, could be predicted in an unique way. As there is no direct experimental hint yet, most work in quantum gravity focuses on the attempt to construct a mathematically and conceptually consistent (and appealing) framework.

There is, of course, no a priori given starting point in the methodological sense. In this context Chris Isham makes a distinction between a 'primary theory of quantum gravity' and a 'secondary theory' [6]. In the primary approach, one starts with a given classical theory and applies heuristic quantization rules. This is the approach usually adopted, and it was successful, for example, in QED. In most cases, the starting point is general relativity, leading to 'quantum general relativity' or 'quantum geometrodynamics', but one could also start from another classical theory such as the Brans-Dicke theory. One usually distinguishes between canonical and covariant approaches. The main advantage of both approaches is that the starting point is given – the classical theory. The main disadvantage is that one does not arrive immediately at a unified theory of all interactions.

The opposite holds for a 'secondary theory'. One starts with a fundamental quantum framework of all interactions and tries to derive (quantum) general relativity in certain limiting situations, e.g. through an energy expansion. The most important example here is string theory (M-theory). The main advantage is that the fundamental quantum theory automatically yields a unification. The main disadvantage is that the starting point is entirely speculative.

Even if quantum general relativity is superseded by a more fundamental theory, such as string theory, it should be valid as an *effective theory* in some appropriate limit. The reason is that far away from the Planck scale, classical general relativity is the appropriate theory, which in turn must be the classical limit of an underlying quantum theory. Except perhaps close to the Planck scale itself, quantum general relativity should be a viable framework (such as QED, which is also supposed to be only an effective theory). It should also be mentioned that string theory automatically implements many of the methods used in the primary approach, such as quantization of constrained systems and covariant perturbation theory.

An important question in the heuristic quantization of a given classical theory is which of the structures in the classical theory should be quantized, i.e. subjected to the superposition principle, and which should remain as classical (or absolute, non-dynamical) structures. Isham distinguishes the following hierarchy of structures [7]:

Point set of events $\longrightarrow$ topological structure $\longrightarrow$ differentiable manifold $\longrightarrow$ causal structure $\longrightarrow$ Lorentzian structure.

Most approaches subject the Lorentzian and the causal structure to quantization, but keep the manifold structure fixed. This is, however, not clear. It might be that even the topological structure is fundamentally quantized. According to the Copenhagen interpretation of quantum theory, all these structures would probably have to stay classical, because they are thought to be necessary ingredients for the measurement process. For the purpose of quantum gravity, such a viewpoint is, however, insufficient and probably inconsistent. In fact, existing approaches to quantum gravity exhibit the absence of a classical time parameter in the fundamental equations (see the 'problem of time' mentioned above).

The history of quantum gravity starts with early perturbative attempts by Leon Rosenfeld in 1929. A brief overview of historical developments can be found in [8]. Currently the most poupular approaches are *canonical quantum gravity* (a direct quantization of general relativity in a Hamiltonian framework) and *string theory*. Within the canonical theory, in particular, the question of initial conditions can be tackled in various ways, leading to tentative answers about the problems of DICE [9, 10].

The following contributions address various important issues concerning the relation between gravity and DICE. Fundamental theories that go beyond general relativity can lead to new effects such as the violation of Lorentz symmetry. This could open a route towards an experimental test of some of the approaches mentioned above. This is discussed in the article by Orfeu Bertolami.

As argued above, decoherence plays a crucial role in the context of quantum gravity and quantum cosmology. Jonathan Halliwell, in his contribution, discusses a general framework – the decoherent histories approach – that has been developed partly to come up with a formulation that can cope with situations which are characterized by the absence of a distinguished time. However, the decoherent histories approach finds also applications in ordinary quantum theory.

In my own contribution, I discuss the information-loss problem for black holes. This problem had been exposed by Hawking in the context of black-hole radiation: if a black hole evaporated completely by the Hawking process and left only thermal radiation behind, any initial state for the full system (not an open system) would develop into a thermal state, in violation of unitary evolution in ordinary quantum theory. I argue that this problem is not present and that Hawking radiation can be solely understood through decoherence.

The contributions by Carlo Rovelli and Julian Barbour are devoted to the problem of time in quantum gravity. Rovelli considers a general Hamiltonian formalism that can be presented in a covariant way. He recovers the basic equations of canonical quantum gravity. Motivated by Machian ideas, Barbour envisages a solution to the problem of time by defining configuration space in purely relational terms. In generalisation of standard classical (and quantum) gravity, the entire dynamics is encoded in pure shape. (While Barbour's work addresses classical gravity in a new and still partly speculative way, it might shed light on the problem of time in quantum gravity as well [note added by editor].)

Finally, Ariel Caticha in his contribution attempts to derive general relativity ('geometrodynamics') from a dynamical formulation for entropy (the concept of an 'information metric'). This would exhibit an unexpected but intriguing connection between statistical physics and gravity.

References

1. E. Joos, H.D. Zeh, C. Kiefer, D. Giulini, J. Kupsch, and I.-O. Stamatescu: *Decoherence and the appearance of a classical world in quantum theory*, 2nd edition (Springer-Verlag, Berlin 2003). See also http://www.decoherence.de

2. H.D. Zeh: *The physical basis of the direction of time*, 4th edition (Springer-Verlag, Berlin 2001). See also `http://www.time-direction.de`

3. R. Penrose: 'Time asymmetry and quantum gravity'. In: *Quantum gravity 2*, ed. by C.J. Isham, R. Penrose, and D.W. Sciama (Clarendon Press, Oxford 1981)

4. *Aspects of quantum gravity*, ed. by D. Giulini, C. Kiefer and C. Lämmerzahl, Lecture Notes in Physics (Springer Verlag, Berlin 2003)

5. M. Rees: *Perspectives in astrophysical cosmology* (Cambridge University Press, Cambridge 1995)

6. C.J. Isham: 'Quantum gravity'. In: *General relativity and gravitation*, ed. by M.A.H. Mac Callum (Cambridge University Press, Cambridge 1987)

7. C.J. Isham: 'Prima facie questions in quantum gravity'. In: *Canonical gravity: From classical to quantum*, ed. by J. Ehlers and H. Friedrich (Springer-Verlag, Berlin 1994)

8. C. Rovelli: 'Notes for a brief history of quantum gravity', gr-qc/0006061

9. J.J. Halliwell: 'Introductory lectures on quantum cosmology'. In: *Quantum cosmology and baby universes*, ed. by S. Coleman et al. (World Scientific, Singapore 1991)

10. C. Kiefer, 'Conceptual issues in quantum cosmology' In: *Towards quantum gravity*, ed. by J. Kowalski-Glikman (Springer-Verlag, Berlin 2000). A related online version is gr-qc/9906100

Dynamics of Pure Shape, Relativity, and the Problem of Time

Julian Barbour

College Farm, South Newington, Banbury, Oxon, OX15 4JG, UK

Abstract. A new approach to the dynamics of the universe based on work by Ó Murchadha, Foster, Anderson and the author is presented. The only kinematics presupposed is the spatial geometry needed to define configuration spaces in purely relational terms. A new formulation of the relativity principle based on Poincaré's analysis of the problem of absolute and relative motion (Mach's principle) is given. The entire dynamics is based on shape and nothing else. It leads to much stronger predictions than standard Newtonian theory. For the dynamics of Riemannian 3-geometries on which matter fields also evolve, implementation of the new relativity principle establishes unexpected links between special relativity, general relativity and the gauge principle. They all emerge together as a self-consistent complex from a unified and completely relational approach to dynamics. A connection between time and scale invariance is established. In particular, the representation of general relativity as evolution of the shape of space leads to a unique dynamical definition of simultaneity. This opens up the prospect of a solution of the problem of time in quantum gravity on the basis of a fundamental dynamical principle.

1   Introduction

In this paper, I wish to discuss the foundations of cosmology and our notion of time. The goal is to contribute to the creation of a quantum theory of the universe. I shall draw attention to some conceptual issues that in my opinion have not hitherto been adequately discussed. Since this is a contribution to an interdisciplinary workshop, I shall keep the discussion as simple as possible. In fact, the model that I shall present, which treats non-relativistic particles in Euclidean space, hardly seems realistic. However, it contains the simplest implementation of a dynamical variational principle that can also be applied in general relativity, as I shall outline in more qualitative terms. It is the principle that I wish to explain. Moreover, the problems to be addressed are so challenging, there is much to be said for attacking them initially in the simplest possible situations. I also happen to believe that anyone competent in quantum mechanics should be interested in the issues raised by the model independently of the quantum cosmological significance. The question is this: Can one do meaningful quantum mechanics with significantly less external kinematic structure than is currently employed in quantum theory? This is a topic ideally suited to this interdisciplinary volume.

J. Barbour, Dynamics of Pure Shape, Relativity, and the Problem of Time, Lect. Notes Phys. **633**, 15–35 (2004)
http://www.springerlink.com/

Some historical background will be helpful. In order to formulate his laws of motion, Newton introduced a rigid external framework: absolute space and time. This framework was simply taken over in its entirety by the creators of quantum mechanics. It was only somewhat modified to accommodate quantum field theory, and a fixed framework is still deeply embedded in that theory. However, ten years before the creation of quantum mechanics, Einstein had created the general theory of relativity. There is universal agreement that Einstein to a very large degree abolished Newton's absolute framework. There is less agreement about what, if anything, replaced it in the classical theory and what kind of framework is appropriate for a quantum theory of gravity. This, I believe, is the main reason why people are still seeking the foundations of quantum cosmology. I want to argue for a new and clearer formulation of the relativity principle. This will lead me to the notion of *dynamics of pure shape*.

It will be helpful to distinguish two different notions of relativity. There is, first, the intuitive idea that position can only be defined relative to observable objects. Suppose we were to contemplate a universe consisting of N point particles in Euclidean space with separations r_{ij} between them. The opponents of Newton's absolute space and time, above all Leibniz [1] and Mach [2], argued that dynamics should be directly formulated in terms of the r_{ij}. Let me call this *kinematic relativity*. In 1902, Poincaré [3] pointed out that in fact Newtonian point-particle dynamics can always be formulated in terms of the r_{ij}, but that then the structure of its initial-value problem is changed compared with the formulation in absolute space (or, in modern terms, an inertial system). This is such an important issue that it needs to be spelled out. If we are given initial positions $\mathbf{x}_i$ and initial velocities $\dot{\mathbf{x}}_i$ in an inertial system (together with the masses m_i and the force law), then Newton's laws determine the past and future motions uniquely. The situation is characteristically different if one is given r_{ij} and $\dot{r}_{ij}$. The fact is that such data contain no information at all about the overall rotation of the system. One cannot determine the angular momentum $\mathbf{M}$ of the system, and for different $\mathbf{M}$ very different evolutions result. Initial data that are identical from the point of view of kinematic relativity give rise to different evolutions. Moreover the defect has an odd structure. Suppose N is large. In three-dimensional Euclidean space, the r_{ij} contain $3N - 6$ independent data, and their time derivatives contain the same number. This number is the order of a million for a globular cluster. Just three more numbers are needed if the evolution is to be uniquely determined. They could be three of the second time derivatives. But why does one need three and not $3n - 6$? Poincaré argued, persuasively in my view, that the only valid objection to Newton's use of absolute space and time resided in this curious need to specify a small number of extra data. He said that if one could formulate a relational dynamics (i.e., one containing only r_{ij} and its time derivatives) free of this defect, the problem of absolute motion would be solved. Poincaré made no attempt to find such a dynamics. Moreover, he pointed out that the empirical evidence seemed to show conclusively that nature did not work in this way.

I shall argue that Poincaré's analysis is very sound and that empirically adequate theories satisfying a criterion along the lines he proposed can be formulated. However, it will be helpful to push his analysis somewhat further, which I shall do in the next section. For the moment, let me merely note that if one takes kinematic relativity seriously one will wish to formulate a dynamics whose initial-value problem satisfies a well-defined criterion. For reasons that will soon become apparent, let me call this the *constructive approach* or *Poincaré's relativity principle*. It can be contrasted with the approach that was adopted by Einstein in creating both special and general relativity and is known as the *principle approach*. It grew out of a generalization of *Galilean relativity*, which Einstein transformed into the restricted relativity principle. In order to construct a theory in which uniform motion could not be detected, Einstein made no attempt to formulate a complete theory of particles interacting with Maxwell's electromagnetic field. Instead, he postulated the existence of a family of distinguished (inertial) frames all in uniform translational motion relative to each other and required the laws of nature to take the identical form in all of them. The dramatic results that he obtained came from combination of this relativity principle with his postulate about the behavior of light. As he explained in his Autobiographical Notes [4] (see also [5], on which this discussion is based), he was encouraged to adopt this approach because of the success of phenomenological thermodynamics based based on 'impotence' principles: the impossibility of constructing perpetual motion machines of the first and second kind. The impotence in his case was the impossibility of detecting uniform motion through the ether (or absolute space) by means of processes that unfold in a closed system.

In 1907, Einstein realized that the equivalence principle enabled him to extend the restricted relativity principle to include another 'impotence' – the inability to detect uniform acceleration. This insight was decisive. Some years earlier, Einstein had read Mach's critique of Newton's absolute space and time and was extremely keen to reformulate dynamics along the broad lines advocated by Mach. He wanted to show that absolute space did not correspond to anything in reality – that it could not be revealed by any experiment. Special relativity had shown that uniform motion – relative to the ether or to absolute space – could not be detected by any physical process. The equivalence principle suggested to him that it might be possible to extend his relativity principle further. If he could extend it so far as to show that the laws of nature could be expressed in identical form in all conceivable frames of reference, this requirement of general covariance would "[take] away from space and time the last remnant of physical objectivity" [6]. He would have achieved his Machian aim. Unfortunately, within two years Einstein had been forced by a critique of Kretschmann [7] to acknowledge that any physical theory must, if it is to have any content, be expressible in generally covariant form. He argued [8] that the principle nevertheless had great heuristic value. One should seek only those theories that are *simple* when expressed in generally covariant form. However, Einstein gave no definition of simplicity. Since then, and especially as a result of quantization attempts, there has been a vast amount of inconclusive discussion about the significance of general covariance [9] and its implications for quantization. A point worth noting is that Einstein treated space and time as a single unit and considered general coordinate transformations on a four-dimensional manifold. The distinction between space and time and the manner in which they are to be treated is very largely erased.

In several recent papers [10–13], my collaborators and I have developed what we call *the 3-space approach*. It is based on a generalization of Poincaré's relativity principle and casts new light on this issue. Above all, it replaces Einstein's vague simplicity requirement by a well-defined constructive principle based on the amount of data needed to formulate initial-value problems. In addition, space and time are treated in completely different ways. This might seem to be a retrogressive step, but actually the approach not only achieves everything that Einstein did by presupposing a four-dimensional unity of space and time but even more. In a real sense it explains why there is a universal light cone (which Einstein presupposed) and why the gauge principle holds. The details of this work goes beyond the scope of the present paper, though I will indicate how these results are obtained in the final section. Readers wishing for full details are referred to the original papers. In this paper, I merely wish to get across the basic ideas and draw attention to some of the interesting possibilities that arise.

2   Basic Ideas

The 3-space approach is a natural modification of the basic scheme employed in the variational principles of mechanics [14]. The main difference is that the 3-space approach uses less kinematic structure. Let us first consider the notion of configuration space. In the Newtonian N-body problem, this is defined relative to an inertial system, which serves two purposes. First, it provides a definition of *equilocality* at different instants of time: relative to the system, one can say that a particle is at the same position at two different times. Displacements are then well defined. This is essentially the reason why Newton introduced absolute space. Second, an inertial system brings with it a notion of time difference. Given a notion of simultaneity, so that one can say that the system has some instantaneous configuration at a given instant, that still does not say 'how much time there is' between two different configurations in a history. This extra information is supplied with an inertial system.

Let us now see how much of this structure can be shed without making dynamics impossible. Suppose a universe consisting of N point particles in Euclidean space. It is rather natural to assume that only the relative configuration counts as physical reality. One would like to say that all configurations that can be carried into exact congruence by Euclidean translations and rotations are the same physical configuration. This amounts to quotienting the Newtonian configuration space Q of $3N$ dimensions by these six symmetries (three translations, three rotations) to obtain the *relative configuration space* (RCS) Q_{RCS} of $3N-6$ dimensions [15]. It is also very natural to go one step further and say that size is relative. Then two configurations that can be made congruent by the action of translations, rotations and dilatations are to be regarded as identical. The corresponding quotient space has $3N - 7$ dimensions and may be called *shape space*: Q_{SS}.

A Newtonian dynamical history is a curve in Q traversed at a certain speed relative to some external measure of time. Now time must always be deduced from the motion of some object in the universe that is itself subject to the laws of nature. Therefore, if the system that we are considering is the entire universe, the information from which we deduce time must in fact already be encoded in the curve in Q. We shall see how this is done shortly. But this observation itself already suggests that, if we are considering the dynamics of the entire universe, it ought to be sufficient to set up a theory of *curves* in the configuration space and dispense with the idea that they are traversed at some speed. The simplest curves are those whose determining law is such that an initial point and initial direction suffice to determine the entire curve. They are geodesics. The key difference from Newtonian curves is that only the initial direction, not the initial direction and the speed in that direction, needs to be specified. This is a natural extension to Poincaré's relativity principle. We seek laws that determine histories with the minimum number of initial data. Such laws are *maximally predictive*.

Going from curves traversed at speed to geodesics is one way of the two ways in which we can reduce the number of initial data. The second is to formulate Poincaré's relativity principle on the smaller quotient spaces Q_{RCS} and Q_{SS}. One of the main points I want to make in this paper is that there is a surprising interconnection between these two ways of reducing initial data. The first may be called the elimination of time and the second elimination of potentially redundant geometrical structure. We shall see that the elimination of time is not truly effective unless the employed geometrical structure is pared down to the absolute minimum. There is an unexpected connection between time and geometry, specifically scale invariance.

An overall characterization of the 3-space approach is here appropriate. The basic idea is to postulate the geometrical structure of space, which is assumed here to have three dimensions but in principle any dimension is possible. This paper will mainly be concerned with Euclidean space, though much more interesting possibilities arise if space is Riemannian. Because of length restrictions, I shall only be able to discuss them briefly and will concentrate on the Euclidean case in order to get the idea across. We assume that the space we consider is occupied by geometrical objects, which may either be point particles or fields. These objects and the space in which they reside define configurations and associated configuration spaces. The key point, as already explained, is that one obtains a hierarchy of configuration spaces by quotienting with respect to the symmetries inherent in the space that is presupposed. The mathematical existence of these quotient spaces seems to reflect a deep property of the world. They seem to be a necessary concomitant of the existence of spatial order. The results so far obtained in the 3-space approach suggest that the structure of the quotient spaces determines much more of the fundamental laws of classical physics than has hitherto been supposed. The point is that it is not easy to construct geodesic principles on quotient spaces. If they are to be consistent, they impose strong restrictions.

There exists more than one way to construct geodesic principles on quotient spaces, but one of them seems to be clearly distinguished compared with all the others on account of its geometrical nature. This is based on a principle that we call *best matching*. If we are to define geodesics on any space, we need to define a metric on it. We need to define a distance between any two pairs of neighboring points of the space. The points in our case are complete configurations of the system that we are considering. They necessarily have some small intrinsic difference. Now the basis of geometry is congruence. However, two intrinsically different configurations cannot be brought to exact congruence. Best matching of two such configurations is based on the idea of bringing them, in a well-defined sense, as close as possible to exact congruence and using the 'mismatch' from it to define the distance between them. Geodesics can then be defined with respect to this best-matching metric. As we shall see, the consistent application of this idea leads to interesting restrictions. Before proceeding with the formal development, let me give an idea of the nature of these restrictions.

The benchmark for considering them is standard Newtonian theory. Suppose we take a generic solution of the Newtonian N-body problem of celestial mechanics and successively 'throw away information'. First, instead of giving the time at which the successive configurations are realized we can simply give their sequence. Next, we can omit the information, characterized by six numbers, that specify the overall position and orientation of the system. We can do this by giving only the inter-particle separations r_{ij}. Finally, we can omit the scale information (one number) contained in the r_{ij}. This is most conveniently done by normalizing them by the square root of the moment of inertia I about the center of mass:

$$I = \sum_i m_i \mathbf{x}_i^2 = \frac{1}{M} \sum_{i<j} m_i m_j r_{ij}^2, M = \sum_i m_i. \tag{1}$$

The resulting information can be plotted as a curve in shape space. If this curve were a geodesic, it would be determined by specification of an initial point and initial direction in shape space. For the 3-body problem, shape space has two dimensions, so in this case one would need three numbers: two to specify the initial point, one to specify the initial direction. This is the ideal that must be met by a dynamics of pure shape. It turns out that a generic Newtonian solution needs no less than five further numbers to be fully specified. It is illuminating to consider what they are.

First, it should be noted that in shape space all dimensional information is lost. We have no knowledge of length scales or clock rates. Now the fundamental dynamical quantities in Newtonian theory such as energy, momentum, and angular momentum, as well as Newton's gravitational constant G, depend on these scales, but only scale-invariant quantities can affect the form of the Newtonian curves projected down to shape space. Let us consider what they are. First, at any instant the angular momentum vector has a certain direction relative to the instantaneous configuration. Two numbers are associated with this information.

Next, the Newtonian kinetic energy can be decomposed into a part associated with overall rotation, a part associated with change of shape, and a part associated with change of size. Two independent scale-invariant ratios can be formed from them. These are four of the five numbers. The final number is in many ways the most enigmatic. It is the instantaneous ratio $H = T/V$ of the kinetic energy T to the potential energy V. Intuitively it exists in Newtonian theory because the external time makes it possible to convert displacements into velocities.

We shall see that the transition from Q to Q_{RCS} ensures that the rotational motion associated with angular momentum no longer plays a role. This eliminates three of the above five numbers. However, two still remain. Rather remarkably, both are eliminated when we take the further step to a dynamics of pure shape on Q_{SS}. One of them measures the kinetic energy associated with change of size, and it is no surprise that this is eliminated in a dynamics of pure shape. But the other, related to the energy, seems intuitively to have something to do with time. After all, time and energy form a canonical pair in Hamiltonian dynamics [14]. It is therefore surprising that a scaling requirement appears to have a bearing on time. I shall come back to this later.

In the next section, I shall discuss the formulation of geodesic principles in a way that highlights the difference from Newtonian theory. In Sect. 4, I shall explain the technique of best matching and in Sect. 5 show how Newtonian theory can be recovered to excellent accuracy from a scale-invariant theory. In Sect. 6, I shall indicate how these ideas can be applied to Riemannian geometry and fields and yield a new perspective on general relativity.

3   Jacobi's Principle

The first step to a dynamics of pure shape is the elimination of time by Jacobi's principle [14], which describes all Newtonian motions of one value E of the total energy as geodesics on configuration space. Further discussion of the implications of Jacobi's principle can be found in [10, 16, 17].

For N particles of masses m_i with potential $U(\mathbf{x}_1, \ldots, \mathbf{x}_N)$ and energy E, the Jacobi action is [14]

$$I_{\text{Jacobi}} = 2 \int \sqrt{E - U} \sqrt{\tilde{T}} \mathrm{d}\lambda, \tag{2}$$

where λ labels the points on trial curves and $\tilde{T} = \sum \frac{m_i}{2} \frac{\mathrm{d}\mathbf{x}_i}{\mathrm{d}\lambda} \cdot \frac{\mathrm{d}\mathbf{x}_i}{\mathrm{d}\lambda}$ is the parametrized kinetic energy. The action (2) is timeless since the label λ could be omitted and the mere displacements $\mathrm{d}\mathbf{x}_i$ employed, as is reflected in the invariance of I_{Jacobi} under the reparametrization

$$\lambda \to f(\lambda). \tag{3}$$

In fact, it is much more illuminating to write the Jacobi action in the form

$$I_{\text{Jacobi}} = 2 \int \sqrt{E - U} \sqrt{T^*}, \qquad T^* = \sum \frac{m_i}{2} \mathrm{d}\mathbf{x}_i \cdot \mathrm{d}\mathbf{x}_i, \tag{4}$$

which makes its timeless nature obvious and dispenses with the label λ.

The characteristic square roots of I_{Jacobi} fix the structure of the canonical momenta:

$$\mathbf{p}_i = \frac{\partial \mathcal{L}}{\partial (\mathrm{d}\mathbf{x}_i / \mathrm{d}\lambda)} = m_i \sqrt{\frac{E - U}{\tilde{T}}} \frac{\mathrm{d}\mathbf{x}_i}{\mathrm{d}\lambda}, \tag{5}$$

which, being homogeneous of degree zero in the velocities, satisfy the constraint [18]

$$\sum \frac{\mathbf{p}_i \cdot \mathbf{p}_i}{2m_i} - E + U = 0. \tag{6}$$

The Euler–Lagrange equations are

$$\frac{\mathrm{d}\mathbf{p}^i}{\mathrm{d}\lambda} = \frac{\partial \mathcal{L}}{\partial \mathbf{x}_i} = -\sqrt{\frac{\tilde{T}}{E - U}} \frac{\partial U}{\partial \mathbf{x}_i}, \tag{7}$$

where λ is still arbitrary. If we choose it such that

$$\frac{\tilde{T}}{E - U} = 1 \Rightarrow \tilde{T} = E - U \tag{8}$$

then (5) and (7) become

$$\mathbf{p}_i = m_i \frac{\mathrm{d}\mathbf{x}_i}{\mathrm{d}\lambda}, \qquad \frac{\mathrm{d}\mathbf{p}_i}{\mathrm{d}\lambda} = -\frac{\partial U}{\partial \mathbf{x}_i},$$

and we recover Newton's second law w.r.t this special λ. However, (8), which is usually taken to express energy conservation, becomes the *definition of time*. Indeed, this emergent time, chosen to make the equations of motion take their simplest form [19], is the astronomers' operational ephemeris time [20]. It is helpful to see how 'change creates time'. The increment δt generated by displacements $\delta\mathbf{x}_i$ is

$$\delta t = \frac{\sqrt{\sum m_i \delta\mathbf{x}_i \cdot \delta\mathbf{x}_i}}{\sqrt{2(E - U)}} \equiv \frac{\delta s}{\sqrt{2(E - U)}}. \tag{9}$$

Each particle 'advances time' in proportion to the square root of its mass and to its displacement, the total contribution δs being weighted by $\sqrt{2(E - U)}$.

In the previous section, I discussed the role of the energy in determining the curves of generic Newtonian solutions when projected down to shape space. The Jacobi action (4) illuminates this issue. Considered purely mathematically, T^* by itself already defines a (Riemannian) metric on Q. It is the kinetic metric [14]. The function $(E - U)$ multiplying T^* is a conformal factor that transforms the original kinetic metric, which describes pure inertial motion, into a conformally related metric. It is this conformal factor that introduces forces and the effect of the energy into Newtonian mechanics. The decomposition of the conformal factor into the constant E and the conventional Newtonian potential $-U$, which is a function of the inter-particle separations, is artificial from this point of view. In the development of best matching in the next section, it will be best to start by allowing the conformal factor to be an arbitrary function on Q.

4   Best Matching

The idea of best matching is simple and arises from a very natural problem: How can one quantify the difference between two nearly identical configurations in an intrinsic manner? No additional structure like an inertial system is to be used. In addition, a universally applicable method is required. This is explained in detail in [12,13]. Here I will explain the gist of the method for the case of the N-body problem. Represent the two configurations in the same coordinate grid. In configuration 1, particle i will have coordinates $\mathbf{x}_i$. In configuration 2, it will have coordinates $\mathbf{x}_i + \mathrm{d}\mathbf{x}_i$. Now consider the quadratic form

$$F \sum_i m_i \mathrm{d}\mathbf{x}_i \cdot \mathrm{d}\mathbf{x}_i, \tag{10}$$

where the conformal factor F, assumed positive since we are going to take the square root of (10) in order to obtain a Jacobi-type action, can in principle be an arbitrary function of the coordinates $\mathbf{x}_i$.[1] We can now use the Euclidean generators of translations, rotations and dilatations separately on each of the configurations, generating different 'placings' of them, and calculate (117) for each change made to the pair of configurations. The points they define in shape space will be unchanged by these operations, which merely affect their mathematical representation. The idea of best matching is to seek the minimum of (117) with respect to all possible placings and to declare this to be the metric distance between the configurations. If a consistent scheme is to be obtained, interesting restrictions arise. It is easier to visualize best matching in the finite-difference form just described. However, calculations are more readily done with continuous variations, which correspond to a Jacobi action of the form

$$I_{\mathrm{BM}} = 2 \int \mathrm{d}\lambda \sqrt{F} \sqrt{T}, \quad T = \sqrt{\frac{1}{2} \sum_i m_i \left(\frac{\mathrm{d}\mathbf{x}_i}{\mathrm{d}\lambda} - \mathbf{c}_i \right) \cdot \left(\frac{\mathrm{d}\mathbf{x}_i}{\mathrm{d}\lambda} - \mathbf{c}_i \right)}, \tag{11}$$

where $\mathbf{c}_i$, which has the dimensions of a velocity, is the correction that arises from λ-dependent transformations on the instantaneous configuration generated by translations, rotations and dilatations. For example, consider a λ-dependent translation $\mathbf{x}_i \rightarrow \mathbf{x}_i + \mathbf{b}(\lambda)$. It generates the velocity transformation $\dot{\mathbf{x}}_i \rightarrow \dot{\mathbf{x}}_i + \dot{\mathbf{b}}(\lambda)$, where $\dot{\mathbf{x}}_i = \mathrm{d}\mathbf{x}_i/\mathrm{d}\lambda$. If we make such transformations, (118) without the correction terms will be changed in an arbitrary manner by the arbitrary vector function $\dot{\mathbf{b}}$. To counteract this effect of translations, we take the correction $\mathbf{c}_i$ to be an arbitrary vector function $\mathbf{a}$ and vary the action (118) with respect to it as a Lagrange multiplier. To counteract the effect of simultaneous arbitrary translations, rotations and dilatations, we take the correction to be

$$\mathbf{c}_i = \mathbf{a} + \omega \times \mathbf{x}_i + D\mathbf{x}_i \tag{12}$$

---

[1] Clearly, a more general, non-diagonal form could be assumed instead of (117). This is one of several issues within the 3-space approach that are currently being studied [21].

and vary with respect to the vector functions $\mathbf{a}$ and ω and the scalar function D as Lagrange multipliers. This variation leads to constraints satisfied by the canonical momenta $\mathbf{p}_i$,

$$\mathbf{p}_i = \sqrt{\frac{F}{T}} m_i \left(\frac{\mathrm{d}\mathbf{x}_i}{\mathrm{d}\lambda} - \mathbf{c}_i \right), \tag{13}$$

of the physical variables $\mathbf{x}_i$. The constraints that arise from the translations, rotations and dilatations are, respectively,

$$\mathbf{P} \equiv \sum_i \mathbf{p}_i = 0, \tag{14}$$

$$\mathbf{M} \equiv \sum_i \mathbf{x}_i \times \mathbf{p}_i = 0, \tag{15}$$

$$v \equiv \sum_i \mathbf{x}_i \cdot \mathbf{p}_i. \tag{16}$$

Now comes a crucial point. Do the Euler–Lagrange equations,

$$\frac{\mathrm{d}\mathbf{p}_i}{\mathrm{d}\lambda} = \sqrt{\frac{T}{F}} \frac{\partial F}{\partial \mathbf{x}_i}, \tag{17}$$

propagate the constraints? The simple calculation shows that (14) will propagate only if F is translationally invariant, (15) will propagate only if F is rotationally invariant, and (16) will propagate only if F is homogeneous of degree -2.[2] Note that, as is described in detail in [12], the linear constraints (14), (15), and (16) owe their existence to the fact that (11) is invariant under λ-dependent translations, rotations, and dilatations provided one defines the transformation law of the three correction terms in $\mathbf{c}_i$ to be the same as that of the velocities but with the opposite sign. Besides the three linear constraints, there is also a quadratic constraint analogous to (6):

$$\sum \frac{\mathbf{p}_i \cdot \mathbf{p}_i}{2m_i} - F = 0. \tag{18}$$

This model, with linear constraints that are uniquely determined by the symmetries of space and a quadratic constraint that follows directly from the idea that time is redundant if the dynamics of the universe (as opposed to subsystems of it) is considered, is interesting from several points of view. Before we discuss them, some preparatory remarks are in order. First, the constraints apply only to the complete system of particles treated as an 'island universe'. Subsystems are

---

[2] These conditions, derived here as consistency requirements, are necessary consequences of the fact that best matching is being used to define a metric on a quotient space. In fact, as is shown in [12], each symmetry with respect to which best matching is performed leads to two conditions: a linear constraint on canonical momenta and a condition on the potential F that ensures its propagation.

not constrained. It is only necessary that the contributions of all subsystems sum to zero. Second, it is always possible to employ a coordinate frame in which the corrections terms c_i vanish and a special 'time' label λ for which $F/T = const.$ Then the Euler–Lagrange equations are identical to Newton's equations in an inertial system. Third, the first two linear constraints tell us that in this preferred system the Newtonian momentum and angular momentum of the universe vanish. Because of the Galilean invariance of Newtonian mechanics, the vanishing of the momentum is not a new physical prediction. It is however derived within the logic of best matching. The vanishing of the angular momentum is not enforced by Newton's equations, the rotational symmetry of which only ensures conservation of angular momentum. Best matching enforces both the symmetry and the exact vanishing of the conserved quantity.

Now we come to consider the third linear constraint. This is by far the most drastic in its consequences and also impacts on the quadratic constraint (18). It introduces a new conserved quantity in Newtonian dynamics, which however, from the point of view of Newtonian dynamics, occurs only under very special circumstances. It has long been recognized by N-body specialists that potentials homogenous of degree -2 in the inter-particle separations represent an interesting special case. This follows from the so-called Lagrange–Jacobi relation [12], which gives an universal expression for the time variation of the moment of inertia (1) in any case in which the the potential is homogenous of degree k:

$$\ddot{I} = 4(E - U) - 2kU. \tag{19}$$

Consider Newtonian celestial mechanics, for which $k = -1$. Then $\ddot{I} = 4E - 2U$, from which Lagrange deduced the first qualitative result in dynamics. Since $U < 0$ for gravity, $E \geq 0$ implies $\ddot{I} > 0$. Thus I is concave upwards and must tend to infinity as $t \longrightarrow +\infty$ and $t \longrightarrow -\infty$. In turn, this means that at least one of the interparticle distances must increase unboundedly, so that any system with $E \geq 0$ is unstable.

Another consequence of (19) is the virial theorem. For suppose that the system has virialized, so that $I \approx 0$. Then $4E = (2k + 4)U$.

For our purposes, the most interesting consequence of (19) arises when $k = -2$. For then

$$\ddot{I} = 4E. \tag{20}$$

Thus, I has the parabolic dependence $I = 2Et^2 + bt + c$ on the time and will tend rapidly to zero or infinity. Such a system is extremely unstable, either imploding or exploding.

However, suppose $E = 0$. Then $\ddot{I} = 0$ by (20), so that

$$\dot{I} = 2 \sum m_i \dot{\mathbf{x}}_i \cdot \mathbf{x}_i = 2 \sum \mathbf{p}_i \cdot \mathbf{x}_i = \text{constant}. \tag{21}$$

Thus, $v = \sum \mathbf{p}_i \cdot \mathbf{x}_i$ is a *new conserved quantity*. I am not aware that it has been given any definite name in the literature (or even that its potential significance in dynamics has been recognized). Since it has the same dimensions (action) as angular momentum and is closely analogous to it, I have called it in [12] the *expansive momentum.* It is precisely the quantity that we have found must vanish if best matching with respect to dilatations is applied. It is especially interesting that vanishing of the energy is simultaneously enforced. The reason for this is the drastic consequence of scale invariance. The point is that the kinetic energy has dimensions of length squared, and even under a λ-independent dilatation $\mathbf{x}_i \to D\mathbf{x}_i$ changes by a factor D^2. This already means that the structure of the Jacobi metric, with kinetic metric that describes pure inertial motion and a conformally related metric that describes inertial motion modified by forces, is not possible on shape space Q_{SS} (though it is still possible on Q_{RCS}). One cannot construct a metric on shape space without a compensating potential term F that is homogeneous of degree -2 and therefore transforms as D^{-2}, thereby compensating the D^2 of the kinetic term. This means that, in contrast to the Jacobi action, one cannot have a conformal factor of the form $E - U$ made up of the constant total energy E and a 'proper' potential that depends on the inter-particle separations.

5   Hidden Scale Invariance

If we are to take scale invariance seriously, we must now confront the problem that the standard potentials in Newtonian dynamics, for gravity and electrostatics, derive from potentials homogenous of degree -1, not -2. Nature would appear to be sending us a strong signal that it is not scale invariant. There is, however, a possibility that scale invariance is realized but hidden remarkably effectively. We have seen above in (21) that the time derivative of the moment of inertia (1) is the expansive momentum v. Therefore, if v vanishes the moment of inertia becomes a conserved quantity. Within standard Newtonian theory, this is an exceptional case, requiring the simultaneous vanishing of the energy and the expansive momentum. It is, however, a necessary consequence of scale invariance as defined here. Let us therefore exploit this fact by converting given Newtonian potentials into scale-invariant analogues that have the necessary homogeneity of degree -2. To do this, we shall use I, or rather $\sqrt{MI}$:

$$\mu = \sqrt{\sum_{i<j} m_i m_j r_{ij}^2}. \tag{22}$$

Just as one passes from special to general relativity (with gravity minimally coupled to matter) by replacing ordinary derivatives in the matter Lagrangians by covariant derivatives, Newtonian potentials can be converted into potentials that respect scale invariance. One simply multiplies by an appropriate power of μ, which has the dimensions of length. This is a rather obvious mechanism. What is perhaps unexpected is that the modified potentials lead to forces *identical* to the originals accompanied by a universal cosmological force with minute local effects. The scale invariance is hidden because μ is conserved.

Let some standard Newtonian potential U consist of a sum of potentials U_k each homogeneous of degree k:

$$U = \sum_{k=-\infty}^{\infty} a_k U_k. \tag{23}$$

The a_k are freely disposable strength constants. The energy E in the Jacobi action (2) will be treated as a constant potential ($k = 0$). (It plays a role like the cosmological constant Λ in GR).

Now replace (23) by

$$\tilde{U} = \sum_{k=-\infty}^{\infty} b_k U_k \mu^{-(2+k)}. \tag{24}$$

The equations of motion for (23) are

$$\frac{d\mathbf{p}^i}{dt} = -\sum_{k=-\infty}^{\infty} a_k \frac{\partial U_k}{\partial \mathbf{x}^i};$$

for (24) they are

$$\frac{d\mathbf{p}^i}{dt} = -\sum_{k=-\infty}^{\infty} b_k \mu^{-(2+k)} \frac{\partial U_k}{\partial \mathbf{x}^i} + \sum_{k=-\infty}^{\infty} (2+k) b_k \mu^{-(2+k)} U_k \frac{1}{\mu} \frac{\partial \mu}{\partial \mathbf{x}^i}.$$

Since μ is constant 'on shell', we can define new strength constants that are truly constant:

$$b_k = a_k \mu^{2+k}. \tag{25}$$

The equations for the modified potential become

$$\frac{d\mathbf{p}^i}{dt} = -\sum_{k=-\infty}^{\infty} a_k \frac{\partial U_k}{\partial \mathbf{x}^i} + \sum_{k=-\infty}^{\infty} (2+k) a_k U_k \frac{1}{\mu} \frac{\partial \mu}{\partial \mathbf{x}^i}. \tag{26}$$

If we define

$$C(t) = \frac{\sum_{k=-\infty}^{\infty} (2+k) a_k U_k}{2 \sum_{i<j} m_i m_j r_{ij}^2} \tag{27}$$

and express μ in terms of r_{ij}, then equations (26) become

$$\frac{d\mathbf{p}^i}{dt} = -\sum_{k=-\infty}^{\infty} a_k \frac{\partial U_k}{\partial \mathbf{x}^i} + C(t) \sum_j m_i m_j \frac{\partial r_{ij}^2}{\partial \mathbf{x}^i}. \tag{28}$$

We recover the original forces exactly together with a universal force. It has an epoch-dependent strength constant $C(t)$ and gives rise to forces between all pairs of particles that, like gravitational forces, are proportional to the inertial mass but increase in strength linearly with the distance.[3] The universal force will be attractive or repulsive depending on the sign of $C(t)$, which is an explicit function of the r_{ij}'s. For small enough r_{ij}, the force will be negligible compared with Newtonian gravity. However, on cosmological scales it will be significant. I refer the reader to [12] for a discussion of the possible cosmological implications. Since this paper is primarily concerned with an alternative formulation of the relativity principle and the problem of time, let me conclude this section with some comments about these two issues and then, in the next section, describe what happens in the context of Riemannian geometry and field theory.

The main significance of the model presented here is, I believe, methodological. It shows that one can formulate a powerful constructive relativity principle and implement it universally. One postulates a spatial geometry and the nature of the objects it. Then the geodesic principle and best matching lead to a highly predictive theoretical framework that is completely relational. Above all, observable effects have genuine observable causes. This is because the kinematic structure that is presupposed is pared down to the bare minimum: spatial geometrical relationships quotiented with respect to the spatial symmetries. No extra kinematics associated with time and inertial systems is assumed. Nevertheless, Newton's laws with extra restrictions that do not appear to be in conflict with observations are recovered. The theory clearly cannot fix everything despite the strong restrictions. The potential can still contain several independent terms with arbitrary relative strengths. However, effects without observable material causes are completely eliminated. Specifically, there are no observable effects that one could attribute to translation, rotation or change of size of the complete universe. These results were to be expected, but, very interestingly, the imposition of scale invariance also eliminates the last vestige of what one might call time kinematics: the possibility of including the constant total energy E in Jacobi's principle. In Newtonian theory, different values of E are possible because the external absolute time allows different kinetic energies for a given spatial configuration. Absolute time does seem to be abolished from Jacobi's principle, but its effect is exactly reproduced by the freely specifiable constant E that is not associated with observable sources. There is still a free constant E in the scale-invariant theory, but it is now the coefficient of a genuine potential.

Scale invariance is a highly intuitive and theoretically desirable attribute of any dynamics of the universe. I find it remarkable that it also 'kills time.' It has even more striking consequences in the context of Riemannian geometry and field theory, to which we now turn.

---

[3] Although $C(t)$ is epoch dependent, this does not mean that the theory contains any fundamental coupling constants with such a dependence. The epoch dependence is an artefact of the decomposition of the forces into Newtonian-type forces and a residue, which is the cosmological force.

6   Riemannian Geometry and Fields

This section merely serves as a summary of the content of the papers [10, 11, 13, 21] with emphasis on the connection between time and scale invariance. I shall start by reviewing the manner in which general relativity, which was, of course, originally formulated as a description of four-dimensional pseudo-Riemannian spacetime, can be interpreted as a dynamical theory of the evolution of three-dimensional Riemannian geometry.

A given Einsteinian spacetime, which for simplicity I shall assume has compact spatial sections (closed universe), can, provided it is globally hyperbolic, be foliated by spacelike hypersurfaces ('leaves'), on which a Riemannian geometry is induced. Such foliations can be generated by laying down some system of coordinates on the original spacetime. Then the surfaces of constant value of the time coordinate are the hypersurfaces of the foliation, on which the induced Riemannian geometry is represented by a Riemannian 3-metric g_{ij} defined relative to the spatial coordinates on the given leaf. Such 3-metrics constitute the dynamical variables when general relativity is treated as Hamiltonian theory. The space of all Riemannian 3-metrics g_{ij} on a given 3-manifold $\mathcal{M}$ is called Riem($\mathcal{M}$). It is the analogue of the Newtonian configuration space. All 3-metrics related to each other by (spatial) coordinate transformations, or equivalently 3-diffeomorphisms, correspond to a given 3-geometry. The space of all 3-geometries (on a given manifold) is called *superspace*. Mathematically, superspace is the quotient of Riem with respect to 3-diffeomorphisms. There is an obvious direct analogy between 3-diffeomorphisms, which 'drag the contents of the universe' around on $\mathcal{M}$, and translations and rotations on Euclidean space. Thus, superspace is the analogue of the relative configuration space (RCS) of the particle model.

Given the spacetime and the foliation, each induced 3-metric will be a point in Riem, and the spacetime will be a curve in Riem. Keeping the foliation unchanged (i.e., the t=constant surfaces the same), but changing the spatial coordinates freely on the different slices, one obtains many different curves in Riem that all represent the same spacetime. On superspace, for the given foliation, there is just one curve. However, if one changes the foliation, one obtains a whole family of curves in superspace, one for each foliation. This multiplicity of curves, all representing the same spacetime, reflects the relativity of time in Einstein's theory and has hitherto presented insuperable obstacles to the attempts at a canonical quantization of general relativity. This is *the problem of time* [22, 23]. It is closely related to the question of the true degrees of freedom of the gravitational field. It is generally accepted that in Einstein's theory there are two at each space point. The question is: can one identify them? When general relativity is treated as a dynamical theory, the $g_{0\nu}, \nu = 0, 1, 2, 3$, components of the spacetime metric turnout to be Lagrange multipliers and are not proper degrees of freedom. The remaining six components g_{ij} in the spatial part of the metric contain three arbitrary functions, corresponding to the possibility of making arbitrary coordinate transformations. This gauge freedom is quotiented out in the passage from Riem to superspace. At this level, one is left with three degrees of freedom per space point. However, the freedom to choose the time coordinate arbitrarily, changing thereby the foliation, represents a further gauge function per space point. But what then is evolving? If one wishes to maintain full general covariance, one cannot quotient away that freedom as one does in the passage from Riem to superspace.

However, if one is prepared to sacrifice general covariance, an obvious step is to perform a further quotienting like the passage from the RCS to shape space. The analogue of shape space is *conformal superspace* (CS), which is obtained by quotienting Riem not only by 3-diffeomorphisms but also by conformal transformations:

$$g_{ij} \to \phi^4 g_{ij}, \tag{29}$$

where the fourth power of the positive function ϕ is chosen for mathematical convenience.

Many years ago, York [24] showed that, indeed, one can parametrize the solutions of the initial-value constraints of general relativity by the two degrees of freedom per space point that reside in conformal superspace. Furthermore, he made effective use of the conformal transformations (29) to find such solutions. This important piece of work involved a *constant-mean-curvature* (CMC) foliation. This is defined as follows. At each point on any leaf of a foliation, the leaf has a certain extrinsic curvature tensor K_{ij} (second fundamental form). This measures the manner in which the leaf is curved in the spacetime in which it is embedded. A CMC foliation is one for which the trace of K_{ij}, $K = g_{ij} K^{ij}$, where indices are raised and lowered by means of g_{ij} and its inverse, is constant on each leaf. For spatially compact solutions, a CMC foliation is unique, and K varies monotonically with the cosmic time. CMC foliations have many useful properties and are uniquely helpful in York's method for finding solutions to the initial-value constraints of general relativity. However, because York did not arrive at his technique through a fundamental variational principle but merely exploited what has proved to be very convenient mathematics, the use of CMC foliations is regarded as a gauge-fixing condition that breaks four-dimensional general covariance. It 'fixes time', introducing a definition of simultaneity, which is anathema to many relativists. However, the 3-space approach suggests an interesting alternative interpretation in which the CMC foliation has a deep physical significance.

The stimulus to the development of the 3-space approach was the Lagrangian reformulation of the Dirac–ADM Hamiltonian representation of general relativity [25, 26] found by Baierlein, Sharp and Wheeler [27]. The key concepts in the ADM formalism are the 3-metrics g_{ij}, the lapse N and the shift N^i. The lapse measures the rate of change of proper time w.r.t. the label time, while the shift determines how the coordinates are laid down on the successive 3-geometries. Prior to the transition to the Hamiltonian, the standard Einstein–Hilbert action for matter-free GR is rewritten, after divergence terms have been omitted, in the 3+1 form

$$I = \int dt \int \sqrt{g} N \left[R + K^{ij} K_{ij} - K^2 \right] d^3 x. \tag{30}$$

Here R is the three-dimensional scalar curvature, and $K_{ij} = -(1/2N)$ $(\partial g_{ij}/\partial t - N_{i;j} - N_{j;i})$ is the extrinsic curvature with trace K. From here the transition made by BSW [27] is trivial. They first replaced K_{ij} in the action by $k_{ij} = \partial g_{ij}/\partial t - N_{i;j} - N_{j;i}$, the unnormalised normal derivative, to give

$$I = \int \mathrm{d}t \int \sqrt{g}\left[NR + \frac{1}{4N}\left(k^{ij}k_{ij} - k^2\right)\right]\mathrm{d}^3x. \tag{31}$$

They varied this action with respect to the lapse and found an algebraic expression for it,

$$N = \sqrt{\frac{k^{ij}k_{ij} - k^2}{4R}}. \tag{32}$$

This expression for N is substituted back into (30) to obtain the BSW Lagrangian

$$I_{\mathrm{BSW}} = \int \mathrm{d}t \int \mathcal{L}\mathrm{d}^3x = \int \mathrm{d}t \int \sqrt{g}\sqrt{R}\sqrt{k^{ij}k_{ij} - k^2}\mathrm{d}^3x. \tag{33}$$

This action is closely analogous to the Jacobi action (2): it has a geodesic-type square root and the Lagrange multiplier N_i generates 3-diffeomorphisms in the same way that the multipliers in the particle model generate the Euclidean symmetry transformations. These properties lead, respectively, to the quadratic and linear constraints

$$-p^{ij}p_{ij} + \frac{1}{2}p^2 + gR = 0, \quad p^{ij}{}_{;i} = 0, \quad p = g_{ab}p^{ab}, \tag{34}$$

where the canonical momenta are

$$p^{ij} = \frac{\delta\mathcal{L}}{\delta\left(\frac{\partial g_{ij}}{\partial\lambda}\right)}, \tag{35}$$

and $p = g_{ij}p^{ij}$ is their trace. It measures the expansion of space and is the analogue of the expansive momentum in the particle model.

The constraints (34), which are the fundamental ADM Hamiltonian and momentum constraints, are evidently like the particle constraints but with a crucial difference: instead of being global constraints that hold for the complete universe, these are infinitely many constraints, one per space point. Note, in particular, that the quadratic constraint is an identity that follows from the mere form of the Lagrangian. It is important that the square root is 'local', i.e., it is taken before the integration over space. This ensures that there is a constraint per space point.

In [10–13], my collaborators and I studied systematically gravity–matter-field Lagrangians that are natural generalizations of the BSW Lagrangian and have the form

$$I = \int \mathrm{d}t \int \sqrt{g}\sqrt{U_{\mathrm{g}} + U_{\mathrm{m}}}\sqrt{T_{\mathrm{g}} + T_{\mathrm{m}}}\mathrm{d}^3x. \tag{36}$$

Here, the gravitational potential term U_{g} depends only on the 3-metric g_{ij}, while the matter potential terms U_{m} for the considered scalar and 3-vector fields depend on g_{ij}, the matter fields and their spatial derivatives. The kinetic terms are quadratic in the velocities, as in conventional Hamiltonian field theory, but with the all-important difference that the velocities are 'corrected' to take into account the effect of time-dependent diffeomorphisms and conformal transformations. The corrections are uniquely determined by the symmetry transformation and introduce corresponding Lagrange multipliers and linear momentum constraints, just as in the particle model. However, we included a free coefficient A in $T_g = k^{ij}k_{ij} - Ak^2$ to reflect that, a priori, two independent scalars can contribute to the kinetic term. I will first discuss our results for the matter-free case and best matching with respect to diffeomorphisms.

They revealed an important difference from the particle case, in which propagation of the constraint linear in the momentum imposes conditions on the potential but the quadratic constraint propagates with any potential. In the case of the Riemannian symmetry and the local square root of the BSW action, the two constraints are 'intertwined' and the simultaneous propagation of them imposes strong restrictions. We found that the only consistent Lagrangians of the form (36) must have the free coefficient A equal to unity, as in general relativity, while the gravitational potential term must have the form $U_g = \Lambda + sR, s = 0, 1, -1$. The freely specifiable constant Λ corresponds to the energy E in Jacobi's principle and is Einstein's cosmological constant. The three possible values of s correspond respectively to so-called strong gravity and general relativity with spacetime signatures -+++ and ++++. This meant that we had found a completely new derivation of general relativity by consistent application of the timeless generalization of Poincaré's relativity principle applied in the case of Riemannian 3-geometries. Spacetime was in no way presupposed. It was derived.

Even more remarkable results came when we tried to couple scalar and 3-vector fields to gravity in the case of the Minkowskian signature -+++. In the standard spacetime approach, the fact that such fields must respect the same light cone as the gravitational field is put in the form of the assumption that spacetime in the small is Minkowskian. A universal light cone is presupposed. However, we found that it is enforced by consistent propagation of the two constraints, both of which are modified by the addition of matter terms. A key point is that the form of the momentum constraint is always completely determined by the tensorial nature of the fields and the fact that one is best matching with respect to diffeomorphisms, which affect all fields in a uniquely determined way. In contrast, the form of the quadratic constraint reflects the particular ansatz made for the Lagrangian. Only very special Lagrangians lead to consistent constraint propagation. Thus, our first result for matter was that there must be a universal light cone. Equally striking is the fact that in the case of 3-vector fields the very same requirement of constraint propagation forces the 3-vector fields to be gauge fields. In fact, the universal light cone and gauge theory are shown to have essentially the same origin. Of course, all derivations in theoretical physics include simplicity assumptions, either explicitly or implicitly. These are discussed by Anderson in [21]. I think it is correct to say that the 3-space approach assumes less and derives more than the standard approach based on Einstein's general principle of covariance.

So far, I have described only the effect of the local square root and best matching with respect to diffeomorphisms. In the more recent paper [13], we constructed Lagrangians in which best matching with respect to conformal transformations is also performed. We obtained two main results. First, if one best matches the BSW Lagrangian as it stands with respect to conformal transformations that preserve the spatial volume $V = \int \sqrt{g} \mathrm{d}^3 x$, then the standard Dirac–ADM constraints are augmented by the constraint

$$\frac{p}{\sqrt{g}} = constant, \tag{37}$$

and one also obtains a condition on the lapse that ensures the propagation of this constraint by the Euler–Lagrange equations. When expressed in terms of the extrinsic curvature K, the constraint (37) is precisely York's CMC slicing condition. (Note that $p/\sqrt{g}$ is constant on each leaf, but its value changes under the evolution.) We therefore have the striking result that such conformal transformations, which change all local scales completely freely subject to the single global restriction on the volume, lead us to general relativity in a distinguished foliation, i.e., to a distinguished definition of simultaneity. Once again, we find a strong connection between time and scale invariance. I believe that this result should be taken seriously. In general relativity, there are four gauge freedoms. Three are associated with 3-diffeomorphisms and the fourth with arbitrary transformations of the time coordinate. In our best-matching approach, there are also four gauge freedoms. The 3-diffeomorphisms are still present, but the freedom in the time gauge is replaced by freedom in the scale gauge. All the four gauge freedoms are now expressed through constraints linear in the canonical momenta. Moreover, all have their origin in the geometry of space.

However, from the point of view of scale invariance, general relativity is frustratingly not quite perfect. I find it extremely puzzling that the solitary volume-preserving restriction on full scale invariance is imposed on the conformal transformations. In fact, this is what permits volume to be a physical degree of freedom in general relativity and allows the expansion of the universe. I already mentioned that the trace p of the canonical momenta measures the expansion of space. We see from the constraint (37) that p does not vanish but that $p/\sqrt{g}$ is equal to an evolving spatial constant. When we best match with respect to all conformal transformations, dropping the volume-preserving restriction, we get

$$p = 0. \tag{38}$$

There is now full agreement with the scale-invariant particle model, for which the vanishing of the expansive momentum ensures constancy of the moment of inertia (the 'size' of the N-particle universe). Here, the vanishing of p means that the volume of the universe cannot change. Because the BSW Lagrangian (33) allows the volume to change, it has to be modified if the constraint $p = 0$ is to propagate. Just as constancy of the moment of inertia was achieved by dividing the potential by a power of the moment of inertia, we achieved propagation of $p = 0$ by dividing the gravitational potential by an appropriate power of the volume. This led us to a consistent fully scale-invariant theory that we call *conformal gravity*. It is a remarkably small modification of general relativity. The single global variable that permits the volume of the universe to change is excised and one obtains a dynamics of the geometry of pure shape. Unfortunately, although conformal gravity should describe the solar-system and binary-pulsar data just as well as general relativity, the cosmology must be quite different. At the time of writing, it seems hard to believe that conformal gravity will be able to supplant general relativity as a cosmological theory. The difficulties are spelled out in [12, 13].

However, this probable failure of conformal gravity does not change the fact that the basic idea of using best-matching geodesics to implement Mach's principle on the basis of Poincaré's relativity principle establishes unexpected links between special relativity, general relativity and the gauge principle. They all emerge together as a self-consistent complex from a unified and completely relational approach to dynamics. We see that all of currently dynamics can be understood in terms of purely spatial geometry. Finally, a deep connection between time and shape is established.

It was a great pleasure to participate in DICE2002, and I hope this contribution to the lecture notes will foster further interdisciplinary workshops.

References

1. H.G. Alexander (ed.): *The Leibniz–Clarke Correspondence* (Barnes and Noble, New York 1956), Sect. 47
2. E. Mach: *Die Mechanik in ihrer Entwicklung historisch-kritsch dargestellt* (Barth, Leipzig 1883); *The Science of Mechanics* (Open Court, Chicago 1893)
3. H. Poincaré: *Science and Hypothesis*, translated from the French edition of 1902 (London 1905)
4. A. Einstein: 'Autobiographical Notes'. In: *Albert Einstein – Philosopher – Scientist*, ed. by P.A. Schilpp (The Library of Living Philosophers, Evanston, Illinois 1949)
5. J. Barbour: 'The part played by Mach's Principle in the genesis of relativistic cosmology'. In: *Modern Cosmology in Retrospect*, ed. by B. Bertotti, R. Balbinot, S. Bergia and A. Messina (Cambridge University Press, Cambridge 1990); (see also: J. Barbour: 'The development of Machian themes in the twentieth century'. In: *The Arguments of Time*, ed. by J. Butterfield (Oxford University Press, Oxford 1999))
6. A. Einstein: Ann. d. Phys. **49**, 769 (1916)
7. E. Kretschmann: Ann. d. Phys. **53**, 575 (1917)
8. A. Einstein: Ann. d. Phys. **55**, 241 (1918)
9. J. Barbour: 'On general covariance and best matching'. In: *Physics Meets Philosophy at the Planck Length*, ed. by C. Callender and N. Huggett (Cambridge University Press, Cambridge 2001)
10. J. Barbour, B.Z. Foster and N. Ó Murchadha: Class. Quant. Grav. **19**, 3217 (2002); 'Relativity without relativity', gr-qc/0012089
11. E. Anderson and J. Barbour: Class. Quant. Grav. **19**, 3249 (2002); 'Interacting vector fields in relativity without relativity', gr-qc/0201092
12. J. Barbour: 'Scale-invariant gravity: particle dynamics', gr-qc/0211021, to be published in Class. Quant. Grav.

13. E. Anderson, J. Barbour, B.Z. Foster, and N. Ó Murchadha: 'Scale-Invariant Gravity: Geometrodynamics', gr-qc/0211022, to be published in Class. Quant. Grav.
14. C. Lanczos: *The Variational Principles of Mechanics* (University of Toronto Press, Toronto 1949)(also available from Dover, New York)
15. J. Barbour and B. Bertotti: Proc. Roy. Soc. A (London) **382**, 295 (1982)
16. J. Barbour: Class. Quant. Grav. **11**, 2875 (1994)
17. J. Barbour: *The End of Time* (Weidenfeld and Nicolson, London 1999; Oxford University Press, New York 1999)
18. P.A.M. Dirac: *Lectures on Quantum Mechanics* (Yeshiva University, New York 1964)
19. H. Poincaré: Rev. Métaphys. Morale **6**, 1 (1898) (English translation: 'The Measure of Time'. In: *The Value of Science* (Science Press, New York 1913)
20. G. Clemence: Rev. Mod. Phys. **29**, 2 (1957)
21. E. Anderson: 'Variations on the seventh route to relativity', gr-qc/0302035
22. K. Kuchař: 'Time and interpretations of quantum gravity'. In: *Proceedings of the 4th Canadian Conference on General Relativity and Relativistic Astrophysics*, ed. by G. Kunstatter, D. Vincent and J. Williams (World Scientific, Singapore 1992)
23. C. Isham: 'Canonical quantum gravity and the problem of time'. In: *Integrable Systems, Quantum Groups, and Quantum Field Theories*, ed. by L. Ibart and M. Rodriguez (Kluwer, Amsterdam 1992)
24. J. York: Phys. Rev. Lett. **26**, 1656 (1971); Phys. Rev. Lett. **28**, 1082 (1972)
25. P. Dirac: Proc. Roy. Soc. A (London) **246**, 333 (1958)
26. R. Arnowitt, S. Deser and C. Misner: 'The dynamics of general relativity'. In: *Gravitation: an Introduction to Current Research*, ed. by L. Witten (Wiley, New York 1972)
27. R. Baierlein, D. Sharp, and J. Wheeler: Phys. Rev. **126**, 1864 (1962)

Dynamics without Time for Quantum Gravity: Covariant Hamiltonian Formalism and Hamilton-Jacobi Equation on the Space G

Carlo Rovelli[1,2]

[1] Centre de Physique Théorique de Luminy, CNRS, Case 907, 13288 Marseille, EU
[2] Perimeter Intitute, 35 King Street North, Waterloo, ON, Canada N2J 2W9

Abstract. Hamiltonian mechanics of field theory can be formulated in a generally covariant and background independent manner over a *finite* dimensional extended configuration space. I study the physical symplectic structure of the theory in this framework. This structure can be defined over a space $\mathcal{G}$ of three-dimensional surfaces without boundary, in the extended configuration space. These surfaces provide a preferred over-coordinatization of phase space. I consider the covariant form of the Hamilton-Jacobi equation on $\mathcal{G}$, and a canonical function S on $\mathcal{G}$ which is a preferred solution of the Hamilton-Jacobi equation. The application of this formalism to general relativity is fully covariant and yields directly the Ashtekar-Wheeler-DeWitt equation, the basic equation of canonical quantum gravity. Finally, I apply this formalism to discuss the partial observables of a covariant field theory and the role of the spin networks –basic objects in quantum gravity– in the classical theory.

1 Introduction

Hamiltonian mechanics is a clean and general formalism for describing a physical system, its states and its observables, and provides a road towards quantum theory. In its traditional formulation, however, the hamiltonian formalism is badly non covariant. This is a source of problems already for finite dimensional systems. For instance, the notions of state and observable are not very clean in the hamiltonian mechanics of the systems where evolution is given in parametric form, especially if the evolution cannot be deparametrized (as in certain cosmological models). But the problem is far more serious in field theory. The conventional field theoretical hamiltonian formalism breaks manifest Lorentz invariance badly. More importantly, in a generally covariant context the conventional hamiltonian formalism is cumbersome and its physical interpretation is far from being transparent.

In my opinion, a proper understanding of the generally covariant structure of mechanics is necessary in order to make progress in quantum gravity. The old notions of observable, state, evolution, hamiltonian, and so on, and the tools that are conventionally employed to relate quantum field theory with experience $-S$ matrix, n-point functions and so on, cease to make sense in a genuinely general covariant and background independent context. Therefore wee need to

C. Rovelli, Dynamics without Time for Quantum Gravity: Covariant Hamiltonian Formalism and Hamilton-Jacobi Equation on the Space G, Lect. Notes Phys. **633**, 36–62 (2004)
http://www.springerlink.com/

understand the the generally covariant version of these notions. In particular, if we want to understand quantum field theory in a truly background independent context, we must find a proper definition of quantum field theoretical transition amplitudes, in a form with a clear operational interpretation that makes sense also when there is no background spacetime.

A covariant formulation of the hamiltonian mechanics of finite dimensional systems is possible. Several versions of this formulation can be found in the literature, with different degrees of developments. Perhaps the first to promote this point of view was Lagrange himself, who first understood that the proper covariant definition of phase space is as the space of the physical motions [1], or the space of the solutions of the equations of motion (modulo gauges). Notable contributions, among many others, are Arnold's identification of the presymplectic space with coordinates (t, q^i, p_i) (time, lagrangian variables and their momenta) as the natural home for mechanics [2], and the beautiful and well developed, but little known, formalism of Souriau [3]. Here, I use the covariant version of hamiltonian mechanics described in reference [4], which builds on previous results. This formalism is based on the physical notion of "partial observable" [5]. The partial observables of a non-relativistic finite dimensional mechanical system are the quantities (t, q^i), treated on the same footing. In particular, the formalism treats the time variable t on the same footing as the lagrangian variables q^i. The space of the partial observables is the extended configuration space $\mathcal{C}$ and the hamiltonian formalism is built over this space. The phase space Γ is identified with a space of one-dimensional curves in $\mathcal{C}$. The elements of $\mathcal{C}$ and Γ provide the proper relativistic generalization of the notions of observable and state, consistent with the modification of the notions of space and time introduced by general relativity [4]. The formalism can be manifestly Lorentz covariant and deals completely naturally with reparametrization invariant systems such as the cosmological models.

The extension of these ideas to field theory require the identification of the partial observables of field theory. These are finite in number. They include the coordinates of the spacetime M and the coordinates of the target space T on which the fields take value. Thus the extended configuration space of a field theory is the finite dimensional space $\mathcal{C} = M \times T$. A hamiltonian formalism for field theory built on the finite dimensional space $\mathcal{C} = M \times T$ has been developed by many authors in several variants, developing classical works by Cartan, Weil [6] and DeDonder [7] on the calculus of variations. See for instance [8] and especially the beautiful work [9] and the extended references therein. Here, I refer to the version of hamiltonian mechanics for field theory described in [10], where the accent is on general covariance and relation to observability. The (infinite dimensional) phase space Γ is identified as a space of four-dimensional surfaces in $\mathcal{C}$. The physical symplectic form of Γ, which determines the Poisson brackets, was not given in [10]. Here, I consider a definition of the symplectic form on Γ, in this context. (On a covariant definition of the symplectic structure on the space of the solutions of the field equations, see also [11], whose relation with this work will be briefly discussed below.)

The key of the construction is to introduce the space $\mathcal{G}$. The space $\mathcal{G}$ is the space of the boundary – "initial and final" –lagrangian data (no momenta). In the finite dimensional case, a typical element γ of $\mathcal{G}$ is a pair of points in $\mathcal{C}$. Generically, two points in $\mathcal{C}$ –say (t_0, q_0^i) and (t, q^i)– identify a motion. In the field theoretical case, $\mathcal{G}$ is a space of three-dimensional surfaces γ, without boundary, in $\mathcal{C}$. Again, γ is a set of Lagrangian data sufficient to identify a motion. Indeed γ defines a closed hypersurface in spacetime and the value of the fields over it. I show below that there is a canonical two-form $\omega_{\mathcal{G}}$ on $\mathcal{G}$, and the physical phase space and its physical symplectic form, namely its Poisson brackets, follow immediately from the pair $(\mathcal{G}, \omega_{\mathcal{G}})$.

I then study the covariant version of the Hamilton-Jacobi equation of the system. This can be defined on $\mathcal{G}$. Furthermore, there exist on $\mathcal{G}$ a canonical solution $S[\gamma]$ of the Hamilton-Jacobi equation.

The Hamilton-Jacobi formalism is a window open towards quantum theory. Schrödinger introduced the Schrödinger equation by interpreting the Hamilton-Jacobi equation as the optical approximation of a wave equation [12]. This means searching for an equation for a wave function ψ, solved to lowest order in $\hbar$ by $\psi = Ae^{i/\hbar S}$, if S solves the Hamilton-Jacobi equation. On the basis of this idea, Schrödinger found his celebrated equation by simply replacing each partial derivative of S in the Hamilton-Jacobi function with (-$i\hbar$ times) a partial derivative operator [13]. This same procedure can be used in the covariant formulation of mechanics. The covariant Hamilton-Jacobi equation yields then directly the quantum dynamical equation of the theory. This is the Schrödinger equation in the case of a non-covariant system, or the appropriate "Wheeler-DeWitt" equation for covariant systems. (In fact, the Wheeler-DeWitt equation as well was first found replacing partial derivatives with partial derivative operators in the Hamilton-Jacobi equation of general relativity [14].) For a parametrized system, this procedure shortcuts Dirac's recipe for the quantization of first class constraints, which is cumbersome and has a very cloudy interpretation when applied to such systems. Furthermore, if $S[\gamma]$ is the canonical solution of the Hamilton-Jacobi equation mentioned above, then $\psi[\gamma] = A[\gamma]e^{i/\hbar S[\gamma]}$ is the propagator of the Schrödinger equation, which was identified in [15] as the quantity providing a direct operational interpretation to a generally covariant quantum system.

I then apply this formalism to general relativity (following also [16] and [10].) I use self-dual variables, which much simplify the equations [17,18]. The extended configuration space is identified with the finite-dimensional space $\tilde{\mathcal{C}} = M \times \mathcal{C}$, where M is the four-dimensional manifold of the spacetime coordinates and $\mathcal{C} = R^4 \times sl(2, C)$ where $sl(2, C)$ is the Lorentz algebra. The formalism is simple and straightforward. It shortcuts the intricacies of the conventional hamiltonian formalism of general relativity, and further simplify the one described in [10]. The Hamilton-Jacobi equation yields immediately to the Ashtekar form of the Wheeler-DeWitt equation. (See extended references in [19]. On early use of Hamilton-Jacobi theory in general relativity, see [20].) In the classical as well as in the quantum theory, the M (spacetime) component of $\tilde{\mathcal{C}}$ ends up playing only an auxiliary role and it disappears from observables and states. This reflects the diffeomorphism invariance of the theory.

Finally, I discuss the physical interpretation of the formalism. This is of interest for the interpretation of quantum gravity and, more in general, any background independent quantum field theory. The physical predictions of the theory are given in terms of correlations between partial observables. These can be obtained directly from $S[\gamma]$. In the context of general relativity, this leads to the introductions of spin networks in the classical context, opening a bridge towards the spin networks used in the quantum theory (spin networks were introduced in quantum gravity in [21] and then developed in [22]. See [23] and extended references therein.)

The paper is structured as follows. The main notions are introduced in Sect. 2 in the finite dimensional context. For concreteness, I exemplify all structures introduced by computing them explicitly in the simple case of a free particle. The field theoretical formalism is developed in Sect. 3. As an example, I describe a self-interacting scalar field in Minkowski space. General relativity is treated in Sect. 4. In this paper, I am not concerned with global issues: I deal only with aspects of the theory which are local in $\mathcal{C}$.

2   Finite Dimensional Systems

The exercise that we perform in this section, following and extending the results of [4], is to carefully reformulate classical hamiltonian mechanics and its interpretation in a form that does not require any of the non-relativistic notions that loose meaning in a generally covariant context. These notions are for instance: instantaneous state of the system, evolution in time, evolution of the observables in time, and so on. The alternative notion that we use (correlations, motions . . .) are introduced in detail in [4]. The motivation of this exercise is to introduce the ideas that will then be used for field theory in Sect. 3 and for general relativity in Sect. 4.

2.1   Relativistic Mechanics

Consider a system with n degrees of freedom governed by a Hamiltonian function $H_0(t, q^i, p_i)$, where q^i with $i = 1, \ldots, n$, are coordinates on the configuration space, p_i the corresponding momenta and t the time variable. Let $\mathcal{C}$ be the $(n+1)$-dimensional extended configuration space, that is, the product of the configuration space with the real line, with coordinates $q^a = (t, q^i)$, with $a = 0, \ldots, n$. From now on we work on this extended configuration space and we treat t on the same footing as the configuration space variables. In this form mechanics is general enough to describe fully relativistic systems.

Let $\Omega = T^*\mathcal{C}$ be the cotangent space to $\mathcal{C}$, and $p_a = (p_0, p_i) \equiv (\pi, p_i)$ the momenta conjugate to q^a. Being a cotangent space, Ω carries the canonical one-form $\theta_\Omega = p_a dq^a$. Dynamics is defined on Ω by the relativistic hamiltonian (or hamiltonian constraint)

$$H(q^a, p_a) = \pi + H_0(q^a, p_i) = 0. \tag{1}$$

This equation defines a surface Σ in Ω. It is convenient to coordinatize Σ with the coordinates $(q^a, p_i) = (t, q^i, p_i)$. We denote θ the restriction of θ_Ω to Σ and $\omega = d\theta$. An orbit of ω is a curve without boundaries $m : \tau \to (q^a(\tau), p_i(\tau))$ in Σ whose tangent $X = (\dot{q}^a \partial_a + \dot{p}_i \partial^i)$ satisfies

$$\omega(X) = 0. \tag{2}$$

Here the dot indicates the derivative with respect to τ while ∂_a and ∂^i form the basis of tangent vectors in Σ associated with the coordinates (q^a, p_i). The last equation is equivalent to the Hamilton equations. The projection $\tilde{m}$ of m to $\mathcal{C}$, that is the curve $\tilde{m} : \tau \to q^a(\tau) = (t(\tau), q^i(\tau))$, is a solution of the equations of motion, given in parametric form.

Let Γ be the space of the orbits. There is a natural projection

$$\Pi : \Sigma \to \Gamma \tag{3}$$

that sends a point to the orbits to which it belongs. There is also a unique two-form ω_Γ on Γ such that its pull back to Σ is ω:

$$\Pi_* \omega_\Gamma = \omega. \tag{4}$$

The symplectic space (Γ, ω_Γ) is the physical phase space with its physical symplectic structure.

This formalism, as well as its interpretation, can be immediately generalized to the case in which the coordinates q^a of $\mathcal{C}$ do *not* split into t and q^i and the relativistic hamiltonian does *not* have the particular form (2). Therefore it remains valid for generally covariant or reparametrization invariant systems [4].

Example: Free Particle. Consider a free particle in one dimension. Then $n = 1$, the extended configuration space has coordinates t (the time) and x (the position of the particle). Denote π and p the corresponding momenta. The constraint (2) that defines the free motion is

$$H = \pi + \frac{1}{2m}p^2 = 0. \tag{5}$$

The restriction of the canonical form $\theta_\omega = \pi dt + pdx$ to the surface Σ defined by (5) is

$$\theta = -\frac{p^2}{2m}dt + pdx \tag{6}$$

where we have taken the coordinates (t, x, p) for Σ. The two form ω on Σ is therefore

$$\omega = d\theta = -\frac{p}{m}dp \wedge dt + dp \wedge dx = dp \wedge \left(dx - \frac{p}{m}dt\right). \tag{7}$$

A curve $(t(\tau), x(\tau), p(\tau))$ on Σ has tangent $X = \dot{t}\partial_t + \dot{x}\partial_x + \dot{p}\partial_p$ and inserting this and (7) in the main equation (2) we obtain

$$\omega(X) = -\dot{t}\frac{p}{m}dp + \dot{x}dp - \dot{p}\left(dx - \frac{p}{m}dt\right) = 0. \tag{8}$$

Equating to zero each component, we have

$$\dot{x} = \frac{p}{m} \dot{t}, \qquad \dot{p} = 0. \tag{9}$$

With the solution $x(\tau) = \frac{p}{m} t(\tau) + Q$ and $p(\tau) = P$ where Q and P are constants and $t(\tau)$ arbitrary. The projection of this orbit on $\mathcal{C}$ gives the motions $x = \frac{P}{m} t + Q$.

The space of the orbits is thus parametrized by the two integration constants Q and P. The point (t, x, p) in Σ, belongs to the orbit $Q = x - \frac{p}{m} t$ and $P = p$. Thus the projection (3) is given by $\Pi(t, x, p) = (Q(t, x, p), P(t, x, p)) = (x - \frac{p}{m} t, p)$. Then $\omega_\Gamma = dP \wedge dQ$, because

$$\begin{aligned} \Pi_* \omega_\Gamma &= dP(t, x, p) \wedge dQ(t, x, p) \\ &= dp \wedge d\left(x - \frac{p}{m} t\right) \\ &= dp \wedge \left(dx - \frac{p}{m} dt\right) \\ &= \omega, \end{aligned} \tag{10}$$

as required by the definition (4). See [4] for examples of relativistic systems.

2.2   The Space $\mathcal{G}$

We now introduce a space which is important for what follows. Let $\mathcal{G}$ be defined as

$$\mathcal{G} = \mathcal{C} \times \mathcal{C}. \tag{11}$$

That is, an element γ of $\mathcal{G}$ is an ordered pair of elements of the extended configuration space $\mathcal{C}$:

$$\gamma = (q^a, q_0^a) = (t, q^i, t_0, q_0^i). \tag{12}$$

We think at γ as initial and final conditions for a physical motion: the motion begins at q_0^i at time t_0 and ends at q^i at time t. Generically, given $\gamma = (q^a, q_0^a)$ there is a unique solution of the equations of motion that goes from q_0^a to q^a.[1] That is, there is a curve $(q^a(\tau), p_i(\tau))$, with boundaries, in Σ, with $\tau \in [0, 1]$ such that

$$q^a(0) = q_0^a, \qquad q^a(1) = q^a, \qquad \omega(X) = 0. \tag{13}$$

We denote m_γ this curve, $\tilde{m}_\gamma$ its projection to $\mathcal{C}$ (namely $q^a(\tau)$). Thus γ is the boundary of $\tilde{m}_\gamma$, which we write as $\gamma = \partial \tilde{m}_\gamma$. We denote s and s_0 the initial and final points of m_γ in Σ. Notice that $s = (q^a, p_i)$ and $s_0 = (q_0^a, p_{0i})$, where in general p_i and p_{0i} depend both on q^a as well as on q_0^a.

There is a natural map $i : \mathcal{G} \to \Gamma$ which sends each pair to the orbit that the pair defines. Thus we can define a two-form $\omega_\mathcal{G}$ on $\mathcal{G}$ as $\omega_\mathcal{G} = i^* \omega_\Gamma$. In other words, $\gamma = (q^a, q_0^a) = (t, q^i, t_0, q_0^i)$ can be taken as a natural (over-)

---

[1] More precisely, we define $\mathcal{G}$ as the set of pairs for which this solution exists. If there is more that one solution, we choose the one with minimal action. See below.

coordinatization of the phase space. Instead of coordinatizing a motion with initial positions and momenta, we coordinatize it with initial and final positions. In these coordinates, the symplectic form is given by $\omega_{\mathcal{G}}$.

For what follows, it is important to notice that there is an equivalent, alternative definition of $\omega_{\mathcal{G}}$, which can be obtained without going first through Γ. Indeed, let $\delta\gamma = (\delta q^a, \delta q_0^a)$ be a vector (an infinitesimal displacement) at γ. Then the following is true:

$$\omega_{\mathcal{G}}(\gamma)(\delta_1\gamma, \delta_2\gamma) = \omega_{\mathcal{G}}(q^a, q_0^a)((\delta_1 q^a, \delta_1 q_0^a), (\delta_2 q^a, \delta_2 q_0^a))$$
$$= \omega(s)(\delta_1 s, \delta_2 s) - \omega(s_0)(\delta_1 s_0, \delta_2 s_0). \tag{14}$$

Notice that $\delta_1 s$, the variation of s, is determined by $\delta_1 q$ as well as by $\delta_1 q_0$, and so on. This equation expresses $\omega_{\mathcal{G}}$ directly in terms of ω. As we shall see, this equation admits an immediate generalization in the field theoretical framework, where ω will be a five-form, but $\omega_{\mathcal{G}}$ is a two-form.

Now fix a pair $\gamma = (q^a, q_0^a)$ and consider a small variation of only one of its elements. Say

$$\delta\gamma = (\delta q^a, 0). \tag{15}$$

This defines a vector $\delta\gamma$ at γ on $\mathcal{G}$, which can be pushed forward to Γ. If the variation is along the direction of the motion, then the push forward vanishes, that is $i_*\delta\gamma = 0$, because γ and $\gamma + \delta\gamma$ define the same motion. It follows that if the variation is along the direction of the motion, $\omega_{\mathcal{G}}(\delta\gamma) = 0$. Thus clearly the equation

$$\omega_{\mathcal{G}}(X) = 0. \tag{16}$$

gives again the solutions of the equations of motion.

Thus, the pair $(\mathcal{G}, \omega_{\mathcal{G}})$ contains all the relevant information on the system. The null directions of $\omega_{\mathcal{G}}$ define the physical motions, and if we divide $\mathcal{G}$ by these null directions, the factor space is the physical phase space, equipped with the physical symplectic structure.

Example: Free Particle. The space $\mathcal{G}$ has coordinates $\gamma = (t, x, t_0, x_0)$. Given this point in $\mathcal{G}$, there is clearly one motion that goes from (t_0, x_0) to (t, x), which is

$$t(\tau) = t_0 + (t - t_0)\tau, \tag{17}$$
$$x(\tau) = x_0 + (x - x_0)\tau. \tag{18}$$

Along this motion,

$$p = m\frac{x - x_0}{t - t_0}, \tag{19}$$

$$\pi = -\frac{(x - x_0)^2}{2m(t - t_0)^2}. \tag{20}$$

The map $i : \mathcal{G} \to \Gamma$ is thus given by

$$P = p = m\frac{x - x_0}{t - t_0}, \tag{21}$$

$$Q = x - \frac{p}{m}t = x - \frac{x - x_0}{t - t_0}t, \tag{22}$$

and therefore the two-form $\omega_{\mathcal{G}}$ is

$$\omega_{\mathcal{G}} = i^* \omega_\Gamma = dP(t, x, t_0, x_0) \wedge dQ(t, x, t_0, x_0)$$

$$= m \, d\frac{x - x_0}{t - t_0} \wedge d\left(x - \frac{x - x_0}{t - t_0}t\right)$$

$$= \frac{m}{t - t_0}\left(dx - \frac{x - x_0}{t - t_0}dt\right) \wedge \left(dx_0 - \frac{x - x_0}{t - t_0}dt_0\right). \tag{23}$$

It is immediate to see that a variation $\delta\gamma = (\delta t, \delta x, 0, 0)$ (at constant (x_0, t_0)) such that $\omega_{\mathcal{G}}(\delta\gamma) = 0$ must satisfy

$$\delta x = \frac{x - x_0}{t - t_0}\,\delta t. \tag{24}$$

This is precisely a variation of x and t along the physical motion (determined by (x_0, t_0)). Therefore $\omega_{\mathcal{G}}(\delta\gamma) = 0$ gives again the equations of motion. The two null directions of $\omega_{\mathcal{G}}$ are thus given by the two vector fields

$$X = \frac{x - x_0}{t - t_0}\partial_x + \partial_t, \tag{25}$$

$$X_0 = \frac{x - x_0}{t - t_0}\partial_{x_0} + \partial_{t_0}, \tag{26}$$

which are in involution (their Lie bracket vanishes), and therefore define a foliation of $\mathcal{G}$ with two-dimensional surfaces. These surfaces are parametrized by P and Q, given in (21,22), and in fact

$$X(P) = X(Q) = X_0(P) = X_0(Q) = 0. \tag{27}$$

In fact, we have simply recovered in this way the physical phase space: the space of these surfaces is the phase space Γ and the restriction of $\omega_{\mathcal{G}}$ to it is the physical symplectic form ω_Γ.

2.3 The Function S on the Space $\mathcal{G}$: A Preferred Solution of the Hamilton-Jacobi Equation

Let us now introduce an important function on $\mathcal{G}$. We define $S(\gamma) = S(q^a, q_0^a)$ by

$$S(\gamma) = \int_{m_\gamma} \theta. \tag{28}$$

It is easy to see that this is the value of the action along the path $\tilde{m}_\gamma$. In fact

$$
\begin{aligned}
S(\gamma) &= \int_{m_\gamma} \theta = \int_{m_\gamma} p_a dq^a = \int_0^1 p_a(\tau)\dot{q}^a(\tau)d\tau = \int_0^1 \left(\pi(\tau)\dot{t}(\tau) + p_i(\tau)\dot{q}^i(\tau)\right) d\tau \\
&= \int_0^1 \left(-H_0(\tau)\dot{t}(\tau) + p_i(\tau)\dot{q}^i(\tau)\right) d\tau = \int_{t_0}^t \left(-H_0(\tau) + p_i(\tau)\frac{dq^i(t)}{dt}\right) dt \\
&= \int_{t_0}^t L\left(q^i, \frac{dq^i(t)}{dt}\right) dt,
\end{aligned}
\tag{29}
$$

where L is the Lagrangian. We have from the definition (28) of S,

$$
\frac{\partial S(q^a, q_0^a)}{\partial q^a} = p_a(q^a, q_0^a)
\tag{30}
$$

where p_a is the value of the momenta in s (which depends on q^a and q_0^a). The derivation of this equation is less obvious than what it looks at first sight: see appendix A.

It follows from (30) that S satisfies the (covariant) Hamilton-Jacobi equation [4]

$$
H\left(q^a, \frac{\partial S(q^a, q_0^a)}{\partial q^a}\right) = 0.
\tag{31}
$$

More precisely, S satisfies the Hamilton-Jacobi equation in both sets of variables, namely it satisfies also

$$
H\left(q_0^a, -\frac{\partial S(q^a, q_0^a)}{\partial q_0^a}\right) = 0,
\tag{32}
$$

where the minus sign comes from the fact that the second set of variable is in the lower integration boundary in (28). $S(\gamma)$, defined in (28), is thus a preferred solution of the Hamilton-Jacobi solution.

In view of the generalization to field theory, it is convenient to extend the definition of $\mathcal{G}$ and $S(\gamma)$ as follow. Define

$$
S(q_f^a, q^a, q_i^a) = \int_{m_{q_f^a, q^a}} \theta + \int_{m_{q^a, q_i^a}} \theta.
\tag{33}
$$

This can be seen as an extension of the definition (28) in the following sense. We can view the path $m = m_{q_f^a, q^a} \cup m_{q^a, q_i^a}$ as a path in $\mathcal{C}$ bounded by the three points (q_f^a, q^a, q_i^a). This form of S will be convenient for extracting physical information from it, and for the extension to field theory.

Example: Free Particle. The function $S(\gamma)$ is easily computed.

$$
S(t, x, t_0, x_0) = \int_0^1 (\pi \dot{t} + p\dot{x}) = \pi \int_{t_0}^t dt + p \int_{x_0}^x dx
$$

$$= -\frac{m(x-x_0)^2}{2(t-t_0)} + m\frac{(x-x_0)^2}{t-t_0}$$

$$= \frac{m(x-x_0)^2}{2(t-t_0)}. \tag{34}$$

We now verify that it satisfies the Hamilton-Jacobi equation of the relativistic hamiltonian (5), which is

$$H\left(q^a, \frac{\partial S}{\partial q^a}\right) = \frac{\partial S}{\partial t} + \frac{1}{2m}\left(\frac{\partial S}{\partial x}\right)^2 = 0. \tag{35}$$

Easily

$$\frac{\partial S}{\partial t} = -\frac{m(x-x_0)^2}{2(t-t_0)^2}, \qquad \frac{\partial S}{\partial x} = \frac{m(x-x_0)}{(t-t_0)}. \tag{36}$$

So that the Hamilton-Jacobi equation (35) is satisfied.

Notice that $S(x, t, x_0, t_0)$ is strictly related to the quantum theory. The Schrödinger equation can be obtained from the relativistic Hamiltonian-Jacobi equation as

$$H\left(q^a, -i\hbar\frac{\partial}{\partial q^a}\right)\psi(q^a) = 0. \tag{37}$$

for a wave function $\psi(q^a)$ on the extended phase space. In fact, this gives immediately

$$\left(-i\hbar\frac{\partial}{\partial t} - \frac{\hbar^2}{2m}\frac{\partial^2}{\partial x^2}\right)\psi(x,t) = 0. \tag{38}$$

which is the conventional time dependent Schrödinger equation. Its propagator $W(x, t, x_0, t_0)$, which satisfies the equation itself

$$\left(-i\hbar\frac{\partial}{\partial t} - \frac{\hbar^2}{2m}\frac{\partial^2}{\partial x^2}\right)W(x,t,x_0,t_0) = 0 \tag{39}$$

is

$$W(x,t,x_0,t_0) = \frac{1}{\sqrt{t-t_0}}\, e^{\frac{i}{\hbar}S(x,t,x_0,t_0)}. \tag{40}$$

Therefore S is the phase of the propagator.

2.4   Physical Predictions

Finally, let us discuss the relation between the formalism described and the physical interpretation of the theory. If the function $S(\gamma)$ on $\mathcal{G}$ is known explicitly, the general solution of the equation of motion is

$$f^a(q^a, q_0^a, p_a^0) = \frac{\partial S(q^a, q_0^a)}{\partial q_0^a} + p_a^0 = 0. \tag{41}$$

We view (41) as an equation for q^a; the quantities p_a^0 and q_0^a are constants determining one solution. The solution defines a curve in $\mathcal{C}$, namely a motion. Physically, once we have determined a motion by means of p_a^0 and q_0^a, equation (41) determines whether or not the correlation q^a can be observed. (See [4] for more details.) In general, there is a redundancy in the system (41): one of the equations ($a = 0$ for non relativistic systems) is a consequence of the others.

For instance, in the case of a free particle, inserting (34) into (41) gives, for $q^a = x$

$$\frac{\partial S(t, x, t_0, x_0)}{\partial x_0} + p^0 = 0. \tag{42}$$

Inserting the explicit form of S given in (34), we obtain

$$x - x_0 = \frac{p^0}{m}(t - t_0). \tag{43}$$

which is the correct relation that relates x and t, at constant values of x_0, t_0, p^0. The other equation, obtained for $q^a = t$, does not give anything new.

There is another way of using the solution of the Hamilton-Jacobi to obtain physical predictions, which is of interest in view of it generalization to quantum field theory. Fix two points q_i^a and q_f^a in $\mathcal{C}$. We can ask if a third point q^a can lay on the same motion as q_i^a and q_f^a. This means asking wether or not we could observe the correlation q^a, given that the correlations q_i^a and q_f^a are observed. A moment of reflection will convince the reader that the answer to this question is positive if and only if

$$\frac{\partial S(q_f^a, q_i^a)}{\partial q_f^a} = \frac{\partial S(q_f^a, q^a)}{\partial q_f^a}, \tag{44}$$

$$\frac{\partial S(q_f^a, q_i^a)}{\partial q_i^a} = \frac{\partial S(q^a, q_1^a)}{\partial q_i^a}. \tag{45}$$

Using the definition (33), this becomes

$$\frac{\partial S(q_f^a, q^a, q_i^a)}{\partial q_\tau^a} = \frac{\partial S(q_f^a, q_i^a)}{\partial q_\tau^a}, \tag{46}$$

with $\tau = i, f$.

Alternatively, we may notice that a first order variation of q^a does not change $S(q_2^a, q^a, q_1^a)$, because the two derivatives of the first and second term in (33) cancel out if m is a motion (and therefore the two momenta at q^a are equal). Thus the condition can be reformulated as

$$\frac{\partial S(q_f^a, q^a, q_i^a)}{\partial q^a} = 0. \tag{47}$$

This equation is easily recognized as a corollary of the principle that the action is an extremum on the motions.

Consider now the same question –whether or not we can observe the correlation q^a- in the quantum theory. The quantum theory does not provide a deterministic yes/no answer to a physical question; it only provides the probability (amplitude) for a positive answer. In the classical theory, the two correlations q_i^a and q_f^a determine a state. In the quantum theory, a (generalized) state of a finite dimensional system is determined by a single point in $\mathcal{C}$, because the detection of the correlation q_i^a is incompatible with the detection of the correlation q_f^a, in the sense of Heisenberg: the detection of the second erases the information on the first. If the state is the generalized state determined by the correlation q_1^a, then the probability density amplitude for detecting the correlation q^a is determined by the propagator $W(q^a, q_i^a)$, defined on $\mathcal{G}$. See [15] for details. Since $S(q^a, q_i^a)$ is the phase of the propagator, equations (45) follow then from the standard optical approximation for the behavior of the wave packets. In this sense the classical interpretation of the formalism can be recovered from the quantum one.

3   Field Theory

3.1   Field Theoretical Relativistic Mechanics

Consider a field theory on Minkowski space M. Let x^μ, where $\mu = 0, 1, 2, 3$, be Minkowski coordinates and call $\phi^A(x^\mu)$ the field, where $A = 1, \ldots, N$. The field is a function $\phi : M \to T$, where $T = \mathbb{R}^N$ is the target space, namely the space in which the field takes values. The extended configuration space of this theory is the finite dimensional space $\mathcal{C} = M \times T$, with coordinates $q^a = (x^\mu, \phi^A)$. In fact, the coordinates of this space correspond to the $(4+N)$ partial observables whose relations are described by the theory [5, 10]. A solution of the equations of motion defines a four-dimensional surface $\tilde{m}$ in $\mathcal{C}$. If we coordinatize this surface using the coordinates x^μ, then this surface is given by $[x^\mu, \phi^A(x^\mu)]$, where $\phi^A(x^\mu)$ is a solution of the field equations. If, alternatively, we use an arbitrary parametrization with parameters $\tau^\rho, \rho = 0, 1, 2, 3$, then the surface is given by $[x^\mu(\tau^\rho), \phi^A(\tau^\rho)]$, and $\phi^A(x^\mu)$ is determined by $\phi^A(x^\mu(\tau^\rho)) = \phi^A(\tau^\rho)$.

In the case of a finite numbe of degrees of freedom (and no gauges), motions are given by one-dimensional curves. At each point of the curve, there is one tangent vector, and momenta coordinatize the one-forms. In field theory, motions are four-dimensional surfaces, and have four independent tangents at each point. Accordingly, momenta coordinatize the four-forms. Let $\Omega = \Lambda^4 T^* \mathcal{C}$, be the bundle of the four-forms $p_{abcd} dq^a \wedge dq^b \wedge dq^c \wedge dq^d$ over $\mathcal{C}$. A point in Ω is thus a pair (q^a, p_{abcd}). The space Ω carries the canonical four-form $\theta_\Omega = p_{abcd} dq^a \wedge dq^b \wedge dq^c \wedge dq^d$. It is convenient to use the notation $p_{\mu\nu\rho\sigma} = \pi \epsilon_{\mu\nu\rho\sigma}$ and $p_{A\nu\rho\sigma} = p_A^\mu \epsilon_{\mu\nu\rho\sigma}$.

The hamiltonian theory can be defined on Ω by the relativistic hamiltonian system

$$p_{ABCD} = p_{ABC\mu} = p_{AB\mu\nu} = 0, \tag{48}$$

$$H = \pi + H_0(x^\mu, \phi^A, p_A^\mu) = 0. \tag{49}$$

where H_0 is DeDonder's covariant Hamiltonian [7] (see below for an example). This system defines a surface Σ in Ω. It is convenient to take coordinates (x^μ, ϕ^A, p^μ_A) on Σ. As before, we denote θ the restriction of θ_Ω to Σ and $\omega = d\theta$. On the surface defined by (48), θ_Ω becomes the canonical four-form

$$\theta = \pi \, d^4x + p^\mu_A \, d\phi^A \wedge d^3x_\mu, \tag{50}$$

where we have introduced the notation $d^4x = dx^0 \wedge dx^1 \wedge dx^2 \wedge dx^3$ and $d^3x_\mu = d^4x(\partial_\mu) = \frac{1}{3!}\epsilon_{\mu\nu\rho\sigma}dx^\nu \wedge dx^\rho \wedge dx^\sigma$. On Σ, defined by (48) and (49),

$$\theta = -H_0(x^\mu, \phi^A, p^\mu_A) \, d^4x + p^\mu_A \, d\phi^A \wedge d^3x_\mu, \tag{51}$$

and ω is the five-form

$$\omega = -dH_0(x^\mu, \phi^A, p^\mu_A) \wedge d^4x + dp^\mu_A \wedge d\phi^A \wedge d^3x_\mu. \tag{52}$$

An orbit of ω is a four-dimensional surface m immersed in Σ, such that at each of its points a quadruplet $X = (X_1, X_2, X_3, X_4)$ of independent tangents to the surface satisfies

$$\omega(X) = 0. \tag{53}$$

The projection of an orbits on $\mathcal{C}$ gives a solution of the field equations.

More in detail, let $(\partial_\mu, \partial_A, \partial^A_\mu)$ be the basis in the tangent space of Σ determined by the coordinates (x^μ, ϕ^A, p^μ_A). Parametrize the surface with arbitrary parameters τ^ρ, so that the surface is given by $[x^\mu(\tau^\rho), \phi^A(\tau^\rho), p^\mu_A(\tau^\rho)]$. Let $\partial_\rho = \partial/\partial\tau^\rho$. Then let

$$X_\rho = \partial_\rho x^\mu(\tau^\rho) \, \partial_\mu + \partial_\rho \phi^A(\tau^\rho) \, \partial_A + \partial_\rho p^\mu_A(\tau^\rho) \, \partial^A_\mu. \tag{54}$$

Then $X = X_0 \otimes X_1 \otimes X_2 \otimes X_3$ is a rank four tensor on Σ. If $\omega(X) = 0$, then $\phi^A(x^\mu)$ determined by $\phi^A(x^\mu(\tau^\rho)) = \phi^A(\tau^\rho)$ is a solution of the equations of motion.

The formalism as well as its interpretation can be immediately generalized to the case in which the coordinates of $\mathcal{C}$ do not split into x^μ and ϕ^A and the relativistic hamiltonian does not have the particular form (48-49) [10].

Example: Scalar Field. Consider a scalar field $\phi(x^\mu)$ on Minkowski space, satisfying the field equations

$$\partial_\mu \partial^\mu \phi(x^\mu) + m^2 \phi(x^\mu) + V'(\phi(x^\mu)) = 0. \tag{55}$$

Here the Minkowski metric has signature $[+, -, -, -]$ and $V'(\phi) = dV(\phi)/d\phi$. The field is a function $\phi : M \to T$, where here $T = \mathbb{R}$. The extended configuration space of this theory is the five dimensional space with coordinates (x^μ, ϕ). The space Ω has coordinates $(x^\mu, \phi, \pi, p^\mu)$ (equation (48) is trivially satisfied) and carries the canonical four-form

$$\theta_\Omega = \pi \, d^4x + p^\mu \, d\phi \wedge d^3x_\mu; \tag{56}$$

The dynamics is defined on this space by the DeDonder relativistic hamiltonian [7]

$$H_0 = \frac{1}{2} \left(p^\mu p_\mu + m^2 \phi^2 + 2V(\phi) \right). \tag{57}$$

The form ω is thus the five-form

$$\omega = - \left(p^\mu dp_\mu + m^2 \phi d\phi + V'(\phi)d\phi \right) \wedge d^4 x + dp^\mu \wedge d\phi \wedge d^3 x_\mu. \tag{58}$$

A straightforward calculation shows that $\omega(X) = 0$ gives

$$\partial_\mu \phi(x^\mu) = p_\mu(x^\mu), \tag{59}$$
$$\partial_\mu p^\mu(x^\mu) = -m^2 \phi(x^\mu) - V'(\phi(x^\mu)). \tag{60}$$

and therefore precisely the field equations (55). Notice that the formalism is manifestly Lorentz covariant, and that no equal time initial data surface has to be chosen.

3.2   The Space $\mathcal{G}$ and the Physical Symplectic Structure

The phase space Γ is defined as the space of the orbits, as in the finite dimensional case. However, notice that now there is no natural projection map π from Σ to Γ, because a point in Σ may belong to many different orbits. It follows that we cannot define a symplectic two-form on the phase space Γ by simply requiring that its pull back with π is ω. As we shall see now, however, the problem can be circumvented.

The key step is to identify the space $\mathcal{G}$. Recall that in the finite dimensional case $\mathcal{G}$ was the Cartesian product of the extended configuration space with itself. The same cannot be true in the field theoretical context, because the proper characterization of $\mathcal{G}$ is as the space of the boundary configuration data that can specify a solution. In field theory, we obviously need an infinite number of data to characterize a solution, therefore $\mathcal{G}$ must be infinite dimensional. The key observation is that in the finite dimensional case $\mathcal{G}$ is the space of the possible *boundaries* of a portion of a motion in $\mathcal{C}$. In the field theoretical context, a portion of a motion is a 4d surface in $\mathcal{C}$ with boundaries. Its boundary is a three-dimensional surface γ. The surface γ bounds a four-dimensional surface $\tilde{m}$, and therefore has no boundaries itself.

Thus, we take $\mathcal{G}$ to be a space of oriented three-dimensional surfaces γ without boundaries in $\mathcal{C}$. The 3d surface γ does not need to be connected. In fact, it is sometimes convenient to think at γ as having two connected components: the initial component and the final component.

Let us coordinatize γ with coordinates $\boldsymbol{\tau} = (\tau^1, \tau^2, \tau^3)$. Then γ is given as $\gamma = [x^\mu(\boldsymbol{\tau}), \phi^A(\boldsymbol{\tau})]$. Notice that $x^\mu(\boldsymbol{\tau})$ defines a 3d surface without boundaries in Minkowski space, which we call γ_M, while $\phi^A(\boldsymbol{\tau})$ determines the value of the field on this surface. The surface in Minkowski space γ_M is the boundary of a connected region V_M of M. A solution of the equation of motion is determined by the value of the field on the boundary. (This is the generic situation, since if two solutions agree on a closed 3d surface, generically they agree in the interior.) Thus, γ determines a solution $\tilde{m}$ of the equations of motion in the interior V_M. Furthermore, let m be the lift of $\tilde{m}$ to Σ. That is, let m be the portion of an orbit of ω that projects down to $\tilde{m}$. Finally, let s_γ be the 3d surface in Σ that bounds m. That is, $s_\gamma = [x^\mu(\tau), \phi^A(\tau), p^\mu_A(\tau)]$, where $p^\mu_A(\tau)$ is determined by the solution of the field equations determined by the entire γ.

We can now define a two-form on $\mathcal{G}$ as follows

$$\omega_\mathcal{G}[\gamma] = \int_{s_\gamma} \omega. \tag{61}$$

The form $\omega_\mathcal{G}$ is a two-form: it is the integral of a five-form over a 3d surface. More precisely, let $\delta\gamma$ be a small variation of γ. This variation can be seen as a vector field $\delta\gamma(\tau)$ defined on γ. This variation determines a corresponding small variation δs_γ, which, in turn, is a vector field $\delta s_\gamma(\tau)$ over s_γ. Then

$$\omega_\mathcal{G}[\gamma](\delta_1\gamma, \delta_2\gamma) = \int_{s_\gamma} \omega(\delta_1 s_\gamma, \delta_2 s_\gamma). \tag{62}$$

Thus, the five-form ω on the finite dimensional space Σ defines the two-form $\omega_\mathcal{G}$ on the infinite dimensional space $\mathcal{G}$.

Now, consider a small local variation $\delta\gamma$ of γ. This means varying the surface γ_M in Minkowski space, as well as varying the value of the field over it. Assume that this variation satisfies the field equations: that is, the variation of the field is the correct one, for the solution of the field equations determined by γ. We have

$$\omega_\mathcal{G}[\gamma](\delta\gamma) = \int_{s_\gamma} \omega(\delta s_\gamma). \tag{63}$$

But the variation δs_γ is by construction along the orbit, namely in the null direction of ω and therefore the right hand side of this equation vanishes. It follows that if $\delta\gamma$ is an infinitesimal physical motion, then

$$\omega_\mathcal{G}(\delta\gamma) = 0. \tag{64}$$

In conclusion, the pair $(\mathcal{G}, \omega_\mathcal{G})$ contains all the relevant information on the system. The null directions of $\omega_\mathcal{G}$ determine the variations of the three-surfaces γ along the physical motions. The space $\mathcal{G}$ divided by these null directions, namely the space of the orbits of these variations is the physical phase space Γ, and the $\omega_\mathcal{G}$, restricted to this space, is the physical symplectic two-form of the system.

Example: Scalar Field. Let us now compute $\omega_\mathcal{G}$ in a slightly more explicit form for the example of the scalar field. From the definition (61),

$$\omega_\mathcal{G}[\gamma] = \int_{s_\gamma} \omega = \int_{s_\gamma} d\pi \wedge d^4x + dp^\mu \wedge d\phi \wedge d^3x_\mu$$

$$= \int_{s_\gamma} (p^\nu dp_\nu + \phi d\phi + V' d\phi) \wedge d^4x + dp^\mu \wedge d\phi \wedge d^3x_\mu$$

$$= \int_{\gamma_M} d^3x_\nu \left((p_\mu - \partial_\mu \phi) dp^\mu \wedge dx^\nu + (\phi + V' + \partial_\mu p^\mu) d\phi \wedge dx^\nu + dp^\nu \wedge d\phi \right)$$

$$= \int_{\gamma_M} d^3x_\nu \ dp^\nu \wedge d\phi. \tag{65}$$

where we have used the x^μ coordinates themselves as integration variables, and therefore the integrand fields are the functions of the x^μ's. Notice that since the integral is on s_γ, the p^μ in the integrand is the one given by the solution of the field equation determined by the data on γ. Therefore it satisfies the equations of motion (59-60), which we have used above. Using (59) again, we have

$$\omega_{\mathcal{G}}[\gamma] = \int_{\gamma_M} d^3x \ n_\nu \ d(\nabla^\nu \phi) \wedge d\phi. \tag{66}$$

In particular, if we consider variations $\delta\gamma$ that do not move the surface and such that the change of the field on the surface is $\delta\phi(x)$, we have

$$\omega_{\mathcal{G}}[\gamma](\delta_1\gamma, \delta_2\gamma) = \int_{\gamma_M} d^3x \ n_\nu \left(\delta_1\phi \nabla^\nu \delta_2\phi - \delta_2\phi \nabla^\nu \delta_1\phi \right). \tag{67}$$

This formula can be directly compared with the expression of the symplectic two-form given on the space of the solutions of the field equations in [11]. The expression is the same, but with a nuance in the interpretation: $\omega_{\mathcal{G}}$ is not defined on the space of the solutions of the field equations – it is defined on the space of the lagrangian data $\mathcal{G}$, and the normal derivative $n_\nu \nabla^\nu \phi$ of these data is determined by the data themselves via the field equations.

3.3 Hamilton-Jacobi

Let us now construct the function S on $\mathcal{G}$. We define as in the finite dimensional case

$$S[\gamma] = \int_{m_\gamma} \theta. \tag{68}$$

Again, it is easy to see that this is in fact the value of the action of the solution $\tilde{m}_\gamma$. For the scalar field, for instance

$$S[\gamma] = \int_{m_\gamma} \theta = \int_{m_\gamma} (\pi d^4x + p^\mu d\phi \wedge d^3x_\mu) = \int_{V_\gamma} (\pi + p^\mu \partial_\mu \phi) \, d^4x$$

$$= \int_{V_\gamma} \left(-\frac{1}{2} p^\mu p_\mu - \frac{1}{2} m^2 \phi^2 - V(\phi) + p^\mu \partial_\mu \phi \right) \, d^4x$$

$$= \int_{V_\gamma} \left(\frac{1}{2} \partial_\mu \phi \partial^\mu \phi - \frac{1}{2} m^2 \phi^2 - V(\phi) \right) \, d^4x$$

$$= \int_{V_\gamma} L(\phi, \partial_\mu \phi) \, d^4x, \tag{69}$$

where L is the Lagrangian density, and we have used the equation of motion $p_\mu = \partial_\mu \phi$.

We have from the definition

$$\frac{\delta S[\gamma]}{\delta x^\mu(\boldsymbol{\tau})} = \pi(\boldsymbol{\tau})\, n_\mu(\boldsymbol{\tau}) + \epsilon_{\mu\nu\rho\sigma}\, p^\nu(\boldsymbol{\tau})\, \partial_i\phi(\boldsymbol{\tau})\, \partial_j x^\rho(\boldsymbol{\tau})\, \partial_k x^\sigma(\boldsymbol{\tau})\, \epsilon^{ijk} \quad (70)$$

where π depends on the full γ, and $n_\mu(\boldsymbol{\tau}) = \frac{1}{3!}\epsilon_{\mu\nu\rho\sigma}\partial_1 x^\nu(\boldsymbol{\tau})\partial_2 x^\rho(\boldsymbol{\tau})\partial_3 x^\sigma(\boldsymbol{\tau})$ is the normal to the three-surface γ_M. Also

$$\frac{\delta S[\gamma]}{\delta\phi(\boldsymbol{\tau})} = p^\mu(\boldsymbol{\tau})n_\mu(\boldsymbol{\tau}). \quad (71)$$

The derivation of these two equations requires steps analogous to the one we used to derive (30). See the appendix for details.

Now, from (49) and (57) we have, for the scalar field

$$\pi + \frac{1}{2}\left(p^\mu p_\mu + m^2\phi^2 + 2V(\phi)\right) = 0. \quad (72)$$

We split p_μ in its normal ($p = p^\mu n_\mu$) and tangential (p^i) components (so that $p^\mu = p^i\partial_i x^\mu + pn^\mu$) and from (70) we have

$$n^\mu(\boldsymbol{\tau})\frac{\delta S[\gamma]}{\delta x^\mu(\boldsymbol{\tau})} = \pi(\boldsymbol{\tau}) - p^i(\boldsymbol{\tau})\,\partial_i\phi(\boldsymbol{\tau}). \quad (73)$$

Using this, (71), and the field equations (70), we obtain

$$\frac{\delta S[\gamma]}{\delta x^\mu(\boldsymbol{\tau})}n_\mu(\boldsymbol{\tau}) + \frac{1}{2}\left[\left(\frac{\delta S[\gamma]}{\delta\phi(\boldsymbol{\tau})}\right)^2 + \partial_j\phi(\boldsymbol{\tau})\partial^j\phi(\boldsymbol{\tau}) + m^2\phi^2(\boldsymbol{\tau}) + 2V(\phi(\boldsymbol{\tau}))\right] = 0. \quad (74)$$

This is the Hamilton-Jacobi equation of the theory. Notice that the function $S[\gamma] = S[x^\mu(\boldsymbol{\tau}), \phi(\boldsymbol{\tau})]$ is a function of the surface, not the way the surface is parametrized. Therefore it is invariant under a change of parametrization. It follows that

$$\frac{\delta S[\gamma]}{\delta x^\mu(\boldsymbol{\tau})}\partial_j x^\mu(\boldsymbol{\tau}) + \frac{\delta S[\gamma]}{\delta\phi(\boldsymbol{\tau})}\partial_j\phi(\boldsymbol{\tau}) = 0. \quad (75)$$

(This equation can be obtained also from the tangential component of (70).) The two equations (74) and (75) govern the Hamilton-Jacobi function $S[\gamma]$.

The connection with the non-relativistic field theoretical Hamilton-Jacobi formalism is the following. We can restrict the formalism to a preferred choice of parameters $\boldsymbol{\tau}$. Choosing $\tau^j = x^j$, we obtain S in the form $S[t(\boldsymbol{x}), \phi(\boldsymbol{x})]$ and the Hamilton-Jacobi equation (74) becomes

$$\frac{\delta S}{\delta t(\boldsymbol{x})} + \frac{1}{2}\left[\left(\frac{\delta S[\gamma]}{\delta\phi(\boldsymbol{x})}\right)^2 + \partial_j\phi\partial^j\phi + m^2\phi^2 + 2V(\phi)\right] = 0. \quad (76)$$

Further restricting the surfaces to the ones of constant t gives the functional $S[t, \phi(\boldsymbol{x})]$, satisfying the Hamilton-Jacobi equation

$$\frac{\partial S}{\partial t} + \frac{1}{2} \int d^3 x \left[\left(\frac{\delta S}{\delta \phi(\boldsymbol{x})} \right)^2 + |\boldsymbol{\nabla} \phi|^2 + m^2 \phi^2 + 2V(\phi) \right] = 0, \qquad (77)$$

which is the usual non-relativistic Hamilton-Jacobi equation

$$\frac{\partial S}{\partial t} + \mathcal{H} \left(\phi, \boldsymbol{\nabla} \phi, \frac{\delta S[\gamma]}{\delta \phi(\boldsymbol{x})} \right) = 0, \qquad (78)$$

where $\mathcal{H}(\phi, \boldsymbol{\nabla} \phi, \partial_t \phi)$ is the non-relativistic hamiltonian.

3.4   Physical Predictions

As in the case of finite dimensional systems, if $S[\gamma] = S[x^\mu(\boldsymbol{\tau}), \phi(\boldsymbol{\tau})]$ is known explicitly, the general solution of the equation of motion can be obtained by derivations. For instance, let γ be formed by two connected components that can be viewed as a past and a future Cauchy surfaces γ_{in} and γ_{out}, parametrized by $\boldsymbol{\tau}_{in}$ and $\boldsymbol{\tau}_{out}$ respectively. Consider the equation

$$\frac{\delta S[\gamma_{out} \cup \gamma_{in}]}{\delta \phi(\boldsymbol{\tau}_{in})} = \frac{\delta S[\tilde{\gamma}_{out} \cup \gamma_{in}]}{\delta \phi(\boldsymbol{\tau}_{in})} \qquad (79)$$

for the variable $\tilde{\gamma}_{out}$, where γ_{in} and γ_{out} are held fix. All the solutions $\tilde{\gamma}_{out}$ of this equation sit on the same motion. That is, this equation determines which are the $\tilde{\gamma}_{out}$ that are compatible with a given state. The situation is completely analogous to the finite dimensional case.

However, these are not the most interesting physical predictions, because operationally well defined observables are local in spacetime. In order to deal with these, fix a surface γ that determines a motion (for instance, formed by two parallel Cauchy surfaces), and consider a single correlation (x^μ, ϕ^A) in $\mathcal{C}$. A well posed question is whether or not the point (x^μ, ϕ^A) sits on the motion defined by γ. That is, whether or not the value of the field at x^μ is ϕ^A, on the solution of the field equations determined by the boundary conditions γ. To answer this question in the Hamilton-Jacobi formalism, observe that the there exist a surface $\gamma \cup (x^\mu, \phi^A)$ in $\mathcal{G}$ and we can consider $S[\gamma \cup (x^\mu, \phi^A)]$. More precisely, pick a small ϵ and let $B_{x^\mu}^\epsilon$ be a 3d surface with radius ϵ surrounding the point x^μ in M. Let $\gamma_{(x^\mu, \phi^A)}^\epsilon$ be the 3d surface in $\mathcal{C}$ defined by the constant value ϕ^A and by $x^\mu \in B_{x^\mu}^\epsilon$. Then

$$S[\gamma \cup (x^\mu, \phi^A)] = \lim_{\epsilon \to 0} S[\gamma \cup \gamma_{(x^\mu, \phi^A)}^\epsilon]. \qquad (80)$$

Using this definition, (x^μ, ϕ^A) is on the motion determined by γ iff

$$\frac{\delta S[\gamma \cup (x^\mu, \phi^A)]}{\delta \phi(\boldsymbol{\tau})} = \frac{\delta S[\gamma]}{\delta \phi(\boldsymbol{\tau})} \qquad (81)$$

where τ parametrizes γ. This is the field theoretical generalization of (46). This can be generalized to an arbitrary number of correlations $(x_1^\mu, \phi_1^A), \ldots, (x_n^\mu, \phi_n^A)$. These are compatible with the initial data γ iff

$$\frac{\delta S[\gamma \cup (x_1^\mu, \phi_1^A) \cup \ldots \cup (x_n^\mu, \phi_n^A)]}{\delta \phi(\tau)} = \frac{\delta S[\gamma]}{\delta \phi(\tau)} \tag{82}$$

In fact, it is clear that if the correlations (x_j^μ, ϕ_j^A) are on m, then the insertion does not change the momenta on γ. Thus, γ determines a state, and equation (81) determines the correlations in the extended configuration space $\mathcal{C}$ that are compatible with this state. Clearly equation (81) is the field theoretical generalization of (46). Alternatively, we can write, as in (47),

$$\frac{\partial S[\gamma \cup (x^\mu, \phi^A)]}{\partial \phi^A} = 0. \tag{83}$$

In conclusion, there are local predictions of the theory. Given a state, the theory can predict whether or not individual correlations (points in $\mathcal{C}$), or sets of correlations, can be observed.

4   General Relativity

4.1   Covariant Hamiltonian Formulation

General relativity can be formulated on the finite dimensional configuration space $\tilde{\mathcal{C}}$ with coordinates (x^μ, A_μ^i). (See [16], [10] and [18].) Here $i = 1, 2, 3$ and A_μ^i is a complex matrix. We raise and lower the $i, j, \ldots$ indices with δ_{ij}.

Assuming (48), the corresponding space Ω has coordinates $(x^\mu, A_\mu^i, \pi, p_i^{\mu\nu})$ and carries the canonical four-form

$$\theta_\Omega = \pi \, d^4x + p_i^{\mu\nu} \, dA_\nu^i \wedge d^3x_\mu. \tag{84}$$

It is convenient to introduce the following notation. We define the gauge covariant differential on all quantities with internal indices as

$$Dv^i = dv^i + \epsilon^i_{jk} A_\mu^j v^k dx^\nu \tag{85}$$

so that, in particular,

$$DA_\mu^i = dA_\mu^i + \epsilon^i_{jk} A_\nu^j A_\mu^k dx^\nu. \tag{86}$$

Using this notation, the canonical form (84) reads

$$\theta_\Omega = p \, d^4x + p_i^{\mu\nu} \, DA_\mu^i \wedge d^3x_\nu. \tag{87}$$

where $p = \pi - p_i^{\mu\nu} A_\nu^j A_\mu^k \epsilon^i_{jk}$. We also define

$$E_{\mu\nu}^i = \epsilon_{\mu\nu\rho\sigma} \, \delta^{ij} \, p_j^{\rho\sigma} \tag{88}$$

and the forms $A^i = A^i_\mu dx^\mu$, $DA^i = dA^i_\mu \wedge dx^\mu + A^j_\nu A^k_\mu \epsilon^i_{jk} dx^\nu \wedge dx^\nu$, $E^i = E^i_{\mu\nu} dx^\mu \wedge dx^\nu$, and so on, on Ω.

General relativity is defined by the hamiltonian system

$$p \qquad\qquad\qquad = 0, \tag{89}$$

$$p^{\mu\nu}_i + p^{\nu\mu}_i \qquad\qquad = 0, \tag{90}$$

$$\bar{E}^i \wedge E^j \qquad\qquad = 0 \tag{91}$$

$$(\delta_{ik}\delta_{jl} - \frac{1}{3}\delta_{ij}\delta_{kl})E^i \wedge E^j = 0. \tag{92}$$

Let me now show that this indeed is general relativity. The key point is that the constraints (91), (92) imply that there exists a real four by four matrix e^I_μ, where $I = 0, 1, 2, 3$, such that $E^i_{\mu\nu}$ is the self-dual part of $e^I_\mu e^J_\nu$. This means the following. Let P^i_{IJ} be the selfdual projector, that is

$$P^i_{jk} \qquad\qquad = \epsilon^i_{jk}, \tag{93}$$

$$P^i_{j0} = -P^i_{0j} = i\,\delta^i_j. \tag{94}$$

Then it is easy to check that (91) and (92) are solved by

$$E^i = P^i_{IJ}\, e^I \wedge e^J. \tag{95}$$

and the counting of degrees of freedom indicates that this is the sole solution.

Therefore we can use the coordinates $(x^\mu, A^i_\mu, e^I_\mu)$ on the constraint surface Σ (where A^i_μ is complex and e^I_μ is real) and the induced canonical four-form is simply

$$\theta = P_{IJi}\, e^I \wedge e^J \wedge DA^i. \tag{96}$$

Indeed, the orbits $(x^\mu, A^i_\mu(x^\mu), e^I_\mu(x^\mu))$ of $\omega = d\theta$ satisfy the Einstein equations, in the form

$$e^I \wedge (de_J + P_{JKi}\, A^i \wedge e^K) = 0, \tag{97}$$

$$P_{IJi}\, e_I \wedge e_J \wedge F^i \qquad\qquad = 0, \tag{98}$$

where $F^i_{\mu\nu}$ is the curvature of A^i_μ. From these equation, it follows that $g_{\mu\nu}(x) \equiv \eta_{IJ} e^I_\mu(x) e^J_\nu(x)$ is Ricci flat. The demonstration is a straightforward calculation.

Thus, rather remarkably, the simple and natural form (96), defined on the finite dimensional space Σ with coordinates $(x^\mu, A^i_\mu, e^I_\mu)$, defines general relativity entirely.

4.2 Hamilton-Jacobi Equation and the S Solution

Let γ be a three-dimensional surface in $\tilde{\mathcal{C}}$. Thus $\gamma = [x^\mu(\tau), A^i_\mu(\tau)]$, where $\tau = (\tau^1, \tau^2, \tau^3) = (\tau^a)$. Define the functional

$$S[\gamma] = \int_{m_\gamma} \theta. \tag{99}$$

as above. That is, m is the four-dimensional surface in Σ which is (part of) an orbit of $d\theta$, and therefore a solution of the field equations, and such that the projection of its boundary to $\tilde{\mathcal{C}}$ is γ. From the definition,

$$\frac{\delta S[\gamma]}{\delta A^i_\mu(\tau)} = P_{iJK} \; \epsilon^{\mu\nu\rho\sigma} \; e^J_\rho(\tau) e^K_\sigma(\tau) n_\nu(\tau). \tag{100}$$

Since from this equation we have immediately

$$n_\mu(\tau)\frac{\delta S[\gamma]}{\delta A^i_\mu(\tau)} = 0, \tag{101}$$

it follows that the dependence of $S[\gamma]$ on $A^i_\mu(\tau)$ is only through the restriction of $A^i(\tau)$ to the three-surface γ_M. That is, only through the components

$$A^i_a(\tau) = \partial_a X^\mu(\tau) A^i_\mu(\tau). \tag{102}$$

Thus $S = S[x^\mu(\tau), A^i_a(\tau)]$ and

$$\frac{\delta S[\gamma]}{\delta A^i_a(\tau)} = P_{iJK} \; \epsilon^{a\nu bc} \; \partial_b X^\rho(\tau)\partial_c X^\sigma(\tau) e^J_\rho(\tau)e^K_\sigma(\tau) n_\nu(\tau) \equiv iE^a_i(\tau). \tag{103}$$

The projection of the field equations (98) on γ_M, written in terms of E^a_i read $D_a E^a_i = 0$, $F^i_{ab} E^{ai} = 0$ and $F^i_{ab} E^{ai} E^{bk}\epsilon_{ijk} = 0$, where D_a and F^i_{ab} are the covariant derivative and the curvature of A^i_a. Using (103) these give the three equations

$$D_a \frac{\delta S[\gamma]}{\delta A^i_a(\tau)} = 0, \tag{104}$$

$$\frac{\delta S[\gamma]}{\delta A^i_a(\tau)} F^i_{ab} = 0, \tag{105}$$

$$\epsilon_{ijk} F^i_{ab}(\tau)\frac{\delta S[\gamma]}{\delta A^j_a(\tau)}\frac{\delta S[\gamma]}{\delta A^k_b(\tau)} = 0. \tag{106}$$

These equations have a well known interpretation. In fact, the first could have been obtained by simply observing that $S[\gamma]$ is invariant under local $SU(2)$ gauge transformations on the three-surface. Under one such transformation generated by a function $f^i(\tau)$ the variation of the connection is $\delta_f A^i_a = D_a f^i$. Therefore S satisfies

$$0 = \delta_f S = \int d^3\tau \; \delta_f A^i_a(\tau) \frac{\delta S[\gamma]}{\delta A^i_a(\tau)}$$
$$= \int d^3\tau \; D_a f^i(\tau) \frac{\delta S[\gamma]}{\delta A^i_a(\tau)} = -\int d^3\tau \; f^i(\tau) \, D_a \frac{\delta S[\gamma]}{\delta A^i_a(\tau)}. \tag{107}$$

This this gives (104). Next, the action is invariant under a change of coordinates on the three surface γ_M. Under one such transformations generated by a function

$f^a(\boldsymbol{\tau})$ the variation of the connection is $\delta_f A_a^i = f^b \partial_b A_a^i + A_b^i \partial_a f^b$. Integrating by parts as in (107) this gives

$$\partial_b A_a^i \frac{\delta S[\gamma]}{\delta A_a^i(\boldsymbol{\tau})} + (\partial_b A_a^i) \frac{\delta S[\gamma]}{\delta A_a^i(\boldsymbol{\tau})} = 0, \tag{108}$$

which, combined with (104) gives (105). Thus, (104) and (105) are simply the requirement that $S[\gamma]$ is invariant under internal gauge and changes of coordinates on the three-surface. The three equations (104),(105) and (106) govern the dependence of S on $A_a^i(\boldsymbol{\tau})$.

On the other hand, it is easy to see that S is independent from $x^\mu(\boldsymbol{\tau})$. A change of coordinates $x^\mu(\boldsymbol{\tau})$ tangential to the surface cannot affect the action, which is independent from the coordinates used. More formally, the invariance under change of parameters $\boldsymbol{\tau}$ implies

$$\frac{\delta S[\gamma]}{\delta x^\mu(\boldsymbol{\tau})} \partial_j x^\mu(\boldsymbol{\tau}) = \frac{\delta S[\gamma]}{\delta A_a^i(\boldsymbol{\tau})} \delta_j A_a^i(\boldsymbol{\tau}), \tag{109}$$

and we have already seen that the right hand side vanishes. The variation of S under a change of $x^\mu(\boldsymbol{\tau})$ normal to the surface is governed by the Hamilton-Jacobi equation proper, (74). In the present case, following the same steps as for the scalar field, we obtain

$$\frac{\delta S[\gamma]}{\delta x^\mu(\boldsymbol{\tau})} n_\mu(\boldsymbol{\tau}) + \epsilon_{ijk} F_{ab}^i \frac{\delta S[\gamma]}{\delta A_a^j(\boldsymbol{\tau})} \frac{\delta S[\gamma]}{\delta A_b^k(\boldsymbol{\tau})} = 0. \tag{110}$$

But the second term vanishes because of (106). Therefore $S[\gamma]$ is independent from tangential as well as normal parts of $x^\mu(\boldsymbol{\tau})$: S depends only on $[A_a^i(\boldsymbol{\tau})]$.

We can thus drop altogether the spacetime coordinates x^μ from the extended configuration space. Define a smaller extended configuration space $\mathcal{C}$ as the 9d complex space of the variables A_a^i. Geometrically, this can be viewed as the space of the linear mappings $A : D \to sl(2,C)$, where $D = R^3$ is a "space of directions" and we have chosen the complex selfdual basis in the $sl(2,C)$ algebra. We then identify the space $\mathcal{G}$ as a space of parametrized three-dimensional surfaces $[A_a^i(\boldsymbol{\tau})]$ without boundaries in $\mathcal{C}$, modulo reparametrizations – where, however, two parametrized surfaces are considered equivalent if $A_a{}^i(\boldsymbol{\tau}) = \frac{\partial \tau'^b}{\partial \tau^a} A_b'{}^i(\tau'(\boldsymbol{\tau}))$. The dynamics of the theory is entirely contained in the equations (104),(105) and (106) for the functional $S[A_a^i(\boldsymbol{\tau})]$ on $\mathcal{G}$.

It is then immediate to obtain the dynamical equation of the quantum theory. This is obtained by replacing the functional derivative of S with a functional derivative operator, and acting over a wave function that has the same argument of S, namely a wave functional $\Psi[A_a^i(\boldsymbol{\tau})]$ on the extended configuration space $\mathcal{C}$. (A different point of view on the use of the covariant formalism in quantum theory is developed in [24].) Equations (104) and (105) do not change and demand that $\Psi[A] = \Psi[A_a^i(\boldsymbol{\tau})]$ is invariant under gauge transformations and changes of coordinates $\boldsymbol{\tau}$, while (106) becomes

$$F_{ab}^{ij}(\boldsymbol{\tau}) \frac{\delta}{\delta A_a^i(\boldsymbol{\tau})} \frac{\delta}{\delta A_b^j(\boldsymbol{\tau})} \Psi[A] = 0. \tag{111}$$

This is the Ashtekar-Wheeler-DeWitt equation, which the basic equation of canonical quantum gravity. See for instance [19].

4.3   Physical Predictions

What are the quantities predicted by the theory in the case of general relativity? Let γ determine a motion m. Notice that we cannot simply ask whether a single point c of $\mathcal{C}$ is in m, because of the non trivial transformation properties of A_a^i under change of parametrization. More precisely, if want to interpret a point in $\mathcal{C}$ as a constant field over the 3d surface of a small ball, as we did for the scalar field case, we run in the difficulty that a constant connection on a three-sphere is trivial. We thus have to look for invariant extended objects in $\mathcal{C}$. These can be defined as follows.

Choose a closed unknotted curve $\alpha : s \mapsto (x^\mu(s), A_\mu^i(s))$ in $\mathcal{C}$. Given a motion m, we can ask whether or not the set of correlations forming the loop α can be realized. More in general, let Γ be a graph (a set of points p_i joined by lines l_{ij}) imbedded in $\mathcal{C}$. We can ask whether the collection of correlations forming Γ is realizable in a given state, determined by a point γ in $\mathcal{G}$. The answer is positive if

$$\frac{\delta S[\gamma]}{\delta A_a^i(\tau)} = \frac{\delta S[\gamma \cup \Gamma]}{\delta A_a^i(\tau)}. \tag{112}$$

Here τ parametrizes γ and $S[\gamma \cup \Gamma]$ is the integral of θ over a motion $m_{\gamma \cup \Gamma}$ which has γ and Γ as boundaries. This means Γ is in $m_{\gamma \cup \Gamma}$. More precisely, we can thicken out the M section of the graph Γ as we did for the scalar field. Here however we do not obtain a ball of radius ϵ, but rather a sort of "tubular structure". The boundary of this tubular structure is a three dimensional surface Γ_ϵ in $\mathcal{G}$ and we pose $S[\gamma \cup \Gamma] = \lim_{\epsilon \to 0} S[\gamma \cup \Gamma_\epsilon]$.

An important observation follows. Consider a simple curve α in $\mathcal{C}$. Define now the "holonomy" of α

$$T_\alpha = TrU_\alpha = TrPe^{\int_\alpha ds \frac{dx^\mu(s)}{ds} A_\mu^i(s)\tau_i}, \tag{113}$$

where τ_i is a basis in the $su(2)$ algebra. Let α' be another closed unknotted curve in $\mathcal{C}$, distinct from α, but with the same holonomy – that is such that $T_{\alpha'} = T_\alpha$. A moment of reflection shows that, due to the internal gauge and diffeomorphism invariance of S, we have

$$S[\gamma \cup \alpha] = S[\gamma \cup \alpha']. \tag{114}$$

Therefore the predictions of the theory do not distinguish α from α', as far as $T_{\alpha'} = T_\alpha \equiv T$. The prediction depend only on T. More in general: to any closed cycle $\alpha = l_{i_1 i_2} \cup l_{i_p i_1}$ in Γ, let the holonomy T_α be defined as in (113). Denote $T_\Gamma = (T_{\alpha_1}, \ldots, T_{\alpha_n})$ the collection of these holonomies. Let $[\Gamma]$ be the knot-class (the equivalence class under diffeomorphisms) to which the restriction of Γ to M belongs, and call $s = ([\Gamma], T_\Gamma)$ a "spin network". Then to graphs Γ and Γ'

cannot be distinguished if they belong to the same spin network s. That is, we have

$$S[\gamma \cup \Gamma] = S[\gamma \cup \Gamma'] \equiv S[\gamma \cup s]. \tag{115}$$

Therefore what the theory predicts is, for a given state, whether or not a spin network s is realizable. A set of correlations determined by a spin network s is realizable iff

$$\frac{\delta S[\gamma \cup s]}{\delta A_a^i(\boldsymbol{\tau})} = \frac{\delta S[\gamma]}{\delta A_a^i(\boldsymbol{\tau})}. \tag{116}$$

The detection of a given s can be realized in principle as follows (see also [10]). Imagine we set up an experience in which we parallel transport a reference system along finite paths l_{ij} in spacetime. This can be realized macroscopically by transporting a gyroscope and a devise keeping track of local acceleration, or, microscopically, by paralleling transport a particle with spin. For instance a left handed neutrino, whose parallel transport is directly described by the self-dual connection A_μ^i. We can then compare the result of the parallel transport at the points p_i, thus effectively measuring the quantities T_α as angles and relative velocities. [2] The gauge and diffeomorphism invariant information provided by such a measurement is then in the topology of the graph formed by the paths and the invariant values of these relative angles and velocities. This is what is contained in the spin network s. In the quantum theory, we expect then a quantum state to determine the probability amplitude for any spin network s to be realizable [23].

5   Conclusions and Open Issues

I think that a proper understanding of the generally covariant structure of mechanics is necessary in order to make progress in quantum gravity. In this paper I have made several steps in this direction. My focus has been on searching a physically viable language for background independent quantum field theory

---

[2] One may object that this setting can be physically realized only if the l_{ij} are all timelike and future oriented. However, this is not a serious experimental limitation. First, we have obviously $U_{l_{ij}^{-1}} = U_{l_{ij}}^{-1}$, therefore future orientation is not a limitation in measurability. Second, the measurement of a spacelike l_{ij} can be obtained in principle as a limit of timelike ones, as in fact we do in practice. That is, divide l_{ij} in a sequence of N (spacelike) segments. For each such spacelike segment, bounded by the points p_1 and p_2, pick a point p in the common past of both p_1 and p_2, and consider the two timelike geodesics that go from p_1 to p and from p to p_2. Then replace the spacelike segment with the union of these two timelike geodesics. It is clear that (dividing l_{ij} and picking the points p appropriately) the parallel transport along the timelike curve obtained in this way converges to the parallel transport along the spacelike curve l_{ij} for large N. That is, spacelike measurements can be seen as bookkeeping for results of measurements obtained by timelike motion.

and not in mathematical completeness. From the mathematical point of view, the structures that I have introduced in this paper certainly can (and need to) be refined.

A result of this paper is the derivation of the symplectic structure and the construction of the Hamilton-Jacobi formalism, in a general covariant hamiltonian formulation of mechanics in field theory. The main ingredient for this is the introduction of the space $\mathcal{G}$. The preferred solution $S[\gamma]$ of the Hamilton-Jacobi equation, defined on $\mathcal{G}$, contains the full dynamical information on the system. In finite dimensional systems, the preferred solution $S[\gamma]$ is also the classical limit of the quantum propagator, which contains the full dynamical information on the quantum system in a form which makes sense in a generally covariant context, and has a direct operational interpretation [15]. In the field theoretical context, operationally realistic measurements are local. Outcome of these can be derived from $S[\gamma]$ as well. I think that the precise relation between $S[\gamma]$ and the Wightman amplitudes in quantum field theory deserves to be explored.

As far as the gravitational field is concerned, the covariant hamiltonian formulation described here is remarkably simple. In fact, general relativity is entirely defined on the finite dimensional space with coordinates $(x^\mu, A^i_\mu, e^I_\mu)$, by the simple and natural form (96) (see also [16]). This formulation leads directly to the basic equation of quantum gravity, and to a classical notion of spin network, which may be of help in clarifying the physical interpretation of the quantum spin networks.

Finally, I think that there should be a proper definition of the extended configuration space $\mathcal{C}$ in which spacetime coordinates x^μ play no role at all, and with a clean operational interpretation. Here and in [10] I have made some steps in this direction, but I think the matter could be further clarified. Ideally, I think, we should get to a clean operational definition of the partial observables of the gravitational field, and of the transition amplitudes between correlations of these. These are the amplitudes that a properly covariant and background independent quantum theory of gravity should allow us to compute.

Appendix

Here we derive equations (30), (70) and (71). The right hand side of (30) is given by the variation of the boundary in the integral (28). However, this boundary variation is not the only variation to be considered, because if q^a changes, the entire curve m_γ may change, becoming a curve $m_{\gamma+\delta\gamma}$. Thus

$$\frac{\partial S(q^a, q^a_0)}{\partial q^a} \delta q^a = \int_{m_{\gamma+\delta\gamma}} \theta - \int_{m_\gamma} \theta. \qquad (117)$$

Consider now the closed line integral of θ along the path α obtained by joining the curves $m_\gamma, \delta s, (m_{\gamma+\delta\gamma})^{-1}, \delta s_0^{-1}$

$$\oint_\alpha \theta = \int_{m_\gamma} \theta + \int_{\delta s} \theta - \int_{m_{\gamma+\delta\gamma}} \theta - \int_{\delta s_0} \theta. \qquad (118)$$

The path α path bounds a surface (a strip) σ which is everywhere tangent to the orbits of ω. Therefore the restriction of $\omega = d\theta$ to σ vanishes. Therefore

$$0 = \int_\sigma d\theta = \int_\alpha \theta = 0. \tag{119}$$

From (117), (118) and (119), we have

$$\frac{\partial S(q^a, q_0^a)}{\partial q^a} \delta q^a = \int_{\delta s} \theta - \int_{\delta s_0} \theta = p_a \delta q^a - p_{0a} \delta q_0^a. \tag{120}$$

And since we are varying q^a at fixed q_0^a, we have (30).

Let us now come to equation (70) and (71). Consider a surface γ and a small variation $\delta\gamma = (\delta x^\mu(\tau), \delta\phi(\tau))$. Consider the five dimensional strip σ in Σ bounded by $m_\gamma, m_{\gamma+\delta\gamma}$, and δs, where s is the boundary of m and δs its variation. The five-form ω vanishes when restricted to σ because σ is tangent to the orbits of ω. Therefore

$$0 = \int_\sigma \omega = \int_{\partial\sigma} \theta = \int_{m_{\gamma+\delta\gamma}} \theta - \int_{m_\gamma} \theta - \int_{\delta s} \theta. \tag{121}$$

By linearity,

$$\int_{m_{\gamma+\delta\gamma}} \theta - \int_{m_\gamma} \theta = \int d^3\tau \left(\frac{\delta S[\gamma]}{\delta x^\mu(\tau)} \delta x^\mu(\tau) + \frac{\delta S[\gamma]}{\delta\phi(\tau)} \delta\phi(\tau) \right). \tag{122}$$

From the last two equations, (70) and (71) follow with a short calculation.

References

1. J.L. Lagrange: *Mémoires de la première classe des sciences mathematiques et physiques* (Institute de France, Paris 1808)
2. V.I. Arnold: *Matematičeskie metody klassičeskoj mechaniki*, (Mir, Moskow 1979). See in particular Chapter IX, Section C
3. J.M. Souriau: *Structure des systemes dynamics* (Dunod, Paris 1969)
4. C. Rovelli: *A note on the foundation of relativistic mechanics. I: Relativistic observables and relativistic states*, gr-qc/0111037
5. C. Rovelli: Phys. Rev. D **65**, 124013 (2002)
6. H. Weil: Ann. Math. (2) **36**, 607 (1935)
7. T. DeDonder: *Theorie Invariantive du Calcul des Variations* (Gauthier-Villars, Paris 1935)
8. J. Kijowski: Commun. Math. Phys. **30**, 99 (1973). M. Ferraris and M. Francaviglia: 'The Lagrangian approach to conserved quantities in General relativity'. In: *Mechanics, Analysis and Geometry: 200 Years after Lagrange*, pf 451-488, ed. by M. Francaviglia (Elsevier Sci.Publ., Amsterdam 1991). I.V. Kanatchikov: Rep. Math. Phys. **41**, 49 (1998). F. Hélein and J. Kouneiher: 'Finite dimensional Hamiltonian formalism for gauge and field theories', math-ph/0010036
9. M.J. Gotay, J. Isenberg and J.E. Marsden: 'Momentum maps and classical relativistic fields', physics/9801019

10. C. Rovelli: 'A note on the foundation of relativistic mechanics. II: Covariant hamiltonian general relativity', gr-qc/0202079
11. Č. Crnković and E. Witten. In: *Newton's tercentenary volume*, ed. by S.W. Hawking and W. Israel (Cambridge University Press, 1987). A. Ashtekar, L. Bombelli and O. Reula. In: *Mechanics, Analysis and Geometry: 200 Years after Lagrange*, ed. by M. Francaviglia (Elsevier, New York 1991)
12. E. Schrödinger: Ann. d. Phys. **79**, 489 (1926), Part 2, English translation in *Collected Papers on Quantum Mechanics* (Chelsea Publ., 1982)
13. E. Schrödinger: Ann. d. Phys. **79**, 361 (1926), Part 1, English translation, op. cit.
14. B. DeWitt, private communication
15. M. Reisenberger and C. Rovelli: Phys. Rev. D **65**, 124013 (2002). D. Marolf and C. Rovelli: 'Relativistic quantum measurement', Phys. Rev. D, to appear; gr-qc/0203056
16. G. Esposito, G. Gionti and C. Stornaiolo: Nuovo Cim. B **110**, 1137 (1995)
17. J.F. Plebanski: J. Math. Phys. **18**, 2511 (1977). A. Sen: Phys. Lett. B **119**, 89 (1982). A. Ashtekar: Phys. Rev. Lett. **57**, 2244 (1986). J. Samuel: Pramana J. Phys. **28**, L429 (1987). T. Jacobson and L. Smolin: Class. Quant. Grav. **5**, 583 (1988). R. Capovilla, J. Dell and T. Jacobson: Phys. Rev. Lett. **63**, 2325 (1991). R. Capovilla, J. Dell, T. Jacobson and L. Mason: Class. Quant. Grav. **8**, 41 (1991)
18. L. Smolin, unpublished notes
19. C. Rovelli: Class. Quant. Grav. **8**, 1613 (1991). C. Rovelli and L. Smolin: Phys. Rev. Lett. **61**, 1155 (1988). C. Rovelli and L. Smolin: Nucl. Phys. B **331**, 80 (1990)
20. A. Peres: Nuovo Cim. **26**, 53 (1962). U. Gerlach: Phys. Rev. **177**, 1929 (1969). K. Kuchar: J. Math. Phys. **13**, 758 (1972). W. Szczyrba: Commun. Math. Phys. **51**, 163 (1976). P. Horava: Class. Quant. Grav. **8**, 2069 (1991). E.T. Newman and C. Rovelli: Phys. Rev. Lett. **69**, 1300 (1992) 1300
21. C. Rovelli and L. Smolin: Phys. Rev. D **52**, 5743 (1995)
22. J.C. Baez: Adv. in Math. **117**, 253 (1996). J.C. Baez: 'Spin networks in nonperturbative quantum gravity'. In: *Interface of Knot Theory and Physics*, ed. by L. Kauffman Am. Math. Soc., Providence, Rhode Island 1996
23. C. Rovelli: 'Loop quantum gravity', Living Reviews in Relativity **1** (1998); T. Thiemann: 'Introduction to Modern Canonical Quantum General Relativity', gr-qc/0110034
24. I.V. Kanatchikov: Phys. Lett. A **283**, 25 (2001); Rep. Math. Phys. **43**, 157 (1999)

Some Recent Developments in the Decoherent Histories Approach to Quantum Theory

Jonathan Halliwell

Blackett Laboratory, Imperial College, London, SW7 2BZ, UK

Abstract. A brief introduction to the decoherent histories approach to quantum theory is given, with emphasis on its role in the discussion of the emergence of classicality from quantum theory. Some applications are discussed, including quantum-classical couplings, the relationship of the histories approach to quantum state diffusion, and the application of the histories approach to situations involving time in a non-trivial way.

1 Introduction

Standard quantum theory is a remarkably successful theory. Indeed, there is at present not one shred of experimental evidence that its basic structure of Hilbert spaces, states, operators *etc.* is in any way wrong. However, in its normal presentation, the Copenhagen interpretation [1], it rests on certain assumptions that may be restrictive. Firstly, it relies on a division into classical and quantum domains. And secondly, it places great emphasis on the notion of measurement [2]. These assumptions are not of course a restriction in the regimes to which it is normally applied. Yet it invites the question, can we do better? Can we formulate standard quantum theory in such a way that it does not rely on these assumptions?

There are a number of reasons why one might want such a more general formulation. Recent experiments have started to probe the traditional border between what is normally called classical and quantum [3]. Furthermore, classical objects are after all built out of atoms, which are quantum mechanical in many of their properties. One can therefore ask how does classical behaviour emerge on large scales for objects made out of small quantum constituents? In addition, some current views of the early universe, and in particular in the area of research known as quantum cosmology, it is supposed that all force and matter fields are subject to the laws of quantum theory. There is no classical domain and there are certainly no measuring devices. How then can we understand, in a truly quantum-mechanical way, how measurements made in the present epoch are related to events in the distant past?

The decoherent histories approach to quantum theory is a reformulation of standard quantum theory for closed quantum systems (such as the entire universe), that removes the usual emphasis on the notion of measurement and of a classical domain. In this contribution I will briefly describe the formalism and

J. Halliwell, Some Recent Developments in the Decoherent Histories Approach to Quantum Theory,
Lect. Notes Phys. **633**, 63–83 (2004)
http://www.springerlink.com/

the way it explains the emergence of classical physics from an underlying quantum theory. I will then also describe a selection of applications of the decoherent histories approach in a variety of different circumstances. This will show what sort of light the approach has been able to shed on other fields. This review is by no means an exhaustive survey of the field, which has by now become quite extensive, and the choice of topics covered largely reflects my own interests in the subject.

2   Interference, Decoherence, and Classicality

Before embarking on a full discussion of the decoherent histories formalism, let us first consider in a simple way the phenomenon of interference in the double-slit experiment, its destruction through decoherence, and the relationship of this to emergent classicality.

Consider then the standard double-slit experiment, in which electrons, say, impinge on a sheet with two slits and are then allowed to fall on a screen. The wave function is assumed to be a tightly peaked wave packet in the direction perpendicular to the screen, but spreads out in the direction parallel to it. In the region of the slits, we may assign a wave function

$$\psi = \psi_1 + \psi_2 \tag{1}$$

to the electrons, where ψ_1 represents a wave emerging from slit 1, and ψ_2 represents a wave emerging from slit 2. The probability of hitting the screen at point x is then

$$p(x) = |\psi(x)|^2$$
$$= |\psi_1|^2 + |\psi_2|^2 + \psi_1\psi_2^* + \psi_1^*\psi_2 \tag{2}$$

That is, the probability of hitting the screen at point x is not the sum of the probabilities of the two separate paths taken by the electron, as it would be in the classical case. The probability sum rules are changed by the presence of quantum interference. Quantum effects are therefore characterized by a failure of usual classical probability sum rules. Or differently put, this corresponds to a failure of Boolean logic.

Note also the appearance of histories at this elementary level. The interference effect arises as a result of the fact that we are trying to talk about the properties of the electron at both the screen *and* the slits, that is, at two moments of time. This simple aspect is rarely mentioned in elementary textbooks, but turns out to be significant, as we shall see.

We see from this simple but important example that for a quantum system to become classical, interference terms must be destroyed. How does this come about? It is known in the double-slit experiment that if we measure the position of the electron close to the slits, then the wave function will collapse into one of the states ψ_1 or ψ_2 and the interference terms will go away. We therefore expect that any physical mechanism that constitutes some kind of physical measurement of the electron will produce a similar effect. This is indeed the case. Suppose, following the original calculation of Joos and Zeh we couple the electrons to an environment [4]. This could be, for example, a bath of photons. If the total wave function of the system (electron) together with its environment is denoted by $|\psi_{S\mathcal{E}}\rangle$, then the density operator of the system only is

$$\rho = \text{Tr}|\psi_{S\mathcal{E}}\rangle\langle\psi_{S\mathcal{E}}| \tag{3}$$

where Tr denotes a trace over the environment. Such a density operator usually obeys a non-unitary master equation. It typically has the form (in one dimension),

$$\frac{\partial\rho}{\partial t} = \frac{i\hbar}{2M}\left(\frac{\partial^2\rho}{\partial x^2} - \frac{\partial^2\rho}{\partial y^2}\right) - D(x-y)^2\rho \tag{4}$$

The important term is the last one on the right, which has the effect of causing the density operator $\rho(x,y)$ to become approximately diagonal in position very rapidly. The means that an initial state of the form

$$\rho = |\psi\rangle\langle\psi| \tag{5}$$

where

$$|\psi\rangle = |\psi_1\rangle + |\psi_2\rangle \tag{6}$$

becomes essentially indistinguishable from the classical mixture

$$\rho = |\psi_1\rangle\langle\psi_1| + |\psi_2\rangle\langle\psi_2| \tag{7}$$

This is the phenomenon of decoherence, in its simplest form.

Consider now the question of how a quantum system, characterized by its positions at a series of times, may become approximately classical. Quantum theory supplies probabilities, so suppose we use it to compute the probability $p(\alpha_1, t_1, \alpha_2, t_2, \cdots)$ that the a particle is in a spatial region α_1 at time t_1, and then in a spatial region α_2 at times t_2, etc. Then, "approximately classical" means at least two things. Firstly, it means that the probability is defined. As we have seen, interference can prevent this from being the case. Secondly, it means that the probability is strongly peaked when the regions $\alpha_1, \alpha_2, \alpha_3 \cdots$ lie along a classical trajectory [5].

From this we see that histories provide an important mode of description for discussing emergent classicality. It is therefore perhaps not surprising that all of the above is most clearly formulated in terms of the decoherent histories approach.

3   The Decoherent Histories Approach

The decoherent (or "consistent") histories approach was put forward by Griffiths in 1984 [6] whose work was substantially developed by Omnès [7–9]. It was also discovered and developed, in part independently, by Gell-Mann and Hartle [10–12]. See [13] for a very extensive bibliography on decoherent histories, and decoherence generally.

3.1   The Formalism

In the decoherent histories approach to quantum theory, a quantum-mechanical history is characterized by an initial (pure or mixed) state ρ, and by a time-ordered string of projection operators,

$$C_{\underline{\alpha}} = P_{\alpha_n}(t_n) \cdots P_{\alpha_1}(t_1) \tag{8}$$

Here, the projectors appearing in (8) are in the Heisenberg picture,

$$P_{\alpha_k}(t_k) = e^{\frac{i}{\hbar} H(t_k - t_0)} P_{\alpha_k} e^{-\frac{i}{\hbar} H(t_k - t_0)} \tag{9}$$

and $\underline{\alpha}$ denotes the string of alternatives $\alpha_1, \cdots \alpha_n$. The projection operators P_α characterize the different alternatives describing the histories at each moment of time. The projectors satisfy

$$\sum_\alpha P_\alpha = 1, \qquad P_\alpha P_\beta = \delta_{\alpha\beta} P_\alpha \tag{10}$$

More generally, it is also of interest to work with histories in which the so-called class operators (8) are given by sums of string of projections (which are not necessarily then equal to strings of projections). Since the class operator (8) is not in fact a projection operator, it is reasonable to ask why this operator should be used to describe a history. This and many other related issues have been considered in detail by Isham and collaborators [14].

Probabilities are assigned to histories of a closed system via the formula,

$$p(\alpha_1, \alpha_2, \cdots \alpha_n) = \mathrm{Tr}\left(C_{\underline{\alpha}} \rho C_{\underline{\alpha}}^\dagger\right) \tag{11}$$

Interference between pairs of histories is measured by the decoherence functional

$$D(\underline{\alpha}, \underline{\alpha}') = \mathrm{Tr}\left(C_{\underline{\alpha}} \rho C_{\underline{\alpha}'}^\dagger\right) \tag{12}$$

Probabilities can be assigned to histories if and only if all pairs of histories in the set obey the condition of *consistency*, which is that

$$\mathrm{Re}D(\underline{\alpha}, \underline{\alpha}') = 0 \tag{13}$$

for $\underline{\alpha} \neq \underline{\alpha}'$. This condition is equivalent to the requirement that all probabilities satisfy the probability sum rules. These, loosely speaking, are that the probability of history $\underline{\alpha}$ or history $\underline{\beta}$ (for two disjoint histories), should be $p(\underline{\alpha}) + p(\underline{\beta})$. Typically, realistic physical mechanisms which bring about the consistency condition (13) also cause the imaginary part of $D(\underline{\alpha}, \underline{\alpha}')$ to vanish for $\underline{\alpha} \neq \underline{\alpha}'$ as well, so it is of interest to work with the stronger condition of *decoherence*, which is

$$D(\underline{\alpha}, \underline{\alpha}') = 0 \tag{14}$$

for $\underline{\alpha} \neq \underline{\alpha}'$. The two conditions of consistency and decoherence have different consequences, and we discuss them below.

When the decoherence condition is satisfied, it is straightforward to see that

$$p(\underline{\alpha}) = \text{Tr}\left(C_{\underline{\alpha}}\rho\right) \tag{15}$$

since we have $\sum_{\underline{\alpha}} C_{\underline{\alpha}} = 1$. Equation (15) is clearly positive when the decoherence condition is satisfied, but not generally not otherwise. Goldstein and Page have turned this around and suggested that positivity of the expression (15) may be used to select physically viable histories [15]. This is clearly weaker than the usual decoherence condition. A variety of decoherence conditions related to those discussed above were also discussed in [16].

3.2   Approximate Decoherence

Typically, the decoherence condition (14) is satisfied only approximately, raising the question of what approximate decoherence actually means and how small the off-diagonal terms really need to be. In this context, it is worth noting that the decoherence functional satisfies the inequality,

$$\left|D(\underline{\alpha}, \underline{\alpha}')\right|^2 \leq D(\underline{\alpha}, \underline{\alpha})D(\underline{\alpha}', \underline{\alpha}') \tag{16}$$

as shown in [17]. A natural approximate decoherence condition is then to insist that the decoherence functional satisfies (15) but with a small number ϵ on the right-hand side. It may be shown that this also guarantees that most of the probability sum rules are satisfied to order $\epsilon^{1/2}$ [17]. On the more general issue of approximate decoherence, it has been suggested that an approximately decoherent set of histories may be turned into an exactly decoherent set by small distortions of the histories, for example, by small distortions of the operators projected onto at each moment of time [18]. A closely related suggestion, which has been worked out in some detail in particular cases, is that approximately decoherent histories are those whose predictions are well-approximated by a hidden variable theory, such as the deterministic quantum theory of 't Hooft [19]. See also [20] for further considerations of approximate decoherence.

3.3   Path Integral Form

It is also useful to note that for histories characterized by projections onto position at different time, the decoherence functional is very usefully expressed in terms of a path integral:

$$D(\underline{\alpha}, \underline{\alpha}') = \int_{\underline{\alpha}} \mathcal{D}x \int_{\underline{\alpha}'} \mathcal{D}y \, \exp\left(\frac{i}{\hbar}S[x(t)] - \frac{i}{\hbar}S[y(t)]\right) \rho_0(x_0, y_0) \tag{17}$$

The path integral is over a pair of paths $x(t), y(t)$ which are folded into the initial state ρ at x_0, y_0, wmeet at the final time (and the final point is summed over), and at intermediate times pass through the series of pairs of regions denoted by $\underline{\alpha}$ and $\underline{\alpha}'$. This form is particularly useful for generalizations to situations which involve time in a non-trivial way.

3.4   Consistency

Consistency, equation (13), is an interesting condition, because, according to a theorem of Omnès, systems described by a set of consistent histories from a representation of classical logic [7]. That is, in a consistent set of histories each history corresponds to a proposition about the properties of a physical system and we can manipulate these propositions without contradiction using ordinary classical logic. In simple terms, we can *talk about* their properties. It is for this reason that the decoherent histories approach may be thought of as supplying a foundation for the application of ordinary reasoning to closed quantum systems. In the Copenhagen interpretation, it was asserted that one could only talk about *measured* quantities in an unambiguous way. In the decoherent histories approach, the idea of a a measured quantity is replaced by the weaker and more general idea of consistency.

An important example is the case of retrodiction of the past from present data. Suppose we have a consistent set of histories. We would say that the alternative α_n (measured present data) implies the alternatives $\alpha_{n-1} \cdots \alpha_1$ (unmeasured past events) if

$$p(\alpha_1, \cdots \alpha_{n-1} | \alpha_n) \equiv \frac{p(\alpha_1, \cdots \alpha_n)}{p(\alpha_n)} = 1 \tag{18}$$

In this way, we can in quantum mechanics build a picture of the history of the universe, given the present data and the initial state, using only logic and the consistency of the histories. We can meaningfully talk about the past properties of the universe even though there was no measuring device there to record them. This is one reason why the histories approach is of interest in quantum cosmology.

3.5   Decoherence and Records

The stronger condition of decoherence is perhaps more interesting since it is related to the existence of records – some physical mechanism existing at a fixed moment of time which is correlated with the past history of the system. An example is a photographic plate showing a particle track. In particular, if the initial state is pure, there exist a set of records at the final time t_n which are perfectly correlated with the alternatives $\alpha_1 \cdots \alpha_n$ at times $t_1 \cdots t_n$ [11]. This follows because, with a pure initial state $|\Psi\rangle$, the decoherence condition implies that the states $C_{\underline{\alpha}}|\Psi\rangle$ are an orthogonal set. It is therefore possible to introduce a projection operator $R_{\underline{\beta}}$ (which is generally not unique) such that

$$R_{\underline{\beta}} C_{\underline{\alpha}} |\Psi\rangle = \delta_{\underline{\alpha}\underline{\beta}} C_{\underline{\alpha}} |\Psi\rangle \tag{19}$$

It follows that the extended histories characterized by the chain $R_{\underline{\beta}} C_{\underline{\alpha}} |\Psi\rangle$ are decoherent, and one can assign a probability to the histories $\underline{\alpha}$ and the records $\underline{\beta}$, given by

$$p(\alpha_1, \alpha_2, \cdots \alpha_n; \beta_1, \beta_2 \cdots \beta_n) = \mathrm{Tr}\left(R_{\beta_1\beta_2\cdots\beta_n} C_{\underline{\alpha}} \rho C_{\underline{\alpha}}^\dagger\right) \tag{20}$$

This probability is then zero unless $\alpha_k = \beta_k$ for all k, in which case it is equal to the original probability $p(\alpha_1, \cdots \alpha_n)$. Hence either the α's or the β's can be completely summed out of (12) without changing the probability, so the probability for the histories can be entirely replaced by the probability for the records at a fixed moment of time at the end of the history:

$$p(\underline{\alpha}) = \mathrm{Tr}\left(R_{\underline{\alpha}}\rho(t_n)\right) = \mathrm{Tr}\left(C_{\underline{\alpha}}\rho C_{\underline{\alpha}}^{\dagger}\right) \tag{21}$$

Conversely, the existence of records $\beta_1, \cdots \beta_n$ at some final time perfectly correlated with earlier alternatives $\alpha_1, \cdots \alpha_n$ at $t_1, \cdots t_n$ implies decoherence of the histories. This may be seen from the relation

$$D(\underline{\alpha}, \underline{\alpha}') = \sum_{\beta_1 \cdots \beta_n} \mathrm{Tr}\left(R_{\beta_1 \cdots \beta_n} C_{\underline{\alpha}}\rho C_{\underline{\alpha}'}^{\dagger}\right) \tag{22}$$

Since each β_k is perfectly correlated with a unique alternative α_k at time t_k, the summand on the right-hand side is zero unless $\alpha_k = \alpha_k'$ (although note that, as we shall see later, a perfect correlation of this type is generally possible only for a pure initial state).

There is, therefore, a very general connection between decoherence and the existence of records. From this point of view, the decoherent histories approach is very much concerned with reconstructing possible past histories of the universe from records at the present time, and then using these reconstructed pasts to understand the correlations amongst the present records [21].

Some explicit models where the records may be explicitly identified have been worked out. Reference [22] showed how the environment stores records of the particle's history in the quantum Brownian motion model (in which the environment is a set of harmonic oscillators). This was repeated for the case of decoherence by a series of scattering processes in [23].

3.6 The Non-uniqueness of Retrodiction

The above scheme contains an important subtlety that we now need to discuss. This is that the retrodicted past is not in fact unique: there are often many sets of consistent histories associated with the same initial state and final measurement, and what's more, can give conditional probability 1 for different complementary observables at an intermediate time. To see how this happens, consider the following example (due to Omnès.) Suppose we have a radioactive atom sitting at the origin which decays at $t = 0$, and therefore emits a particle in an outgoing spherical wave state, and then hits a detector at time $t = t_2$. It is then reasonable to ask if we can say anything about the system at an intermediate time t_1.

We may analyze this using the decoherent histories approach. The appropriate decoherence functional is,

$$D(\alpha_1, \alpha_2 | \alpha_1', \alpha_2) = \mathrm{Tr}\left(P_{\alpha_2}(t_2)P_{\alpha_1}(t_1)\rho P_{\alpha_1'}(t_1)\right) \tag{23}$$

Here, P_{α_2} is a projector onto the position of the detector, $\rho = |\psi\rangle\langle\psi|$ is the spherical wave state, and P_{α_1} is a projector onto possible properties we may measure at the intermediate time t_1. We consider two cases.

First, suppose that we project at time t_1 onto a spatial region lying between the detector and the origin. Then, one finds quite easily that the histories are approximately decoherent, and secondly, that the conditional probability $p(\alpha_1|\alpha_2)$ of finding the particle at α_1, given that it was at α_2 is approximately equal to 1. That is, if the detector clicks, then we may logically deduce that the particle followed a trajectory in the past, along the direct line from the origin to the detector.

Second, suppose we project instead onto a completely different intermediate state. The initial state is an outgoing spherical wave, and we there expect that there will be consistent histories reflecting this fact. Suppose we therefore project onto this possibility, using the projector,

$$P_{\alpha_1}(t_1) = |\psi(t_1)\rangle\langle\psi(t_1)| \tag{24}$$

(together with its negation). Then we again get consistency, and that $p(\alpha_1|\alpha_2) = 1$, from which we would be inclined to say that the particle was in a spherical wave state at time t_1. A third possibility would be to project onto the momentum at t_1, which would lead to yet another set of consistent histories, in which the value of momentum is predicted with near certainty.

This illustrates that there are different sets of consistent histories for the same physical situation, depending on which variables one would like to talk about. Moreover, these different choices for $P_{\alpha_1}(t_1)$ do not commute, and this means that if we attempted to combine them in a single set, the consistency condition would no longer be satisfied, and we would not be allowed to make any logical deductions. This feature of the formalism is essentially quantum-mechanical complementarity, although it appears in a form which is for some quite disconcerting. It raises the question as to whether one would be able to assign definite values to non-commuting observables. This, however, is excluded by a rule proposed by Griffiths [6], which states that all logical deductions about the system must be made from the framework of a single consistent set.

The non-uniqueness of the retrodicted past then raises some questions as to the value of the whole formalism, and what practical use it is. It means that quantum theory, in the decoherent histories version of it, does not uniquely tell us what "actually happened", although to be sure, this feature of quantum theory is certainly known already in various ways. See [9, 18, 24, 25] for discussions of these issues.

The value of the formalism becomes clear when we ask what do we actually do with the retrodicted past. The answer is that we use it to make other predictions about the present. That is, we start from some present data (measurements or cosmological observations), and, using consistency or decoherence, we retrodict the past. We then use the retrodicted past to make more predictions about the present. The role of the histories, therefore, is that they are an intermediate tool which helps us to identify correlations between data sets at a fixed moment of time. For example, when we look at a photograph of a particle track, we see a series of dots, existing at a fixed moment of time, which appear to be correlated. The explanation of their correlation is to be found in appealing to the past history of the system that produced the dots. Differently put, by looking, for example, at just three dots, we can use decoherence and retrodiction to deduce that a particle passed through on an approximately classical history. We then use the history to correctly deduce the location of the remaining dots on the photographic plate. In brief, therefore, histories are a useful tool to help us understand present records [21].

4   Quantum Brownian Motion Model

We now briefly consider a particular model, namely the quantum Brownian motion model. This model has been extensively studied in the literature so only the briefest of accounts will be given here [17]. The model consists of a particle of mass M in a potential $V(x)$ linearly coupled to an environment consisting of a large bath of harmonic oscillators in a thermal state at temperature T [26]. We consider histories of position samplings of the distinguished system. The samplings are continuous in time and Gaussian sampling functions are used (corresponding to approximate projection operators). The decoherence functional for the model is most conveniently given in path-integral form:

$$D[\bar{x}(t), \bar{y}(t)] = \int \mathcal{D}x\mathcal{D}y \; \delta(x_f - y_f) \; \rho(x_0, y_0)$$
$$\times \exp\left(\frac{i}{\hbar}S[x(t)] - \frac{i}{\hbar}S[y(t)] + \frac{i}{\hbar}W[x, y]\right)$$
$$\times \exp\left(-\int dt \; \frac{(x(t) - \bar{x}(t))^2}{2\sigma^2} - \int dt \; \frac{(y(t) - \bar{y}(t))^2}{2\sigma^2}\right) \quad (25)$$

Here, S is the action for a particle in a potential $V(x)$, $\bar{x}(t)$, $\bar{y}(t)$ are the sampled positions and x_f and x_0 denote the final and initial values respectively. The effects of the environment are summarized entirely by the Feynman-Vernon influence functional phase, $W[x, y]$, given by,

$$W[x(t), y(t)] = -\int_0^t ds \int_0^s ds'[x(s) - y(s)] \; \eta(s - s') \; [x(s') + y(s')]$$
$$+ i\int_0^t ds \int_0^s ds'[x(s) - y(s)] \; \nu(s - s') \; [x(s') - y(s')] \quad (26)$$

The explicit forms of the non-local kernels η and ν may be found, for example, in [27]. Here it is assumed, as is typical in these models, that the initial density matrix of the total system is simply a product of the initial system and environment density matrices, and the initial environment density matrix is a thermal state at temperature T. Considerable simplifications occur in a purely ohmic environment in the Fokker-Planck limit (a particular form of the high-temperature limit), in which one has

$$\eta(s - s') = M\gamma\, \delta'(s - s') \qquad (27)$$

$$\nu(s - s') = \frac{2M\gamma kT}{\hbar}\, \delta(s - s') \qquad (28)$$

where γ is the dissipation. For convenience we will work in this limit. One can see almost immediately that the imaginary part of W, together with the Gaussian samplings in (25), will have the effect of suppressing widely differing paths $\bar{x}(t)$, $\bar{y}(t)$. Indeed, the suppression factor will be of order

$$\exp\left(-\frac{2M\gamma kT\sigma^2}{\hbar^2}\right) \qquad (29)$$

In cgs units $\hbar \sim 10^{-27}$ and $k \sim 10^{-16}$, so $kT/\hbar^2 \sim 10^{40}$ if T is room temperature. Values of order 1 for M, γ and σ therefore lead to an astoundingly small suppression factor. Decoherence through interaction with a thermal environment is thus a very effective process indeed.

More precisely, one can approximately evaluate the functional integral (25). Let $X = (x + y)/2$, $\xi = x - y$, and use the smallness of the suppression factor to expand about $\xi = 0$. Then the ξ functional integral may be carried out with the result,

$$D[\bar{x}(t), \bar{y}(t)] = \int \mathcal{D}X\; W(M\dot{X}_0, X_0)\; \exp\left(-\int dt\; \frac{(X - \frac{\bar{x}+\bar{y}}{2})^2}{\sigma^2}\right)$$

$$\times \exp\left(-\int dt\; \frac{F[X]^2}{2(\Delta F)^2} - i\hbar \int dt\; \frac{(\bar{x} - \bar{y})F[X]}{4\sigma^2(\Delta F)^2}\right)$$

$$\times \exp\left(-\int dt\; \frac{(\bar{x} - \bar{y})^2}{2\ell^2}\right) \qquad (30)$$

where

$$F[X] = M\ddot{X} + M\gamma\dot{X} + V'(X) \qquad (31)$$

are the classical field equations with dissipation, and

$$(\Delta F)^2 = \frac{\hbar^2}{\sigma^2} + 4M\gamma kT \qquad (32)$$

$$\ell^2 = 2\sigma^2 + \frac{\hbar^2}{4M\gamma kT} \qquad (33)$$

$W(M\dot{X}_0, X_0)$ is the Wigner transform of the initial density operator.

The decoherence width (33) does not, in fact, immediately indicate the expected suppression of interference, because the temperature-dependent term will typically be utterly negligible compared to the σ^2 term. The point, however, is that more precise notions of decoherence need to be employed. One should check some of the probability sum rules, or use the approximate decoherence condition discussed in [17], in which the sizes of the off and on-diagonal terms are compared. This has not been carried out for the general expression (30), and in fact seems to be rather hard. Satisfaction of the approximate decoherence condition was checked for some special cases in [17]. Still, one expects the standard to which decoherence is attained to be of the order of the suppression factor (29), *i.e.*, very good indeed.

Now consider the diagonal elements of the decoherence function, representing the probabilities for histories.

$$p[\bar{x}(t)] = \int \mathcal{D}X \, W(M\dot{X}_0, X_0)$$
$$\times \, \exp\left(-\int dt \, \frac{(X - \bar{x})^2}{\sigma^2} - \int dt \, \frac{F[X]^2}{2(\Delta F)^2}\right) \tag{34}$$

The distribution is peaked about configurations $\bar{x}(t)$ satisfying the classical field equations with dissipation; thus approximate classical predictability is exhibited. The width of the peak is given by (32). Loosely speaking, a given classical history occurs with a weight given by the Wigner function of its initial data. This cannot be strictly correct, because the Wigner function is not positive in general, although it is if coarse-grained over an $\hbar$-sized region of phase space [28].

The width (32) has clearly identifiable contributions from quantum and thermal fluctuations. The thermal fluctuations dominate the quantum ones when $8M\gamma kT\sigma^2 \gg \hbar^2$, which, from (29), is precisely the condition required for decoherence, as previously noted. Environmentally-induced fluctuations are therefore inescapable if one is to have decoherence. This means that there is a tension between the demands of decoherence and classical predictability, both of which are necessary (although generally not sufficient) for the emergence of a quasiclassical domain [11]. This tension is due to the fact that the degree of decoherence (29) improves with increasing environment temperature, but predictability deteriorates, because the fluctuations (32) grow. However, the smallness of Boltzmann's constant ensures that the fluctuations (32) will be small compared to $F[X]$ for a wide range of temperatures if M is sufficiently large. Moreover, the efficiency of decoherence as evidenced through (29) is largely due to the smallness of $\hbar$, and will hold for a wide range of temperatures. So although there is some tension, there is a broad compromise regime in which decoherence and classical predictability can each hold extremely well.

One might wonder what sort of restrictions the uncertainty principle places on the degree to which probabilities for histories may be peaked about a particular history. For clearly it is not possible to be perfectly peaked because that would mean definite values for positions at different times, which are non-commuting operators. It turns out that the Shannon information,

$$I = \sum_{\underline{\alpha}} p(\underline{\alpha}) \ln p(\underline{\alpha}) \tag{35}$$

provides a useful measure of the degree of peaking, and the uncertainty principle appears as a lower bound on I, as shown in [29].

5   Quantum Classical Couplings

An interesting further development of the previous section concerns the construction of consistent theories describing the interaction of classical and quantum systems. This section primarily follows [30]. (See also [31].) This is a question of interest in a variety of different areas, in particular, in quantum field theory in curved spacetime, where one is interesting in assessing the effect a quantum field has on a classical gravitational field. To be more precise, let us consider the simpler case of a massive particle of mass M, position X and velocity $\dot{X}$. Suppose this particle, which we take to be well-described by classical mechanics, comes into interaction with a particle of mass m, which is sufficiently light that it is predominantly quantum-mechanical and described by a state vector $|\psi\rangle$. What happens to the classical particle under these conditions?

With a simple linear coupling between these systems, the classical particle obeys the equation of motion,

$$M\ddot{X} + V'(X) + \lambda x = 0 \tag{36}$$

and the state of the quantum particle obeys the equation

$$i\hbar \frac{d|\psi\rangle}{dt} = H_X|\psi\rangle \tag{37}$$

where H_X denotes the quantum particle's Hamiltonian in the presence of a classical external field X. The problem with this system, however, is how to interpret the quantity x in (36), which clearly should be an operator because it describe a quantum system. The simplest suggested resolution to this is to insert a quantum-mechanical expectation value, and to use instead of (36) the equation

$$M\ddot{X} + V'(X) + \lambda\langle\psi|x|\psi\rangle = 0 \tag{38}$$

However, one would expect this prescription to yield reasonable results only in a limited set of circumstances. Indeed, it gives physically unreasonable results in the interesting case when the quantum particle is in a superposition of localized position states. Clearly in that case the correct physical answer is that the classical particle should "see" one or other of the localized superposition states, and not some kind of averaged position, which is what (38) indicates.

An elementary extension of the results of the previous section may be used to construct a more sensible coupled classical-quantum system. One may start from the assumption that there are no truly classical systems, only quantum systems that are approximately classical under certain circumstances, as discussed in the previous section. Consider therefore, a large particle of mass M and position X coupled to an environment, to make it approximately classical. Suppose also that the massive particle is again coupled to the light particle, as in (36). We now look for decoherent histories of the massive particle.

It is easily shown that there are decoherent histories of positions of the massive particle, whose probabilities are strongly peaked about the equation of motion,

$$M\ddot{X} + M\gamma\dot{X} + V'(X) + \lambda\bar{x} = 0 \tag{39}$$

These are the classical equations of motion with dissipation, as expected (and also with fluctuations, encoded in the width of the peak). The interesting extra ingredient, however, is the quantity $\bar{x}$ which is now not an operator, but a stochastic c-number, for which the decoherent histories approach supplies a probability distribution function $p[\bar{x}(t)]$. This is a physically sensible result: the classical particle responds in a stochastic way to the quantum system. Furthermore, this construction gives the expected sensible results for superposition states of the quantum particle.

6 Decoherent Histories and Quantum State Diffusion

For the particular, yet commonly realized situation in which there is a natural split into system and environment, it turns out that the decoherent histories approach is closely related to the quantum state diffusion approach to open quantum systems. We begin be briefly explaining what this is.

Very many open quantum systems are accurately described by the Lindblad master equation for the reduced density operator ρ [32]. This is

$$\frac{d\rho}{dt} = -\frac{i}{\hbar}[H, \rho] - \frac{1}{2}\sum_{j=1}^{n}\left(\{L_j^\dagger L_j, \rho\} - 2L_j\rho L_j^\dagger\right) \tag{40}$$

Here, H is the Hamiltonian of the open system in the absence of the environment (sometimes modified by terms depending on the L_j) and the n operators L_j model the effects of the environment. The Lindblad form is the most general possible evolution equation preserving positivity, hermiticity and trace, subject only to the physically useful assumption that the evolution is Markovian. For example, in the quantum Brownian motion model, the master equation has a single non-hermitian L which is a linear combination of position and momentum operators, and the Markovian approximation is valid for reasonably high temperatures.

Whilst the master equation is the correct quantum-mechanical description for many open systems, its solutions do not easily yield the expected physical picture that may be directly compared with experiments. Take again the case of quantum Brownian motion. There, coarse observations of the system yield trajectories in phase space following approximately classical paths with dissipation and fluctuations. Yet an initial localized wavepacket will spread indefinitely as a result of quantum and thermal fluctuations under evolution according to (40). Differently put, the density operator really corresponds to an ensemble of trajectories, yet an experiment measures an individual trajectory of the system.

The quantum state diffusion picture, introduced by Gisin and Percival [33], aimed to expose the individual trajectories contained in the density operator equation. In this picture, the density operator ρ satisfying (40) is regarded as a mean over a distribution of pure state density operators,

$$\rho = M|\psi\rangle\langle\psi| \tag{41}$$

where M denotes the mean (defined below), with the pure states evolving according to the non-linear stochastic Langevin-Ito equation,

$$|d\psi\rangle = -\frac{i}{\hbar}H|\psi\rangle dt + \frac{1}{2}\sum_j \left(2\langle L_j^\dagger\rangle L_j - L_j^\dagger L_j - \langle L_j^\dagger\rangle\langle L_j\rangle\right)|\psi\rangle\, dt$$
$$+ \sum_j \left(L_j - \langle L_j\rangle\right)|\psi\rangle\, d\xi_j(t) \tag{42}$$

for the normalized state vector $|\psi\rangle$. Here, the $d\xi_j$ are independent complex differential random variables representing a complex Wiener process. Their linear and quadratic means are,

$$M[d\xi_j d\xi_k^*] = \delta_{jk}\, dt, \quad M[d\xi_j d\xi_k] = 0, \quad M[d\xi_j] = 0 \tag{43}$$

This equation is very similar to the explicit modified versions of quantum theory, such as that due to Ghirardi *et al.* [34], but here the motivation is different, in that the stochastic equation is not proposed as a fundamental equation.

An interesting feature of (42) is that its solutions tend to undergo some kind of localization in time, quite the opposite of the master equation. For example, it may be shown that in the quantum Brownian motion model, all initial states become localized around a wave packet tightly peaked in phase space after a very short time, and thereafter remain localized and follow the classical equations of motion [33, 35]. Hence the quantum state diffusion approach naturally provides the intuitive appealing and experimentally correct picture of an individual trajectory. Furthermore, one can also calculate a probability for each trajectory.

It is therefore easily seen that both of the intuitive picture and physically predictions of quantum state diffusion are in fact essentially the same as those provided by the decoherent histories approach. For the histories approach also naturally provides the picture of an individual trajectory, and a probability for those trajectories. This is argued in much greater detail in [36]. Differently put, the solutions to quantum state diffusion in some sense represent a single history from a decoherent set. Also, QSD was put forward as a phenomenological picture of open quantum systems. This connection with the decoherent histories approach may be thought of as a more fundamental justification of QSD phenomenology.

More general versions of this connection, involving other types of stochastic unravelings have been discovered by Brun [37].

7   Hydrodynamic Equations

Most models of decoherence, such as the quantum Brownian motion model of Section 4, rely on an obvious separation of the whole closed system in a distinguished subsystem, and the rest (the environment). This is a reasonable assumption for a wide variety of physically interesting situations. The decoherent histories approach, however, does not rely on such a separation, and this is important, because there are situations or regimes where such a split does not necessarily exist. This then raises the question, in a large and possibly complex quantum system, with no obvious system-environment split, what are the variables that naturally become classical and what sorts of classical equations of motion emerge from the underlying quantum theory? Differently put, what is the most general possible derivation of emergent classicality?

Gell-Mann and Hartle have argued that one particular set of variables that are strong candidates for the "habitually decohering" variables are the integrals over small volumes of locally conserved densities [11]. These variables are distinguished by the existence of conservation laws for total energy, momentum, charge, particle number, *etc.* Associated with such conservation laws are local conservation laws of the form

$$\frac{\partial \rho}{\partial t} + \nabla \cdot \mathbf{j} = 0 \tag{44}$$

The candidate quasiclassical variables are then

$$Q_V = \int_V d^3x \; \rho(\mathbf{x}) \tag{45}$$

If the volume V over which the local densities are smeared is infinite, Q_V will be an exactly conserved quantity. In quantum mechanics it will commute with the Hamiltonian, and, as is easily seen, histories of Q_V's will then decohere exactly [38]. If the volume is finite but large compared to the microscopic scale, Q_V will be slowly varying compared to all other dynamical variables. This is because the local conservation law (44) permits Q_V to change only by redistribution, which is limited by the rate at which the locally conserved quantity can flow out of the volume. Because these quantities are slowly varying, histories of them should therefore approximately decohere. Furthermore, the fact that the Q_V's are slowly varying may also be used, at least classically, to derive an approximately closed set of equations involving only those quantities singled out by the conservation laws. These equations are, for example, the Navier-Stokes equations, and the derivation of them is a standard (although generally non-trivial) exercise in non-equilibrium statistical mechanics [39].

One of the current goals of the decoherent histories approach is to carry this programme through in detail. A more detailed sketch of how this works was put forward in [40, 41] and a variety of models and related aspects are described in [42].

8   Spacetime Coarse Grainings

Another class of questions to which the decoherent histories approach adapts very well are those that involve time in a non-trivial way. In particular, the arrival time and tunneling time problems have been the subject of considerable recent interest, that interesting stemming from the fact that quantum mechanics does not obviously supply a unique and physically reasonable prescription for addressing those problems. The decoherent histories approach offers yet another way to analyzing these questions.

To focus ideas, consider the question, what is the probability that a particle enters the spatial region Δ at any time, during the time interval $[0, \tau]$? This has been address by Yamada and Takagi [43], Hartle [44] and Micanek and Hartle [45], most of these concentrating on the case where Δ is the region $x < 0$. The decoherence functional can be constructed using the path integral form (17) as a starting point. The fact that the question is non-trivial in time means that there is interference, and the decoherence or consistency conditions are satisfied only for very special choices of initial state, and the resultant probabilities are rather trivial. For example, in the case where Δ is $x < 0$, the initial state must be antisymmetric about $x = 0$, and the probability of entering the region is zero. In particular, this analysis does not admit the interesting case of a wave packet starting in $x > 0$ and approaching the origin. Moreover, it does not even give results that have a sensible classical limit.

One can develop this approach further, and include an environment to produce decoherence, along the lines of the quantum Brownian motion model of Section 4 [46]. This then allows the assignment of probabilities for a wide variety of initial states, with a sensible and expected classical limit. However, they are essentially the probabilities one might anticipate on the basis of a classical Fokker-Planck equation for classical Brownian motion, and moreover, the probabilities seem to depend on the features of the environment.

At the present state of play, therefore, the results of the decoherent histories analysis are not very striking. There is, however, space for much more work to be done in this area. Furthermore, they appear to be compatible with other approaches to the arrival time problem, which involves, for example, time operators or detector models. The fact that most initial states do not satisfy the consistency condition (in the case of no environment) appears to be related to the fact that paths in the path integral move in and out of the region Δ many times, which in turn appears to be related to the fact that the time operators proposed in other approaches are not self-adjoint. Similarly, the dependence of the result on the details of the environment seems to be related to the fact that, in detector model approaches, the result depends on the form of the detector. See [47] for a more detailed review of these issues.

Finally, on the subject of temporal issues, it is worth noting that the so-called temporal Bell inequalities [48] may play in interesting role in the decoherent histories approach. The temporal Bell inequalities are a set of inequalities on certain probabilities or correlation functions which are derived by assuming that there exists a valid probability distribution for certain variables at a series of times. From the point of view of the decoherent histories approach, this assumption is only true when the histories are decoherent. The temporal Bell inequalities may therefore provide an interesting way of characterizing decoherence of histories, or the lack of decoherence, although this possibility does not seem to have received much attention to date.

9   Quantum Theory without Time

An important extension of the ideas outlined in the previous section is the further extension of the decoherent histories approach to situations which do no involve specifying time in any way whatsoever. For example, in classical mechanics whether a particle follows a particular classical path. Or we can ask whether a particle's path enters a given region of space at stage along the entire trajectory. Such questions are interesting and relevant for quantum cosmology. There, the wave function of the system obeys not a Schrödinger equation, but the Wheeler-DeWitt equation, which has the form,

$$H\Psi = 0 \tag{46}$$

That is, the state function of the system is a zero energy eigenstate. The form of this equation arises as a result of reparametrization invariance, or more generally, the four-dimensional diffeomorphism invariance of general relativity. The problems of interpreting the solutions to this equation are similar to the problems of interpreting the Klein-Gordon equation as a first-quantized wave equation, and have attracted a considerable amount of interest over the years. Like the arrival time problem, there are a variety of different approaches to it. The decoherent histories approach is well-adapted to this problem. This is partly because, as we shall see, the natural reparametrization-invariant notion is an entire classical trajectory, and the histories approach handles this notion straightforwardly. We will briefly review the decoherent histories analysis of the question. The detailed analysis of the Klein-Gordon equation has been carried out in [49] and more general timeless models have been studied in [50]. We follow the latter quite closely.

To be precise, suppose we have an n dimensional configuration space with coordinates $\mathbf{x} = (x_1, x_2, \cdots x_n)$, and suppose the wave function $\psi(\mathbf{x})$ of the system is in an eigenstate of the Hamiltonian, as in (46). What is the probability of finding the system in a region Δ of configuration space *without reference to time*? The classical case contains almost all the key features of the problem and we concentrate on this.

We will consider a classical system described by a $2n$-dimensional phase space, with coordinates and momenta $(\mathbf{x}, \mathbf{p}) = (x_k, p_k)$, and Hamiltonian

$$H = \frac{\mathbf{p}^2}{2M} + V(\mathbf{x}) \tag{47}$$

We assume that there is a classical phase space distribution function $w(\mathbf{p}, \mathbf{x})$, which is normalized according to

$$\int d^n p \, d^n x \, w(\mathbf{p}, \mathbf{x}) = 1 \tag{48}$$

and obeys the evolution equation

$$\frac{\partial w}{\partial t} = \sum_k \left(-\frac{p_k}{M} \frac{\partial w}{\partial x_k} + \frac{\partial V}{\partial x_k} \frac{\partial w}{\partial p_k} \right) = \{H, w\} \tag{49}$$

where $\{ \, , \, \}$ denotes the Poisson bracket. The interesting case is that in which w is the classical analogue of an energy eigenstate, in which case $\partial w/\partial t = 0$, so the evolution equation is simply

$$\{H, w\} = 0 \tag{50}$$

It follows that

$$w(\mathbf{p}^{cl}(t), \mathbf{x}^{cl}(t)) = w(\mathbf{p}(0), \mathbf{x}(0)) \tag{51}$$

where $\mathbf{p}^{cl}(t), \mathbf{x}^{cl}(t)$ are the classical solutions with initial data $\mathbf{p}(0), \mathbf{x}(0)$, so w is constant along the classical orbits.

Given a set of classical solutions $(\mathbf{p}^{cl}(t), \mathbf{x}^{cl}(t))$, and a phase space distribution function w, we are interested in the probability that a classical solution will pass through a region Δ of configuration space. To see whether the classical trajectory $\mathbf{x}^{cl}(t)$ intersects this region, consider the phase space function

$$A(\mathbf{x}, \mathbf{p}_0, \mathbf{x}_0) = \int_{-\infty}^{\infty} dt \, \delta^{(n)}(\mathbf{x} - \mathbf{x}^{cl}(t)) \tag{52}$$

(In the case of periodic classical orbits, the range of t is taken to be equal to the period). This function is positive for points $\mathbf{x}$ on the classical trajectory labeled by $\mathbf{p}_0, \mathbf{x}_0$ and zero otherwise. It also has the property that

$$\{H, A\} = 0 \tag{53}$$

so is a reparametrization-invariant observable. This is the mathematical expression of the statement above that an entire classical trajectory is a reparametrization-invariant notion. Intersection of the classical trajectory with the region Δ means,

$$\int_{\Delta} d^n x \, A(\mathbf{x}, \mathbf{p}_0, \mathbf{x}_0) > 0 \tag{54}$$

The quantity on the left is essentially the amount of parameter time the trajectory spends in the region Δ, so we are simply requiring it to be positive. We may now write down the probability for a classical trajectory entering the region Δ. It is,

$$p_\Delta = \int d^n p_0 d^n x_0 \, w(\mathbf{p}_0, \mathbf{x}_0) \, \theta \left(\int_\Delta d^n x \, A - \epsilon \right) \tag{55}$$

In this construction, ϵ is a small positive number that is eventually sent to zero, and is included to avoid possible ambiguities in the θ-function at zero argument. The θ-function ensures that the phase space integral is over all initial data whose corresponding classical trajectories spend a time greater than ϵ in the region Δ. Equation (55) is the desired formula for the probability of entering the region Δ without regard to time. All elements of it are reparametrization invariant.

At the meeting, a possible difficulty with (54) was pointed out by T.Brun. This is that for chaotic systems, the trajectories visit every region in phase space, and it therefore appears that every trajectory will intersect the region Δ. At present it therefore looks like the analysis presented here is valid only for integrable systems. A more general version of this analysis avoiding this difficulty is currently being sought, and will be described elsewhere. A possible solution is that asking whether the system intersects a given region of configuration space is not in fact a useful observable for chaotic systems, although there is no reason why there should not exist other types of observables that are useful.

Turning now to the quantum theory, a decoherence functional of the general form (12) may be constructed. Special attention is required for the inner product (since the solutions to (46) are typically not normalizable in the usual sense), and also for the class operators C_α, which must be reparametrization invariant and describe paths which pass through the region Δ. Also, an environment is typically required in order to obtain decoherence. These non-trivial aspects were discussed at length in [50]. The final result, in essence, is that in the semiclassical limit, one obtains a formula of the form (55) in which the probability function w is replaced by the Wigner function of the quantum state. Furthermore, this formula may be rewritten in such a way that it coincides with earlier heuristic analyses of the Wheeler-DeWitt equation (such as the so-called WKB interpretation).

In summary, the decoherent histories approach successfully adapts to genuinely timeless systems, such as the Wheeler-DeWitt equation, although there is much more work to be done on this question.

10   Summary

I have described the decoherent histories approach, its properties and some recent applications and developments. These may be summarized as follows:
• The decoherent histories approach is essentially the Copenhagen approach to quantum mechanics, but relies on a smaller set of assumptions. In essence, it formalizes intuition.
• The approach gives a very comprehensive account of emergent classicality in a variety of situations, plus fluctuations about it, and couplings to quantum systems.
• It naturally accommodates situations where there is no system-environment split.
• It readily generalizes to situations which are non-trivial in time, or which do not involve time at all, as we expect to be the case in quantum gravity.

Acknowledgements

I am very grateful to Hans-Thomas Elze to inviting me to take part in this most stimulating meeting. I would also like to thank Todd Brun for useful comments on my talk at the meeting.

References

1. A useful source of literature on the Copenhagen interpretation is: *Quantum Theory and Measurement*, ed. by J.A. Wheeler and W. Zurek (Princeton University Press, Princeton, NJ, 1983)

2. J. Bell: Phys. World **3**, 33 (1990).
3. A.J. Leggett: Suppl. Prog. Theor. Phys. **69**, 80 (1980)
4. E. Joos and H.D. Zeh: Z. Phys. B **59**, 223 (1985)
5. J.B.Hartle. In: *Proceedings of the Cornelius Lanczos International Centenary Con-ference*, ed. by J.D. Brown, M.T. Chu, D.C. Ellison and R.J. Plemmons (SIAM, Philadelphia 1994), also available as e-print gr-qc/9404017
6. R.B. Griffiths: J. Stat. Phys. **36**, 219 (1984)
7. R. Omnès: J. Stat. Phys. **53**, 893 (1988)
8. R. Omnès: J. Stat. Phys. **53**, 933 (1988); **53**, 957 (1988); **57**, 357 (1989); Phys. Lett. A **138**, 157 (1989); Ann. Phys. **201**, 354 (1990); Rev. Mod. Phys. **64**, 339 (1992)
9. R. Omnès: Phys. Lett. A **187**, 26 (1994); J. Stat. Phys. **62**, 841 (1991)
10. M. Gell-Mann and J.B. Hartle. In: *Complexity, Entropy and the Physics of In-formation, SFI Studies in the Sciences of Complexity*, Vol. VIII, ed. by W. Zurek (Addison Wesley, Reading, MA, 1990); and in: *Proceedings of the Third Interna-tional Symposium on the Foundations of Quantum Mechanics in the Light of New Technology*, ed. by S. Kobayashi, H. Ezawa, Y. Murayama and S. Nomura (Physical Society of Japan, Tokyo 1990)
11. M. Gell-Mann and J.B. Hartle: Phys. Rev. D **47**, 3345 (1993)
12. J.B. Hartle. In: *Proceedings of the 1992 Les Houches Summer School, Gravitation et Quantifications*, ed. by B. Julia and J. Zinn-Justin (Elsevier Science B.V., 1995)
13. T. Brun: 'Decoherence and Consistent Histories Reference List', www.sns.ias.edu/~tbrun.
14. C. Isham: J. Math. Phys. **23**, 2157 (1994); C. Isham and N. Linden: J. Math. Phys. **35**, 5452 (1994); **36**, 5392 (1995); C. Isham, N. Linden and S. Schreckenberg: J. Math. Phys. **35**, 6360 (1994)
15. S. Goldstein and D.N. Page: Phys. Rev. Lett. **74**, 3715 (1995)
16. J. Finkelstein: Phys.Rev. D **47**, 5430 (1993).
17. H.F. Dowker and J.J. Halliwell: Phys. Rev. D **46**, 1580 (1992)
18. H.F. Dowker and A. Kent: J. Stat. Phys. **82**, 1575 (1996); Phys. Rev. Lett. **75**, 3038 (1995)
19. J.J. Halliwell: Phys. Rev. D **63** 085013 (2001)
20. J.N. McElwaine: Phys. Rev. A **53**, 2021 (1996).
21. J.B. Hartle: gr-qc/9712001
22. J.J. Halliwell: Phys. Rev. D **60**, 105031 (1999)
23. P.J. Dodd and J.J. Halliwell: quant-ph/0301104
24. I. Giardina and A. Rimini: Found. Phys. **26**, 973 (1996)
25. C. Anastopoulos: Int. J. Theor. Phys. **37**, 2261 (1998)
26. A. Caldeira and A. Leggett: Physica A **121**, 587 (1983)
27. C. Anastopoulos and J.J. Halliwell: Phys. Rev. D **51**, 6870 (1995)
28. J.J. Halliwell: Phys. Rev. D **46**, 1610 (1992)
29. J.J. Halliwell: Phys. Rev. D **48**, 2739 (1993) See also: Phys. Rev. D **48**, 4785 (1993)
30. J.J. Halliwell: Phys. Rev. D **57**, 2337 (1998)
31. L. Diosi and J.J. Halliwell: Phys. Rev. Lett. **81**, 2846 (1998)
32. G. Lindblad: Commun. Math. Phys. **48**, 119 (1976)
33. N. Gisin and I.C. Percival: J. Phys. A **25**, 5677 (1992); J. Phys. A **26**, 2233 (1993); J. Phys. A **26**, 2245 (1993); Phys. Lett. A **167**, 315 (1992)
34. G.C. Ghirardi, A. Rimini and T. Weber: Phys. Rev. D **34**, 470 (1986); G.C. Ghir-ardi, P. Pearle, and A. Rimini: Phys. Rev. A **42**, 78 (1990)
35. J.J. Halliwell and A. Zoupas, Phys. Rev. D **52**, 7294 (1995)

36. L. Diósi, N. Gisin, J. Halliwell and I.C. Percival: Phys. Rev. Lett. **74**, 203 (1995)
37. T. Brun: Phys. Rev. Lett. **78**, 1833 (1997); Phys. Rev. A **61**, 042107 (2000)
38. J.B. Hartle, R. Laflamme and D. Marolf: Phys. Rev. D **51**, 7007 (1995)
39. D. Forster: *Hydrodynamic Fluctuations, Broken Symmetry and Correlation Functions* (Benjamin, Reading, MA, 1975)
40. J.J. Halliwell: Phys. Rev. Lett. **83**, 2481 (1999)
41. J.J. Halliwell: Phys. Rev. D **58**, 105015 (1998)
42. C. Anastopoulos: gr-qc/9805074; T. Brun and J.J. Halliwell: Phys. Rev. D **54**, 2899 (1996); E. Calzetta and B.L. Hu: Phys. Rev. D **59**, 065018 (1999); and in: *Directions in General Relativity*, ed. by B.L. Hu and T.A. Jacobson (Cambridge University Press, Cambridge 1993); T. Brun and J.B. Hartle: Phys. Rev. D **60**, 123503 (1999). See quant-ph/9912037 for a review of the decoherent histories approach applied to the hydrodynamic problem
43. N. Yamada and S. Takagi: Prog. Theor. Phys. **85**, 985 (1991); Prog. Theor. Phys. **86**, 599 (1991); Prog. Theor. Phys. **87**, 77 (1992); N. Yamada: Sci. Rep. Tôhoku Uni., Series 8, **12**, 177 (1992); Phys. Rev. A **54**, 182 (1996)
44. J.B. Hartle: Phys. Rev. D **44**, 3173 (1991)
45. R.J. Micanek and J.B. Hartle: Phys. Rev. A **54**, 3795 (1996)
46. J.J. Halliwell and E. Zafiris, Phys. Rev. D **57**, 3351 (1998)
47. J.J. Halliwell. In: *Time in Quantum Mechanics*, ed. by J.G. Muga, R. Sala Mayato and I.L. Egususquiza (Springer Verlag, Berlin 2001); also available as quant-ph/0101099
48. A.J. Leggett and A. Garg: Phys. Rev. Lett. **54**, 857 (1985); J.P. Paz and G. Mahler, Phys. Rev. Lett. **71**, 3235 (1993)
49. J.J. Halliwell and J. Thorwart: Phys. Rev. D **64**, 124018 (2001)
50. J.J. Halliwell and J. Thorwart: Phys. Rev. D **65**, 104009 (2002)

Is There an Information-Loss Problem for Black Holes?

Claus Kiefer

Institut für Theoretische Physik, Universität zu Köln, Zülpicher Str. 77, 50937 Köln, Germany

Abstract. Black holes emit thermal radiation (Hawking effect). If after black-hole evaporation nothing else were left, an arbitrary initial state would evolve into a thermal state ('information-loss problem'). Here it is argued that the whole evolution is unitary and that the thermal nature of Hawking radiation emerges solely through decoherence – the irreversible interaction with further degrees of freedom. For this purpose a detailed comparison with an analogous case in cosmology (entropy of primordial fluctuations) is presented. Some remarks on the possible origin of black-hole entropy due to interaction with other degrees of freedom are added. This might concern the interaction with quasi-normal modes or with background fields in string theory.

1   The Information-Loss Problem

Black holes are amazing objects. According to general relativity, stationary black holes are fully characterised by just three numbers: Mass, angular momentum, and electric charge. This "no-hair theorem" holds within the Einstein-Maxwell theory in four spacetime dimensions. It reminds one at the properties of a macroscopic gas which can be described by only few variables such as energy, entropy, and pressure. In fact, there exist laws of black-hole mechanics which are analogous to the laws of thermodynamics (see e.g. [1] for a detailed review). The temperature is proportional to the surface gravity, κ, of the black hole, and the entropy is proportional to its area, A. That this correspondence is not only a formal one, but possesses physical significance, was shown by Hawking in his seminal paper [2]. Considering quantum field theory on the background of a collapsing star (see Fig. 1), it is found that the black hole radiates with a temperature proportional to $\hbar$,

$$T_{\mathrm{BH}} = \frac{\hbar\kappa}{2\pi k_{\mathrm{B}}} \; . \tag{1}$$

The origin of this temperature is the presence of a horizon. Due to the high gravitational redshift in its vicinity (symbolised in Fig. 1 by the dashed line γ near the horizon γ_{H}), the vacuum modes are excited for a very long time, until the black hole has evaporated. The entropy connected with this temperature is given by the 'Bekenstein-Hawking' formula,

$$S_{\mathrm{BH}} = \frac{k_{\mathrm{B}} A}{4G\hbar} \; . \tag{2}$$

C. Kiefer, Is There an Information-Loss Problem for Black Holes?, Lect. Notes Phys. **633**, 84–95 (2004)
http://www.springerlink.com/

In the case of a spherically-symmetric ('Schwarzschild') black hole, one has $\kappa = (4GM)^{-1}$, and the Hawking temperature is given by

$$T_{\mathrm{BH}} = \frac{\hbar}{8\pi G k_{\mathrm{B}} M} \approx 6.2 \times 10^{-8} \frac{M_{\odot}}{M} \ \mathrm{K} \ . \tag{3}$$

If the quantum field on the black-hole background is a massless scalar field, the expectation value of the particle number for a mode with wave number $\mathbf{k}$ is given by

$$\langle n_k \rangle = \frac{1}{e^{8\pi\omega GM} - 1} \ , \tag{4}$$

with $\omega = |\mathbf{k}| = k$. This is a Planck distribution with temperature (3). The usual interpretation is 'particle creation': a mode with pure positive frequency will evolve (along the dashed line in Fig. 1) into a superposition of positive and negative frequences. Therefore, an initial vacuum will evolve into a superposition of excited states ('particles'). In the Heisenberg picture, this is described by the occurrence of a non-vanishing Bogoliubov coefficient β [1,2].

An alternative point of view arises through the use of the Schrödinger picture. Taking the scalar field in Fourier space, $\phi(k)$, the initial vacuum state can be expressed as a Gaussian wave functional. One assumes for simplicity that the full wave functional can be written as a product over independent modes, $\Psi = \prod_k \psi_k$ (one can imagine putting the whole system into a box). Then,

$$\psi_k \propto \exp\left[-k|\phi(k)|^2\right] \ , \tag{5}$$

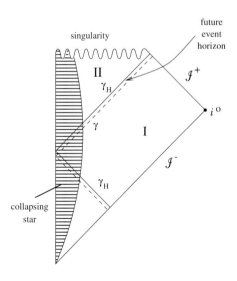

Fig. 1. Penrose diagram of a star to form a black hole

see e.g. [3] for details about the functional Schrödinger picture. With this initial state, the functional Schrödinger equation can be solved exactly [4]. Due to the dynamical gravitational background, the various ψ_k, albeit always of Gaussian form, develop a complex function in the exponent [4,5],

$$\psi_k \propto \exp\left[-k \coth(2\pi kGM + ikt)|\phi(k)|^2\right] . \tag{6}$$

This is still a *pure state*, but the expectation value of the particle number operator with respect to this state is of the same Planckian form as (4),

$$\langle\psi_k|n_k|\psi_k\rangle = \frac{1}{e^{8\pi\omega GM} - 1} . \tag{7}$$

As a side remark I note that such a result can also be obtained from the wave functional solving the Wheeler-DeWitt equation in the WKB approximation [6]. A state such as (6) is well known from quantum optics and called a *two-mode squeezed state*. That Hawking radiation can be described in this terminology was first recognised – using the Heisenberg picture – in [7].

For spatial hypersurfaces that enter the horizon one must trace out the degrees of freedom which reside inside the horizon. This results in a thermal density matrix in the outside region [8], with the temperature being equal to the Hawking temperature (3). It is a general property of two-modes squeezed states that one gets a thermal density matrix if half of the modes is being traced out [9]. The emergence of a density matrix is not surprising. Taking into account only part of the degrees of freedom one is dealing with an open quantum system [10]. Such systems do not obey a unitary dynamics, but are described by master equations. This is not related to any information-loss in the full system (including the interior of the black hole), but only to an effective information loss (or gain) for the reduced system.

The black-hole entropy (2) is much bigger than the entropy of a collapsing star. The entropy of the Sun, for example, is $S_\odot \approx 10^{57}$, but the entropy of a solar-mass black hole is $S_{\mathrm{BH}} \approx 10^{77}$, i.e. twenty orders of magnitudes larger (all entropies are measured in units of k_{B}). If all matter in the observable Universe were in a single gigantic black hole, its entropy would be $S_{\mathrm{BH}} \approx 10^{123}$. Black holes thus seem to be the most efficient objects for swallowing information.

Due to Hawking radiation, black holes have a finite lifetime. It is given by

$$\tau_{\mathrm{BH}} \approx \left(\frac{M_0}{m_{\mathrm{p}}}\right)^3 t_{\mathrm{p}} \approx 10^{65} \left(\frac{M_0}{M_\odot}\right)^3 \text{years} , \tag{8}$$

where m_{p} and t_{p} denote Planck mass and Planck time, respectively. The question now arises what happens at the end of black-hole evaporation. If only thermal radiation were left behind, an arbitrary initial state (for the star collapsing to form a black hole) would evolve into a mixed state. This process does not happen in the standard quantum theory for closed systems. There, the entropy

$$S = -k_{\mathrm{B}}\mathrm{Tr}(\rho\ln\rho) \tag{9}$$

is conserved for the full system. Since a thermal state contains least information, one would be faced with the *information-loss problem*. This is, in fact, what Hawking speculated to happen. The calculations in [2] are, however, restricted to the semiclassical approximation (gravity classical and matter quantum), which breaks down when the black hole approaches the Planck mass. The final answer can only be obtained within quantum gravity. The options are [11]

- Information is indeed *lost* during the evaporation, i.e. the evolution is non-unitary,

$$\rho \to \$\rho \neq S\rho S^\dagger \ .$$

- The full evolution is *unitary*, but this cannot be seen in the semiclassical approximation.
- The black hole leaves a *remnant* carrying all the information.

The state (6) also holds on a spatial surface that in Fig. 1 would start at the intersection of the collapsing star with the horizon and extend to spatial infinity i^0, i.e. a surface that does not enter the horizon. In fact, such surfaces seem quite natural, since they correspond to constant Schwarzschild time far away from the black hole. The question then arises where the thermal nature of Hawking radiation comes from; although (7) is Planckian, the state (6) is pure and the differences to a thermal distribution can be recognised in higher-order correlation functions.

I shall argue in Sect. 3 that the thermal appearance of Hawking radiation can be understood, even for the pure state (6), through *decoherence* – the irreversible and unavoidable interaction with the environment [10]. For this purpose it will be appropriate to rewrite (6) in a form where the two-mode squeezed nature becomes explicit. In fact, one can rewrite ψ_k in the form

$$\psi_k \propto \exp\left[-k\frac{1 + e^{2i\varphi_k}\tanh r_k}{1 - e^{2i\varphi_k}\tanh r_k}|\phi(k)|^2\right]$$
$$\equiv \exp\left[-(\Omega_R + i\Omega_I)|\phi(k)|^2\right] \ , \tag{10}$$

where the *squeezing parameter* r_k is given by

$$\tanh r_k = \exp(-4\pi\omega GM) \ , \tag{11}$$

and the *squeezing angle* φ_k reads

$$\varphi_k = -kt \ . \tag{12}$$

Thus, $r_k \to 0$ for $k \to \infty$ and $r_k \to \infty$ for $k \to 0$: modes with bigger wavelength are more squeezed than modes with smaller wavelength. At the maximum of the Planck spectrum one has $r \approx 0.25$. This corresponds to $\langle n_k \rangle = \sinh^2 r_k \approx 0.06$ for the expectation value of the particle number.

At $kt = 0$ the squeezing is in ϕ, whereas at $kt = \pi/2$ the squeezing is in p_ϕ, the momentum conjugate to ϕ. The ratio of the corresponding widths is $\tanh^2(2\pi kGM)$ (≈ 0.37 at the maximum of the Planck spectrum). Before I apply this to the study of decoherence, it is appropriate to review the analogous situation in inflationary cosmology. This will serve to understand the similarities to and the differences from the black-hole case.

2   Entropy of Cosmological Fluctuations

One of the most important advantages of an inflationary scenario for the early Universe is the possibility to obtain a dynamical explanation of structure formation (see e.g. [12] for a review). Quantum vacuum fluctuations are amplified by inflation, leading to a squeezed state (corresponding to particle creation in the Heisenberg picture). The scalar field(s) and the scalar part of the metric describe primordial density fluctuations, while the tensor part of the metric describes gravitons. The generation of gravitons is, in fact, an effect of linear quantum gravity. The primordial fluctuations can at a later stage serve as seeds for structure (galaxies and clusters of galaxies). They exhibit themselves in the CMB temperature anisotropy spectrum. The gravitons may lead to a stochastic gravitational-wave background that might in principle be observable with space-based experiments.

An important issue in the theoretical understanding of the above process is the exact way in which these quantum fluctuations[1] become classical stochastic variables, see e.g. [13] for discussion and references. Two ingredients are responsible for the emergence of their classical behaviour. Firstly, inflation leads to a huge squeezing of the quantum state for the fluctuations and, therefore, to a huge particle creation (the number of generated particles for a mode with wave number k is $N_k = \sinh^2 r_k$, where r_k is the squeezing parameter.) But in the limit of large r_k the quantum state becomes an approximate WKB state, corresponding in the Heisenberg picture to neglecting the part of the solution which goes as e^{-r_k} and which is called the "decaying mode". Secondly, interactions with other, "environmental", degrees of freedom lead to the field-amplitude basis (here called y_k) as the classical basis ("pointer basis") with respect to which interferences become unobservable. This process of *decoherence* transforms the y_k into effective classical stochastic quantities. On the largest cosmological scales one finds for the squeezing parameter $r_k \approx 100$, far beyond any values which can be attained in the laboratory.

Whereas the quantum theory therefore does not lead to deviations from the usual predictions (based on a phenomenological classical stochastic theory) of the inflationary scenario, the entropy of the fluctuations depends on their quantum nature (on the presence of the decaying mode, albeit small), see [14]. The entropy of the squeezed quantum state is of course zero, because it is a pure state. Due to the interaction with other degrees of freedom, however, the fluctuations have to be described by a density matrix ρ. The relevant quantity is then the von Neumann entropy (9).

In the context of primordial fluctuations, different mechanisms of coarse-graining have been investigated in order to calculate the local entropy. It was found that the maximal value for the entropy is $S_{\mathrm{max}} = 2r_k$, resulting from a coarse-graining with respect to the particle-number basis, i.e. non-diagonal elements of the reduced density matrix ρ are neglected in this basis. Consideration of the corresponding contour of the Wigner ellipse in phase space shows that this

---

[1] In the following, "fluctuations" refers to primordial density fluctuations as well as gravitons.

would smear out the thin elongated ellipse of the squeezed state (corresponding to the high values of r_k) into a big circle. Does such a coarse-graining reflect the actual process happening during inflation? Since the pointer basis in the quantum-to-classical transition is the field-amplitude basis and *not* the particle-number basis (which mixes the field variable with its canonical momentum), one would expect that coarse-graining should be done with respect to $y(k)$. This would then lead to $S = S_{\max}/2$, which is noticeably different from maximal entropy [14].

A crucial observation for the calculation of (9) is to note that the wavelength of the amplified fluctuations during inflation is bigger than the horizon scale, i.e bigger than H_{I}^{-1}, where H_{I} is the Hubble parameter of inflation (here taken to be approximately constant for simplicity). This prevents a direct causal interaction with the other, environmental, fields. However, nonlocal quantum correlations can still develop due to interaction terms in the total Hamiltonian. Since, as remarked above, the interaction is local in $y(k)$ (as opposed to its momentum), the density matrix will be of the form (suppressing k for simplicity)

$$\rho_\xi(y, y') = \rho_0(y, y') \exp\left(-\frac{\xi}{2}(y - y')^2\right) , \tag{13}$$

where $\rho_0(y, y')$ denotes the density matrix referring to the squeezed state, which is a Gaussian state, see e.g. [10] for a discussion. Equation (13) guarantees that interferences in the y-basis become small, while the probabilities (diagonal elements) remain unchanged. The details of the interaction are encoded in the phenomenological parameter ξ and are not needed for our discussion. Decoherence is efficient if the Gaussian in (13) dominates over the Gaussian from the squeezed state described by ρ_0. This leads to the condition [14]

$$\frac{\xi e^{2r}}{k} \gg 1 , \tag{14}$$

which is called the *decoherence condition*.

The entropy of the fluctuations is now calculated by setting $\rho = \rho_\xi$ in (9). For an arbitrary Gaussian density matrix, the result has been obtained in [15], see also Appendix A2.3 in [10]. With the abbreviation

$$\chi = \frac{\xi}{k}(1 + 4\sinh^2 r) , \tag{15}$$

the result is in our case given by the expression

$$S = -\ln 2 + \frac{1}{2}\ln\chi - \frac{\sqrt{1+\chi}}{2}\ln\frac{\sqrt{1+\chi}-1}{\sqrt{1+\chi}+1} . \tag{16}$$

The decoherence condition means $\chi \gg 1$, which leads for arbitrary r to

$$S \approx 1 + \frac{1}{2}\ln\frac{\xi(1 + 4N)}{4k} . \tag{17}$$

In the high-squeezing limit, $e^r \to \infty$, this yields

$$S \approx 1 - \ln 2 + \frac{1}{2} \ln \frac{e^{2r}\xi}{k} = 1 + \frac{1}{2} \ln \frac{N\xi}{k} . \tag{18}$$

In phase space this corresponds to $S \approx \ln A$, where A is the area of the Wigner ellipse. Application of the decoherence condition (14) leads to

$$S \gg 1 - \ln 2 \approx 0.31 , \tag{19}$$

where $\gg$ holds here in a logarithmic sense (it directly holds for the number of states e^S). Therefore, decoherence already occurs after few bits of information are lost. This is much less than the maximal entropy, which is obtained if the ellipse is smeared out to a big circle, corresponding to the choice $\xi/k = e^{2r}$ in (18), and leading to $S_{max} = 2r$, as has been remarked above. That only few bits of information loss can be sufficient for decoherence is well known from quantum optics [10]. The correlation between y and its canonical momentum is only preserved if $S \ll S_{max} = 2r$ because otherwise the squeezed Wigner ellipse is no longer recognizable.

One would expect that the maximal possible entropy due to quantum entanglement alone (i.e. without dynamical back reaction) is obtained if the coarse-graining is performed exactly with respect to the *field-amplitude basis* y. As remarked above, this would lead to $S = r$. Inspecting (18), this would correspond to the choice $\xi = k$. Therefore, as long as the modes are outside the horizon, one would expect to have $\xi < k$ which has a very intuitive interpretation: the coherence length $\xi^{-1/2}$ is larger than the width of the ground state ($r = 0$), so that the environment does not spoil the property of the quantum state being squeezed in some direction compared to the ground state. It can be shown that the correlation between y and the conjugate momentum remains for a sufficiently long time after the second horizon crossing (in the postinflationary phase), so that it really leads to the observed acoustic peaks (B-polarisation for gravitational waves) in the CMB [16]. The information contained in these peaks can be interpreted as a measure for the deviation of the entropy from the maximal entropy. Therefore, coarse-grainings that lead to maximal entropy would prevent the occurrence of such peaks and would thus be *in conflict with observation.*

This analysis has also borne out an interesting analogy of the primordial fluctuations with a chaotic system: the Hubble parameter corresponds to a Lyapunov exponent, although our system is not chaotic, but only classically unstable [14]. In the next section the comparison of the cosmological case with the black-hole case will be made.

3   Hawking Radiation from Decoherence

As in the cosmological case, the quantum state corresponding to Hawking radiation is a two-mode squeezed state. There are, however, pronounced differences in the black-hole case. As has been remarked at the end of Sect. 1, the squeezing parameter for the maximum of the Planck distribution is only $r_k \approx 0.25$, which is far below the values attained from inflation. High squeezing values are only obtained for very big wavelengths.

The quantum state can again be represented by the contour of the Winger ellipse in phase space. In the cosmological case the rotation of this ellipse is very slow, the corresponding time being about the age of the Universe [16]. This reflects the fact that it is not allowed to coarse-grain this ellipse into a big circle (Sect. 2). What about the situation for black holes? Again, the Wigner ellipse rotates around the origin, and the typical timescale is given by

$$t_k = \frac{\pi}{2k} \cdot$$

(20)

This corresponds to the exchange of squeezing between ϕ and its conjugate momentum p_ϕ, cf. the end of Sect. 1. Evaluated at the maximum of the Planck spectrum, one has

$$t_k(\max) \approx 14GM \approx 7 \times 10^{-5} \frac{M}{M_\odot} \text{ s} ,$$

(21)

which is much smaller than typical observation times. It is for this reason that a coarse-graining with respect to the squeezing angle *can* be performed. Squeezed states are extremely sensitive to interactions with environmental degrees of freedom [10]. In the present case of a quickly rotating squeezing angle this interaction leads to a diagonalisation of the reduced density matrix with respect to the particle-number basis [17], not the field-amplitude basis. Thereby the local entropy is maximised, corresponding to the coarse-graining of the Wigner ellipse into a circle. The value of this entropy can be calculated along the lines of Sect. 2. In contrast to the cosmological case one finds the standard expression for a thermal ensemble,

$$S_k = (1 + n_k) \ln(1 + n_k) - n_k \ln n_k \xrightarrow{r_k \gg 1} 2r_k .$$

(22)

The integration over all modes gives $S = (2\pi^2/45)T_{\mathrm{BH}}^3 V$, which is just the entropy of the Hawking radiation with temperature $T_{\mathrm{BH}} = (8\pi GM)^{-1}$. In this way, the pure squeezed state becomes indistinguishable from a canonical ensemble with temperature T_{BH} [5].

Independent of this practical indistinguishability from a thermal ensemble, the state remains a pure state. In fact, for timescales smaller than t_k the above coarse-graining is not allowed and the difference to a thermal state could be seen in principle. For the case of a primordial black hole with mass $M \approx 5 \times 10^{14}$ g one has at the maximum of the Planck spectrum $t_k(\max) \approx 1.7 \times 10^{-23}$ s. The observation of Hawking radiation at smaller times could then reveal the difference between the (true) pure state and a thermal state. It is also clear from (20) that large wavelengths (much larger than the wavelength corresponding to the maximum of the Planck spectrum) have a much longer rotation time for the Wigner ellipse. This would also offer, in principle, the possibility to distinguish observationally between pure and mixed states – provided, of course, that primordial black holes exist and can be observed.

To summarise, no mixed state for the total system has appeared at any stage of this discussion. This indicates that the full quantum evolution of collapsing star plus scalar field evolves unitarily. The thermal nature of Hawking radiation thus only emerges through coarse-graining. For hypersurfaces entering the horizon, this is achieved by tracing out degrees of freedom referring to the interior [8]. As has been emphasised here, however, a similar result holds for hypersurfaces that stay outside the horizon. Such hypersurfaces are a natural choice for asymptotic observers. The mixed appearance of the pure state for the quantum field is due to its squeezed nature. Squeezed states are very sensitive to interactions with other fields, even for very weak coupling. Such interactions lead to decoherence for the Hawking radiation. The thermal nature of the reduced density matrix is a consequence of the presence of the horizon as encoded in the particular squeezed state (6), cf. [4].

Similar conclusions hold for hypersurfaces entering the horizon [5]. A general problem is, however, that the evolution along different foliations is in general not unitarily equivalent [18]. Observations at spatial infinity cannot thus be correctly recovered from an arbitrary foliation. The precise relationship between observation and choice of hypersurfaces is not yet properly understood and deserves investigation.

Since no information loss occurs in the first place, there does not seem to be any case for an information-loss problem. The above line of argument holds, of course, only within the semiclassical regime, neglecting quantum effects of the gravitational field itself. Only a full quantum theory of gravity can give an exact description of black-hole evaporation. One would, however, not expect that a unitary evolution during the semiclassical phase would be followed by a sudden information loss at the final stage.

4   Bekenstein-Hawking Entropy through Decoherence?

The above discussion was concerned with Hawking radiation. I have argued that its mixed appearance is due to the decohering influence of other fields. The entropy of the Hawking radiation, (22), is the result of a coarse-graining and application of von Neumann's formula (9). But what about the entropy of the black hole itself? In other words, can the Bekenstein-Hawking formula (2) be recovered, within the semiclassical approximation, along the same lines?

For an answer, one should be able to specify microscopic degrees of freedom of the gravitational field itself. This has been achieved, in certain situations, within tentative approaches to quantum gravity: loop quantum gravity (see e.g. [19]) and string theory (see e.g. [20]). In string theory, the derivation of (2) in achieved in an indirect way: using duality and the properties of 'BPS states', a black hole in the limit of strong coupling ($g \gg 1$) can be related to a bound collection of 'D-branes' in flat space in the limit of weak coupling ($g \ll 1$). Both configurations should have the same number of quantum states – this is the property of BPS states. For the D-branes, standard formulas of string theory give a definite answer which coincides with the expression (2) for the duality-related black hole.

It was argued in [21] that black holes are inherently associated with mixed states, and that pure D-brane states rapidly develop entanglement with other degrees of freedom, leading to decoherence. Only the decohered D-branes should be associated with black holes. In fact, the derivation of (2) employs, at least implicitly, the use of decohered D-branes.

A related situation was investigated in [22]. There, the decay of a massive string state (i.e. a string state with high excitation number n) and small coupling was addressed. The mass is given by

$$M^2 = \frac{n}{l_{\mathrm{s}}^2} , \quad n \gg 1 , \tag{23}$$

where l_{s} is the string length. The degeneracy N_n of a level n is given by

$$N_n \sim \mathrm{e}^{a\sqrt{n}} = \mathrm{e}^{aMl_{\mathrm{s}}} , \quad a = 2\pi\sqrt{\frac{D-2}{6}} , \tag{24}$$

where D is the number of spacetime dimensions in which the string moves. The decay spectrum of a single excited state does not exhibit any thermal properties. However, if one averages over all the degenerate states with the same mass M, the decay spectrum is of Planckian form, with the temperature given by the Hagedorn temperature $T_{\mathrm{H}} = (al_{\mathrm{s}})^{-1}$ [22]. It is speculated that the reason is decoherence generated by the entanglement with quantum background fields being present in the string spectrum.

It might well be that (2) can generally be justified in an analogous manner, either in loop quantum gravity or string theory. Quantitative calculations in this direction have still to be performed. They should in particular reveal the universal nature of the Bekenstein-Hawking entropy. One might expect that therein the *quasi-normal modes* of a black hole could play a crucial role in serving as an environment. For the Schwarzschild black hole the maximal value for the real part of the mode frequency (energy) is [23]

$$\hbar\omega = \frac{\hbar\ln 3}{8\pi GM} = (\ln 3)k_{\mathrm{B}}T_{\mathrm{BH}} , \tag{25}$$

This frequency seems to play a crucial role in the calculation of (2) from loop quantum gravity [24]. In quantum cosmology, the global structure of spacetime assumes classical properties through interaction with higher modes [1, 25]. In a similar way one could envisage the quasi-normal modes to produce a classical behaviour for black holes – the entropy (2) would then result as the entanglement entropy of the correlated state between black hole and quasi-normal modes. Since in the corresponding calculation a sum over all modes has to be performed, one might expect the maximal frequency (25) to play a crucial role. The details, however, are far from being explored. To understand black-hole entropy as an entanglement entropy has been tried before, see e.g. [26] and the references therein. A special example is induced gravity where (2) was recovered in the presence of non-minimally coupled fields [27]. (This result might be of relevance to string theory.) Here it is suggested that the role of the environment is played by the quasi-normal modes, which are (for large mode number) characteristic of the black hole itself and therefore should be able to yield a universal result.

It is known that spacetime as such is a classical concept, arising from decoherence in quantum gravity (see Sect. 5 in [10]). The event horizon of a black hole is a spacetime concept and should therefore have no fundamental meaning in quantum gravity. It should arise from decoherence in the semiclassical limit, together with (2).

References

1. C. Kiefer: 'Thermodynamics of black holes and Hawking radiation'. In: *Classical and quantum black holes*, ed. by P. Fré, V. Gorini, G. Magli and U. Moschella (IOP Publishing, Bristol 1999), p. 17
2. S.W. Hawking: Commun. Math. Phys. **43**, 199 (1975)
3. R. Jackiw: *Diverse topics in theoretical and mathematical physics* (World Scientific, Singapore 1995), Sect. IV.4
4. J.-G. Demers and C. Kiefer: Phys. Rev. D **53**, 7050 (1996)
5. C. Kiefer: Class. Quant. Grav. **18**, L151 (2001)
6. C. Vaz, C. Kiefer, T.P. Singh and L. Witten, Phys. Rev. D **67**, 024014 (2003)
7. L.P. Grishchuk and Y.V. Sidorov: Phys. Rev. D **42**, 3413 (1990)
8. W. Israel: Phys. Lett. A **57**, 107 (1976)
9. D.F. Walls and G.J. Milburn: *Quantum optics* (Springer Verlag, Berlin 1994), Sect. 5.2.5
10. E. Joos, H.D. Zeh, C. Kiefer, D. Giulini, J. Kupsch, and I.-O. Stamatescu: *Decoherence and the appearance of a classical world in quantum theory*, 2nd edit. (Springer-Verlag, Berlin 2003)
11. D.N. Page: 'Black hole information'. In: *Proceedings of the 5th Canadian conference on general relativity and relativistic astrophysics*, ed. by R. Mann and R. McLenaghan (World Scientific, Singapore 1994), p. 1
12. A.R. Liddle and D.H. Lyth: *Cosmological inflation and large-scale structure* (Cambridge University Press, Cambridge 2000)
13. C. Kiefer and D. Polarski: Ann. Phys. (Leipzig) **7**, 137 (1998)
14. C. Kiefer, D. Polarski and A.A. Starobinsky: Phys. Rev. D **62**, 043518 (2000)
15. E. Joos and H.D. Zeh: Z. Phys. B **59**, 223 (1985)
16. C. Kiefer, J. Lesgourgues, D. Polarski and A.A. Starobinsky: Class. Quant. Grav. **15**, L67 (1998)
17. T. Prokopec: Class. Quant. Grav. **10**, 2295 (1993)
18. C.G. Torre and M. Varadarajan: Class. Quant. Grav. **16**, 2651 (1999)
19. A. Ashtekar, J.C. Baez, and K. Krasnov: Adv. Theor. Math. Phys. **4**, 1 (2000)
20. G.T. Horowitz: 'Quantum states of black holes.' In: *Black holes and relativistic stars*, ed. by R.M. Wald (The University of Chicago Press, Chicago 1998), p. 241
21. R. Myers: Gen. Rel. Grav. **29**, 1217 (1997)
22. D. Amati and J.G. Russo: Phys. Lett. B **454**, 207 (1999)
23. S. Hod: Phys. Rev. Lett. **81**, 4293 (1998); L. Motl: gr-qc/0212096
24. O. Dreyer: Phys. Rev. Lett. **90**, 081301 (2003)
25. H.D. Zeh: Phys. Lett. A **116**, 9 (1986); C. Kiefer: Class. Quant. Grav. **4**, 1369 (1987)

26. C. Kiefer: 'Towards a full quantum theory of black holes'. In: *Black Holes: Theory and Observation*, ed. by F.W. Hehl, C. Kiefer and R. Metzler (Springer-Verlag, Berlin 1998), p. 416

27. V.P. Frolov, D.V. Fursaev and A.I. Zelnikov: Nucl. Phys. B **486**, 339 (1997); D.V. Fursaev: Nucl. Phys. B (Proc. Suppl.) **88**, 277 (2000)

Threshold Effects and Lorentz Symmetry

Orfeu Bertolami

Instituto Superior Técnico, Departamento de Física, Av. Rovisco Pais,
1049-001 Lisboa, Portugal

Abstract. Evidence on the violation of Lorentz symmetry arises from the observation of cosmic rays with energies beyond the GZK cutoff, $E_{GZK} \simeq 4 \times 10^{19} \ eV$, from the apparent transparency of the Universe to the propagation of high energy gamma radiation, and from the stability of pions in air showers. These three paradoxes can be explained through deformations of the relativistic dispersion relation. Theoretical ideas aimed to understand how Lorentz symmetry may be broken and how phenomenologically interesting deformations of the relativistic dispersion relation may arise are briefly discussed.

1 Introduction

Invariance under Lorentz transformations is one of the most fundamental symmetries of physics and is a key feature of all known physical theories. However, recently, evidence has emerged that this symmetry may not be respected in at least three different phenomena:

i) Observation of ultra-high energy cosmic rays with energies [1–4] beyond the Greisen-Zatsepin-Kuzmin (GZK) cutoff, $E_{GZK} \simeq 4 \times 10^{19} \ eV$ [5]. These events, besides challenging our knowledge of mechanisms that allow accelerating cosmic particles to such high energies, may imply in a violation of Lorentz invariance as only through this violation that new threshold effects may arise and the resonant scattering reactions with photons of the Cosmic Microwave Background (CMB), e.g. $p + \gamma_{2.73K} \to \Delta_{1232}$, are suppressed [6–9]. An astrophysical solution to this paradox is possible and imply identifying viable sources at distances within, $D_{Source} \lesssim 50 - 100 \ Mpc$ [10, 11], so that the travelling time of the emitted particles is shorter than the attenuation time due to particle photoproduction on the CMB. Given the energy of the observed ultra-high energy cosmic rays (UHECRs) and the Hilla's criteria [12] on the energy, size and intensity of the magnetic field to accelerate protons, $E_{18} \leq \frac{1}{2}\beta B(\mu G)L(kpc)$ - where E_{18} is the maximum energy measured in units of $10^{18} \ eV$, β the velocity of the shock wave relative to c - it implies that, within a volume of radius $50 - 100 \ Mpc$ about the Earth, only neutron stars, active galactic nuclei, gamma-ray bursts and cluster of galaxies are feasible acceleration sites [12,13]. This type of solution has been recently suggested in [14], where it was also argued that the near isotropy of the arrival directions of the observed UHECRs can be attributed to

O. Bertolami, Threshold Effects and Lorentz Symmetry, Lect. Notes Phys. **633**, 96–102 (2004)
http://www.springerlink.com/   © Springer-Verlag Berlin Heidelberg 2004

extragalactical magnetic fields near the Milky Way that are strong enough to deflect and isotropize the incoming directions of UHECRs from sources within D_{Source}. This is a debatable issue and it is worth bearing in mind that scenarios for the origin of UHECRs tend to generate anisotropies (see e.g. [15]). A further objection against this proposal is the mismatch in the energy fluxes of observed UHECRs and of the potential sources as well as the lack of spatial correlations between observed UHECRs and candidate sources (see e.g. [16] and references therein).

ii) Observations of gamma radiation with energies beyond 20 TeV from distant sources such as Markarian 421 and Markarian 501 blazars [17–19]. These observations suggest a violation of Lorentz invariance as otherwise, due to pair creation, there should exist a strong attenuation of fluxes beyond 100 Mpc of γ-rays with energies higher than 10 TeV by the diffuse extragalactic background of infrared photons [20–23].

iii) Studies of longitudinal evolution of air showers produced by ultra high-energy hadronic particles seem to suggest that pions are more stable than expected [24].

Violations of the Lorentz symmetry may lead to other threshold effects associated to asymmetric momenta in pair creation, photon stability, alternative Čerenkov effects, etc. [25, 26].

On the theoretical front, work in the context of string/M-theory has shown that Lorentz symmetry can be spontaneously broken due to non-trivial solutions in string field theory [27], from interactions that may arise in braneworld scenarios where our 3-brane is dynamical [28], in loop quantum gravity [29, 30], in noncommutative field theories [1] [31], and in quantum gravity inspired spacetime foam scenarios [33]. The resulting novel interactions may have striking implications at low-energy [34, 35, 37–39]. Putative violations of the Lorentz invariance may also lead to the breaking of CPT symmetry [40]. An extension of the Standard Model (SM) that incorporates violations of Lorentz and CPT symmetries was developed in [41].

2 Possible Solutions for the Observational Paradoxes and Experimental Bounds

Potential violations of fundamental symmetries naturally raise the question of how to experimentally verify them. In the case of CPT symmetry, its violation can be experimentally tested by various methods, such as for instance, via neutral-meson experiments [42], Penning-trap measurements and hydrogen-antihydrogen spectroscopy [43]. The breaking of CPT symmetry also allows for a mechanism to generate the baryon asymmetry of the Universe [44]. In what

---

[1] We mention however, that in a model where a scalar field is coupled to gravity, Lorentz invariance may still hold, at least at first non-trivial order in perturbation theory of the noncommutative parameter [32].

concerns Lorentz symmetry, astrophysics plays, as we have already seen, an essential role. Moreover, it will soon be possible to make correlated astrophysical observations involving high-energy radiation and, for instance, neutrinos, which will make viable direct astrophysical tests of Lorentz invariance [9, 20, 45] .

The tighest experimental limit on the extent of which Lorentz invariance is an exact symmetry arises from measurements of the time dependence of the quadrupole splitting of nuclear Zeeman levels along Earth's orbit. Experiments of this nature can yield an impressive upper limit on deviations from the Lorentz invariance, $\delta < 3 \times 10^{-22}$ [46], and even more stringent bounds according to [47].

On very broad terms, proposals to explain the three abovementioned paradoxes rely on deformations of the relativistic dispersion relation, that can be written, for a particle species a, as:

$$E_a^2 = p_a^2 c_a^2 + m_a^2 c_a^4 + F(E_a, p_a, m_a, c_a) \quad , \tag{1}$$

where c_a is the maximal attainable velocity for particle a and F is a function of c_a and of the relevant kinematical variables .

For instance, Coleman and Glashow [7] proposed to explain the observation of cosmic rays beyond the GZK limit assuming that each particle has its own maximal attainable velocity and a vanishing function F. This is achieved studying the relevant interaction between a CMB photon and a proton primary yielding the $\Delta(1224)$ hadronic resonance. A tiny difference between the maximal attainable velocities, $c_p - c_\Delta \equiv \epsilon_{p\Delta} \simeq 1.7 \times 10^{-25}$ c, can explain the events beyond the GZK cutoff. This bound is three orders of magnitude more stringent than the experimental one. A bound from the search of neutrino oscillations can also be found, even though less stringent, $|\epsilon| \lesssim few \times 10^{-22}$ c [48]. Interestingly these limits can be turned into bounds on parameters of the Lorentz-violating extension of the SM [9]. As discussed in [9], a characteristic feature of the Lorentz violating extension of the SM of [41] is that it gives origin to a time delay, Δt, in the arrival of signals brought by different particles that is energy independent, in opposition to what is expected from other models (see [37] and references therein), and has a dependence on the chirality of the particles involved as well:

$$\Delta t \simeq \frac{D}{c}[(c_{00} \pm d_{00})_i - (c_{00} \pm d_{00})_j] \quad , \tag{2}$$

where c_{00} and d_{00} are the time-like components of the CPT-even flavour-dependent parameters that have to be added to the fermion sector of the SM so to exhbit Lorentz-violating interactions [41]:

$$\mathcal{L}_{\text{Fermion}}^{\text{CPT}-\text{even}} = \tfrac{1}{2}ic_{\mu\nu}\overline{\psi}\gamma^\mu \overset{\leftrightarrow}{\partial^\nu} \psi + \tfrac{1}{2}id_{\mu\nu}\overline{\psi}\gamma_5\gamma^\mu \overset{\leftrightarrow}{\partial^\nu} \psi \quad . \tag{3}$$

The $\pm$ signs in (2) arise from the fact that parameter $d_{\mu\nu}$ depends on the chirality of the particles in question, and D is the proper distance of the source.

The function F arising from this SM Lorentz violating extension is [9]:

$$F = -2c_{00}E^2 \pm 2d_{00}Ep \quad ,$$ (4)

with $c_a = c$ for all particles.

It has been argued that some quantum gravity and stringy inspired models (see [30, 37–39]) lead to modifications of the dispersion relation of the following form:

$$F = -k_a \frac{p_a^3}{M_P} \quad ,$$ (5)

where k_a is a constant and M_P is Planck's mass. This deformation can explain the three discussed paradoxes [25, 30, 37–39].

At very high energies, deformation (5) can be approximately written as

$$F \simeq -\frac{E^3}{E_{QG}} \quad ,$$ (6)

where E_{QG} is a quantum gravity scale. For photons this leads to the following dispersion relation

$$pc = E\sqrt{1 + \frac{E}{E_{QG}}} \quad ,$$ (7)

from which bounds on the quantum gravity scale [20, 37, 38, 45] can be astrophysically determined, the most stringent being [49]

$$E_{QG} > 4 \times 10^{18} \ GeV \quad .$$ (8)

Another aspect of the problem of violating a fundamental symmetry like Lorentz invariance concerns gravity. A putative violation of Lorentz symmetry renews the interest in gravity theories that have intrinsically built in this feature. From the point of the post-Newtonian parametrization the theory that most closely resembles general relativity is Rosen's bimetric theory [2] [50]. Indeed, this theory shares with general relativity the same values for all post-Newtonian parameters [51]

$$\beta = \gamma = 1 \ ; \ \alpha_1 = \alpha_3 = \zeta_1 = \zeta_2 = \zeta_3 = \zeta_4 = \xi = 0 \quad ,$$ (9)

except for parameter α_2 that signals the presence preferred-frame effects (Lorentz invariance violation) in the g_{00} and g_{0i} components of the metric. Naturally, this parameter vanishes in General Relativity, but in Rosen's bimetric theory it is

$$\alpha_2 = \frac{f_0}{f_1} - 1 \quad ,$$ (10)

where f_0 and f_1 are the asymptotic values of the components of the metric in the Universe rest frame, i.e. $g_{\mu\nu}^{(0)} = diag(-f_0, f_1, f_1, f_1)$, which must be close to the

---

[2] This theory has some difficulties as in its simplest form it does not admit black hole solutions and it is unclear to which extent it is compatible with cosmology.

Minkowski metric. A non-vanishing α_2 implies that angular momentum is not conserved. Bounds on this parameter are obtained from the resulting anomalous torques on the Sun, whose absence reveals that $\alpha_2 < 4 \times 10^{-7}$ [52]. It is worth remarking that the other parameters leading to preferred-frame effects are bound by the pulsar PSR J2317+1439 data, that lead to $\alpha_1 < 2\times10^{-4}$ [51], and from the average on the pulse period of millisecond pulsars, which gives $\alpha_3 < 2.2 \times 10^{-20}$ [53]. It is clear that Rosen's theory deserves a closer examination.

3   Conclusions and Outlook

Lorentz and CPT symmetries may be spontaneously broken in string theory and in some quantum gravity inspired models. Modifications to the relativistic dispersion relation arising from these models allow for explaning the three paradoxes associated to threshold effects in ultra high-energy cosmic rays, pair creation in the propagation of TeV photons and its interaction with the diffuse gamma radiation background, and the longitudinal evolution of high energy hadronic particles in extensive air showers. Confirmation that these phenomena signal the breaking of Lorentz symmetry is an exciting prospect as it would constitute in an inequivocal indication of physics beyond the SM. Near future observations that will be carried by extensive detectors such as by the Auger Observatory [54] may unfold interesting questions and challenges to theory.

References

1. N. Hayashida et al. (AGASA Collab.): Phys. Rev. Lett. **73**, 3491 (1994); M. Takeda et al. (AGASA Collab.): Phys. Rev. Lett. **81**, 1163 (1998)
2. D.J. Bird et al. (Fly's Eye Collab.): Phys. Rev. Lett. **71**, 3401 (1993); Astrophys. J. **424**, 491 (1994); **441**, 144 (1995)
3. M.A. Lawrence, R.J.O. Reid, and A.A. Watson (Haverah Park Collab.): Journ. Phys. G **17**, 733 (1991)
4. N.N. Efimov et al. (Yakutsk Collab.): In *ICRR Symposium on Astrophysical Aspects of the Most Energetic Cosmic Rays*, ed. by N. Nagano and F. Takahara (World Scientific, Singapore 1991)
5. K. Greisen: Phys. Rev. Lett. **16**, 748 (1966); G.T. Zatsepin, V.A. Kuzmin: JETP Lett. **41**, 78 (1966)
6. H. Sato and T. Tati: Prog. Theor. Phys. **47**, 1788 (1972)
7. S. Coleman and S.L. Glashow: Phys. Lett. B **405**, 249 (1997); Phys. Rev. **D59**, 116008 (1999)
8. L. Gonzales-Mestres: hep-ph/9905430
9. O. Bertolami and C.S. Carvalho: Phys. Rev. D **61**, 103002 (2000)
10. F.W. Stecker: Phys. Rev. Lett. **11**, 1016 (1968)
11. C.T. Hill, D. Schramm and T. Walker: Phys. Rev. D **36**, 1007 (1987)
12. A.M. Hillas: Ann. Rev. Astron. Astrophys. **22**, 425 (1984)
13. J.W. Cronin: Rev. Mod. Phys. **71**, S165 (1999)
14. G.R. Farrar and T Piran: Phys. Rev. Lett. **84**, 3527 (2000)
15. A.V. Olinto: astro-ph/0003013

16. O. Bertolami: Gen. Rel. Grav. **34**, 707 (2002)
17. F. Krennrich et al.: Astrophys. J. **560**, L45 (2001)
18. F.A. Aharonian et al.: Astron. and Astrophys. **349**, 11A (1999)
19. A.N. Nikishov: Sov. Phys. JETP **14**, 393 (1962);
    J. Gould and G. Schreder: Phys. Rev. **155**, 1404 (1967);
    F.W. Stecker, O.C. De Jager, and M.H. Salmon: Astrophys. J. **390**, L49 (1992)
20. G. Amelino-Camelia, J. Ellis, N.E. Mavromatos, D.V. Nanopuolos, and S. Sarkar:
    Nature **393**, 763 (1998)
21. P.S. Coppi and F.A. Aharonian: Astropart. Phys. **11**, 35 (1999)
22. T. Kifune: Astrophys. J. Lett. **518**, L21 (1999)
23. W. Kluzniak: astro-ph/9905308;
    R.J. Protheroe, H. Meyer: Phys. Lett. B **493**, 1 (2000);
    G. Amelino-Camelia and T. Piran: Phys. Rev. D **64**, 036005 (2001)
24. E.E. Antonov et al.: Pisma ZhETF **73**, 506 (2001)
25. T.J. Konopka and S.A. Major: New J. Phys. **4**, 57 (2002)
26. T. Jacobsen, S. Liberati, and D. Mattingly: hep-ph/0209264
27. V.A. Kostelecký and S. Samuel: Phys. Rev. D **39**, 683 (1989); Phys. Rev. Lett. **63**,
    224 (1989)
28. G. Dvali and M. Shifman: hep-th/9904021
29. R. Gambini and J. Pullin: Phys. Rev. D **59**, 124021 (1999)
30. J. Alfaro, H.A. Morales-Tecotl, and L.F. Urrutia: Phys. Rev. Lett. **84**, 2183 (2000)
31. S.M. Carroll, J.A. Harvey, V.A. Kostelecký, C.D. Lane, and T. Okamoto: Phys.
    Rev. Lett. **87**, 141601 (2001)
32. O. Bertolami and L. Guisado: Phys. Rev. D **67**, 025001 (2003)
33. L.J. Garay: Phys. Rev. Lett. **80**, 2508 (1998)
34. O. Bertolami: Class. Quant. Grav. **14**, 2748 (1997)
35. O. Bertolami and D.F. Mota: Phys. Lett. B **455**, 96 (1999)
36. H. Sato: astro-ph/0005218
37. G. Amelino-Camelia and T. Piran: Phys. Lett. B **497**, 265 (2001)
38. N. Mavromatos: gr-qc/0009045
39. R. Aloisio, P. Blasi, P.L. Ghia, and A.F. Grillo: astro-ph/0001258
40. V.A. Kostelecký and R. Potting: Phys. Rev. D **51**, 3923 (1995); Phys. Lett. B **381**,
    389 (1996)
41. D. Colladay and V.A. Kostelecký: Phys. Rev. D **55**, 6760 (1997); Phys. Rev. D **58**,
    116002 (1998)
42. D. Colladay and V.A. Kostelecký: Phys. Lett. B **344**, 259 (1995); Phys. Rev. D
    52, 6224 (1995);
    V.A. Kostelecký and R. Van Kooten: Phys. Rev. D **54**, 5585 (1996)
43. R. Bluhm: hep-ph/0006033.
44. O. Bertolami, D. Colladay, V.A. Kostelecký, and R. Potting: Phys. Lett. B **395**,
    178 (1997)
45. S.D. Biller et al.: Phys. Rev. Lett. **83**, 2108 (1999)
46. S.K. Lamoreaux, J.P. Jacobs, B.R. Heckel, F.J. Raab, and E.N. Fortson: Phys.
    Rev. Lett. **57**, 3125 (1986)
47. V.A. Kostelecký and C.D. Lane: Phys. Rev. D **60**, 116010 (1999)
48. E.B. Brucker et al.: Phys. Rev. D **34**, 2183 (1986);
    S.L. Glashow, A. Halprin, P.I. Krastev, C.N. Leung, and J. Pantaleone: Phys. Rev.
    D **56**, 2433 (1997)
49. G. Amelino-Camelia: gr-qc/0212002
50. N. Rosen: Gen. Rel. Grav. **4**, 435 (1973); Ann. Phys. (New York) **84**, 455 (1974);
    Gen. Rel. Grav. **9**, 339 (1978)

51. C.M. Will: *Theory and Experiment in Gravitational Physics* (Cambridge University Press, Cambridge 1993); 'The Confrontation between General Relativity and Experiment: A 1998 Update', gr-qc/9811036
52. K. Nordtvedt: Astrophys. J. **320**, 871 (1987)
53. J.F. Bell and T. Damour: Class. Quant. Grav. **13**, 3121 (1996)
54. L. Anchordoqui, T. Paul, S. Reucroft, and J. Swain: astro-ph/0206072

Towards a Statistical Geometrodynamics

Ariel Caticha

Department of Physics, University at Albany, SUNY, Albany, NY 12222, USA

Abstract. Can the spatial distance between two identical particles be explained in terms of the extent that one can be distinguished from the other? Is the geometry of space a macroscopic manifestation of an underlying microscopic statistical structure? Is geometrodynamics derivable from general principles of inductive inference? Tentative answers are suggested by a model of geometrodynamics based on the statistical concepts of entropy, information geometry, and entropic dynamics.

1 Introduction

The purpose of dynamical theories is to predict or explain the changes observed in physical systems on the basis of information that is codified into what one calls the states of the system. One common view is that these dynamical theories – the laws of physics – are successful because they happen to reflect the true laws of nature.

Here I wish to follow an alternative path: perhaps once the relevant information has been identified the question of predicting changes is just a matter of careful consistent manipulation of the available information. If this turns out to be the case, then the laws of physics should follow directly from rules for processing information, that is, the rules of probability theory [1] and the method of maximum entropy (ME) [2–4].[1]

There are some indications that this point of view is worth pursuing. Indeed, thermodynamics is a prime example of a fundamental physical theory that can be derived from general principles of inference [2]. Quantum mechanics provides a second, less trivial, and less well known example [6]. Both theories follow from a correct specification of the subject matter, that is, an appropriate choice of variables – this is the truly difficult step – plus probabilistic and entropic arguments.

---

[1] On terminology: The ME method is designed for processing information to update from a prior probability distribution to a posterior distribution. (The terms 'prior' and 'posterior' are used with similar meanings in the context of Bayes' theorem.) The ME method is usually understood in the restricted sense that one updates from a prior distribution that happens to be uniform – this is the usual postulate of equal a priori probabilities. Here we adopt a broader meaning that includes updates from arbitrary priors and which involves the maximization of *relative* entropy. Since all entropies are relative to some prior, be it uniform or not, the qualifier 'relative' is redundant and will henceforth be omitted. For a brief account of the ME method in a form that is convenient for our current purposes see [5].

A. Caticha, Towards a Statistical Geometrodynamics, Lect. Notes Phys. **633**, 103–116 (2004)
http://www.springerlink.com/ © Springer-Verlag Berlin Heidelberg 2004

A third independent clue is found when one attempts to derive classical dynamical theories from purely entropic arguments. The surprising outcome is that the resulting "entropic" dynamics (ED) shows remarkable similarities with the general theory of relativity – geometrodynamics (GD). The general purpose of this paper is to take the first tentative steps towards explaining geometrodynamics as a form of entropic dynamics.

The procedure to derive an ED involves three steps [7]. The first step is to identify the subject matter and the corresponding space of observable states or, perhaps more appropriately, the space of macrostates. This is not easy because there exists no systematic way to search for the right macrovariables; it is a matter of taste and intuition, trial and error.

The second step is to define a quantitative measure of the change or the "distance" from one state to another. Although in general the choice of distance is not unique an exception occurs when the macrostates can be interpreted as probability distributions over some appropriate space of microstates. Then there is a natural distance which is given by the Fisher-Rao information metric [8,9] (its uniqueness is discussed in [10,11]; for a brief heuristic derivation see [12]). It measures the extent to which one probability distribution can be distinguished from another. This second step – assigning a statistical distance – is not straightforward either: more inspired guesswork is needed unless the right microstates happen to be known beforehand.[2]

The third and final step is easier. We ask: Given the initial and the final states, what trajectory is the system expected to follow? The question implicitly assumes that there is a trajectory, that in moving from one state to another the system will pass through a continuous set of intermediate states, and that information about the initial and final states is sufficient to determine them. The answer follows from a principle of inference, the ME principle, and not from any additional "physical" postulates.

The resulting ED is elegant and not trivial: the system moves along a geodesic but the geometry of the space of states is curved and possibly quite complicated. Since the only available clock is the system itself there is no reference to an external physical time. The natural intrinsic time is defined by the change of the system itself – in ED time is change – and can only be obtained after the equations of motion are solved. ED is a timeless Machian dynamics and its features resemble those advocated by Barbour [19]: it is reversible; it can be derived from a Jacobi action principle rather than the more familiar action principle of Hamilton; and its canonical Hamiltonian formulation is an example of a dynamics driven by constraints.

The similarities to GD are striking. For example, in GD there is no reference to an external physical time. The proper time interval along any curve between an initial and a final three-dimensional geometries of space is determined only after solving the Einstein equations of motion [20]. The absence of an external

---

[2] The recognition that spaces of probability distributions are metric spaces has nevertheless been fruitful in statistics, where the subject is known as Information Geometry [13,14], and in physics [15–18].

time has been a serious impediment in understanding GD because it is not clear which variables represent the true gravitational degrees of freedom [21–24]. GD is also derived from a Jacobi action principle [25, 26] and its canonical Hamiltonian formulation is an example of a dynamics driven by constraints [27–29]. The question, therefore, is whether GD is an example of ED. The answer requires identifying those variables that describe the true degrees of freedom of the gravitational field.

The tentative steps of making assumptions about the subject matter, the macrostates, and about how to associate a probability distribution to each of them are taken in Sect. 2. We want to predict the evolution of the three-dimensional geometry of space. The problem is that space is invisible. What we see is not space, but matter in space and we do not quite know how to disentangle which properties should be attributed to the matter and which to space. The best one can do is to choose the simplest form of matter: a substance that is neutral to all interactions and is itself describable by a minimal number of attributes. This ideal form of matter is a dust of identical particles; being neutral they will only interact gravitationally, and being identical the issue of what it is that distinguishes them – size, mass, flavor – does not arise. Thus we assume there is nothing to space beyond what can be learned from observing the evolving distribution of dust particles. The geometry of space is just the geometry of all the distances between dust particles. Furthermore, we assume this geometry is of statistical origin. Identical particles that are close together are easy to confuse, those that are far apart are easy to distinguish. The distance between two neighboring particles is the distinguishability distance given by a Fisher-Rao metric. Notice that the Fisher-Rao metric is used in two conceptually different ways. One is to distinguish successive states of the same system, the other is to distinguish different neighboring particles. The first is related to time, the second to space.

Having decided what system is under study and how it is statistically described we can proceed to define its ED. In Sect. 3, as a warm up problem, we develop the ED of a single point, and then, in Sect. 4, we generalize to the whole dust cloud. Although the resulting statistical GD is not Einstein's GD of space-time – an indication that the states and variables we have chosen do not accurately describe the gravitational degrees of freedom – it is close enough to be encouraging. The model GD developed here corresponds to what is called an ultralocal or strong gravity theory [30–32]. We do not recover the notion of space-time but we do find an embryonic form of Lorentz invariance in that simultaneity is relative. Finally, in Sect. 5 we summarize our conclusions.

2   The Geometry of a Dust Cloud

Consider a cloud of identical specks of dust suspended in an otherwise empty space. And there is nothing else; in particular, there are no rulers and no clocks, just dust. Our goal is to study how the cloud evolves. We do this by keeping track of individual specks of dust.

Being identical the particles are easy to confuse. The only distinction between two of them is that one happens to be here while the other is over there. To distinguish one speck of dust from another we assign labels or coordinates to each particle. We assume that three real numbers (y^1, y^2, y^3) are sufficient.

Of course, particles can be mislabeled. Then the "true" coordinates y are unknown and one can only provide an estimate, x. Let $p(y|x)\mathcal{D}y$ be the probability that the particle labeled x should have been labeled y. The labels x are introduced to distinguish one particle from another, but can we distinguish a particle at x from another at $x + \mathcal{D}x$? If $\mathcal{D}x$ is small enough the corresponding probability distributions $p(y|x)$ and $p(y|x + \mathcal{D}x)$ overlap considerably and it is easy to confuse them. We seek a quantitative measure of the extent to which these two distributions can be distinguished.

The following crude argument is intuitively appealing. Consider the relative difference,

$$\frac{p(y|x+\mathcal{D}x)-p(y|x)}{p(y|x)} = \frac{\partial \log p(y|x)}{\partial x^i} \mathcal{D}x^i. \tag{1}$$

The expected value of this relative difference does not provide us with the desired measure of distinguishability: it vanishes identically. However, the variance does not vanish,

$$d\lambda^2 = \int \mathcal{D}y\, p(y|x) \frac{\partial \log p(y|x)}{\partial x^i} \frac{\partial \log p(y|x)}{\partial x^j} \mathcal{D}x^i \mathcal{D}x^j \overset{def}{=} \gamma_{ij}(x)\mathcal{D}x^i\mathcal{D}x^j. \tag{2}$$

This is the measure of distinguishability we seek. Except for an overall multiplicative constant, the Fisher-Rao metric γ_{ij} is the only Riemannian metric that adequately reflects the underlying statistical nature of the abstract manifold of the distributions $p(y|x)$ [10, 11].

We take the further step of interpreting $d\lambda$ as the *spatial* distance of the three-dimensional space the dust inhabits. Indeed, one would normally say that the reason it is easy to confuse two particles is that they happen to be too close together. We argue in the opposite direction and *explain* that the reason the particles at x and at $x + \mathcal{D}x$ are close together is *because* they are difficult to distinguish.

The origin of the uncertainty will be left unspecified; perhaps it is due to a limit on the ultimate resolution of observation devices, or perhaps, as with a particle undergoing Brownian motion, the uncertainty might be caused by a fluctuating physical agent. It is required, however, that two particles at the same location in space must be affected by the same uncertainty, the same irreducible noise. Then the noise is not linked to the particle, but to the place, and we might as well say that the source of the irreducible noise is space itself. This is somewhat analogous to the principle of equivalence: it is the fact that all particles irrespective of their mass move along the same trajectories in a gravitational field that allows us to eliminate the notion of a gravitational field and attribute their common behavior to a single universal agent, the curvature of space-time.

To assign an explicit $p(y|x)$ and explore the geometry it induces we will consider what is perhaps the simplest possibility. We assume that the uncertainty in the coordinate x is small so that $p(y|x)$ is sharply localized in a neighborhood about x and within this very small region curvature effects can be neglected. Further, we assume that particles are labeled by the expected values $\langle y^i \rangle = x^i$ and that the information that happens to be necessary for the purpose of prediction of future behavior is given by the second moments $\langle (y^i - x^i)(y^j - x^j) \rangle = C^{ij}(x)$. This is physically reasonable: for each particle we have estimates for its position and of the small margin of error. Then $p(y|x)$ can be determined maximizing entropy relative to an appropriate prior. To the extent that curvature effects are negligible, the underlying space is flat and translationally invariant. Thus, symmetry suggests a uniform prior and the resulting ME distribution is Gaussian,

$$p(y|x) = \frac{C^{1/2}}{(2\pi)^{3/2}} \exp\left[-\tfrac{1}{2} C_{ij}(y^i - x^i)(y^j - x^j)\right], \tag{3}$$

where C_{ij} is the inverse of the covariance coefficients C^{ij}, $C^{ik}C_{kj} = \delta^i_j$, and $C \equiv \det C_{ij}$. The corresponding metric is obtained substituting into (2). For small uncertainties $C_{ij}(x)$ is constant within the region where $p(y|x)$ is appreciable and the result is

$$\gamma_{ij}(x) = C_{ij}(x). \tag{4}$$

The metric changes smoothly over space and, in general, space is curved. The connection, the curvature, and other aspects of its Riemannian geometry can be computed in the standard way. The probability distributions,

$$p(y|x) = \frac{\gamma^{1/2}(x)}{(2\pi)^{3/2}} \exp\left[-\tfrac{1}{2} \gamma_{ij}(x)(y^i - x^i)(y^j - x^j)\right], \tag{5}$$

also vary smoothly with x.

To summarize, we have succeeded in describing the information geometry that derives from considerations of distinguishability among particles. The idea is rather general but was developed explicitly only for the special case of small uncertainties, that is, for particles that can be localized within regions much smaller than those where curvature effects become appreciable. An interesting question that will not be addressed here concerns the extension to those situations of extreme curvature found near singularities.

Before discussing dynamics we mention that there is one very peculiar feature of the distance $d\lambda$, (2), that may be very significant: $d\lambda^2$ is dimensionless. The metric $\gamma_{ij}(x)$ allows one to measure spatial lengths in terms of a local standard, the local uncertainty width. This immediately raises the question of how to compare the uncertainty widths, and therefore lengths, at two distant locations. One possibility, which we pursue in the rest of this paper, is that γ_{ij} describes the Riemannian geometry of space. This amounts to asserting that the uncertainty widths are the same everywhere, they provide us with a universal standard of length. A second, more intriguing possibility, which we will explore elsewhere, is that all the information metric γ_{ij} allows us to do is to compare the lengths of small segments in different orientations at the same location; it allows one to measure angles. Then γ_{ij} does not describe the geometry of space completely, it only describes its conformal geometry.

3   Entropic Dynamics of a Single Point

In this section we develop the ED of a single Gaussian distribution, an analogue of GD in zero spatial dimensions. Let Γ be the space of states. The points in Γ are Gaussian distributions with zero mean $\langle y \rangle = 0$,

$$p(y|\gamma) = \tfrac{\gamma^{1/2}}{(2\pi)^{3/2}} \exp\left(-\tfrac{1}{2}\gamma_{ij}y^i y^j\right),\qquad (6)$$

where $\gamma = \det \gamma_{ij}$ and $y = (y^1, y^2, y^3)$ are points in R^3. Whether γ denotes the matrix γ_{ij} or its determinant should, in what follows, be clear from the context. Since $\gamma_{ij} = \gamma_{ji}$ is symmetric, Γ is a six dimensional space.

The following notation is convenient: the derivative $\partial/\partial\gamma_{ij}$ of a function $F(\gamma)$ is defined so that $\mathcal{D}F$ takes the simple form

$$\mathcal{D}F \overset{def}{=} \tfrac{\partial F}{\partial\gamma_{ij}}\mathcal{D}\gamma_{ij}.\qquad (7)$$

$\partial F/\partial\gamma_{ij}$ coincides with the usual partial derivative times $(1+\delta_{ij})/2$. To operate with $\partial/\partial\gamma_{ij}$ we only need to find out how it acts on γ_{kl} and on its inverse γ^{kl}. We find

$$\tfrac{\partial\gamma_{kl}}{\partial\gamma_{ij}} = \tfrac{1}{2}\left(\delta_k^i\delta_l^j + \delta_l^i\delta_k^j\right) \overset{def}{=} \delta_{kl}^{ij}\quad\text{and}\quad \tfrac{\partial\gamma^{kl}}{\partial\gamma_{ij}} = -\tfrac{1}{2}\left(\gamma^{ki}\gamma^{lj} + \gamma^{kj}\gamma^{li}\right).\qquad (8)$$

Note that $\delta_{kl}^{ij}\gamma_{ij} = \gamma_{kl}$ and $\delta_{kl}^{ij}\gamma^{kl} = \gamma^{ij}$. We will also need to differentiate the determinant $\gamma = \det\gamma_{ij}$,

$$\mathcal{D}\gamma = \gamma\gamma^{ij}\mathcal{D}\gamma_{ij}\quad\text{or}\quad \tfrac{\partial\gamma}{\partial\gamma_{ij}} = \gamma\gamma^{ij}.\qquad (9)$$

The Fisher-Rao metric $g^{ij\,kl}$ on the space Γ is

$$g^{ij\,kl} = \int \mathcal{D}y\, p(y|\gamma)\tfrac{\partial\log p(y|\gamma)}{\partial\gamma_{ij}}\tfrac{\partial\log p(y|\gamma)}{\partial\gamma_{kl}} = \tfrac{1}{4}\left(\gamma^{ki}\gamma^{lj} + \gamma^{kj}\gamma^{li}\right),\qquad (10)$$

and its inverse metric, defined by $g^{ij\,kl}g_{kl\,mn} = \delta_{mn}^{ij}$, is

$$g_{kl\,mn} = \gamma_{km}\gamma_{ln} + \gamma_{kn}\gamma_{lm}.\qquad (11)$$

Now we can tackle the dynamics. The key to the question "Given initial and final states, what trajectory is the system expected to follow?" lies in the implicit assumption that there exists a continuous trajectory. This means that large changes are the result of a continuous succession of very many small changes; the problem of studying large changes is reduced to the simpler problem of studying small changes.

We want to determine the states along a short segment of the trajectory as the system moves from an initial state γ to a neighboring final state $\gamma + \Delta\gamma$. To find the intermediate states we reason that in going from the initial to the final state the system must pass through a halfway point, that is, an intermediate state that is equidistant from γ and $\gamma + \Delta\gamma$. Finding the halfway point clearly determines the trajectory: first find the halfway point, and use it to determine 'quarter of the way' points, and so on. But there is nothing special about halfway states. In general, we can assert that the system must pass through intermediate states γ_ω such that, having already moved a distance $\mathcal{D}\ell$ away from the initial γ, there remains a distance $\omega\mathcal{D}\ell$ to be covered to reach the final $\gamma + \Delta\gamma$; ω is any positive number.

The basic dynamical question can be rephrased as follows: The system is initially described by the probability distribution $p(y|\gamma)$ and we are given the new information that the system has moved to one of the neighboring states in the family $p(y|\gamma_\omega)$. Which $p(y|\gamma_\omega)$ do we select? Phrased in this way it is clear that this is precisely the kind of problem to be tackled using the ME method.[1] The selected distribution is that which maximizes the relative entropy of $p(y|\gamma_\omega)$ relative to a prior distribution p_{old}. Since in the absence of new information there is no reason to change one's mind, when there are no constraints the selected posterior distribution should coincide with the prior distribution. Therefore the prior p_{old} is the initial state $p(y|\gamma)$. Thus, to determine the intermediate state $\gamma_\omega = \gamma + \mathcal{D}\gamma$ one varies over $\mathcal{D}\gamma_{ij}$ to maximize

$$S\left[p(y|\gamma_\omega), p(y|\gamma)\right] = -\int \mathcal{D}y\, p(y|\gamma + \mathcal{D}\gamma) \log \frac{p(y|\gamma+\mathcal{D}\gamma)}{p(y|\gamma)}$$
$$= -\tfrac{1}{2} g^{ij\,kl} \mathcal{D}\gamma_{ij} \mathcal{D}\gamma_{kl} = -\tfrac{1}{2}\mathcal{D}\ell^2\,, \tag{12}$$

subject to the constraint $\mathcal{D}\ell_f = \omega\mathcal{D}\ell$ where

$$\mathcal{D}\ell_f^2 = g^{ij\,kl}\left(\Delta\gamma_{ij} - \mathcal{D}\gamma_{ij}\right)\left(\Delta\gamma_{kl} - \mathcal{D}\gamma_{kl}\right). \tag{13}$$

Introducing a Lagrange multiplier $\lambda/2$,

$$0 = \delta\left[-\tfrac{1}{2} g^{ij\,kl}\mathcal{D}\gamma_{ij}\mathcal{D}\gamma_{kl} + \tfrac{\lambda}{2}\left(\omega^2\mathcal{D}\ell^2 - \mathcal{D}\ell_f^2\right)\right], \tag{14}$$

then, the selected $\mathcal{D}\gamma_{ij}$ is given by

$$\mathcal{D}\gamma_{ij} = \chi\Delta\gamma_{ij} \quad \text{where} \quad \chi = \tfrac{\lambda}{1+\lambda(1-\omega^2)}. \tag{15}$$

Substituting $\mathcal{D}\gamma_{ij}$ into $\mathcal{D}\ell$ and $\mathcal{D}\ell_f$ we get $\mathcal{D}\ell = \chi\Delta\ell$ and $\mathcal{D}\ell_f = (1-\chi)\Delta\ell$, so that $\chi = (1+\omega)^{-1}$ with $0 < \chi < 1$ and

$$\mathcal{D}\ell + \mathcal{D}\ell_f = \Delta\ell. \tag{16}$$

The interpretation is clear: the three states γ, γ_ω and $\gamma + \Delta\gamma$ lie on a straight line. The expected trajectory is the geodesic that passes through the given initial and final states.

Note that each different value of ω provides a different criterion to select the trajectory and an inconsistency would arise if these criteria led to different trajectories. It is reassuring to find that indeed the ED trajectory is independent of the value ω.

ED determines the vector tangent to the trajectory $\mathcal{D}\gamma/\mathcal{D}\ell$, but not the actual velocity $\mathcal{D}\gamma/\mathcal{D}t$. In conventional forms of dynamics the distance ℓ along the trajectory is related to an external time t through a Hamiltonian which fixes the evolution relative to external clocks. But here the only clock available is the system itself which can only provide an internal, intrinsic time. It is best to define the intrinsic time so that motion looks simple. A natural definition consists in stipulating that the system moves with unit velocity, then the intrinsic time is given by the distance ℓ itself. The intrinsic time interval is the amount of change. A peculiar feature of this notion of time is that intervals are not a priori known, they are determined only after the equations of motion are solved and the actual trajectory is determined.

The geodesics in the space Γ are obtained minimizing the Jacobi action

$$J[\gamma] = \int_{\eta_i}^{\eta_f} \mathcal{D}\eta\, L(\gamma, \dot{\gamma})\,, \tag{17}$$

where η is an arbitrary parameter along the trajectory and $\dot{\gamma}_{ij} = \mathcal{D}\gamma_{ij}/\mathcal{D}\eta$. The Lagrangian is just the arc length

$$L(\gamma, \dot{\gamma}) = \left(g^{ij\,kl}\dot{\gamma}_{ij}\dot{\gamma}_{kl}\right)^{1/2} = \left(\tfrac{1}{2}\gamma^{ik}\gamma^{jl}\dot{\gamma}_{ij}\dot{\gamma}_{kl}\right)^{1/2}\,. \tag{18}$$

The canonical momenta are

$$\pi^{mn} = \frac{\partial L}{\partial \dot{\gamma}_{mn}} = \frac{1}{2L}\gamma^{ik}\gamma^{jl}\dot{\gamma}_{ij}\delta^{mn}_{kl} = \frac{1}{2L}\gamma^{mi}\gamma^{nj}\dot{\gamma}_{ij}\,, \tag{19}$$

and have a fixed magnitude

$$g_{ij\,kl}\pi^{ij}\pi^{kl} = 1\,. \tag{20}$$

The canonical Hamiltonian vanishes identically,

$$H_{\text{can}}(\gamma, \pi) = \dot{\gamma}_{ij}\pi^{ij} - L(\gamma, \dot{\gamma}) \equiv 0\,, \tag{21}$$

because the Lagrangian is homogeneous of first degree in the velocities. The manifest reparametrization invariance of the action $J[\gamma]$ conveniently reflects the absence of an external time with respect to which the system could possibly evolve.

Since variations of the momenta are constrained to preserve their magnitude the action principle is

$$I[\gamma, \pi, N] = \int_{\eta_i}^{\eta_f} \mathcal{D}\eta \left[\dot{\gamma}_{ij}\pi^{ij} - N(\eta)h(\gamma, \pi)\right]\,, \tag{22}$$

where

$$h(\gamma, \pi) \overset{def}{=} \tfrac{1}{2}g_{ij\,kl}\pi^{ij}\pi^{kl} - \tfrac{1}{2}\,, \tag{23}$$

and $N(\eta)$ are Lagrange multipliers that at each instant η enforce the constraints

$$h(\gamma, \pi) = 0\,. \tag{24}$$

Equations of motion are obtained varying with respect to γ and π with γ fixed at the endpoints $\delta\gamma_{ij}(\eta_i) = \delta\gamma_{ij}(\eta_f) = 0$. Then

$$\dot{\gamma}_{mn} = N\frac{\partial h}{\partial \pi^{mn}} = 2N\gamma_{mi}\gamma_{nj}\pi^{ij} \tag{25}$$

$$\dot{\pi}^{mn} = -N\frac{\partial h}{\partial \gamma_{mn}} = -2N\gamma_{ij}\pi^{mi}\pi^{nj}. \tag{26}$$

There is no equation of motion for N. Comparing (19) and (25) we get

$$N(\eta) = L(\gamma, \dot{\gamma}) = \frac{\mathcal{D}\ell}{\mathcal{D}\eta}, \tag{27}$$

which is recognized as the "lapse" function which gives the increase of intrinsic time ℓ per unit increase of the parameter η. Then the equations of motion simplify to

$$\frac{\mathcal{D}\gamma_{mn}}{\mathcal{D}\ell} = \frac{\partial h}{\partial \pi^{mn}} = 2\gamma_{mi}\gamma_{nj}\pi^{ij} \tag{28}$$

$$\frac{d\pi^{mn}}{\mathcal{D}\ell} = -\frac{\partial h}{\partial \gamma_{mn}} = -2\gamma_{ij}\pi^{mi}\pi^{nj}. \tag{29}$$

One can check that $dh/\mathcal{D}\eta = 0$. Therefore if $h = 0$ initially, the constraint will be consistently preserved by the evolution. One can also check that the action $I[\gamma, \pi, N]$ is invariant under the gauge transformations

$$\delta\gamma_{mn} = \varepsilon(\eta)\frac{\partial h}{\partial \pi^{mn}}, \quad \delta\pi^{mn} = -\varepsilon(\eta)\frac{\partial h}{\partial \gamma_{mn}}, \quad \text{and} \quad \delta N = \dot{\varepsilon}(\eta) \tag{30}$$

provided $\varepsilon(\eta)$ vanishes at the end points, $\varepsilon(\eta_i) = \varepsilon(\eta_f) = 0$. The invariance $\delta I = 0$ holds for any path $\gamma(\eta)$, $\pi(\eta)$ and not just for those paths at which the action is stationary. In addition, as is evident in the action $J[\gamma]$, there is an additional invariance under global (η-independent) "conformal" transformations, $\gamma_{ij} \to \psi^4\gamma_{ij}$. The corresponding conserved quantity is $\mathrm{tr}\pi$. To appreciate the significance of this conserved quantity note that

$$\mathrm{tr}\pi = \gamma^{mn}\pi_{mn} = \frac{\gamma^{mn}\dot{\gamma}_{mn}}{2N} = \frac{1}{2N\gamma}\frac{\mathcal{D}\gamma}{\mathcal{D}\eta} = \frac{1}{2\gamma}\frac{\mathcal{D}\gamma}{d\tau}, \tag{31}$$

so that the determinant γ expands or contracts at a constant relative rate. In particular, if the initial velocity happens to be such that $\mathrm{tr}\pi = 0$, then γ remains fixed at its constant initial value.

4   Geometrodynamics: The Ultralocal Case

The system we study is a *single* dust cloud. To the dust cloud we associate a probability distribution P given by a product of the distributions (5) of the individual particles,

$$P[y|\gamma] = \prod_x p\left(y(x)|x, \gamma_{ij}(x)\right)$$

$$= \left[\prod_x \frac{\gamma^{1/2}(x)}{(2\pi)^{3/2}}\right]\exp\left[-\frac{1}{2}\sum_x \gamma_{ij}(x)(y^i - x^i)(y^j - x^j)\right]. \tag{32}$$

It was the necessity to quantify whether we can distinguish a test particle at x from its neighbor at $x + \mathcal{D}x$ that led us to introduce the metric γ_{ij} in the first place. When we consider the change from an earlier state γ to a later state $\gamma + \Delta\gamma$ the distinguishability problem manifests itself yet again. Even if we had managed to distinguish a test particle at x from a neighboring test particle at $x + \mathcal{D}x$, there is no guarantee that the particle that earlier had coordinates x will be the *same* particle that will later be found at x. Particles do not just need to be identified, they need to be re-identified. For the invisible points of space this difficulty is only exacerbated because the re-identification of points depends on the state of motion of the test particles. If we allow for the possibility of particles moving past each other we conclude that the points of space cannot be treated as enduring things. And this is precisely where the model discussed in this section becomes unrealistic: we maintain such a strict correspondence between a test particle and the point it occupies that we end up treating the individual points of space as if they were real enduring objects. A more realistic model of space should deal with *several* potentially coexisting dust clouds in relative motion.

Once a dust particle in the earlier state γ is identified with the label x, we will assume that this particle can be assigned the same label x as it evolves into the later state $\gamma + \Delta\gamma$. These are comoving coordinates. Then we can write the change $\Delta\ell$ between $P[y|\gamma + \Delta\gamma]$ and $P[y|\gamma]$, (32), from their relative entropy,

$$S[\gamma + \Delta\gamma, \gamma] = -\int \left(\prod_x \mathcal{D}y(x)\right) P[y|\gamma + \Delta\gamma] \log \frac{P[y|\gamma+\Delta\gamma]}{P[y|\gamma]} = -\tfrac{1}{2}\Delta\ell^2 \quad (33)$$

Since $P[y|\gamma]$ and $P[y|\gamma + \Delta\gamma]$ are products $S[\gamma + \Delta\gamma, \gamma]$ can be written as a sum over the individual particles,

$$S[\gamma + \Delta\gamma, \gamma] = \sum_x S[\gamma(x) + \Delta\gamma(x), \gamma(x)] = -\tfrac{1}{2}\sum_x \Delta\ell^2(x), \quad (34)$$

where

$$\Delta\ell^2(x) = g^{ij\,kl}(x)\Delta\gamma_{ij}(x)\Delta\gamma_{kl}(x), \quad (35)$$

with $g^{ij\,kl}$ given by (10). Therefore, the overall change in going from γ to $\gamma + \Delta\gamma$ is

$$\Delta\ell^2 = \sum_x \Delta\ell^2(x) = \int \mathcal{D}x\,\rho(x)\Delta\ell^2(x), \quad (36)$$

where we have written the discrete sum as an integral – the number of dust particles within $\mathcal{D}x$ is $\mathcal{D}x\rho(x)$.

Having given a sufficient specification of what we mean by a state of the system we can now proceed to formulate its ED. Once again we ask, 'Given initial and final states, what trajectory is the system expected to follow?' and the answer follows from the implicit assumption that there exists a continuous trajectory, but here we must pay closer attention to what precisely we mean by 'trajectory'. Indeed, if predicting changes is just a matter of careful consistent manipulation of the available information, then we must recognize that we know more than just that the product state (32) must evolve through a continuous sequence of intermediate states. We also know that each and every one of the individual factors $p(y|x, \gamma)$ must also evolve continuously through a sequence of intermediate states to reach the corresponding final state. This means that instead of one parameter ω there are many such parameters, one for each position x, and there is no reason why they should all take the same value. In other words, the intermediate states γ_ω should be labeled by a local function $\omega(x)$ rather than a single global parameter ω. A continuous sequence of states γ_ω interpolating between the initial γ and the final $\gamma + \Delta\gamma$ can be defined by imposing $\omega(x) = \zeta f(x)$ where $f(x)$ is a fixed positive function and the parameter ζ varies from 0 to ∞. There is no single trajectory; each choice of the function $f(x)$ defines one possible trajectory. In a sense, the cloud follows many alternative paths "simultaneously". To guarantee consistency we should check that physical predictions are independent of the choice of the arbitrary function $f(x)$.

Before we formulate the ED we should remark on the significance of invariance under choices of $f(x)$. The product state $P[y|\gamma]$ provides the only definition of what an instant is, of which states $p(y|x', \gamma')$ at distant points x' we can agree to call simultaneous with a certain state $p(y|x, \gamma)$ at the point x. Therefore, if there is no unique sequence of intermediate states, then there is no unique, absolute definition of simultaneity. We see here a kind of foliation invariance, a rudimentary, and yet extreme form of local Lorentz invariance. Since the metric γ_ω of the intermediate states $P[y|\gamma_\omega]$ remains positive for arbitrary choices of the function $\omega(x)$ the analogues of the light cones are collapsed into light lines. The invariant speed – the speed of light – is zero. The GD model described here resembles the so-called ultralocal or strong gravity theories [30–32] more closely than it resembles general relativity.

Now we address the question: Given initial and final states, γ and $\gamma + \Delta\gamma$, what are the possible trajectories? Let η be an arbitrary time parameter labeling successive intermediate states. The initial state is $\gamma_{ij}(\eta, x) = \gamma_{ij}(x)$, the final state is $\gamma_{ij}(\eta + \Delta\eta, x) = \gamma_{ij}(x) + \Delta\gamma_{ij}(x)$, and the intermediate states are $\gamma_{ij}(\eta + \mathcal{D}\eta, x) = \gamma_{ij}(x) + \mathcal{D}\gamma_{ij}(x)$. To determine the intermediate state $\gamma + \mathcal{D}\gamma$ one varies over $\mathcal{D}\gamma_{ij}$ to maximize the entropy

$$S[\gamma + \mathcal{D}\gamma, \gamma] = -\int \left(\prod_x \mathcal{D}y(x) \right) P[y|\gamma + \mathcal{D}\gamma] \log \frac{P[y|\gamma + \mathcal{D}\gamma]}{P[y|\gamma]} = -\tfrac{1}{2}\mathcal{D}\ell^2 , \quad (37)$$

where

$$\mathcal{D}\ell^2 = \int \mathcal{D}x \, \rho(x)\mathcal{D}\ell^2(x) \quad \text{with} \quad \mathcal{D}\ell^2(x) = g^{ij\,kl}(x)\mathcal{D}\gamma_{ij}(x)\mathcal{D}\gamma_{kl}(x), \quad (38)$$

subject to independent constraints at each point x,

$$\mathcal{D}\ell_f(x) = \omega(x)\mathcal{D}\ell(x) \quad (39)$$

where

$$\mathcal{D}\ell_f^2(x) = g^{ij\,kl}(x) \left(\Delta\gamma_{ij}(x) - \mathcal{D}\gamma_{ij}(x) \right) \left(\Delta\gamma_{kl}(x) - \mathcal{D}\gamma_{kl}(x) \right) \quad (40)$$

Introducing Lagrange multipliers $\lambda(x)/2$,

$$0 = \delta \left[\int \mathcal{D}x \, \rho(x) \left\{ -\tfrac{1}{2}\mathcal{D}\ell^2(x) + \tfrac{\lambda(x)}{2} \left(\omega^2(x)\mathcal{D}\ell^2(x) - \mathcal{D}\ell_f^2(x) \right) \right\} \right] \qquad (41)$$

the result, $\mathcal{D}\gamma_{ij}(x) = \chi(x)\Delta\gamma_{ij}(x)$, coincides with the single point result (15) for each value of x. Substituting $\mathcal{D}\gamma_{ij}$ into $\mathcal{D}\ell(x)$ and $\mathcal{D}\ell_f(x)$ we get $\mathcal{D}\ell(x) = \chi\Delta\ell(x)$ and $\mathcal{D}\ell_f(x) = [1 - \chi(x)]\Delta\ell(x)$, so that

$$\mathcal{D}\ell(x) + \mathcal{D}\ell_f(x) = \Delta\ell(x). \qquad (42)$$

The conclusion is that the states of the individual particles evolve independently of each other along geodesics in the single point configuration space given by (28-29). The dynamics of the cloud is independent of the choice of $\omega(x)$ as desired – this is foliation invariance.

The ultralocal statistical GD deduced in the previous paragraphs is the dynamics of a large or perhaps infinite number of independent subsystems. The action for the whole cloud can be written as the sum of the individual particle actions given in (17). Thus, the proposed action is

$$J[\gamma, \dot\gamma] = \int_{\eta_i}^{\eta_f} \mathcal{D}\eta \int \mathcal{D}x \, \rho \left(g^{ij\,kl}\dot\gamma_{ij}\dot\gamma_{kl} \right)^{1/2}, \qquad (43)$$

where $\dot\gamma_{ij} = \partial\gamma_{ij}/\partial\eta$. In commoving coordinates $\dot\rho = \partial\rho/\partial\eta = 0$. It is straightforward to develop the constrained Hamiltonian formalism and recover the single particle equations of motion.

Notice that the actual distance from the initial state to the final state along a certain path is given by (36),

$$\ell = \int_i^f \mathcal{D}\eta \left(\dot\ell^2 \right)^{1/2} = \int_{\eta_i}^{\eta_f} \mathcal{D}\eta \left[\int \mathcal{D}x \, \rho \, g^{ij\,kl}\dot\gamma_{ij}\dot\gamma_{kl} \right]^{1/2}. \qquad (44)$$

Therefore, unlike the action for a single point (17), the action (43) is not the natural arc length. The dust cloud does not evolve along a geodesic. The reason for this can be traced to the additional constraint that individual particles evolve continuously, which allows a multitude of different trajectories and leads to foliation invariance.

5   Conclusions

One idea explored in this work is whether it is possible to establish a connection between ordinary spatial distances and the information metric of Fisher and Rao – whether one can explain the notion of spatial distance. We succeeded in describing the information geometry that derives from considerations of distinguishability among particles; particles that are easily confused are said to be near, those that are easily distinguished are farther apart. The idea is that distances between particles are not distances between structureless points but distances between probability distributions.

According to Euclid, a point is that which has no size. General relativity was founded upon a revision of Euclid's fifth postulate. Statistical geometrodynamics is founded upon the further revision of Euclid's first definition, the notion of structureless points.

The second idea we explored is whether Einsteinian macroscopic geometrodynamics is derivable from an underlying microscopic statistical theory purely on the basis of principles of inference, without additional postulates of a more "physical" nature. We can only claim a partial success; the result is close enough to be promising. The model GD we obtained satisfies the main requirement, it describes the dynamics of a geometry; it is related to gravity because it describes an ultralocal gravity theory; and it exhibits foliation invariance. Moreover, the somewhat puzzling fact that space and time are so different and yet enter the formalism in such a symmetric way receives a natural explanation: a time interval refers to the extent we can distinguish an earlier state from a later state of the same system, while a spatial distance refers to the extent we can distinguish two different systems.

Einstein's GD might be recovered by making a different choice of the states and variables that describe the gravitational degrees of freedom. Two possible alternative choices were suggested. First, one should avoid a too strict correspondence between a test particle and the point it occupies because this treats the individual points of space as if they were real objects. Second, it may be that the Fisher-Rao metric does not describe the full geometry of space, as we assumed in this work, but only describes its conformal geometry.

Should the ideas proposed here prove successful one can further expect that the currently popular approaches to a quantum theory of gravity will require revision.

References

1. R.T. Cox: Am. J. Phys. **14**, 1 (1946); *The Algebra of Probable Inference* (Johns Hopkins, Baltimore 1961)
2. E.T. Jaynes: Phys. Rev. **106**, 620 and **108**, 171 (1957); *E.T. Jaynes: Papers on Probability, Statistics and Statistical Physics*, ed. by R.D. Rosenkrantz (Reidel, Dordrecht 1983)
3. J.E. Shore and R.W. Johnson: IEEE Trans. Inf. Theory **IT-26**, 26 (1980)
4. J. Skilling: 'The Axioms of Maximum Entropy'. In: *Maximum-Entropy and Bayesian Methods in Science and Engineering*, ed. by G.J. Erickson and C.R. Smith (Kluwer, Dordrecht 1988)
5. A. Caticha: 'Maximum Entropy, Fluctuations and Priors'. In: *Bayesian Methods and Maximum Entropy in Science and Engineering*, ed. by A. Mohammad-Djafari, AIP Conf. Proc. **568**, 94 (2001); math-ph/0008017
6. A. Caticha: Phys. Lett. A **244**, 13 (1998); Phys. Rev. A **57**, 1572 (1998); Found. Phys. **30**, 227 (2000); quant-ph/9810074
7. A. Caticha: 'Entropic Dynamics'. In: *Bayesian Inference and Maximum Entropy Methods in Science and Engineering*, ed. by R. L. Fry, AIP Conf. Proc. **617**, 302 (2002); gr-qc/0109068

8. R.A. Fisher: Proc. Cambridge Philos. Soc. **122**, 700 (1925)
9. C.R. Rao: Bull. Calcutta Math. Soc. **37**, 81 (1945)
10. N.N. Čencov: *Statistical Decision Rules and Optimal Inference*, Transl. Math. Monographs, vol. 53 (Am. Math. Soc., Providence 1981)
11. L.L. Campbell: Proc. Am. Math. Soc. **98**, 135 (1986)
12. A. Caticha: 'Change, Time and Information Geometry'. In: *Bayesian Methods and Maximum Entropy in Science and Engineering*, ed. by A. Mohammad-Djafari, AIP Conf. Proc. **568**, 72 (2001); math-ph/0008018
13. S. Amari: *Differential-Geometrical Methods in Statistics* (Springer Verlag, Berlin 1985)
14. C.C. Rodríguez: 'Entropic Priors for Discrete Probabilistic Networks and for Mixtures of Gaussian Models'. In: *Bayesian Inference and Maximum Entropy Methods in Science and Engineering*, ed. by R.L. Fry, AIP Conf. Proc. **617**, 410 (2002)
15. F. Weinhold: J. Chem. Phys. **63**, 2479 (1975)
16. G. Ruppeiner: Phys. Rev. A **20**, 1608 (1979); Phys. Rev. A **27**, 1116 (1983)
17. L. Diósi and B. Lukács: Phys. Rev. A **31**, 3415 (1985) and Phys. Lett. A **112**, 13 (1985)
18. R. Balian, Y. Alhassid and H. Reinhardt: Phys. Rep. **131**, 1 (1986)
19. J. Barbour: Class. Quant. Grav. **11**, 2853(1994)
20. R.F. Baierlein, D.H. Sharp and J.A. Wheeler: Phys. Rev. **126**, 1864 (1962)
21. J.W. York: 'Kinematics and Dynamics of General Relativity'. In: *Sources of Gravitational Radiation*, ed. by L. Smarr (Cambridge Univ. Press, Cambridge 1979)
22. C. Rovelli: 'Strings , Loops and the Others: a Critical Survey of the Present Approaches to Quantum Gravity'. In: *Gravitation and Relativity: At the Turn of the Millenium*, ed. by N. Dadhich and J. Narlikar (Poona Univ. Press, Poona 1998)
23. K. Kuchar: 'Time and Interpretations of Quantum Gravity'. In: *Proc. of the 4th Canadian Conf. on General Relativity and Astrophysics*, ed. by G. Kunsatter, D. Vincent, and J. Williams (World Scientific, Singapore 1992)
24. C. Isham: 'Canonical Quantum Gravity and the Problem of Time'. In: *Integrable Systems, Quantum Groups and Quantum Field Theories*, ed. by L.A. Ibort and M.A. Rodriguez (World Scientific, Singapore 1993)
25. J.D. Brown and J.W. York: Phys. Rev. D **40**, 3312 (1989)
26. J.B. Hartle: Phys. Rev. D **38**, 2985 (1988)
27. P. Dirac: Proc. Roy. Soc. A (London), **246**, 333 (1958)
28. R. Arnowitt, S. Deser and C.W. Missner: 'The Dynamics of General Relativity'. In: *Gravitation: an Introduction to Current Research*, ed. by L. Witten (Wiley, New York 1962)
29. S.A. Hojman, K. Kuchar and C. Teitelboim: Ann. Phys. **96**, 88 (1976)
30. C.J. Isham: Proc. Roy. Soc. A (London), **351**, 209 (1976)
31. C. Teitelboim: Phys. Rev. D **25**, 3159 (1982)
32. M. Pilati: Phys. Rev. D **26**, 2645 (1982)

Part III

Quantum Mechanics in Between
– Decoherence and Determinism

Introduction:
Quantum Theory and Beneath?

Hans-Thomas Elze

Instituto de Física, Universidade Federal do Rio de Janeiro, C.P. 68.528,
21941-972 Rio de Janeiro, RJ, Brazil

The question raised in the title is old, as old as quantum theory itself. An answer has not been given. Leading physicists of the first "quantum century" have taken a stand on the mostly philosophical issues which were surrounding it before. Can the uncomfortable unfamiliarity of quantum mechanics be overcome or is it here to stay?

In view of the uninterrupted success story of the applications of quantum theory, in the laboratory or otherwise, a pragmatic character might tend to discard these questions, until conflicting evidence from experiment will be reported. However, the issue is pressing from the theoretical side, perhaps as much as the blackbody spectrum and spectral lines were generally not understood a hundred years ago.

It is worth while to recall some aspects of the unresolved clash between quantum theory and general relativity motivating the contemporary surge of interest in the foundations of quantum mechanics.

In recent work 't Hooft argues that physically realistic models for the unification of general relativity with quantum theory necessarily will have to restore determinism in the latter, in order to account for the so-called holographic principle [1,2]. A complete and accordingly unified theory of physical reality should produce an upper bound for the total number of independent quantum states of a system which scales like the exponential of its surface area. This is in marked contrast to the usual behavior of quantum field theories (QFT), where the exponential scaling is expected to be governed by the volume instead. The difference is conjectured to be due to the intrinsic nonlocality of such a unified theory.

Some indications that a truly quantum and general relativistic theory possibly has to be nonlocal, which implies a modification of the generally accepted formulation of quantum theory, have been discussed in the context of the black-hole information paradox first discovered by Hawking [3]. When a quantum system collapses gravitationally from a pure initial state to form a black hole, which subsequently evaporates via Hawking radiation, then it appears to make a transition to a mixed state. This is accompanied by an increase of the von Neumann entropy or, equivalently, a loss of information, which cannot be accounted for in local QFT.

The pre-holographic discussion of these issues is summarized in [4], where three remaining possible resolutions of the information paradox were pointed

H.-T. Elze, Introduction: Quantum Theory and Beneath?, Lect. Notes Phys. **633**, 119–124 (2004)
http://www.springerlink.com/   © Springer-Verlag Berlin Heidelberg 2004

out: *i*) Metastable black-hole remnants may possess an unexpected capacity for information storage; *ii*) unitary quantum theory has to be given up; *iii*) the apparently lost information is restored by nonlocal effects other than quantum correlations. While *i*) seems unlikely (see further references in [4]), the options *ii*) and *iii*) found some surprising correspondents in recent developments in string theory, as further discussed in [2,5], for example. However, it is pointed out that basic concepts of the interpretation of physical reality, such as space, time, matter, causality, and the role of the observer become increasingly questionable, up to the point of not being operationally well-defined, when quantum theory is accepted in its present form in this context.

A new proposal to resolve the paradox is described by Kiefer in Part II of this volume [6]. There, several potential sources of decoherence are identified which were not considered previously. They may account for the entropy increase or information loss, respectively.

Instead, 't Hooft proposes to attack these problems in a radically different way. For example, fundamentally discrete classical degrees of freedom may exist, with a deterministic evolution determined by the rules of a cellular automaton [7]. An auxiliary Hilbert space can then be introduced, in which the corresponding evolution of probabilities is indeed determined, in the continuum limit, by the Schrödinger equation.

Generalizing, the underlying deterministic degrees of freedom are assumed to evolve dissipatively, i.e. incorporating information loss at the fundamental classical level, residing at the Planck scale [2,8]. Under fairly general conditions such systems evolve into stable limit cycles, i.e. equivalence classes of states which are defined by the same asymptotic behavior. Then, if a reasonable continuum limit exists, an effective unitary quantum theory may emerge based on the identification of the equivalence classes with the physical states.

This idea has been studied in the context of a concrete model of two dissipatively coupled oscillators by Blasone, Jizba and Vitiello, which is presented in the following [9].

To round off the sketch of 't Hooft's pioneering efforts, we mention that recently, instead of invoking dissipation, a mapping between equivalence classes of states of several classical field theories and the quantum states of corresponding free QFT has been constructed, based on (the fixing of) a large group of classical gauge transformations [10,11].

The tentative conclusion here is that quantum theory can and should be reconstructed as an "effective" theory describing the large-scale statistical behavior of fundamental deterministic degrees of freedom. The prospective consequences of such an approach, where quantum states do not represent the primordial degrees of freedom, are provocative and might lead to a paradigm change.

However, a major obstacle has been encountered in the works referred to above. So far, no interacting models could successfully be constructed. Here, three contributions to this part of the present volume may open new vistas.

Based on extensive numerical studies of classical Yang-Mills theories, Biró, Müller and Matinyan show that these classical models surprisingly tend to "quantize themselves" [12, 13].

More specifically, they consider the classical Hamiltonian dynamics of Yang-Mills fields on a spatial lattice. Their recent simulations of such non-Abelian lattice gauge theories have demonstrated strong chaoticity properties, following from calculations of the spectrum of Lyapunov exponents and analysis of the fluctuations of observables sampled along ergodic trajectories, which even indicate mixing behavior.

It is argued furthermore that this chaotic dynamics, due to the color degrees of freedom interacting nonlinearly at each space-time point, is sufficient to effectively thermalize the system. From there the reasoning runs to some extent analogous to the stochastic quantization introduced by Parisi and Wu [14–16]. This leads from a thermal classical gauge theory in d dimensions to a Euclidean zero-temperature QFT in $d - 1$ dimensions.

What is very appealing in this approach is that no extraneous sources of noise with corresponding unknown dynamics and additional parameters need to be introduced. The necessary equilibrium state is reached automatically on a sufficiently short time scale, in which observables are distributed according the microcanical ensemble.

Interesting questions, which arise and are discussed in this work, concern the dependence on dimensionality of the system, whether the mechanism could be modified in such a way as to lead to the quantized Yang-Mills theory in Minkowski space, and whether the mechanism could possibly work for gravity.

The question whether this "self-quantization" approach could also lead to QFT in spaces with Lorentzian signature must again be raised in the context of the work contributed to this volume by Wetterich [17].

Based on the concept of "incomplete statistics" [18], it is shown that under very general assumptions a succinct description of a classical statistical field theory which is only incompletely known, or only partially accessible to observation, is the one provided by the Euclidean QFT of the particular model under consideration. This statement actually implies that no other description of the system can be more complete than this quantum mechanical description.

In the example considered in this work, the incompleteness of the statistics consists in the spatial separation of all degrees of freedom into those "inside", which are observable or whose statistical properties are known in terms of an appropriate probability distribution, and the inaccessible rest "outside", with interactions coupling both across the boundary. This also suggests that transfer of information between "inside" and "outside" could possibly be associated with real-time evolution.

This last point actually forms the basis of the work contributed by Diósi [19]. With the help of Bayesian inference from the observation of the random process associated with irreversible evolution, it is shown how probabilistic statements about the sign of the intrinsic time arrow of a given system can be made. The sign of the flow of time corresponds to the sign of information loss and a phenomenological expression quantifying this is proposed. Interestingly, it applies to the thermodynamic, microscopic, and quantum mechanical examples studied here in a unfied way.

Furthermore, the method of Diósi is related to the one applied by Caticha in his ambitious attempt to base a geometrodynamics on the Bayesian analysis of what can be known locally about a system [20]. Related recent developments pointing further into this exciting direction, attempting even to bridge the gap to quantum gravity on a statistical basis, can be found in [21].

Coming back to the emergence of quantum theory, two observations suggest themselves. First, abstracting from the dynamical details of the respective models, the situation described by Wetterich seems to have intrinsically much in common with the one described in the work by Biró, Müller and Matinyan, and even with the dissipation model of Blasone, Jizba and Vitiello. Apparently, the lack of complete information about the system bundles up classically unresolvable configurations into equivalence classes which appear as quantum states. This was anticipated on general grounds by 't Hooft. In the Yang-Mills case it is the exponential sensitivity to variations of initial conditions that leads to the rapid loss of information about the trajectory in the extradimension, i.e. rapid increase of the Kolomogorov-Sinai entropy. Thus, a case of "incomplete statistics" must arise in the remaining $d - 1$ dimensions. However, seen from afar, we would like to understand better why this is realized precisely by dimensional reduction.

Second, it is intriguing to note a similarity with the separation of system and environment degrees of freedom that lies at the heart of decoherence. Could it be that, starting with fundamentally deterministic degrees of freedom, the *emergence of quantum mechanics is a dual phenomenon to decoherence?*

In the classical-to-quantum transition we loose information about the classical system due to inaccessibility of the environment, which apparently can be realized in a variety of ways (extradimensions, etc.). While in the quantum-to-classical transition the environment provides the access to additional information about the quantum system that is sufficient to undo the quantization.

In my own contribution I follow yet another route [22]. Previously, I studied two simple models, a nonrelativistic and a relativistic particle, respectively, which are "timeless" in the sense of being reparametrization invariant. In references [23] it is shown that one can define a reparametrization invariant physical time in terms of local observables. Basically, it is related to counting incidents, i.e. coincidences of the ergodic (by construction) trajectory with a suitably defined detector. Since such a "clock" is related to a Poincaré section of the full trajectory, it is not surprising that the resulting time is discrete. Again, a case of missing information results, concerning the full trajectory.

More generally, reparametrization invariant classical Hamiltonian systems have been considered in which physical time is discrete and related by a statistical distribution to proper time of the Hamiltonian flow. Then, it is shown that a natural description of such systems is the one furnished by real-time unitary quantum mechanics [24].

Most appropriately, one may call this "stroboscopic quantization", since it originates from the discreteness of physical time, like reading a clock (the system) under a stroboscopic light.

Many questions may arise here. I discuss the construction of interacting models and the need to regularize the emergent quantum Hamiltonian, in order to arrive at a well-defined Schrödinger equation. The extension of these considerations from reparametrization to diffeomorphism invariant models still remains to be seen.

Completing this part is the contribution by Haba and Kleinert [25]. This comprises detailed derivations of master equations from the forward-backward path integral for a particle in the background of electromagnetic and gravitational radiation, respectively. Among other effects, these equations describe the dissipation and decoherence that such a particle experiences and the ensuing entropy production or information loss in the respective environments.

Particularly interesting in the present context is the following point raised by Haba and Kleinert. A very massive object which collapses gravitationally should be described by their theory incorporating the graviton environment, once its size becomes sufficiently small. A detailed analysis of this process, starting classically and leading into the quantum regime dynamically, apparently is still lacking. However, its outcome could provide additional insight complementing the contribution by Kiefer in Part II on the information-loss problem for black holes [6].

In conclusion, some of the threads should have become visible here which interweave issues of the following contributions, even though the underlying pattern has not been reveiled.

References

1. G. 't Hooft: In *Salamfestschrift: a collection of talks*, ed. by A. Ali, J. Ellis and S. Randjbar-Daemi, World Scientific Series in 20th Century Physics, vol. 4 (World Scientific, Singapore 1993), gr-qc/9310026; in *The Oskar Klein Centenary*, ed. by U. Lindström (World Scientific, Singapore 1995);
   L. Susskind: J. Math. Phys. **36**, 6377 (1995)
2. G. 't Hooft: Class. Quant. Grav. **16**, 3263 (1999)
3. S. Hawking: Phys. Rev. D **14**, 2460 (1976)
4. U.H. Danielsson and M. Schiffer: Phys. Rev. D **48**, 4779 (1993)
5. J. Ellis, N.E. Mavromatos, and D.V. Nanopoulos: Gen. Rel. Grav. **31**, 1257 (1999)
6. C. Kiefer: Lecture in this volume; gr-qc/0304102
7. G. 't Hooft: J. Stat. Phys. **53**, 323 (1988); Nucl. Phys. B **342**, 471 (1990);
   G. 't Hooft, K. Isler and S. Kalitzin: Nucl. Phys. B **386** (1992) 495
8. G. 't Hooft: Found. Phys. Lett. **10**, 105 (1997)
9. M. Blasone, P. Jizba, and G. Vitiello: Lecture in this volume;
   Phys. Lett. A **287**, 205 (2001)
10. G. 't Hooft: 'Quantum Mechanics and Determinism'. In *Particles, Strings and Cosmology*, ed. by P. Frampton and J. Ng (Rinton Press, Princeton 2001), p. 275; hep-th/0105105; 'Determinism Beneath Quantum Mechanics', quant-ph/0212095
11. P. Pouliot: JHEP 0111, (2001) 028

12. T.S. Biró, S.G. Matinyan, and B. Müller: *Chaos and Gauge Field Theory* (World Scientific, Singapore 1994)
13. T.S. Biró, B. Müller, and S.G. Matinyan: Lecture in this voulme;
    T.S. Biró, S.G. Matinyan, and B. Müller: Found. Phys. Lett. **14**, 471 (2001)
14. G. Parisi and Y.S. Wu: Sci. Sin. **24**, 483 (1981)
15. P.H. Damgaard and H. Hüffel: Phys. Rep. **152**, 227 (1987)
16. C. Beck: *Spatio-temporal Chaos and Vacuum Fluctuations of Quantized Fields* (World Scientific, Singapore 2002)
17. C. Wetterich: Lecture in this volume; quant-ph/0212031
18. C. Wetterich: Nucl. Phys. B **14**, 40 (1989); ibid. **397**, 299 (1993)
19. L. Diósi: Lecture in this volume; quant-ph/0302183
20. A. Caticha: Lecture in this volume; gr-qc/0301061
21. T. Padmanabhan: Mod. Phys. Lett. A **17**, 1147 (2002); 'Gravity from Spacetime Thermodynamics'. Invited talk at the conference *Fred Hoyle's Universe*, Cardiff, June 24–26, 2002, gr-qc/0209088
22. H.-T. Elze: Lecture in this volume; gr-qc/0307014
23. H.-T. Elze and O. Schipper: Phys. Rev. D **66**, 044020 (2002);
    H.-T. Elze: Phys. Lett. A **310**, 110 (2003)
24. H.-T. Elze: 'Quantum mechanics emerging from "timeless" classical dynamics', submitted to Phys. Rev. D; quant-phys/0306096
25. Z. Haba and H. Kleinert: Lecture in this volume

Probability of Intrinsic Time-Arrow from Information Loss

Lajos Diósi

Research Institute for Particle and Nuclear Physics, P.O. Box 49,
1525 Budapest 114, Hungary

Abstract. Time-arrow $s = \pm$, intrinsic to a concrete physical system, is associated with the direction of information loss ΔI displayed by the random evolution of the given system. When the information loss tends to zero the intrinsic time-arrow becomes uncertain. We propose the heuristic relationship $1/[1 + \exp(-s\Delta I)]$ for the probability of the intrinsic time-arrow. The main parts of the present work are trying to confirm this heuristic equation.

The probability of intrinsic time arrow is defined by Bayesian inference from the observed random process. From irreversible thermodynamic systems, the proposed heuristic probabilities follow via the Gallavotti-Cohen relations between time-reversed random processes. In order to explore the underlying microscopic mechanism, a trivial microscopic process is analyzed and an obvious discrepancy is identified. It can be resolved by quantum theory. The corresponding trivial quantum process will exactly confirm the proposed heuristic time-arrow probability.

1 Introduction

Both experiment and theory confirm that physical processes are time-reversal invariant in 'simple' systems. This invariance may eventually be lost if the system is chaotic, singular, of many degrees of freedom, or not isolated [1]. It seems plausible now that time-reversal asymmetry (irreversibility) is always accompanied by some information loss. Yet, little is known quantitatively. The present work discusses an elementary informatic mechanism of irreversibility. It leads to a simple analytic expression for the asymmetric probability of the two possible directions of time.

Suppose we use *reference-time t* to label the order of events but we leave open whether *physical-time st* is passing with increasing or decreasing t, according to the respective *time-arrow $s = \pm$*. We make no a priori (extrinsic) assignment for s. The ambiguity is to be resolved by analyzing irreversible physical processes. We consider informatic irreversibility in a sense that the Shannon information changes by ΔI along the process. We call the resulting a posteriori time-arrow *intrinsic*. It belongs to the given irreversible process. It would not exist in 'empty space' at all. In the spirit of the second law of thermodynamics, the physical entropy production $s\Delta I$ must be positive, hence the intrinsic time-arrow is unique:

$$s = \text{sign}(\Delta I) . \tag{1}$$

L. Diósi, Probability of Intrinsic Time-Arrow from Information Loss, Lect. Notes Phys. **633**, 125–135 (2004)
http://www.springerlink.com/ © Springer-Verlag Berlin Heidelberg 2004

This assignment is only valid if the magnitude $|\Delta I|$ is macroscopic which means that it is much bigger than 1 bit. If, however, the irreversibility is weak then we have to be content with a probabilistic intrinsic time-arrow. The main suggestion of our work is that this probabilistic time-arrow is a relevant concept and, furthermore, the probability $P(s)$ depends on the Shannon information change ΔI under rather general conditions. I will consider the following relationship:

$$P(s) = \frac{1}{1+e^{-s\Delta I}} \, . \tag{2}$$

If we change the sign of the reference time ($t \to -t$) then also the sign of the information loss will change ($\Delta I \to -\Delta I$). Hence the above expression is *covariant* against time-reversal of the reference frame. Asymptotically it yields the unique thermodynamic arrow (1) if the information loss $|\Delta I|$ is much greater than 1 bit. On the contrary, the two time-arrows become equally probable for a reversible process where ΔI is much less than 1 bit [2]. The suggested relationship is heuristic and lacks a general proof. It is intimately related to the fluctuation theorem [3,4] proved for a particular class of irreversible processes [5]. On the other hand, it intends to reflect a fundamental meaning of the time-arrow in terms of information flow. I am going to prove that the relationship (2) follow from elementary statistical considerations provided we assume some further conditions to fulfill.

Section 2 presents the mathematical steps of Bayesian statistical inference adapted to the estimation of the time-arrow from the observed data. In Sect. 3 we discuss the inference from irreversible thermodynamic process, in Sect. 4 from microscopic process. The time-arrow is derived from quantum irreversibility in Sect. 5. The Appendix offers a short proof of the fluctuation theorem.

2   Bayesian Time-Arrow

Given a statistical system, let X denote a certain random process in a given interval of reference-time t. Let $\tilde{X}$ denote the time-reversal of X. Assume that from the principles of statistical physics we can calculate the probability $\mathcal{P}(X)$ in physical-time! We also introduce the probability distribution $\tilde{\mathcal{P}}(X)$ of the same random process seen from a reference frame with reversed time,

$$\tilde{\mathcal{P}}(X) \equiv \mathcal{P}(\tilde{X}) \, . \tag{3}$$

The conditional probability distribution of X takes the form

$$P(X|s) = \begin{cases} \mathcal{P}(X) & s = + \\ \tilde{\mathcal{P}}(X) & s = - \end{cases} , \tag{4}$$

where s is the a priori time-arrow. Prior to the irreversible process X, the distribution of s is symmetric: $P_0(s) = 1/2$. Hence the joint distribution of X and s is the following,

$$P(X,s) = P(X|s)P_0(s) = \frac{1}{2} \begin{cases} \mathcal{P}(X) & s = + \\ \tilde{\mathcal{P}}(X) & s = - \end{cases} . \tag{5}$$

According to the Bayes rule, the conditional a posteriori distribution of the time-arrow reads

$$P(s|X) = \frac{1}{\mathcal{P}(X)+\tilde{\mathcal{P}}(X)} \times \begin{cases} \mathcal{P}(X) & s = + \\ \tilde{\mathcal{P}}(X) & s = - \end{cases}, \tag{6}$$

which can be cast into the following covariant form:

$$P(s|X) = \frac{1}{1+e^{-sD(X)}}, \tag{7}$$

where

$$D(X) = -\log\frac{\tilde{\mathcal{P}}(X)}{\mathcal{P}(X)}. \tag{8}$$

This Bayesian estimate means that if 1) we know the a priori distribution $\mathcal{P}(X)$ of the random process X in physical-time but 2) experimentally we observe either X or $\tilde{X}$ with equal probability since we have no a priori information regarding the relationship of our reference-time to the physical-time then 3) learning X in the reference-time will lead us to the Bayesian probabilistic estimate $P(s)$ of the times-arrow.

Let us calculate the mean fidelity of the estimated time-arrow: from (4-8) we shall obtain the following closed form:

$$F \equiv \sum_X P(+|X)P(X|+) = \left\langle \frac{1}{1+e^{-D(X)}} \right\rangle_{\mathcal{P}}. \tag{9}$$

The expectation value should refer to $\mathcal{P}(X)$ which is the distribution in the physical frame. We can easily derive an ultimate covariant expression of the average Bayesian estimate,

$$P(s) = \left\langle \frac{1}{1+e^{-sD(X)}} \right\rangle, \tag{10}$$

where the average refers already to the observed statistics and the form is valid in time-reversed reference frames as well.

3   Thermodynamic Case

Let X be a coarse-grained macroscopic random process in a given statistical system in the period $[-T, +T] \equiv [t_1, t_2]$ and let $\tilde{X}$ be the same process seen from the time-reversed reference frame,

$$\begin{aligned} X &= \{X(t); \ t_1 \le t \le t_2\}, \\ \tilde{X} &= \{X(-t); \ t_1 \le t \le t_2\}. \end{aligned} \tag{11}$$

Typically, X can be an irreversible thermodynamic process $X(t)$. Assume that we know the irreversible entropy $\Delta I(X)$ produced by the process X. Obviously, the time-reversed process 'produces' the same entropy with the opposite sign,

$$\Delta I(\tilde{X}) = -\Delta I(X). \tag{12}$$

Let us introduce the following conditional distributions:

$$P(X|\xi) = \mathcal{P}(X)/P_1(\xi) \,,$$
$$P(\tilde{X}|\eta) = \tilde{\mathcal{P}}(X)/P_2(\eta) \,, \tag{13}$$

where $P_1(\xi), P_2(\eta)$ are the probability distributions of the extreme values $\xi = X(t_1)$ and $\eta = X(t_2)$, respectively. In the Appendix the reader finds an elementary proof of the fluctuation theorem [3–5] encoding the asymmetry of the time-reversal $X \leftrightarrow \tilde{X}$ into covariant equation,

$$P(\tilde{X}|\eta) = e^{-\Delta I(X)} P(X|\xi) \,. \tag{14}$$

Accordingly, the violation of the time-reversal symmetry is exponentially increasing with the magnitude $|\Delta I|$ of the irreversible entropy. We are going to show that, via the Bayesian statistics of Sect. 2, the relationship (14) reproduces the heuristic probabilities (2) for the thermodynamic time-arrow.

Let us express the r.h.s. of (8) from (13,14),

$$D(X) = \Delta I(X) - \log \frac{P_2(\eta)}{P_1(\xi)} \,. \tag{15}$$

For long enough periods, the r.h.s. is dominated by the information loss $\Delta I(X)$, the second (boundary) term can be ignored (cf. [5]). In this limit we can write the covariant Bayesian estimate (7) into this form,

$$P(s|X) = \frac{1}{1+e^{-s\Delta I(X)}} \,, \tag{16}$$

which on average leads to the covariant distribution

$$P(s) = \left\langle \frac{1}{1+e^{-s\Delta I(X)}} \right\rangle \,. \tag{17}$$

Finally, this yields the heuristic form (2) provided we can ignore the statistical fluctuations of the entropy production around its expectation value $\Delta I = \langle I(X) \rangle$. This is justified for common macroscopically irreversible processes where $|\Delta I| \gg 1$.

4   Microscopic Case

Let us consider a statistical ensemble of $n \gg 1$ independent d–state systems characterized by the probability distribution $\rho^i, i = 1, 2, \ldots, d$. Let X be an abstract random process as trivial as the transition from an initial microscopic ensemble state ξ into a final one η, the time-reversed process $\tilde{X}$ will be the opposite transition,

$$X = (\xi, \eta) \,,$$
$$\tilde{X} = (\eta, \xi) \,. \tag{18}$$

Let ρ_1^i and ρ_2^i be the probability distributions of the systems within the ensembles ξ and η, respectively. Then the change of Shannon information along the process X reads:

$$\Delta I \equiv nI_2 - nI_1 = -n\sum_{i=1}^{d} \rho_2^i \log \rho_2^i + n\sum_{i=1}^{d} \rho_1^i \log \rho_1^i \ . \tag{19}$$

The process X is irreversible if $\Delta I \neq 0$ and we should assign the time-arrow s so that $s\Delta I$ be positive (1). The point is that the two samples ξ and η may, by chance, not realize the asymmetry especially when the shapes of their probability distributions ρ_1^i and ρ_2^i do not much differ from each other.

Let us characterize the two constituting configurations of $X = (\xi, \eta)$ by the multiplicities n_1^i and n_2^i,

$$\xi = (n_1^i; i = 1, 2, \ldots, d) \ ,$$
$$\eta = (n_2^i; i = 1, 2, \ldots, d) \ , \tag{20}$$

which follow independent multinomial distributions with the respective mean values

$$\langle n_1^i \rangle = n\rho_1^i \ ,$$
$$\langle n_2^i \rangle = n\rho_2^i \ . \tag{21}$$

For large n we can approximate the multinomial distributions by Gaussian functions,

$$\mathcal{P}(\xi, \eta) = C \exp\left(-\sum_{i=1}^{d} \frac{[n_1^i - n\rho_1^i]^2}{2n\rho_1^i} - \sum_{i=1}^{d} \frac{[n_2^i - n\rho_2^i]^2}{2n\rho_2^i}\right) \ ,$$

$$\mathcal{P}(\eta, \xi) = C \exp\left(-\sum_{i=1}^{d} \frac{[n_2^i - n\rho_1^i]^2}{2n\rho_1^i} - \sum_{i=1}^{d} \frac{[n_1^i - n\rho_2^i]^2}{2n\rho_2^i}\right) \ . \tag{22}$$

We substitute these expressions into (8) to calculate $D(\xi, \eta)$, then we calculate the mean value,

$$D = -\frac{n}{2}\sum_{i=1}^{d} \left((\rho_2^i)^2 - (\rho_1^i)^2\right)\left(\frac{1}{\rho_2^i} - \frac{1}{\rho_1^i}\right) \ . \tag{23}$$

Suppose that $D(\xi, \eta)$ is, for very large n, dominated by the mean value D and fluctuations will thus be ignored. Hence the average Bayes estimate (10) reads

$$P(s) = \frac{1}{1 + e^{-sD}} \ . \tag{24}$$

This could become equivalent with our heuristic proposal provided $D = \Delta I$ which is apparently not true in general. I was looking for further conditions at least to achieve the asymptotic equivalence of D and ΔI. I concluded to the following elementary assumptions. First, the shapes ρ_1^i and ρ_2^i must be close to each other so that the lowest nontrivial order in $\Delta\rho^i = \rho_2^i - \rho_1^i$ will be sufficient. Second, the statistics of *either* ξ *or* η must be totally random. This sets the a priori time-arrow for $s = -$ or $s = +$, respectively. For concreteness, I consider the case $s = +$ and adopt flat distribution for η [6],

$$\rho_2^i = \tfrac{1}{d} \ . \tag{25}$$

This second assumption is a necessary one, otherwise ΔI contains a linear term in $\Delta\rho^i$ while D does not. From (19) and (23) the above two assumptions lead to the following results:

$$\Delta I = \tfrac{nd}{2} \sum_{i=1}^{d} (\Delta\rho^i)^2 \ , \tag{26}$$

and

$$D = nd \sum_{i=1}^{d} (\Delta\rho^i)^2 \ . \tag{27}$$

The result is surprising: D has come out twice the information loss.

Mathematically, D is the Kullback divergence between two neighboring ensembles ξ and η and it should asymptotically coincide with the information loss between them. The reason of the anomalous factor 2 is that we happened to use the Kullback divergence between the composite ensembles (ξ, η) and (η, ξ) instead of ξ and η. This gives a hint how the factor 2 would go away. It is interesting to note that the physical resolution has a typical quantum mechanical motivation. In microphysics it is conceptually impossible to observe the full quantity $X = (\xi, \eta)$. If, e.g., the time-arrow is positive ($s = +$) then η is testable and ξ is not because its observation would significantly perturb the initial preparation. And vice versa, when $s = -$ then η is testable and ξ is not. Accordingly, we are going to change the concept of experimental data. In the concrete case, we forbid the observation of ξ. In this sense, we have to redefine the distribution of the observed quantities,

$$\mathcal{P}(\xi, \eta) \to \mathcal{P}(\eta) \equiv \sum_{\xi} \mathcal{P}(\xi, \eta) \ ,$$

$$\tilde{\mathcal{P}}(\xi, \eta) \to \tilde{\mathcal{P}}(\eta) \equiv \sum_{\xi} \tilde{\mathcal{P}}(\xi, \eta) \ . \tag{28}$$

Repeating the calculation of the Kullback divergence in the leading order, inserting the flat values (25) for ρ_2^i, D turns out to be half of the previous value (26). Thus in the given approximation we have obtained the identity

$$D = \Delta I \ , \tag{29}$$

and confirmed the heuristic relationship (2).

5   Quantum Case

Let us consider the statistical ensemble of $n \gg 1$ independent $d-$state quantum systems where each one has the same density matrix $\hat{\rho}$. Let X be an abstract random process as trivial as the transition from an initial ensemble $\hat{\xi}$ into a final one $\hat{\eta}$, the time-reversed process $\tilde{X}$ will be the opposite transition,

$$X = (\hat{\xi}, \hat{\eta}) \, ,$$
$$\tilde{X} = (\hat{\eta}, \hat{\xi}) \, , \tag{30}$$

where

$$\hat{\xi} = \hat{\rho}_1 \otimes \hat{\rho}_1 \otimes \ldots \otimes \hat{\rho}_1 \equiv \hat{\rho}_1^{\otimes n} \, ,$$
$$\hat{\eta} = \hat{\rho}_2 \otimes \hat{\rho}_2 \otimes \ldots \otimes \hat{\rho}_2 \equiv \hat{\rho}_2^{\otimes n} \, , \tag{31}$$

if $\hat{\rho}_1$ and $\hat{\rho}_2$ stand for the density matrices of the systems within the ensembles $\hat{\xi}$ and $\hat{\eta}$, respectively. The change of von Neumann information during the process X reads

$$\Delta I \equiv n I_2 - n I_1 = -n \operatorname{Tr} (\hat{\rho}_2 \log \hat{\rho}_2) + n \operatorname{Tr} (\hat{\rho}_1 \log \hat{\rho}_1) \, . \tag{32}$$

The process X is irreversible if $\Delta I \neq 0$ and we should assign the time-arrow s so that $s\Delta I$ be positive (1). In order to ΔI have a definite sign the two ensembles $\hat{\xi}$ and $\hat{\eta}$ should display experimentally significant asymmetry.

Quantum theory says that if the reference-time is the physical time $(s = +)$ then we cannot test the ensemble $\hat{\xi}$ but the ensemble $\hat{\eta}$. And in the opposite case $(s = -)$ the ensemble $\hat{\xi}$ is testable and $\hat{\eta}$ is not. We see that the estimation of the time-arrow s boils down to the statistical decision whether the actually observed ensemble is $\hat{\xi} = \hat{\rho}_1^{\otimes n}$ or $\hat{\eta} = \hat{\rho}_2^{\otimes n}$ whereas both alternatives have equal a priori likelihoods.

We can mechanically follow the Bayes method of the previous chapters. Note, however, the typical quantum informatic arguments: this is the way I approached the issue originally.

The two collective states (31) reside in a Hilbert space of dimension d^n. According to the quantum counterpart of Shannon's code theory [7], in the large n limit such collective states become asymptotically equivalent with totally random states restricted for given subspaces. Our states (31) become random states in subspaces $\hat{E}_1$ and $\hat{E}_2$,

$$\hat{\xi} = \hat{\rho}_1^{\otimes n} \sim e^{-n I_1} \hat{E}_1 \, ,$$
$$\hat{\eta} = \hat{\rho}_2^{\otimes n} \sim e^{-n I_2} \hat{E}_2 \, , \tag{33}$$

where $\hat{E}_1$ and $\hat{E}_2$ are Hermitian projectors of dimensions $e^{n I_1}$ and $e^{n I_2}$, respectively. The dimensions depend on the von Neumann entropies. We are interested in the situations where the experimental distinguishability of the above two ensembles would exclusively depend on the difference $\Delta I = n I_2 - n I_1$ of the informations. This is obviously not true in general because the distinguishability will depend e.g. on the overlap $\mathrm{Tr}(\hat{E}_1\hat{E}_2)$. Nonetheless, the asymptotic forms (33) suggest simple conditions to achieve our goal. Suppose that one of the ensembles, say $\hat{\eta}$, is of minimal information,

$$\hat{\rho}_2 = \tfrac{\hat{1}}{d} \, , \tag{34}$$

which means that $I_2 = \log d$ and $\hat{E}_2 = \hat{1}^{\otimes n}$. The ensemble $\hat{\eta}$ is totally random over the whole collective Hilbert space of dimension d^n. The overlap between $\hat{E}_1$ and $\hat{E}_2$ becomes trivial. The information loss is always positive and the true time-arrow is thus $s = +$. But we have to find it by deciding whether we have tested the ensemble $\hat{\xi}$ or $\hat{\eta}$ which are of equal a priori likelihoods.

Now the experimental distinguishability of $\hat{\xi}$ and $\hat{\eta}$ is already trivial. All we have to do is to define $\hat{E}_1$ as observable and to observe it! If the tested ensemble is $\hat{\xi}$ itself then we get 1 with certainty since $\mathrm{Tr}(\hat{E}_1\hat{\xi}) = 1$. If the observed ensemble is the fully random $\hat{\eta}$ then we get 1 with probability $\mathrm{Tr}(\hat{E}_1\hat{\eta}) = e^{-\Delta I}$ and we get 0 with the complementary probability. As we see, the complete experimental statistics is determined by the information loss ΔI.

Let us turn to the Bayes method of Sect. 2 to estimate the time-arrow s. As we suggested above, the observed data is the value $E_1 = \{0, 1\}$ of the quantum observable $\hat{E}_1$. The probability $\mathcal{P}(E_1)$ stands for its distribution in the reference time with time-arrow $s = +$ and $\tilde{\mathcal{P}}(E_1)$ stands for its distribution in the reversed time $s = -$. In the preceding paragraph we established their values,

$$\mathcal{P}(E_1) = E_1 e^{-\Delta I} + (1 - E_1)(1 - e^{-\Delta I}) \, ,$$
$$\tilde{\mathcal{P}}(E_1) = E_1 \, . \tag{35}$$

Applying the steps of Sect. 2 mechanically, first we write (8) into this form:

$$e^{-D(E_1)} = \frac{\tilde{\mathcal{P}}(E_1)}{\mathcal{P}(E_1)} \, , \tag{36}$$

which is then substituted into the expression (9) of the mean fidelity, yielding

$$F = \left\langle \frac{1}{1 + \tilde{\mathcal{P}}(E_1)/\mathcal{P}(E_1)} \right\rangle_{\mathcal{P}} = \frac{1}{1 + e^{-\Delta I}} \tag{37}$$

We have used (35) to calculate the average. The result implies exactly the form (2) for the probability of intrinsic time-arrow in function of the information loss.

6   Concluding Remarks

I proposed a heuristic probability distribution (2) for the time-arrow intrinsic to a given irreversible process. The proposed probability is solely a function of the information loss ΔI. The idea itself comes from the phenomenological fluctuation theorem. Indeed, the concrete form of my proposal can easily be confirmed for the intrinsic time-arrow of standard irreversible processes, at least in the limit of macroscopic entropy production $\Delta I \gg 1$. My basic goal, however, was the construction of whatever trivial microscopic process which could underly the proposed dependence on ΔI. I analyze the irreversible process of the simplest possible structure in classical and quantum versions. The quantum version confirmed the proposed probabilistic time-arrow. Let me summarize this central result.

1) Suppose we know that (in physical time) a quantum ensemble $\hat{\xi}$ of $n \gg 1$ identical systems of given (also known to us) state transforms into an ensemble $\hat{\eta}$ of n totally random systems. 2) Suppose we do not know at all whether our reference-time is the physical-time or not, and whether the 'resulting ensemble' of the above process has been $\hat{\eta}$ or $\hat{\xi}$. 3) We test the 'resulting ensemble' and Bayesian inference will give us the time-arrow with fidelity

$$F = \frac{1}{1+e^{-|\Delta I|}} \ .$$

An infite number of conceptual issues could be raised against the presented ideas. I mention and discuss only two. First, the assignment of a non-trivial intrinsic time-arrow to a local irreversible process is a speculation. Nature might retain the same universal time-arrow for the whole Universe independently of the measure or direction of local information flows. Yet, we do not know if Nature is that conservative indeed. We learned from Einstein that Nature delegates the issues of local geometry to local physical systems. I adopted the hypothesis that this happens with time-arrow as well. Second, the proposed confirmation of the time-arrow probability includes Bayesian inference. Many would say that inference is subjective. The obtained probability is also subjective. Nonetheless, famous arguments using inference have been used earlier to confirm objective statistics of quantized fields [11]. It is, furthermore, a common knowledge that the maximum-likelihood inference of the intensive thermodynamic parameters confirms their true equilibrium fluctuations in Gibbs ensembles.

The present work is an attempt to find universal expressions for the hypothetic intrinsic time-arrow. There is a hint of the information loss to play the key role. This does not mean that we can already claim an experimental significance which should, of course, be inevitable after all. But theory of intrinsic time opens a series of natural questions to study in the future and there is apparently a promise of further analytic results.

Acknowledgments

I am grateful to Hans-Thomas Elze and to the sponsors of the workshop "DICE 2002" for the invitation to talk in Piombino and to contribute to this volume. My research was also supported by the Hungarian OTKA Grant No. 032640.

Appendix

Let $X(t)$ denote a thermodynamic variable of equilibrium value $\bar{X}$, where λ is the relaxation rate, and γ is the Onsager kinetic coefficient. The time-dependent fluctuations of $X(t)$ are governed by the phenomenological Langevin equation,

$$\frac{dX(t)}{dt} = -\lambda(X(t) - \bar{X}) + \sqrt{2\gamma}\, w(t) \tag{38}$$

with the standard white-noise $w(t)$. The expression

$$\frac{dI}{dt} = \frac{\lambda}{\gamma}(\bar{X} - X)\frac{dX}{dt} \tag{39}$$

will be the local rate of irreversible entropy production (information loss) along the process $X(t)$ (see, e.g., in Landau-Lifshitz [8], or in [9]). According to Onsager and Machlup [10], the conditional probability distribution of the process $X = \{X(t); -T \le t \le T\}$ at fixed initial value $\xi = X(-T)$ and for equilibrium value $\bar{X}$ takes this functional Gaussian form,

$$P(X|\xi; \bar{X}) = \exp\left(-\frac{1}{4\gamma}\int_{-T}^{T}\left[\frac{dX(t)}{dt} + \lambda(X(t) - \bar{X})\right]^2 dt\right). \tag{40}$$

We shall consider *driven* thermodynamic processes which can be described by the (38-40) with time-dependent equilibrium values $\{\bar{X}(t); -T \le t \le T\}$. For convenience of forthcoming calculations let us write down the distribution functional of the driven process,

$$P(X|\xi; \bar{X}) = \exp\left(-\frac{1}{4\gamma}\int_{-T}^{T}\left[\frac{dX(t)}{dt} + \lambda(X(t) - \bar{X}(t))\right]^2 dt\right). \tag{41}$$

Obviously the above equations assume physical time t. Let us express the conditional distribution of the time-reversed process $\tilde{X}$ starting from $\tilde{X}(-T) = \eta$, driven by the time-reversed function $\tilde{\bar{X}}$. Namely, we replace $X, \xi, \bar{X}$ in (41) by $\tilde{X}, \eta, \tilde{\bar{X}}$, respectively,

$$P(\tilde{X}|\eta; \tilde{\bar{X}}) = \exp\left(-\frac{1}{4\gamma}\int_{-T}^{T}\left[\frac{d\tilde{X}(t)}{dt} + \lambda(\tilde{X}(t) - \tilde{\bar{X}}(t))\right]^2 dt\right). \tag{42}$$

Now we change the variable t in the integrand for $-t$ and insert the relations

$$\tilde{X}(t) \equiv X(-t),$$
$$\tilde{\bar{X}}(t) \equiv \bar{X}(-t), \tag{43}$$

leading to

$$P(\tilde{X}|\eta; \tilde{\bar{X}}) = \exp\left(-\frac{1}{4\gamma}\int_{-T}^{T}\left[\frac{dX(t)}{dt} - \lambda(X(t) - \bar{X}(t))\right]^2 dt\right). \tag{44}$$

(Recall that this expression would be the conditional distribution of the process had we observed it in the time-reversed frame.) The logarithm of the physical distribution (41) over the time-reversed one (44) will result in a remarkable expression,

$$\log \frac{P(X|\xi;\bar{X})}{P(\tilde{X}|\eta;\bar{X})} = \frac{\lambda}{\gamma} \int_{-T}^{T} (\bar{X}(t) - X(t)) dX(t) .$$ (45)

It follows from (39) that the r.h.s. is equal to the total entropy production (information-loss) of the driven process,

$$\Delta I(X;\bar{X}) = \frac{\lambda}{\gamma} \int_{-T}^{T} (\bar{X}(t) - X(t)) dX(t) .$$ (46)

This and the preceding equation yield the fluctuation theorem [3–5],

$$P(\tilde{X}|\eta;\tilde{\bar{X}}) = e^{-\Delta I(X;\bar{X})} P(X|\xi;\bar{X}) .$$ (47)

References

1. H.D. Zeh: *The Direction of Time* (Springer Verlag, Berlin 1989)
2. Especially in thermodynamic context, the Shannon and von Neumann information (loss) will be deliberately identified with the entropy (production). While information is the fundamental quantity in the present work, I use Boltzmann units of entropy. In Shannon units (bits) the central equation (2) would read:

$$P(s) = \frac{1}{1+2^{-s\Delta I}}$$

3. D.J. Evans, E.G.D. Cohen, and G.P. Morriss: Phys. Rev. Lett. **71**, 2401 (1993)
4. G. Gallavotti and E.G.D. Cohen: Phys. Rev. Lett. **74**, 2694 (1995)
5. C. Maes, F. Redig, and A. Van Moffaert: J. Math. Phys. **41**, 1528 (2000)
6. The choice only means that, for the sake of simplicity, the forthcoming calculations are done in the physical reference time. The covariance for $t \to -t$ will only be restored on the final equations
7. B. Schumacher: Phys. Rev. A **51**, 2738 (1995)
8. L.D. Landau and E.M. Lifshitz: *Statistical Physics* (Clarendon, Oxford 1982)
9. L. Diósi, Katalin Kulacsy, B.Lukács, and A.Rácz: J. Chem. Phys. **105**, 11220 (1996)
10. L. Onsager and S. Machlup: Phys. Rev. **91**, 1505 (1953)
11. N. Bohr and L. Rosenfeld: Kgl. Danske Videnskab S. Mat. Fys. Medd. **12** 1 (1933)

Master and Langevin Equations for Interaction with Electromagnetic and Gravitational Environments

Zbigniew Haba[1,2] and Hagen Kleinert[2]

[1] Institute of Theoretical Physics, University of Wroclaw, Poland
[2] Institut für Theoretische Physik, Freie Universität Berlin, Germany

Abstract. We set up a forward–backward path integral for a point particle in a bath of photons and gravitons and derive from it a master equation for the density matrix which describes electromagnetic and gravitational dissipation. The associated Langevin equations are discussed.

1 Introduction

The Feynman-Vernon influence functional [1–3] is a useful tool for investigating the role of an environment in decoherence and dissipation phenomena. In the cosmos, there exists a well-studied universal environment of the cosmic background radiation. In addition, we expect also an environment of gravitational waves as remnant of the big bang. Here we study a Markovian approximation to the quantum mechanics in either environment leading to a master equation correcting the Hamiltonian equation of the conventional quantum mechanics.

2 Photonic Environment

Consider first an interaction with a quantized electromagnetic field. The time evolution of a quantum-mechanical density matrix $\rho(\mathbf{x}_{+a}, \mathbf{x}_{-a}; t_a)$ of a particle coupled to an external electromagnetic vector potential $\mathbf{A}(\mathbf{x}, t)$ is determined by a forward–backward path integral [1–3]

$$(\mathbf{x}_{+b}, t_b | \mathbf{x}_{+a}, t_a)(\mathbf{x}_{-b}, t_b | \mathbf{x}_{-a}, t_a)^* \equiv U(\mathbf{x}_{+b}, \mathbf{x}_{-b}, t_b | \mathbf{x}_{+a}, \mathbf{x}_{-a}, t_a)$$

$$= \int \mathcal{D}\mathbf{x}_+ \mathcal{D}\mathbf{x}_- \, \exp\left\{ \frac{i}{\hbar} \int \left[\frac{M}{2}\left(\dot{\mathbf{x}}_+^2 - \dot{\mathbf{x}}_-^2 \right) - V(\mathbf{x}_+) + V(\mathbf{x}_-) \right. \right.$$
$$\left. \left. - \frac{e}{c}\dot{\mathbf{x}}_+ \mathbf{A}(\mathbf{x}_+, t) + \frac{e}{c}\dot{\mathbf{x}}_- \mathbf{A}(\mathbf{x}_-, t) \right] \right\}, \tag{1}$$

where $\mathbf{x}_+(t)$ and $\mathbf{x}_-(t)$ are two fluctuating paths connecting the initial and final points $\mathbf{x}_{+a}$ and $\mathbf{x}_{+b}$, and $\mathbf{x}_{-a}$ and $\mathbf{x}_{-b}$, respectively. Then, the density matrix $\rho(\mathbf{x}_{+b}, \mathbf{x}_{-b}; t_b)$ at a time t_b is found from that at an earlier time t_a by the integral

$$\rho(\mathbf{x}_{+b}, \mathbf{x}_{-b}; t_b) = \int d\mathbf{x}_{+a} \, d\mathbf{x}_{-a} \, U(\mathbf{x}_{+b}, \mathbf{x}_{-b}, t_b | \mathbf{x}_{+a}, \mathbf{x}_{-a}, t_a) \rho(\mathbf{x}_{+a}, \mathbf{x}_{-a}; t_a). \tag{2}$$

Z. Haba and H. Kleinert, Master and Langevin Equations for Interaction with Electromagnetic and Gravitational Environments, Lect. Notes Phys. **633**, 136–150 (2004)
http://www.springerlink.com/

The vector potential $\mathbf{A}(\mathbf{x}, t)$ is a superposition of oscillators $\mathbf{X_k}(t)$ of frequency $\Omega_\mathbf{k} = c|\mathbf{k}|$ in a volume V:

$$\mathbf{A}(\mathbf{x}, t) = \sum_\mathbf{k} f_\mathbf{k}(\mathbf{x}) \mathbf{X_k}(t), \quad f_\mathbf{k}(\mathbf{x}) = \frac{e^{i\mathbf{kx}}}{\sqrt{2V\Omega_\mathbf{k}/c}} \quad . \tag{3}$$

These oscillators are assumed to be in equilibrium at a finite temperature T, where we shall write their time-ordered correlation functions as $G_{\mathbf{kk}'}^{ij}(t, t') = \langle \hat{T} \hat{X}_\mathbf{k}^i(t), \hat{X}_{-\mathbf{k}'}^j(t') \rangle = \delta_{\mathbf{kk}'}^{ij \, \mathrm{tr}} G_{\Omega_\mathbf{k}}(t, t') \equiv \delta_{\mathbf{kk}'}(\delta^{ij} - k^i k^j/\mathbf{k}^2) G_{\Omega_\mathbf{k}}(t, t')$, the transverse Kronecker symbol resulting from the sum over the two polarization vectors $\sum_{h=\pm} \epsilon^i(\mathbf{k}, h)\epsilon^{j*}(\mathbf{k}, h)$ of the vector potential $\mathbf{A}(\mathbf{x}, t)$. The thermal average of the evolution kernel (1) over the electromagnetic field is given by the forward–backward path integral

$$U(\mathbf{x}_{+b}, \mathbf{x}_{-b}, t_b | \mathbf{x}_{+a}, \mathbf{x}_{-a}, t_a) = \int \mathcal{D}\mathbf{x}_+(t) \int \mathcal{D}\mathbf{x}_-(t)$$

$$\times \exp\left\{ \frac{i}{\hbar} \int dt \left[\frac{M}{2}(\dot{\mathbf{x}}_+^2 - \dot{\mathbf{x}}_-^2) - (V(\mathbf{x}_+) - V(\mathbf{x}_-)) \right] + \frac{i}{\hbar} \mathcal{A}^{\mathrm{FV}}[\mathbf{x}_+, \mathbf{x}_-] \right\}, \tag{4}$$

where $\exp\{i\mathcal{A}^{\mathrm{FV}}[\mathbf{x}_+, \mathbf{x}_-]/\hbar\}$ is the Feynman-Vernon *influence functional*. The influence action $\mathcal{A}^{\mathrm{FV}}[\mathbf{x}_+, \mathbf{x}_-]$ is the sum of a dissipative and a fluctuating part $\mathcal{A}_D^{\mathrm{FV}}[\mathbf{x}_+, \mathbf{x}_-]$ and $\mathcal{A}_F^{\mathrm{FV}}[\mathbf{x}_+, \mathbf{x}_-]$, respectively, whose explicit forms are

$$\mathcal{A}_D^{\mathrm{FV}}[\mathbf{x}_+, \mathbf{x}_-] = \frac{ie^2}{2\hbar c^2} \int dt \int dt' \, \Theta(t - t') \Big[\dot{\mathbf{x}}_+ \mathbf{C}_\mathrm{b}(\mathbf{x}_+ t, \mathbf{x}_+' t') \dot{\mathbf{x}}_+'$$

$$-\dot{\mathbf{x}}_+ \mathbf{C}_\mathrm{b}(\mathbf{x}_+ t, \mathbf{x}_-' t') \dot{\mathbf{x}}_-' - \dot{\mathbf{x}}_- \mathbf{C}_\mathrm{b}(\mathbf{x}_- t, \mathbf{x}_+' t') \dot{\mathbf{x}}_+' - \dot{\mathbf{x}}_- \mathbf{C}_\mathrm{b}(\mathbf{x}_- t, \mathbf{x}_-' t') \dot{\mathbf{x}}_-' \Big] \tag{5}$$

and

$$\mathbf{A}_F^{\mathrm{FV}}[\mathbf{x}_+, \mathbf{x}_-] = \frac{ie^2}{2\hbar c^2} \int dt \int dt' \, \Theta(t - t') \Big[\dot{\mathbf{x}}_+ \mathbf{A}_\mathrm{b}(\mathbf{x}_+ t, \mathbf{x}_+' t') \dot{\mathbf{x}}_+'$$

$$-\dot{\mathbf{x}}_+ \mathbf{A}_\mathrm{b}(\mathbf{x}_+ t, \mathbf{x}_-' t') \dot{\mathbf{x}}_-' - \dot{\mathbf{x}}_- \mathbf{A}_\mathrm{b}(\mathbf{x}_- t, \mathbf{x}_+' t') \dot{\mathbf{x}}_+' + \dot{\mathbf{x}}_- \mathbf{A}_\mathrm{b}(\mathbf{x}_- t, \mathbf{x}_-' t') \dot{\mathbf{x}}_-' \Big], \tag{6}$$

where $\mathbf{x}_\pm$, $\mathbf{x}_\pm'$ are short for $\mathbf{x}_\pm(t)$, $\mathbf{x}_\pm(t')$, and $\mathbf{C}_\mathrm{b}(\mathbf{x}_- t, \mathbf{x}_-' t')$, $\mathbf{A}_\mathrm{b}(\mathbf{x}_- t, \mathbf{x}_-' t')$ are 3×3 commutator and anticommutator functions of the bath of photons,

$$C_\mathrm{b}^{ij}(\mathbf{x} t, \mathbf{x}' t') = -ic^2\hbar \int \frac{d\omega' d^3k}{(2\pi)^4} \sigma_\mathbf{k}(\omega') \delta_{\mathbf{kk}}^{ij \, \mathrm{tr}} e^{i\mathbf{k}(\mathbf{x}-\mathbf{x}')} \sin\omega'(t - t'), \tag{7}$$

$$A_\mathrm{b}^{ij}(\mathbf{x} t, \mathbf{x}' t') = c^2\hbar \int \frac{d\omega' d^3k}{(2\pi)^4} \sigma_\mathbf{k}(\omega') \delta_{\mathbf{kk}}^{ij \, \mathrm{tr}} \coth\frac{\hbar\omega'}{2k_BT} e^{i\mathbf{k}(\mathbf{x}-\mathbf{x}')} \cos\omega'(t - t'), \tag{8}$$

where T is the temperature, k_B is the Boltzmann constant, $\beta \equiv 1/k_BT$, and $\sigma_\mathbf{k}(\omega')$ is the spectral density contributed by the oscillator of momentum $\mathbf{k}$,

$$\sigma_\mathbf{k}(\omega') \equiv \frac{2\pi}{2\Omega_\mathbf{k}} [\delta(\omega' - \Omega_\mathbf{k}) - \delta(\omega' + \Omega_\mathbf{k})]. \tag{9}$$

We assume now that a system is so small that the effects of retardation can be neglected. Then we can ignore the $\mathbf{x}$-dependence in (7) and (8) and find

$$C_b^{ij}(\mathbf{x}\,t, \mathbf{x}'\,t') \approx C_b^{ij}(t, t') = i\frac{\hbar}{2\pi c}\frac{2}{3}\delta^{ij}\partial_t\delta(t - t'). \tag{10}$$

Inserting this into (5) we find (after a renormalization)

$$\mathcal{A}_D^{FV}[\mathbf{x}_+, \mathbf{x}_-] = -\gamma\frac{M}{2}\int_{t_a}^{t_b} dt\,(\dot{\mathbf{x}}_+ - \dot{\mathbf{x}}_-)(t)(\ddot{\mathbf{x}}_+ + \ddot{\mathbf{x}}_-)(t), \tag{11}$$

with the friction constant

$$\gamma \equiv \frac{e^2}{6\pi c^3 M} = \frac{2}{3}\frac{\alpha}{\omega_M}, \tag{12}$$

where $\alpha \equiv e^2/\hbar c \approx 1/137$ is the fine-structure constant and $\omega_M \equiv Mc^2/\hbar$ the Compton frequency associated with the mass M.

We consider next the anticommutator function. Inserting (9) and the friction constant γ from (12), it becomes

$$\frac{e^2}{c^2}A_b(\mathbf{x}\,t, \mathbf{x}'\,t') \approx 2\gamma k_B T K(t, t'), \tag{13}$$

where

$$K(t, t') = K(t - t') \equiv \int_{-\infty}^{\infty} \frac{d\omega'}{2\pi}K(\omega')e^{-i\omega'(t-t')}, \tag{14}$$

and

$$K(\omega') \equiv \frac{\hbar\omega'}{2k_B T}\coth\frac{\hbar\omega'}{2k_B T}. \tag{15}$$

With the function $K(t, t')$, the fluctuation part of the influence functional in (6), (5), (4) becomes

$$\mathcal{A}_F^{FV}[\mathbf{x}_+, \mathbf{x}_-] = i\frac{w}{2\hbar}\int_{t_a}^{t_b} dt \int_{t_a}^{t_b} dt'\,(\dot{\mathbf{x}}_+ - \dot{\mathbf{x}}_-)(t)\,K(t, t')\,(\dot{\mathbf{x}}_+ - \dot{\mathbf{x}}_-)(t'). \tag{16}$$

Here we have introduced the constant

$$w \equiv 2Mk_B T\gamma, \tag{17}$$

At very high temperatures an expansion in $\frac{1}{T}$ leads to the time evolution amplitude for the density matrix

$$U(\mathbf{x}_{+b}, \mathbf{x}_{-b}, t_b | \mathbf{x}_{+a}, \mathbf{x}_{-a}, t_a) = \tag{18}$$

$$\int \mathcal{D}\mathbf{x}_+(t) \int \mathcal{D}\mathbf{x}_-(t) \exp\left\{\frac{i}{\hbar}\int dt\,\left[\frac{M}{2}(\dot{\mathbf{x}}_+^2 - \dot{\mathbf{x}}_-^2) - (V(\mathbf{x}_+) - V(\mathbf{x}_-))\right]\right\}$$

$$\times \exp\left\{-\frac{i}{2\hbar}M\gamma\int dt\,(\dot{\mathbf{x}}_+ - \dot{\mathbf{x}}_-)(\ddot{\mathbf{x}}_+ + \ddot{\mathbf{x}}_-) - \frac{w}{2\hbar^2}\int_{t_a}^{t_b} dt\,(\dot{\mathbf{x}}_+ - \dot{\mathbf{x}}_-)^2\right\}.$$

3   Master Equation for Time Evolution of Density Matrix

We now derive a differential equation describing the evolution of the density matrix $\rho(x_{+a}, x_{-a}; t_a)$ in (2). For this purpose we consider first a canonical formulation of the quantum system without electromagnetism, and include the effect of the latter recursively. For simplicity, we shall treat only the local limiting form of the last term in (18). In this limit, we define a Hamilton-like operator as follows:

$$\hat{\mathcal{H}} \equiv \tfrac{1}{2M}\left(\hat{\mathbf{p}}_+^2 - \hat{\mathbf{p}}_-^2\right) + V(\mathbf{x}_+) - V(\mathbf{x}_-)$$
$$+ \tfrac{M\gamma}{2}(\hat{\dot{\mathbf{x}}}_+ - \hat{\dot{\mathbf{x}}}_-)(\hat{\dot{\mathbf{x}}}_+ + \hat{\dot{\mathbf{x}}}_-) - i\tfrac{w}{2\hbar}(\hat{\dot{\mathbf{x}}}_+ - \hat{\dot{\mathbf{x}}}_-)^2. \tag{19}$$

Here $\hat{\dot{\mathbf{x}}}$, $\hat{\ddot{\mathbf{x}}}$ are abbreviations for the commutators

$$\hat{\dot{\mathbf{x}}} \equiv \tfrac{i}{\hbar}[\hat{\mathcal{H}}, \hat{\mathbf{x}}], \qquad \hat{\ddot{\mathbf{x}}} \equiv \tfrac{i}{\hbar}[\hat{\mathcal{H}}, \hat{\dot{\mathbf{x}}}]. \tag{20}$$

A direct differentiation of (18) over time leads to the conclusion that the density matrix $\rho(x_+, x_-; t_a)$ satisfies the time evolution equation

$$i\hbar\partial_t\rho(x_+, x_-; t_a) = \hat{\mathcal{H}}\rho(x_+, x_-; t_a). \tag{21}$$

At moderately high temperatures, we also include a term coming from (16) through an expansion of $K(\omega)$ in $\frac{1}{T}$

$$\mathcal{H}_1 \equiv i\tfrac{w\hbar}{24(k_B T)^2}(\hat{\ddot{\mathbf{x}}}_+ - \hat{\ddot{\mathbf{x}}}_-)^2. \tag{22}$$

For systems with friction caused by a conventional heat bath of harmonic oscillators as discussed by Caldeira and Leggett [4], the analogous extra term was shown by Diosi [5] to bring the master equation to the general Lindblad form [6] which ensures positivity of the probabilities resulting from the solutions of (21).

It is useful to re-express (21) in the standard quantum-mechanical operator form where the density matrix has a bra–ket representation $\hat{\rho}(t) = \sum_{mn}\rho_{nm}(t)$ $|m\rangle\langle n|$. Let us denote the initial Hamilton operator of the system in (1) by $\hat{H} = \hat{\mathbf{p}}^2/2M + \hat{V}$, then (21) with the term (22) takes the operator form

$$i\hbar\partial_t\hat{\rho} = \hat{\mathcal{H}}\hat{\rho} \equiv [\hat{H}, \hat{\rho}] + \tfrac{M\gamma}{2}\left(\hat{\dot{\mathbf{x}}}\hat{\ddot{\mathbf{x}}}\hat{\rho} - \hat{\rho}\hat{\ddot{\mathbf{x}}}\hat{\dot{\mathbf{x}}} + \hat{\dot{\mathbf{x}}}\hat{\rho}\hat{\ddot{\mathbf{x}}} - \hat{\ddot{\mathbf{x}}}\hat{\rho}\hat{\dot{\mathbf{x}}}\right)$$
$$- \tfrac{iw}{2\hbar}[\hat{\dot{\mathbf{x}}}, [\hat{\dot{\mathbf{x}}}, \hat{\rho}]] - \tfrac{iw\hbar}{24(k_B T)^2}[\hat{\ddot{\mathbf{x}}}, [\hat{\ddot{\mathbf{x}}}, \hat{\rho}]]. \tag{23}$$

The positivity of $\hat{\rho}$ is ensured by the observation, that (23) can be written in the extended *Lindblad form* [6]

$$\partial_t\hat{\rho} = -\tfrac{i}{\hbar}[\hat{H}_\gamma, \hat{\rho}] - \sum_{n=1}^{2}\left(\tfrac{1}{2}\hat{L}_n\hat{L}_n^\dagger\hat{\rho} + \tfrac{1}{2}\hat{\rho}\hat{L}_n\hat{L}_n^\dagger - \hat{L}_n^\dagger\hat{\rho}\hat{L}_n\right), \tag{24}$$

where

$$\hat{H}_\gamma = \tfrac{\hat{\mathbf{p}}^2}{2M} + V + \tfrac{M\gamma}{4}(\hat{\dot{\mathbf{x}}}\hat{\ddot{\mathbf{x}}} + \hat{\ddot{\mathbf{x}}}\hat{\dot{\mathbf{x}}}). \tag{25}$$

Here, the two Lindblad operators are of the form

$$L_1 = a_1\hat{\mathbf{x}} + ib_1\dot{\hat{\mathbf{x}}}, \qquad L_2 = a_2\hat{\mathbf{x}} + ib_2\dot{\hat{\mathbf{x}}} \qquad (26)$$

where a and b satisfy the equations

$$-\tfrac{M\gamma}{2\hbar} = b_1a_1 + a_2b_2, \qquad \tfrac{w}{\hbar^2} = a_1^2 + a_2^2, \qquad \tfrac{w\beta^2}{12} = b_1^2 + b_2^2. \qquad (27)$$

The solution is not unique. In the case $b_1 = 0$ one has

$$\hat{L}_1 \equiv \tfrac{\sqrt{w}}{2\hbar}\hat{\mathbf{x}}, \qquad \hat{L}_2 \equiv \tfrac{\sqrt{3w}}{2\hbar}\left(\hat{\mathbf{x}} - i\tfrac{\hbar}{6k_BT}\dot{\hat{\mathbf{x}}}\right). \qquad (28)$$

Another interesting solution is found for the choice $b_1 = \mu a_1$ and $b_2 = \mu a_2$, with $\mu = \hbar\beta/2\sqrt{3}$. Then $L_1 = a_1 A$ and $L_2 = a_2 A^\dagger$ with $A = \hat{\mathbf{x}} + i\mu\dot{\hat{\mathbf{x}}}$, and

$$a_1^2 = \tfrac{1}{2\hbar^2}w\left(1 - \tfrac{\sqrt{3}}{2}\right), \qquad a_1^2 = \tfrac{1}{2\hbar^2}w\left(1 + \tfrac{\sqrt{3}}{2}\right). \qquad (29)$$

For small γ we could set in (23) $\dot{\hat{\mathbf{x}}} = -i\hbar\nabla/M$ and $\ddot{\hat{\mathbf{x}}} = -\nabla V(\hat{\mathbf{x}})/M$ and obtain a proper Lindblad equation. In the case of the harmonic oscillator, such an equation coincides with the master equation describing an exchange of quanta with the heat bath at intermediate temperatures. Then, it is known that the solution tends to equilibrium [7], which is the Gibbs state. We expect also the solution of (23) to tend to equilibrium. The special case of a quadratic potential is discussed in detail in [8].

4   Line Width

Let us apply the master equation (23) to an atom assuming it to be initially in an eigenstate $|i\rangle$ of H, with a density matrix $\hat{\rho}(0) = |i\rangle\langle i|$. Since atoms decay rather slowly, we may treat the γ-term in (23) perturbatively. It leads to a time derivative of the density matrix

$$\partial_t\langle i|\hat{\rho}(t)|i\rangle = -\tfrac{\gamma}{\hbar M}\langle i|[\hat{H},\hat{\mathbf{p}}]\,\hat{\mathbf{p}}\,\hat{\rho}(0)|i\rangle = \tfrac{\gamma}{M}\sum_{f\neq i}\omega_{if}\langle i|\mathbf{p}|f\rangle\langle f|\mathbf{p}|i\rangle$$

$$= -M\gamma\sum_f \omega_{if}^3\,|\mathbf{x}_{fi}|^2, \qquad (30)$$

where $\hbar\omega_{if} \equiv E_i - E_f$, and $\mathbf{x}_{fi} \equiv \langle f|\mathbf{x}|i\rangle$ are the matrix elements of the dipole operator.

An extra width comes from the last two terms in (23):

$$\partial_t\langle i|\hat{\rho}(t)|i\rangle = -\tfrac{w}{M^2\hbar^2}\langle i|\mathbf{p}^2|i\rangle - \tfrac{w}{12M^2(k_BT)^2}\langle i|\dot{\mathbf{p}}^2|i\rangle$$

$$= -w\sum_n \omega_{if}^2\left[1 + \tfrac{\hbar^2\omega_{if}^2}{12(k_BT)^2}\right]|\mathbf{x}_{fi}|^2. \qquad (31)$$

This time dependence is caused by spontaneous emission and induced emission and absorption. To identify the different contributions, we rewrite the spectral decompositions (7) and (8) in the **x**-independent approximation,

$$C_b(t,t') + A_b(t,t') = \frac{4\pi}{3}\hbar \int \frac{d\omega' d^3k}{(2\pi)^4} \frac{\pi}{2M\Omega_{\mathbf{k}}} \left\{1 + \coth\frac{\hbar\omega'}{2k_BT}\right\}$$
$$\times \left[\delta(\omega' - \Omega_{\mathbf{k}}) - \delta(\omega' + \Omega_{\mathbf{k}})\right] e^{-i\omega'(t-t')}, \quad (32)$$

as

$$C_b(t,t') + A_b(t,t') = \frac{4\pi}{3}\hbar \int \frac{d\omega' d^3k}{(2\pi)^4} \frac{\pi}{2M\Omega_{\mathbf{k}}} \left\{2\delta(\omega' - \Omega_{\mathbf{k}}) + \frac{2}{e^{\hbar\Omega_{\mathbf{k}}/k_BT} - 1}\right.$$
$$\left. \times \left[\delta(\omega' - \Omega_{\mathbf{k}}) + \delta(\omega' + \Omega_{\mathbf{k}})\right]\right\} e^{-i\omega'(t-t')}. \quad (33)$$

Following Einstein's intuitive interpretation, the first term in curly brackets is due to spontaneous emission, the other two terms accompanied by the Bose occupation function account for induced emission and absorption. For high and intermediate temperatures, the right-hand side of (33) has the expansion

$$\frac{4\pi}{3}\hbar \int \frac{d\omega' d^3k}{(2\pi)^4} \frac{\pi}{2M\Omega_{\mathbf{k}}} \left\{2\delta(\omega' - \Omega_{\mathbf{k}}) + \left(\frac{2k_BT}{\hbar\Omega_{\mathbf{k}}} - 1 + \frac{1}{6}\frac{\hbar\Omega_{\mathbf{k}}}{k_BT}\right)\right.$$
$$\left. \times \left[\delta(\omega' - \Omega_{\mathbf{k}}) + \delta(\omega' + \Omega_{\mathbf{k}})\right]\right\} e^{-i\omega'(t-t')}. \quad (34)$$

The first term in curly brackets corresponds to the spontaneous emission. It contributes to the rate of change $\partial_t\langle i|\hat{\rho}(t)|i\rangle$ a term $-2M\gamma\sum_{f<i}\omega_{if}^3|\mathbf{x}_{fi}|^2$. This differs from the right-hand side of (30) in two important respects. First, the sum is restricted to the lower states $f < i$ with $\omega_{if} > 0$, since the δ-function allows only for decays. Second, there is an extra factor 2. Indeed, by comparing (32) with (34) we see that the spontaneous emission receives equal contributions from the 1 and the $\coth(\hbar\omega'/2k_BT)$ in the curly brackets of (32), i.e., from dissipation and fluctuation terms $C_b(t,t')$ and $A_b(t,t')$.

Thus our master equation yields for the natural line width of atomic levels the equation

$$\Gamma = 2M\gamma \sum_{f<i} \omega_{if}^3 |\mathbf{x}_{fi}|^2, \quad (35)$$

in agreement with the historic *Wigner-Weisskopf formula*.

In terms of Γ, the rate (30) can therefore be written as

$$\partial_t\langle i|\hat{\rho}(t)|i\rangle = -\Gamma + M\gamma \sum_{f<i} \omega_{if}^3 |\mathbf{x}_{fi}|^2 + M\gamma \sum_{f>i} |\omega_{if}|^3 |\mathbf{x}_{fi}|^2. \quad (36)$$

The second and third terms do not contribute to the total rate of change of $\langle i|\hat{\rho}(t)|i\rangle$ since they are canceled by the induced emission and absorption terms associated with the -1 in the big parentheses of the fluctuation part of (34). The finite lifetime changes the time dependence of the state $|i,t\rangle$ from $|i,t\rangle = |i,0\rangle e^{-iEt}$ to $|i,0\rangle e^{-iEt-\Gamma t/2}$.

5   Langevin Equation for a Photonic Environment

For high γT, the last term in the forward–backward path integral (18) makes the size of the fluctuations in the difference between the paths $\mathbf{y}(t) \equiv \mathbf{x}_+(t) - \mathbf{x}_-(t)$ very small. It is then convenient to introduce the average of the two paths as $\mathbf{x}(t) \equiv [\mathbf{x}_+(t) + \mathbf{x}_-(t)]/2$, and expand

$$V\left(\mathbf{x} + \tfrac{\mathbf{y}}{2}\right) - V\left(\mathbf{x} - \tfrac{\mathbf{y}}{2}\right) \sim \mathbf{y} \cdot \boldsymbol{\nabla} V(\mathbf{x}) + \mathcal{O}(\mathbf{y}^3) \dots \; , \tag{37}$$

keeping only the first term.

Then the path integral (18) takes the form (see [9] for more details)

$$U(\mathbf{x}_{+b}, \mathbf{x}_{-b}, t_b | \mathbf{x}_{+a}, \mathbf{x}_{-a}, t_a) = \int \mathcal{D}\mathbf{x}\mathcal{D}\mathbf{y}\,\delta\Big(\mathbf{x}(t_b) - \mathbf{x}_b\Big)\delta\Big(\mathbf{y}(t_b) - \mathbf{y}_b\Big)$$

$$\times \exp\left\{ \frac{i}{\hbar} \int_{t_a}^{t_b} dt \left[\dot{\mathbf{y}}\Big(M\dot{\mathbf{x}} - M\gamma\ddot{\mathbf{x}} + i\frac{w}{2\hbar}\dot{\mathbf{y}} - i\frac{w\hbar}{24(k_B T)^2}\,\dddot{\mathbf{y}} + \dots \Big) \right. \right.$$

$$\left. \left. -\mathbf{y}\boldsymbol{\nabla} V(\mathbf{x}) \right] \right\}. \tag{38}$$

At t_a, the paths $\mathbf{x}(t)$ and $\mathbf{y}(t)$ start from $\mathbf{x}_a \equiv (\mathbf{x}_{+a} + \mathbf{x}_{-a})/2$ and $\mathbf{y}_a \equiv \mathbf{x}_{+a} - \mathbf{x}_{-a}$, respectively. Representing the δ-function $\delta\left(\mathbf{y}(t_b) - \mathbf{y}_b\right)$ in (38) as a Fourier integral, and inserting for $\mathbf{y}(t_b)$ the equation

$$\mathbf{y}(t_b) = \int_{t_a}^{t_b} dt' \,\dot{\mathbf{y}}(t') + \mathbf{y}_a, \tag{39}$$

we rewrite the evolution kernel in the form

$$U\Big(\mathbf{x}_{+b}, \mathbf{x}_{-b}, t_b | \mathbf{x}_{+a}, \mathbf{x}_{-a}, t_a\Big) = \int \frac{d^3 p}{(2\pi)^3} \int \mathcal{D}\mathbf{x}\mathcal{D}\mathbf{y}\,\delta\Big(\mathbf{x}(t_b) - \mathbf{x}_b\Big)$$

$$\times \exp\left\{ \frac{i}{\hbar} \int_{t_a}^{t_b} dt \left[\dot{\mathbf{y}}\Big(\boldsymbol{\eta} + \frac{iw}{2\hbar}\dot{\mathbf{y}} + \dots \Big) - \mathbf{y}_b \boldsymbol{\nabla} V(\mathbf{x}) \right] - \frac{i\mathbf{P}}{\hbar}(\mathbf{y}_b - \mathbf{y}_a) \right\}, \tag{40}$$

where we introduced the new variable

$$\boldsymbol{\eta}(t) \equiv M\dot{\mathbf{x}}(t) - M\gamma\ddot{\mathbf{x}}(t) + \int_{t_a}^{t} dt'\boldsymbol{\nabla} V\left(\mathbf{x}(t')\right) - \mathbf{p}. \tag{41}$$

In (40), the variables $\dot{\mathbf{y}}(t)$ at different points $t_a < t < t_b$ are independent of each other, and we choose at the end points $\dot{\mathbf{y}}(t_a) = \dot{\mathbf{y}}(t_b) = 0$. This may be justified in a time sliced formulation, where the integrations over the variables next to the end points give only a trivial factor with respect to the product of integrals in which these variables are held fixed at the endpoint values. Furthermore, the path integral (40) does not depend on $\boldsymbol{\eta}(t)$ outside the interval $t \in (t_a, t_b)$, and is independent of $\boldsymbol{\eta}(t_b)$ and $\boldsymbol{\eta}(t_a)$. Hence we may choose $\boldsymbol{\eta}(t_a) = \boldsymbol{\eta}(t_b) = 0$, for convenience, and (40) can be solved as a differential equation on the interval $t_a \leq t \leq t_b$

$$M\ddot{\mathbf{x}} - M\gamma\,\dddot{\mathbf{x}} + \boldsymbol{\nabla} V(\mathbf{x}) = \dot{\boldsymbol{\eta}}(t), \tag{42}$$

with the initial conditions

$$\mathbf{x}(t_a) = \mathbf{x}_a, \quad M\dot{\mathbf{x}}(t_a) - M\gamma\ddot{\mathbf{x}}(t_a) = \mathbf{p}. \tag{43}$$

We now perform the integral over $\dot{\mathbf{y}}$ in (40). We shall, from now on, neglect the expansion terms indicated by the dots, and obtain the path integral

$$U\left(\mathbf{x}_{+b}, \mathbf{x}_{-b}, t_b | \mathbf{x}_{+a}, \mathbf{x}_{-a}, t_a\right) = \int \frac{d^3p}{(2\pi)^3} \int \mathcal{D}\mathbf{x} \, \delta\left(\mathbf{x}(t_b) - \mathbf{x}_b\right) \tag{44}$$

$$\times \exp\left[-\frac{1}{2w} \int_{t_a}^{t_b} dt\, \boldsymbol{\eta}^2(t)\right] \exp\left\{-\frac{i}{\hbar}\left[\mathbf{y}_b \int_{t_a}^{t_b} dt\, \boldsymbol{\nabla}V(x) + \mathbf{p}(\mathbf{y}_b - \mathbf{y}_a)\right]\right\},$$

where $\boldsymbol{\eta}(t)$ depends on $\mathbf{x}(t)$ via (42). There are some virtues of this representation in comparison with the path integral (40), in particular, if forward and backward paths start out and end at the same points, such that $\mathbf{y}_a = \mathbf{y}_b = \mathbf{0}$: the oscillatory integral in (40) is transformed into a Gaussian integral (44) which converges exponentially fast. Let us note that from (42) with zero noise, we obtain the equation for the energy change of the particle

$$\frac{d}{dt}\left[\frac{M}{2}\dot{\mathbf{x}}^2 + V(\mathbf{x}) - M\gamma\dot{\mathbf{x}}\ddot{\mathbf{x}}\right] = -M\gamma\,\ddot{\mathbf{x}}^2. \tag{45}$$

The right-hand side is the classical electromagnetic power radiated by an accelerated particle. The extra term in the brackets is known as *Schott term* [10].

As an application of the Langevin equation we show how to express the evolution of the density matrix (53) in terms of the Wigner function defined by the Fourier transform

$$W(\mathbf{x}, \mathbf{p}; t) = \left(\frac{1}{2\pi\hbar}\right)^3 \int d^3y\, e^{i\mathbf{p}\mathbf{y}/\hbar} \rho(\mathbf{x}, \mathbf{y}; t). \tag{46}$$

Here and in the sequel, we omit subscripts b from t for brevity. Then (44) yields the time evolution equation for the Wigner function as a functional integral

$$W(\mathbf{x}, \mathbf{p}; t) = \int \mathcal{D}\mathbf{x}\, \exp\left[-\frac{1}{2w} \int_{t_a}^{t} dt'\, \boldsymbol{\eta}^2(t')\right]$$

$$\times W\left(\mathbf{x}(t), \mathbf{p} - \int_{t_a}^{t} dt'\, \boldsymbol{\nabla}V(\mathbf{x}(t')); t_a\right). \tag{47}$$

This equation has a simple physical interpretation. In the limit $T \to 0$, the functional integral (47) is concentrated around $\boldsymbol{\eta}(t') \equiv \mathbf{0}$, corresponding to a deterministic solution of (42) with $\boldsymbol{\eta}(t') \equiv \mathbf{0}$. In this limit, we obtain from (47) the Wigner function

$$W(\mathbf{x}, \mathbf{p}; t) = W\left(\mathbf{x}(t), \mathbf{p} - \int_{t_a}^{t} dt'\, \boldsymbol{\nabla}V(\mathbf{x}(t')); t_a\right). \tag{48}$$

If the particle is decoupled from the bath, $\gamma = 0$, we have $\mathbf{p} = M\dot{\mathbf{x}}(t_a)$ and $M\dot{\mathbf{x}}(t) = \mathbf{p} - \int_{t_a}^{t} dt\, \boldsymbol{\nabla}V(\mathbf{x}(t))$, and we see that the time evolution of the Wigner function is given by the Liouville equation

$$\partial_t W + \frac{1}{M}\mathbf{p} \cdot \boldsymbol{\nabla}_\mathbf{x}W - \boldsymbol{\nabla}_\mathbf{x}V \cdot \boldsymbol{\nabla}_\mathbf{p}W = 0. \tag{49}$$

The time evolution kernel can also be expressed as a path integral over the noise variable $\boldsymbol{\eta}(t)$. For this simply change the integration variable in (47) from $\mathbf{x}(t)$ to $\boldsymbol{\eta}(t)$. It can be shown that the functional determinant for a change of variables $\mathbf{x}(t) \to \boldsymbol{\eta}(t)$ is a constant (see [11]). Hence, we can rewrite (47) as

$$W(\mathbf{x}, \mathbf{p}; t) = \left\langle W\left(\mathbf{x}(t), \mathbf{p} - \int_{t_a}^{t} dt'\, \boldsymbol{\nabla} V(\mathbf{x}(t')), t_a\right)\right\rangle_{\boldsymbol{\eta}}, \tag{50}$$

where the average with respect to $\boldsymbol{\eta}(t)$ fluctuations is performed with the functional integral

$$\langle \dots \rangle_{\boldsymbol{\eta}} \equiv \int \mathcal{D}\mathbf{x} \dots \exp\left[-\frac{1}{2w}\int_{t_a}^{t_b} dt\, \boldsymbol{\eta}^2(t)\right]. \tag{51}$$

The calculation of (50) proceeds by solving first the Langevin equation (42) with the boundary conditions (43) to obtain the solution $\mathbf{x}(t)$, and subsequently take the expectation value with respect to the white noise with the correlation function

$$\langle \eta^i(t)\eta^j(t')\rangle = w\,\delta^{ij}\delta(t-t'). \tag{52}$$

6   Path Integral for Density Matrix in a Gravitational Environment

It is believed that quantum effects are negligible for objects of a large mass which consist of a great number of particles. However, in some extreme conditions such objects can be of microscopic size. Then the weak gravitational forces may sum up coherently. The process of a collapse of a star will be accompanied by an emission of gravitational waves which should be treated quantum mechanically in the final stages. Here, we discuss a gravitational object of this kind when interacting with an equilibrium state of gravitons.

Consider the Schrödinger equation

$$i\hbar\partial_t \langle \mathbf{x}|\psi(t)\rangle = \hat{H}(t)\langle \mathbf{x}|\psi(t)\rangle, \tag{53}$$

where $\hat{H}(t) = -\hbar^2\Delta/2M + V(\mathbf{x})$ is the Hamilton operator formed from the Laplace-Beltrami operator $\Delta = g^{-1/2}\left(\partial_\mu g^{\mu\nu} g^{1/2}\partial_\nu\right)$ containing the inverse $g^{\mu\nu}$ of the metric $g_{\mu\nu}(\mathbf{x}, t)$ and the determinant $g = \det(g_{\mu\nu})$. In empty space, small fluctuations $2\varepsilon\, u_{\mu\nu}(x)$ of the metric around the Minkowski metric $\eta_{\mu\nu}$ describe gravitational waves which may be quantized as usual. In the linear approximation, the Einstein-Hilbert action reads

$$\mathcal{A}_{\mathrm{E}} = -\frac{1}{2\varepsilon^2}\int d^4x \sqrt{-g}\, R$$

$$= -\frac{1}{2}\int d^4x\, G^{\mu\nu} u_{\mu\nu}, \quad G^{\mu\nu} \equiv \epsilon_{\mu\lambda\sigma\tau}\epsilon_{\nu\lambda\kappa\delta}\partial_\sigma\partial_\kappa u_{\tau\delta}, \tag{54}$$

where $G^{\mu\nu}$ is the linearized Einstein tensor $R^{\mu\nu} - g^{\mu\nu}R/2$, and ε a gravitational coupling constant related to Newton's gravitational constant $\varepsilon^2 = 8\pi G/c^3$. In the radiation gauge with $u^{0a} \equiv 0$, the free fields may be expanded into creation and annihilation operators for gravitons as

$$\hat{u}_{\mu\nu}(x) = \sum_{\mathbf{k}} \sum_{h=-2,2} \frac{1}{\sqrt{2V\Omega_{\mathbf{k}}}} \left[\epsilon_{\mu\nu}(\hat{\mathbf{k}}, h)\hat{a}(\hat{\mathbf{k}}, h)e^{-ikx} + \text{h.c.} \right], \tag{55}$$

where $\epsilon_{\mu\nu}(\hat{\mathbf{k}}, \pm 2)$ are the traceless divergence-free polarization tensors of helicity ± 2. The correlation functions at temperature T of these fields are:

$$\langle \hat{T}\hat{u}_{ij}(\mathbf{x}, t)\hat{u}_{kl}(\mathbf{x}', t') \rangle \equiv G_{ij;kl}(x, x')$$
$$= c\hbar \int \frac{d^3k}{(2\pi)^3\Omega_{\mathbf{k}}} \Lambda_{ij;kl}(\hat{\mathbf{k}}) \cos \mathbf{k}(\mathbf{x}-\mathbf{x}')$$
$$\times \left[\cos \Omega_{\mathbf{k}}(t-t') \coth \frac{\beta\hbar\Omega_{\mathbf{k}}}{2} - i \sin \Omega_{\mathbf{k}}(t-t') \right], \tag{56}$$

where $\Omega_k = c|\mathbf{k}|$ and

$$\Lambda_{ij,kl}(\hat{\mathbf{k}}) \equiv \sum_{h=-2,2} \epsilon(\hat{\mathbf{k}}, h)\epsilon(\hat{\mathbf{k}}, h)^*$$
$$= \tfrac{1}{2} \left[\Lambda_{ik}(\hat{\mathbf{k}})\Lambda_{jl}(\hat{\mathbf{k}}) + \Lambda_{il}(\hat{\mathbf{k}})\Lambda_{jk}(\hat{\mathbf{k}}) - \Lambda_{ij}(\hat{\mathbf{k}})\Lambda_{kl}(\hat{\mathbf{k}}) \right], \tag{57}$$

is the projection tensor to the physical polarization states, expressed in terms of transverse projection matrices $\Lambda_{ij}(\hat{\mathbf{k}}) = \delta_{ij} - k_i k_j/\mathbf{k}^2$ of electromagnetism. Here, the operator $\hat{T}$ is the time-ordering operator.

We can solve the evolution equation for the density matrix by means of the influence functional (as in Sect. 2),

$$U\left(\mathbf{x}_{+b}\mathbf{x}_{-b}, t_b | \mathbf{x}_{+a}\mathbf{x}_{-a}, t_a\right) =$$
$$\int \mathcal{D}\mathbf{x}_+ \mathcal{D}\mathbf{x}_- \exp\left\{ \frac{i}{\hbar} \int_{t_a}^{t_b} dt' \left[\frac{M}{2}\left[\dot{\mathbf{x}}_+^2 - \dot{\mathbf{x}}_-^2 \right] - V(\mathbf{x}_+) + V(\mathbf{x}_-) \right] \right\}$$
$$\times \exp\left[-\frac{\epsilon^2 M^2}{2\hbar^2} \int_{t_a}^{t_b} dt \int_{t_a}^{t} dt' \left(\dot{x}_+^k \dot{x}_+^l - \dot{x}_-^k \dot{x}_-^l \right) A_{kl;mn} \left(\dot{x}_+^{'m} \dot{x}_+^{'n} - \dot{x}_-^{'m} \dot{x}_-^{'n} \right) \right.$$
$$\left. - \frac{\epsilon^2 M^2}{2\hbar^2} \int_{t_a}^{t_b} dt \int_{t_a}^{t} dt' \left(\dot{x}_+^k \dot{x}_+^l - \dot{x}_-^k \dot{x}_-^l \right) C_{kl;mn} \left(\dot{x}_+^{'m} \dot{x}_+^{'n} - \dot{x}_-^{'m} \dot{x}_-^{'n} \right) \right], \tag{58}$$

where x' stands for $x(t')$ and $A_{kl;mn}$ and $C_{kl;mn}$ are twice the real and imaginary parts of (56). We are assuming the wavelengths to be much larger than the size of the object, which allows us to neglect the $\mathbf{x}$-dependence in $A_{kl;mn}$ and $C_{kl;mn}$ [12]. Then we take the angular average

$$\langle \Lambda_{kl;mn}(\hat{\mathbf{k}}) \rangle = \Lambda_{kl;mn} \equiv \tfrac{2}{5} \left[\tfrac{1}{2}(\delta_{km}\delta_{ln} + \delta_{kn}\delta_{lm}) - \tfrac{1}{3}\delta_{kl}\delta_{mn} \right], \tag{59}$$

and find

$$C_{kl;mn} \approx i\frac{\hbar}{2\pi c^2} \Lambda_{kl;mn} \, \partial_t \delta(t - t'), \tag{60}$$

and at high temperature:

$$A_{kl;mn} = \frac{\hbar}{2\pi c^2} \frac{1}{\beta\hbar} \Lambda_{kl;mn} \, \delta(t - t'). \tag{61}$$

Introducing the traceless dimensionless tensor

$$q^{kl} \equiv \frac{1}{c^2} \left(\dot{x}^k \dot{x}^l - \frac{1}{3} \delta^{kl} \dot{x}_i \dot{x}_i \right), \tag{62}$$

the second exponent can be rewritten as

$$-i\frac{\gamma}{2} \int_{t_a}^{t_b} dt \, (q_+^{kl} - q_-^{kl}) \partial_t (q_+^{kl} + q_-^{kl}) - \frac{w}{2} \int_{t_a}^{t_b} dt \, (q_+^{kl} - q_-^{kl})^2, \tag{63}$$

where

$$\gamma \equiv \frac{\varepsilon^2 M^2 c^2}{10\pi\hbar}, \qquad w \equiv 2\gamma \frac{k_B T}{\hbar}. \tag{64}$$

7 Master Equation for Density Matrix in an Environment of Gravitons

As in Sect. 2 we can now derive an operator differential equation for the time evolution of the density matrix. We denote the Hamilton operator without gravitational field by

$$\hat{H} = \frac{\hat{\mathbf{p}}^2}{2M} + V(\mathbf{x}). \tag{65}$$

The resulting master equation has the form

$$\partial_t \hat{\rho} = -\frac{i}{\hbar} \mathcal{H} \hat{\rho} \equiv -\frac{i}{\hbar}[\hat{H}, \hat{\rho}] - i\frac{\gamma}{2} \left(\hat{q}^{kl} \hat{\dot{q}}^{kl} \hat{\rho} - \hat{\rho} \, \hat{\dot{q}}^{kl} \hat{q}^{kl} + \hat{q}^{kl} \hat{\rho} \, \hat{\dot{q}}^{kl} - \hat{\dot{q}}^{kl} \hat{\rho} \, \hat{q}^{kl} \right)$$
$$- \frac{w}{2} \left\{ [\hat{q}^{kl}, [\hat{q}^{kl}, \hat{\rho}]] + \frac{\beta^2 \hbar^2}{12} [\hat{\dot{q}}^{kl}, [\hat{\dot{q}}^{kl}, \hat{\rho}]] \right\}. \tag{66}$$

where $\hat{\dot{x}}^k$, $\hat{\ddot{x}}^k$, $\hat{\dot{q}}^{kl}$, ... are defined recursively by the commutators $\hat{\dot{x}}^k \equiv i[\hat{\mathcal{H}}, \hat{x}^k]/\hbar$, $\hat{\ddot{x}}^k \equiv i[\hat{\mathcal{H}}, \hat{\dot{x}}^k]/\hbar$, $\hat{\dot{q}}^{kl} \equiv i[\hat{\mathcal{H}}, \hat{q}^{kl}]/\hbar$, ...

This equation will conserve the positivity of the probability as can be assured by bringing it to the extended *Lindblad form* [6] (as in the electromagnetic case the choice of the Lindblad operators L_n is not unique; we make a particular choice here)

$$\partial_t \hat{\rho} = -\frac{i}{\hbar}[\hat{H}_\gamma, \hat{\rho}] - \sum_{n=1}^{2} \sum_{kl} \left(\frac{1}{2} \hat{L}_n^{kl} \hat{L}_n^{kl\dagger} \hat{\rho} + \frac{1}{2} \hat{\rho} \, \hat{L}_n^{kl} \hat{L}_n^{kl\dagger} - \hat{L}_n^{kl\dagger} \hat{\rho} \, \hat{L}_n^{kl} \right). \tag{67}$$

with the Lindblad operators

$$\hat{L}_1^{kl} \equiv \frac{\sqrt{w}}{2} \hat{q}^{kl}, \qquad \hat{L}_2^{kl} \equiv \frac{\sqrt{3w}}{2} \left(\hat{q}^{kl} - i\frac{\hbar}{6k_B T} \hat{\dot{q}}^{kl} \right), \tag{68}$$

and

$$\hat{H}_\gamma = \frac{\hat{\mathbf{p}}^2}{2M} + V(\mathbf{x}) + \frac{M\gamma}{4} (\hat{q}^{kl} \hat{\dot{q}}^{kl} + \hat{\dot{q}}^{kl} \hat{q}^{kl} \sum_{kl}). \tag{69}$$

The master equation (66) allows us to calculate the rate of change of an initial eigenstate $|i\rangle$ of the Schrödinger equation $H|i\rangle = E_i|i\rangle$ by emission and absorption of gravitons to lowest order in G. It is a purely classical rate, determined by the matrix elements of the γ-terms in (66) (see [9] for more details):

$$\partial_t\tilde\rho_i$$
$$= -\tfrac{\gamma}{2}\Big[\langle i|\hat q_{kl}|f\rangle\langle f|\hat q_{kl}|i\rangle - \langle i|\hat q_{kl}|f\rangle\langle f|\hat q_{kl}|i\rangle - \langle i|\hat q_{kl}|i\rangle\langle i|\hat q_{kl}|i\rangle + \langle i|\hat q_{kl}|i\rangle\langle i|\hat q_{kl}|i\rangle\Big]$$
$$= -\gamma\sum_{f\neq i}\omega_{if}\langle i|\hat q_{kl}|f\rangle\langle f|\hat q_{kl}|i\rangle. \tag{70}$$

The part of the sum with $f < i$, where the final energy is lower than the initial energy, yields the spontaneous decay rate of the system:

$$\Gamma = 2\gamma\sum_{f<i}\omega_{if}\langle i|\hat q_{kl}|f\rangle\langle f|\hat q_{kl}|i\rangle. \tag{71}$$

For more details of this identification, in particular the origin of the factor 2, see [9]. This decay rate coincides with the perturbative quantum-mechanical result obtained from Fermi's golden rule and the interaction operator $\frac{1}{M}\int_{t_q}^{t_b} u_{kl}\hat p_k\hat p_l$:

$$\Gamma = \frac{\varepsilon^2}{8\pi^2\hbar}\frac{1}{c^2M^2}\int d\Omega \sum_{h=-2,2}\sum_{f<i}\omega_{if}\left|\epsilon^{kl}(\hat{\mathbf{k}},h)\langle f|\hat p_k\hat p_l|i\rangle\right|^2$$
$$= \frac{\varepsilon^2}{2\pi\hbar}\frac{2}{5c^2M^2}\sum_{f<i}\omega_{if}\left|\langle f|\hat p_k\hat p_l - \tfrac{1}{3}\mathbf{p}^2\delta_{kl}|i\rangle\right|^2. \tag{72}$$

In terms of Γ, the right-hand side of (70) can be written as

$$\partial_t\tilde\rho_i = -\Gamma + \gamma\sum_{f<i}\omega_{if}\langle i|\hat q_{kl}|f\rangle\langle f|\hat q_{kl}|i\rangle - \gamma\sum_{f>i}\omega_{if}\langle i|\hat q_{kl}|f\rangle\langle f|\hat q_{kl}|i\rangle. \tag{73}$$

By spontaneous emission, $\hat\rho_i$ decays like $e^{-\Gamma t}$. The rest is due to induced emissions and absorptions contributing to the rate of change of $\hat\rho_i$.

The total level width is increased further by the induced emissions and absorptions proportional to w in (66):

$$\partial_t\rho_i = -w\Big\{\sum_f\langle i|\hat q_{kl}|f\rangle\langle f|\hat q_{kl}|i\rangle - \langle i|\hat q_{kl}^2|i\rangle^2$$
$$+\frac{\beta^2\hbar^2}{12}\Big[\sum_f\langle i|\hat q_{kl}|f\rangle\langle f|\hat q_{kl}|i\rangle - \langle i|\hat q_{kl}^2|i\rangle^2\Big]\Big\}. \tag{74}$$

These are due to further induced emission and absorption processes.

Only the last term in this expressions is caused by the quantum nature of the emitted gravitons. In fact, (72) agrees with the classical result for the emitted power by inserting an energy per graviton $\hbar\omega_{if}$ into the sum and replacing ω_{if}^2 times the square of the matrix elements by the classical expression $(M^{-2}\partial_t q_{kl})^2/2$.

8   Langevin Equation in a Gravitational Environment

At high and moderately high temperatures, the fluctuations of $\mathbf{y}$ are small due to the last term in the exponent of (58). We therefore introduce average and difference variables of the forward and backward paths in (58):

$$\mathbf{x} \equiv (\mathbf{x}_+ + \mathbf{x}_-)/2, \qquad \mathbf{y} = \mathbf{x}_+ - \mathbf{x}_-, \tag{75}$$

and expand $V(\mathbf{x}_\pm) = V(\mathbf{x} \pm \mathbf{y}/2)$ in $\mathbf{y}$ up to the third order. Then the exponent in (58) becomes

$$\frac{i}{\hbar} \int_{t_a}^{t_b} dt \left\{ M\dot{\mathbf{y}}\dot{\mathbf{x}} - \mathbf{y}\boldsymbol{\nabla} V(\mathbf{x}) - \tfrac{1}{24} y_i y_j y_k \nabla_i \nabla_j \nabla_k V(\mathbf{x}) \right.$$
$$-2\tfrac{\gamma\hbar}{c^4} \left[\dot{\mathbf{x}}^2(\ddot{\mathbf{x}}\mathbf{y}) + \tfrac{1}{3}(\dot{\mathbf{y}}\dot{\mathbf{x}})(\ddot{\mathbf{x}}\mathbf{x}) + \tfrac{1}{3}(\ddot{\mathbf{y}}\mathbf{y})(\dot{\mathbf{x}}\dot{\mathbf{y}}) + \tfrac{1}{4}\dot{\mathbf{y}}^2(\ddot{\mathbf{y}}\mathbf{x}) \right]$$
$$\left. +2i\,\tfrac{w\hbar}{c^4} \left[\dot{\mathbf{y}}^2\dot{\mathbf{x}}^2 + \tfrac{1}{3}(\dot{\mathbf{y}}\dot{\mathbf{x}})^2 \right] \right\}, \tag{76}$$

where fourth and higher orders in $\mathbf{y}(t)$ are neglected. Neglecting also the quadratic terms $\mathbf{y}(t)$, extremization with respect to $\mathbf{y}(t)$ gives the classical equation of motion with radiation damping

$$M\ddot{\mathbf{x}} + \boldsymbol{\nabla} V - 2\tfrac{\gamma\hbar}{c^4} \tfrac{d}{dt} \left[\dddot{\mathbf{x}}\,\dot{\mathbf{x}}^2 + \tfrac{1}{3}\dot{\mathbf{x}}\,(\ddot{\mathbf{x}}\dot{\mathbf{x}}) \right] = 0. \tag{77}$$

From this equation we can calculate the energy dissipation of the gravitational body due to the coupling to the bath of gravitons:

$$\frac{d}{dt} \left\{ \left[\frac{M\dot{\mathbf{x}}^2}{2} + V(\mathbf{x}) \right] - \tfrac{2}{3}\tfrac{\gamma\hbar}{c^4}\tfrac{d}{dt}(\dot{\mathbf{x}}^2)^2 \right\} = -2\tfrac{\gamma\hbar}{c^4} \left[\ddot{\mathbf{x}}^2\dot{\mathbf{x}}^2 + \tfrac{1}{3}(\ddot{\mathbf{x}}\dot{\mathbf{x}})^2 \right]. \tag{78}$$

Expressed in terms of q_{kl}, the right-hand side yields a rate of energy loss

$$\dot{E} = -\gamma\hbar\,\dot{q}_{kl}^2 = \tfrac{4G}{5}\,\dot{q}_{kl}^2. \tag{79}$$

This is the formula derived for classical gravitational radiation. It explains the energy loss which causes the shrinking of the orbit of a binary system seen in astronomical observations [13]. The extra term in brackets is the gravitational analog of Schott's term in the energy balance of electromagnetic radiation damping discussed in [9, 10]. We can conclude that the time-average of the energy decreases owing to the dissipative term on the r.h.s. of (78).

The quadratic fluctuations of $\mathbf{y}(t)$ in the path integral (44) are governed by the quadratic exponent (76). Let us introduce the noise η with the correlation function

$$\langle \eta_i(t)\eta_j(t') \rangle = \tfrac{4w\hbar^2}{3c^4}\delta_{ij}\delta(t - t'). \tag{80}$$

Then we obtain the following Langevin equation describing the effect of the thermal graviton fluctuations upon the motion of a body in a potential V,

$$M\ddot{\mathbf{x}} + \boldsymbol{\nabla}V(\mathbf{x}) - 2\frac{\gamma\hbar}{c^4}\frac{d}{dt}\left[\ddot{\mathbf{x}}\dot{\mathbf{x}}^2 + \frac{1}{3}\dot{\mathbf{x}}(\ddot{\mathbf{x}}\dot{\mathbf{x}})^2\right] = \frac{d}{dt}\Omega^{1/2}\boldsymbol{\eta}. \tag{81}$$

For a definition of Ω and a derivation of this equation we refer to [14].

Our derivation of the classical equation (78) from the quantum path integral is much simpler and more direct than the purely classical derivations in the literature. There one always starts from extended objects to find the forces exerted upon a body by its own gravitational radiation. The subsequent limit of a point particle is quite delicate. For reviews on the subject see [10, 15, 16].

Let us finally remark that similar results have been derived by other methods in [17, 18].

References

1. R.P. Feynman and F.L. Vernon: Ann. Phys. (N.Y.) **24**, 118 (1963)
2. R.P. Feynman and A.R. Hibbs: *Quantum Mechanics and Path Integrals* (McGraw Hill, New York 1965)
3. H. Kleinert: *Path Integrals in Quantum Mechanics, Statistics and Polymer Physics* (World Scientific, Singapore 1995);
   http://www.physik.fu-berlin.de/~kleinert/re0.html#b5
4. A.O. Caldeira and A.J. Leggett: Physica A **121**, 587 (1983)
5. L. Diosi: Europhys. Lett. **22**, 1 (1993)
6. G. Lindblad: Commun. Math. Phys. **48**, 119 (1976)
7. C.W. Gardiner and P. Zoller: *Quantum Noise*, 2nd ed. (Springer-Verlag, Berlin 2000)
8. A. Sandulescu and H. Scutaru: Ann. Phys. (N.Y.) **173**, 277 (1987)
9. Z. Haba and H. Kleinert: Eur. Phys. J. B **21**, 553 (2001)
10. F. Rohrlich: *Classical and Charged Particles* (Addison-Wesley, Reading, MA 1963); Phys. Rev. D **60**, 084017 (1999); Am. J. Phys. **68**, 1109 (2000)
11. Z. Haba and H. Kleinert: 'Langevin Equation for Particle in Thermal Photon Bath', Berlin preprint (2000)
12. A discussion of the **x**-dependence can be found in Z. Haba: Mod. Phys. Lett. A **15**, 1519 (2000)
13. R.A. Hulse and J.H. Taylor, Ap. J. **195**, L51 (1975)
14. Z. Haba and H. Kleinert: Int. Journ. Mod. Phys. A **17**, 3729 (2002)
15. Y. Mino, M. Sasaki and T. Tanaka: Phys. Rev. D **55**, 3457 (1997)
16. T.C. Quinn and R.M. Wald: Phys. Rev. D **56**, 3381 (1997)
17. S. Adler: Phys. Rev. D **62**, 117901 (2000)
18. E. Lisi, A. Marrone and D. Montanino: hep-ph/0002053
19. For more details on dimensional regularization in the present context see H. Kleinert: Ann. Phys. (N.Y.), in press; quant-ph/0008109
20. H. Kleinert: J. Math. Phys. **27**, 3003 (1986)
21. H. Kleinert: *Gauge Fields in Condensed Matter*, op.cit., Vol. II (World Scientific, Singapore 1989)
22. U. Weiss: *Quantum Dissipative Systems* (World Scientific, Singapore 1993)
23. C. Cohen-Tannoudji, J. Dupont-Roc and G. Grynberg: *Photons and Atoms : Introduction to Quantum Electrodynamics* (Wiley, New York 1992)

24. W.E. Cook and T.F. Gallagher: Phys. Rev. A **21**, 588 (1980);
T. Nakajima, P. Lambropoulos and H. Walther: Phys. Rev. A **56**, 5100 (1997);
G. Barton: J. Phys. B **20**, 879 (1987);
G.W. Ford, J.T. Lewis and R.F. O'Connell: Phys. Rev. A **34**, 2001 (1986)
25. H. Kleinert and A. Chervyakov: Phys. Lett. A **245**, 345 (1998), quant-ph/9803016;
J. Math. Phys. B **40**, 6044 (1999), physics/9712048. For more details see
Sects. 3.3.1, 3.5, 3.21, and 4.3 in the third edition of [3], which can be downlo-
aded from http://www.physik.fu-berlin.de/~kleinert/b3
26. See equation (4.99) in the third edition of [3] on the www

Dissipation, Emergent Quantization, and Quantum Fluctuations

Massimo Blasone[1,2], Petr Jizba[3], and Giuseppe Vitiello[2]

[1] Blackett Lab., Imperial College, Prince Consort Rd., London SW7 2BW, UK
[2] Dipartimento di Fisica "E.R. Caianiello", INFN – INFM, Università di Salerno, 84100 Salerno, Italy
[3] Institute of Theoretical Physics, University of Tsukuba, Ibaraki 305-8571, Japan

Abstract. We review some aspects of the quantization of the damped harmonic oscillator. We derive the exact action for a damped mechanical system in the frame of the path integral formulation of the quantum Brownian motion problem developed by Schwinger and by Feynman and Vernon. The doubling of the phase space degrees of freedom for dissipative systems and thermal field theories is discussed and the doubled variables are related to quantum noise effects. The 't Hooft proposal, according to which the loss of information due to dissipation in a classical deterministic system manifests itself in the quantum features of the system, is analyzed and the quantum spectrum of the harmonic oscillator is shown to originate from the dissipative character of the original classical deterministic system.

1 Introduction

Our purpose in this report is to derive the action for a damped mechanical system in the path integral formalism, to discuss the role of quantum fluctuations and to show that the loss of information due to dissipation manifests itself in the form of quantum noise effects.

A microscopic theory for a dissipative system must include the details of the processes responsible for dissipation. One would then start with a Hamiltonian that describes the system, the bath, and the system-bath interaction. The description of the original dissipative system is recovered by the reduced density matrix obtained by eliminating the bath variables which originate the damping and the fluctuations. In quantum mechanics canonical commutation relations are not preserved by time evolution due to damping terms. The role of fluctuating forces is in fact the one of preserving the canonical structure. However, the knowledge of the details of the processes inducing the dissipation may not always be available; these details may not be explicitly known and the dissipation mechanisms are sometimes globally described by such parameters as friction, resistance, viscosity etc. In some sense, such parameters are introduced in order to compensate for the information loss caused by dissipation.

Our discussion in the present paper is aimed at considering, from one side, the description of dissipative systems in the frame of the quantum Brownian motion as described by Schwinger [1] and by Feynman and Vernon [2], and

M. Blasone, P. Jizba, and G. Vitiello, Dissipation, Emergent Quantization, and Quantum Fluctuations, Lect. Notes Phys. **633**, 151–163 (2004)
http://www.springerlink.com/

from the other side, the suggestion put forward by 't Hooft in a recent series of papers [3, 4], according to which Quantum Mechanics may be an effective theory resulting from a more fundamental deterministic theory after a process of information loss has taken place. Some results recently obtained by us point to an intriguing underlying connection between our approach to dissipative systems, framed, as said, in the Schwinger and Feynman and Vernon formalism, and with a strong relation with the thermal field theory formalism of Takahashi and Umezawa [5–9], and 't Hooft's proposal.

In the next section we derive the exact action for a damped mechanical system (and the special case of the linear oscillator) from the path integral formulation of the quantum Brownian motion problem developed by Schwinger and by Feynman and Vernon. We will closely follow the references [10,11] in our discussion. The doubling of the phase space degrees of freedom for dissipative systems and thermal field theories is discussed and the doubled variables are related to quantum noise effects. In Sect. 3, the 't Hooft proposal is discussed and the loss of information due to dissipation in a classical deterministic system is shown to manifest in the quantum character of the spectrum of the harmonic oscillator. The geometric (Berry-Anandan-like) phase arising in dissipative systems is recognized and related to the zero-point energy of the quantum oscillator spectrum. The results obtained in [12] are there reported. In Sect. 4 we consider the connection with thermal observables recognizing the role played by the system free energy. Section 5 is devoted to further remarks and to conclusions.

For the sake of shortness we do not report on the coherent structure of the states of the quantum system. Details on that part and on other features of the approach here presented can be found in the literature (see, e.g. [7–9,13]). Also, we do not report on several applications of our formalism, ranging from the study of topologically massive gauge theories in the infrared region in $2 + 1$ dimensions [8], the Chern-Simons-like dynamics of Bloch electrons in solids [8], the expanding geometry model in inflationary cosmology [14], to the study of the quantum brain model [15–17] and non-commutative geometry [18].

2   The Exact Action for Damping

Our aim in this section is to obtain the exact action for a particle of mass m, damped by a mechanical resistance γ, moving in a potential V. To be definite, we consider the damped harmonic oscillator (dho)

$$m\ddot{x} + \gamma\dot{x} + \kappa x = 0, \tag{1}$$

as a simple prototype for dissipative systems. Our discussion and our results also apply, however, to more general systems than the one represented in (1).

The damped oscillator (1) is a non-hamiltonian system and the canonical formalism, which one needs in order to proceed to its quantization, cannot be set up [19]. Let us see, however, how one can face the problem by resorting to well known tools such as the density matrix and the Wigner function.

We start with the preliminary consideration of the special case of zero mechanical resistance. The Hamiltonian for an isolated particle reads

$$H = -\frac{\hbar^2}{2m} \left(\frac{\partial}{\partial x}\right)^2 + V(x). \tag{2}$$

On the other hand, it is useful to consider the Wigner function, whose standard expression is, see e.g. [20, 21],

$$W(p, x, t) = \frac{1}{2\pi\hbar} \int \psi^* \left(x - \tfrac{1}{2}y, t\right) \psi \left(x + \tfrac{1}{2}y, t\right) e^{\left(-i\frac{py}{\hbar}\right)} dy , \tag{3}$$

with the associated density matrix function

$$W(x, y, t) = \langle x + \tfrac{1}{2}y|\rho(t)|x - \tfrac{1}{2}y\rangle = \psi^* \left(x - \tfrac{1}{2}y, t\right) \psi \left(x + \tfrac{1}{2}y, t\right). \tag{4}$$

For an isolated particle one obtains the density matrix equation of motion

$$i\hbar\frac{\partial}{\partial t}\rho = [H, \rho]. \tag{5}$$

In the coordinate representation, employing

$$x_\pm = x \pm \tfrac{1}{2}y, \tag{6}$$

(5) reads

$$i\hbar\frac{\partial}{\partial t}\langle x_+|\rho(t)|x_-\rangle =$$
$$\left\{-\frac{\hbar^2}{2m}\left[\left(\frac{\partial}{\partial x_+}\right)^2 - \left(\frac{\partial}{\partial x_-}\right)^2\right] + [V(x_+) - V(x_-)]\right\} \langle x_+|\rho(t)|x_-\rangle, \tag{7}$$

which, in terms of x and y, is

$$i\hbar\frac{\partial}{\partial t}W(x, y, t) = \mathcal{H}_o W(x, y, t) \tag{8}$$

$$\mathcal{H}_o = \tfrac{1}{m}p_x p_y + V\left(x + \tfrac{1}{2}y\right) - V\left(x - \tfrac{1}{2}y\right), \tag{9}$$

$$p_x = -i\hbar\frac{\partial}{\partial x}, \quad p_y = -i\hbar\frac{\partial}{\partial y}. \tag{10}$$

Of course, the "Hamiltonian" (9) may be constructed from the "Lagrangian"

$$\mathcal{L}_o = m\dot{x}\dot{y} - V\left(x + \tfrac{1}{2}y\right) + V\left(x - \tfrac{1}{2}y\right). \tag{11}$$

Now let us suppose that the particle interacts with a thermal bath at temperature T. The interaction Hamiltonian between the bath and the particle is taken as

$$H_{\text{int}} = -fx, \tag{12}$$

where f is the random force on the particle at the position x due to the bath.

In the Feynman-Vernon formalism, the effective action for the particle has the form

$$\mathcal{A}[x, y] = \int_{t_i}^{t_f} dt \, \mathcal{L}_o(\dot{x}, \dot{y}, x, y) + \mathcal{I}[x, y], \tag{13}$$

where $\mathcal{L}_o$ is defined in (11) and

$$e^{\frac{i}{\hbar}\mathcal{I}[x,y]} = \langle (e^{-\frac{i}{\hbar}\int_{t_i}^{t_f} f(t)x_-(t)dt})_- (e^{\frac{i}{\hbar}\int_{t_i}^{t_f} f(t)x_+(t)dt})_+\rangle. \tag{14}$$

In (14), the average is with respect to the thermal bath; "$(.)_+$" and "$(.)_-$" denote time ordering and anti-time ordering, respectively; the c-number coordinates $x_\pm$ are defined as in (6). We observe that if the interaction between the bath and the coordinate x (i.e $H_{\text{int}} = -fx$) were turned off, then the operator f of the bath would develop in time according to $f(t) = e^{iH_\gamma t/\hbar} f e^{-iH_\gamma t/\hbar}$ where H_γ is the Hamiltonian of the isolated bath (decoupled from the coordinate x). $f(t)$ is the force operator of the bath to be used in (14).

The reduced density matrix function in (4) for the particle which first makes contact with the bath at the initial time t_i is given at a final time by

$$W(x_f, y_f, t_f) = \int_{-\infty}^{\infty} dx_i \int_{-\infty}^{\infty} dy_i\, K(x_f, y_f, t_f; x_i, y_i, t_i) W(x_i, y_i, t_i), \tag{15}$$

with the path integral representation for the evolution kernel

$$K(x_f, y_f, t_f; x_i, y_i, t_i) = \int_{x(t_i)=x_i}^{x(t_f)=x_f} \mathcal{D}x(t) \int_{y(t_i)=y_i}^{y(t_f)=y_f} \mathcal{D}y(t)\, e^{\frac{i}{\hbar}\mathcal{A}[x,y]}. \tag{16}$$

The evaluation of $\mathcal{I}[x, y]$ for a linear passive damping thermal bath requires the use of several Greens functions all of which have been discussed by Schwinger [1] and for shortness here we only mention that the fundamental correlation function for the random force on the particle due to the thermal bath is given by (see [10])

$$G(t - s) = \frac{i}{\hbar}\langle f(t)f(s)\rangle. \tag{17}$$

The retarded and advanced Greens functions are defined by

$$G_{\text{ret}}(t - s) = \theta(t - s)[G(t - s) - G(s - t)], \tag{18}$$

$$G_{\text{adv}}(t - s) = \theta(s - t)[G(s - t) - G(t - s)] . \tag{19}$$

The mechanical impedance $Z(\zeta)$ (analytic in the upper half complex frequency plane $\mathcal{I}m\,\zeta > 0$) is given by

$$-i\zeta Z(\zeta) = \int_0^{\infty} dt\, G_{\text{ret}}(t)e^{i\zeta t}. \tag{20}$$

and the quantum noise in the fluctuating random force is given by

$$N(t - s) = \frac{1}{2}\langle f(t)f(s) + f(s)f(t)\rangle, \tag{21}$$

and is distributed in the frequency domain in accordance with the Nyquist theorem

$$N(t - s) = \int_0^{\infty} d\omega\, S_f(\omega) \cos[\omega(t - s)], \tag{22}$$

$$S_f(\omega) = \frac{\hbar\omega}{\pi} \coth \frac{\hbar\omega}{2kT} \mathcal{R}eZ(\omega + i0^+). \tag{23}$$

The mechanical resistance is defined by

$$\gamma = \lim_{\omega \to 0} \mathcal{R}eZ(\omega + i0^+). \tag{24}$$

Equation (14) may be now evaluated following Feynman and Vernon [10] as,

$$\mathcal{I}[x, y] = \frac{1}{2} \int_{t_i}^{t_f} \int_{t_i}^{t_f} dt ds \, [G_{ret}(t - s) + G_{adv}(t - s)][x(t)y(s) + x(s)y(t)]$$

$$+ \frac{i}{2\hbar} \int_{t_i}^{t_f} \int_{t_i}^{t_f} dt ds N(t - s) y(t) y(s). \tag{25}$$

By defining the retarded force on y and the advanced force on x as

$$F_y^{\text{ret}}(t) = \int_{t_i}^{t_f} ds \, G_{ret}(t - s) y(s), \tag{26}$$

$$F_x^{\text{adv}}(t) = \int_{t_i}^{t_f} ds \, G_{adv}(t - s) x(s), \tag{27}$$

respectively, the interaction between the bath and the particle is then

$$\mathcal{I}[x, y] = \frac{1}{2} \int_{t_i}^{t_f} dt \, [x(t)F_y^{\text{ret}}(t) + y(t)F_x^{\text{adv}}(t)]$$

$$+ \frac{i}{2\hbar} \int_{t_i}^{t_f} \int_{t_i}^{t_f} dt ds \, N(t - s) y(t) y(s). \tag{28}$$

Thus the real and the imaginary part of the action are finally given by

$$\mathcal{R}e\mathcal{A}[x, y] = \int_{t_i}^{t_f} dt \, \mathcal{L}, \tag{29}$$

$$\mathcal{L} = m\dot{x}\dot{y} - \left[V(x + \tfrac{1}{2}y) - V(x - \tfrac{1}{2}y)\right] + \tfrac{1}{2}\left[xF_y^{\text{ret}} + yF_x^{\text{adv}}\right], \tag{30}$$

and

$$\mathcal{I}m\mathcal{A}[x, y] = \frac{1}{2\hbar} \int_{t_i}^{t_f} \int_{t_i}^{t_f} dt ds \, N(t - s) y(t) y(s). \tag{31}$$

respectively. Equations (29),(30),(31) are *rigorously exact* for linear passive damping due to the bath when the path integral (16) is employed for the time development of the density matrix.

The lesson we learn from our result is that in the classical limit "$\hbar \to 0$" nonzero y yields an "unlikely process" in view of the large imaginary part of the action implicit in (31). On the contrary, at quantum level nonzero y may allow quantum noise effects arising from the imaginary part of the action [10].

In conclusion, we can consider the approximation to (30) with $F_y^{\text{ret}} = \gamma\dot{y}$ and $F_x^{\text{adv}} = -\gamma\dot{x}$, i.e.

$$\mathcal{L}(\dot{x}, \dot{y}, x, y) = m\dot{x}\dot{y} - V\left(x + \tfrac{1}{2}y\right) + V\left(x - \tfrac{1}{2}y\right) + \tfrac{\gamma}{2}(x\dot{y} - y\dot{x}), \tag{32}$$

which implies the classical equations of motion

$$m\ddot{x} + \gamma\dot{x} + \tfrac{1}{2}\left[V'(x + \tfrac{1}{2}y) + V'(x - \tfrac{1}{2}y)\right] = 0, \tag{33}$$

$$m\ddot{y} - \gamma\dot{y} + V'(x + \tfrac{1}{2}y) - V'(x - \tfrac{1}{2}y) = 0. \tag{34}$$

It is easy to see that, for $V\left(x \pm \tfrac{1}{2}y\right) = \tfrac{1}{2}\kappa(x \pm \tfrac{1}{2}y)^2$, (35),(36) give the dho equation (1) and its complementary equation for the y coordinate

$$m\ddot{x} + \gamma\dot{x} + \kappa x = 0, \tag{35}$$

$$m\ddot{y} - \gamma\dot{y} + \kappa y = 0. \tag{36}$$

The y-oscillator is the time–reversed image of the x-oscillator. If from the manifold of solutions to (35),(36) we choose those for which the y coordinate is constrained to be zero, then (35),(36) simplify to

$$m\ddot{x} + \gamma\dot{x} + \kappa x = 0, \quad y = 0. \tag{37}$$

Thus we obtain a classical damped equation of motion from a Lagrangian theory at the expense of introducing an "extra" coordinate y, later constrained to vanish. Note that $y(t) = 0$ is a true solution to (35),(36) (and (33),(34)) so that the constraint is *not* in violation of the equations of motion.

We stress, however, that *the role of the "doubled" y coordinate is absolutely crucial in the quantum regime since there it accounts for the quantum noise as shown above*. This result leads us to consider carefully 't Hooft's proposal [3,4]. When, as customary, one adopts the classical (legitimate) solution $y = 0$, the x system appears to be open, "incomplete"; the loss of information due to dissipation essentially amounts to neglecting the bath and/or to the ignorance of specific features of the bath-system interaction, i.e. the ignorance of "where" and "how" energy flows out of the system. According to our result, reverting from the classical level to the quantum level, the loss of information occurring at the classical level due to dissipation manifests itself in terms of "quantum" noise effects arising from the imaginary part of the action, to which the y contribution is indeed crucial. In the next sections we analyze in some details our approach to dissipation in connection with 't Hooft proposal.

3   The Damped/Amplified Harmonic Oscillator System

To establish a link with 't Hooft's quantization scenario it is important to dwell a bit on some formal aspects of the damped–amplified harmonic oscillator system (35)–(36) [19]. To do this let us note first that by defining $x^\alpha = (x, y)$ the equations of motion (35)–(36) can be written in the compact form

$$m\ddot{\mathbf{x}} + \gamma\sigma_3\dot{\mathbf{x}} + \kappa\mathbf{x} = 0. \tag{38}$$

This suggest that under appropriate boundary conditions the y–oscillator is the time–reversed image of the x–oscillator. Introducing the metric tensor $g_{\alpha\beta} = (\sigma_1)_{\alpha\beta}$ the corresponding Lagrangian reads

$$\mathcal{L} = \tfrac{m}{2}\,\dot{\mathbf{x}}\dot{\mathbf{x}} + \tfrac{\gamma}{2}\,\dot{\mathbf{x}} \wedge \mathbf{x} - \tfrac{\kappa}{2}\,\mathbf{x}\mathbf{x}, \tag{39}$$

with an obvious notation $\mathbf{ab} = g_{\alpha\beta} a^\alpha b^\alpha$ and $\mathbf{a} \wedge \mathbf{b} = \varepsilon^{\alpha\beta} a_\alpha b_\beta$ $(\varepsilon^{\alpha\beta} = -\varepsilon_{\alpha\beta})$. It is convenient to reformulate the former in the rotated coordinate system, i.e.

$$x_1 = \tfrac{x+y}{\sqrt{2}}, \quad x_2 = \tfrac{x-y}{\sqrt{2}}.$$

In these coordinates the Lagrangian has the form

$$\mathcal{L} = \tfrac{m}{2}\ \dot{\mathbf{x}}\dot{\mathbf{x}} + \tfrac{\gamma}{2}\ \mathbf{x} \wedge \dot{\mathbf{x}} - \tfrac{\kappa}{2}\ \mathbf{xx}. \tag{40}$$

Here $x^\alpha = (x_1, x_2)$ and the metric tensor $g_{\alpha\beta} = (\sigma_3)_{\alpha\beta}$ (note the change of sign in the wedge product). Introducing the canonical momenta $p^\alpha = \partial\mathcal{L}/\partial\dot{x}_\alpha$ and $p_\alpha \equiv (p_1, p_2)$ we obtain

$$\mathbf{p} = m\ \dot{\mathbf{x}} - \tfrac{\gamma}{2}\ \sigma_1 \mathbf{x}, \tag{41}$$

and thus the corresponding Hamiltonian reads

$$H = \tfrac{1}{2m}\ \mathbf{p}^2 + \tfrac{\gamma}{2m}\ \mathbf{p} \wedge \mathbf{x} + \tfrac{1}{2}\left(\kappa - \tfrac{\gamma^2}{4m}\right)\mathbf{x}^2. \tag{42}$$

The key observation is that with the system (42) we can affiliate the $su(1,1)$ algebraic structure. Indeed, from the dynamical variables p_α and x^α one may construct the functions

$$J_1 = \tfrac{1}{2m\Omega}\ p_1 p_2 - \tfrac{m\Omega}{2}\ x_1 x_2,$$
$$J_2 = -\tfrac{1}{2}\ \mathbf{p} \wedge \mathbf{x} = -\tfrac{1}{2}\varepsilon^\alpha_\beta\ p_\alpha x^\beta = \tfrac{1}{2}\left(p_1 x_2 + p_2 x_1\right),$$
$$J_3 = \tfrac{1}{4m\Omega}\left(p_1^2 + p_2^2\right) + \tfrac{m\Omega}{4}\left(x_1^2 + x_2^2\right). \tag{43}$$

Here $\Omega = \sqrt{\tfrac{1}{m}(\kappa - \tfrac{\gamma^2}{4m})}$, with $\kappa > \tfrac{\gamma^2}{4m}$. Applying now the canonical Poisson brackets $\{p_\alpha, x^\beta\} = g^\beta_\alpha = \delta^\beta_\alpha$ we obtain Poisson's subalgebra

$$\{J_2, J_3\} = J_1, \quad \{J_3, J_1\} = J_2, \quad \{J_1, J_2\} = -J_3. \tag{44}$$

The algebraic structure (44) corresponds to $su(1,1)$ algebra [22]. The quadratic Casimir for the algebra (44) is defined as [22]

$$\mathcal{C} = \tfrac{1}{2}\left(J_1^2 + J_2^2 - J_3^2\right) = -\tfrac{1}{2}\left(\tfrac{\mathbf{p}^2 + m^2\Omega^2}{4m\Omega}\right)^2 \equiv -\tfrac{1}{2}\ C^2. \tag{45}$$

In terms of J_2 and the Casimir $\mathcal{C}$ the Hamiltonian (42) can be formulated as

$$H = 2(\Omega\mathcal{C} - \Gamma J_2), \tag{46}$$

with $\Gamma = \gamma/2m$. It might be shown [8] that when (42) is quantized then $SU(1,1)$ is the dynamical group of the system. A formal simplification occurs when the hyperbolic coordinates are introduced, i.e.

$$x_1 = r\cosh u, \quad x_2 = r\sinh u. \tag{47}$$

Then J_2 and $\mathcal{C}$ have a particularly simple structure [8], namely

$$\mathcal{C} = \tfrac{1}{4\Omega m}\left[p_r^2 - \tfrac{1}{r^2}p_u^2 + m^2\Omega^2 r^2\right], \quad J_2 = \tfrac{1}{2}p_u, \tag{48}$$

Let us finally note that the dynamical system described by the Lagrangian (39) is sometimes called Bateman's dual system [19].

4   Deterministic Dissipative Systems and Quantization

In a recent series of papers [3, 4], G. 't Hooft has put forward the idea that
Quantum Mechanics may result from a more fundamental deterministic theory,
after that a process of information loss has taken place. He has found a class of
Hamiltonian systems which remain deterministic, even when they are described
by means of Hilbert space techniques. The truly quantum systems are obtained
when constraints are imposed on the original Hilbert space: these constraints
implement the information loss.

More specifically, the Hamiltonian for such systems is of the form

$$H = \sum_i p_i \, f_i(q) \,, \tag{49}$$

where $f_i(q)$ are non–singular functions of the canonical coordinates q_i. The crucial point is that equations for the q's (i.e. $\dot{q}_i = \{q_i, H\} = f_i(q)$) are decoupled
from the conjugate momenta p_i and this implies [3] that the system can be described deterministically even when expressed in terms of operators acting on
the Hilbert space. The condition for the deterministic description is the existence
of a complete set of observables commuting at all times, called *beables* [23] - a
condition which is guaranteed for the systems of (49), for which such a set is
given by the $q_i(t)$ [3].

A problem with the above mentioned class of Hamiltonians is that they are
not bounded from below. This might be cured by splitting H in (49) as [3]:

$$H = H_1 - H_2 \quad , \quad H_1 = \tfrac{1}{4\rho}\left(\rho + H\right)^2 \quad , \quad H_2 = \tfrac{1}{4\rho}\left(\rho - H\right)^2 \,, \tag{50}$$

with ρ a certain time–independent, positive function of q_i. As a result, H_1 and
H_2 are positively (semi)definite and $\{H_1, H_2\} = \{\rho, H\} = 0$.

To get the lower bound for the Hamiltonian one thus imposes the constraint
condition onto the Hilbert space:

$$H_2|\psi\rangle = 0 \,, \tag{51}$$

which projects out the states responsible for the negative part of the spectrum.
In the deterministic language this means that one gets rid of the unstable trajectories [3]. In the line of 't Hooft's proposal, it has been shown [24] that a
reparametrization-invariant time technique in a specific model also leads to a
quantum dynamics emerging from a deterministic classical evolution. Deterministic models with discrete time evolution have been recently studied in [25].

We now show that the above discussed system of damped-antidamped oscillators does provide indeed an explicit realization of 't Hooft's mechanism. Furthermore, we shall see that there is a connection between the zero-point energy of
the quantum harmonic oscillator and the geometric phase of the (deterministic)
system of damped/antidamped oscillators.

The first thing we need to realize is that the Hamiltonian of our model belongs
to the same class of the Hamiltonians considered by 't Hooft. Indeed (46) can

be then rewritten as [12]:

$$H = \sum_{i=1}^{2} p_i \, f_i(q) , \qquad (52)$$

with $f_1(q) = 2\Omega$, $f_2(q) = -2\Gamma$, provided we use the canonical transformation:

$$q_1 = \int \frac{dz \; m\Omega}{\sqrt{4J_2^2 + 4m\Omega C z - m^2\Omega^2 z^2}} , \qquad (53)$$

$$q_2 = 2u + \int \frac{dz}{z} \frac{2J_2}{\sqrt{4J_2^2 + 4m\Omega C z - m^2\Omega^2 z^2}} , \qquad (54)$$

$$p_1 = C , \qquad p_2 = J_2 , \qquad (55)$$

with $z = r^2$. One has $\{q_i, p_i\} = 1$, and the other Poisson brackets vanishing. Thus J_2 and C are beables. Yet also q_1 and q_2 are beables as it can be directly seen from the Hamiltonian (52).

Now we set

$$H = H_I - H_{II} \quad , \quad H_I = \tfrac{1}{2\Omega C}(2\Omega C - \Gamma J_2)^2 \quad , \quad H_{II} = \tfrac{\Gamma^2}{2\Omega C} J_2^2 . \qquad (56)$$

Of course, only nonzero r^2 should be taken into account in order for C to be invertible. Note that C is a constant of motion (being the Casimir operator): this ensures that once it has been chosen to be positive, as we do from now on, it will remain such at all times.

We then implement the constraint

$$J_2|\psi\rangle = 0 , \qquad (57)$$

which defines the physical states. Although the system (52), i.e.(46), is deterministic, $|\psi\rangle$ is not an eigenvector of u (u does not commute with p_u). Of course, if one does not use the operatorial formalism to describe our system, then $p_u = 0$ implies $u = -\frac{\gamma}{2m}t$. Equation (57) implies

$$H|\psi\rangle = H_I|\psi\rangle = 2\Omega C|\psi\rangle = \left(\tfrac{1}{2m}p_r^2 + \tfrac{K}{2}r^2\right)|\psi\rangle , \qquad (58)$$

where $K \equiv m\Omega^2$. H_I thus reduces to the Hamiltonian for the linear harmonic oscillator $\ddot{r} + \Omega^2 r = 0$. The physical states are even with respect to time-reversal ($|\psi(t)\rangle = |\psi(-t)\rangle$) and periodical with period $\tau = \frac{2\pi}{\Omega}$.

Having denoted with $|\psi\rangle$ the physical states, we now introduce the states $|\psi(t)\rangle_H$ and $|\psi(t)\rangle_{H_I}$ satisfying the equations:

$$i\hbar \tfrac{d}{dt}|\psi(t)\rangle_H = H \, |\psi(t)\rangle_H , \qquad (59)$$

$$i\hbar \tfrac{d}{dt}|\psi(t)\rangle_{H_I} = 2\Omega C|\psi(t)\rangle_{H_I} . \qquad (60)$$

Equation (60) describes the 2D "isotropic" (or "radial") harmonic oscillator. $H_I = 2\Omega C$ has the spectrum $\mathcal{H}_I^n = \hbar\Omega n$, $n = 0, \pm 1, \pm 2, \dots$ According to our choice for C to be positive, only positive values of n will be considered.

The generic state $|\psi(t)\rangle_H$ can be written as

$$|\psi(t)\rangle_H = \hat{T}\left[\exp\left(\frac{i}{\hbar}\int_{t_0}^t 2\Gamma J_2 dt'\right)\right]|\psi(t)\rangle_{H_I} , \qquad (61)$$

where $\hat{T}$ denotes time-ordering. Note that here $\hbar$ is introduced on purely dimensional grounds and its actual value cannot be fixed by the present analysis.

We obtain [12]:

$$_H\langle\psi(\tau)|\psi(0)\rangle_H = {}_{H_I}\langle\psi(0)|\exp\left(i\int_{C_{0\tau}} A(t')dt'\right)|\psi(0)\rangle_{H_I} \equiv e^{i\phi} , \qquad (62)$$

where the contour $C_{0\tau}$ is the one going from $t' = 0$ to $t' = \tau$ and back and $A(t) \equiv \frac{\Gamma m}{\hbar}(\dot{x}_1 x_2 - \dot{x}_2 x_1)$. Note that $(\dot{x}_1 x_2 - \dot{x}_2 x_1)dt$ is the area element in the (x_1, x_2) plane enclosed by the trajectories (see Fig. 1). Notice also that the evolution (or dynamical) part of the phase does not enter in ϕ, as the integral in (62) picks up a purely geometric contribution [26].

Because the physical states $|\psi\rangle$ are periodic ones, let us focus our attention on those. Following [26], one may generally write

$$|\psi(\tau)\rangle = e^{i\phi - \frac{i}{\hbar}\int_0^\tau \langle\psi(t)|H|\psi(t)\rangle dt}|\psi(0)\rangle = e^{-i2\pi n}|\psi(0)\rangle , \qquad (63)$$

i.e. $\frac{\langle\psi(\tau)|H|\psi(\tau)\rangle}{\hbar}\tau - \phi = 2\pi n$, $n = 0, 1, 2, \ldots$, which by using $\tau = \frac{2\pi}{\Omega}$ and $\phi = \alpha\pi$, gives

$$\mathcal{H}^n_{I,\text{eff}} \equiv \langle\psi_n(\tau)|H|\psi_n(\tau)\rangle = \hbar\Omega\left(n + \frac{\alpha}{2}\right) , \qquad (64)$$

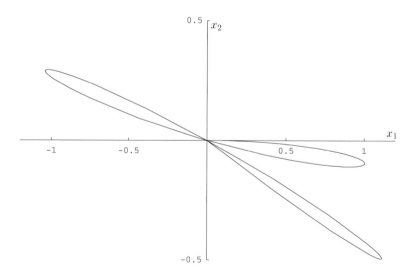

Fig. 1. Trajectories for $r_0 = 0$ and $v_0 = \Omega$, after three half-periods for $\kappa = 20$, $\gamma = 1.2$ and $m = 5$. The ratio $\int_0^{\tau/2}(\dot{x}_1 x_2 - \dot{x}_2 x_1)dt/\mathcal{E} = \pi\frac{\Gamma}{m\Omega^3}$ is preserved. $\mathcal{E}$ is the initial energy: $\mathcal{E} = \frac{1}{2}mv_0^2 + \frac{1}{2}m\Omega^2 r_0^2$

where the index n has been introduced to exhibit the n dependence of the state and the corresponding energy. $\mathcal{H}^n_{i,\text{eff}}$ gives the effective nth energy level of the physical system, namely the energy given by $\mathcal{H}^n_i$ corrected by its interaction with the environment. We thus see that the dissipation term J_2 of the Hamiltonian is actually responsible for the "zero-point energy" ($n = 0$): $E_0 = \frac{\hbar}{2}\Omega\alpha$.

As well known, the zero-point energy is the "signature" of quantization since in Quantum Mechanics it is formally due to the non-zero commutator of the canonically conjugate q and p operators. Thus dissipation manifests itself as "quantization". In other words, E_0, which appears as the "quantum contribution" to the spectrum of the conservative evolution of physical states, signals the underlying dissipative dynamics. If we want to match the Quantum Mechanics zero-point energy, we have to fix $\alpha = 1$, which gives [12] $\Omega = \frac{\gamma}{m}$.

5   Thermodynamics

In order to better understand the dynamical role of J_2 we rewrite (61) as

$$|\psi(t)\rangle_H = \hat{T}\left[\exp\left(i\frac{1}{\hbar}\int_{u(t_0)}^{u(t)} 2J_2 du'\right)\right]|\psi(t)\rangle_{H_I},\qquad (65)$$

by using $u(t) = -\Gamma t$. Accordingly, we have

$$-i\hbar\frac{\partial}{\partial u}|\psi(t)\rangle_H = 2J_2|\psi(t)\rangle_H.\qquad (66)$$

We thus see that $2J_2$ is responsible for shifts (translations) in the u variable, as is to be expected since $2J_2 = p_u$ (cf. (48)). In operatorial notation we can write indeed $p_u = -i\hbar\frac{\partial}{\partial u}$. Then, in full generality, equation (57) defines families of physical states, representing stable, periodic trajectories (cf. (58)). $2J_2$ implements transition from family to family, according to (66). Equation (59) can be then rewritten as

$$i\hbar\frac{d}{dt}|\psi(t)\rangle_H = i\hbar\frac{\partial}{\partial t}|\psi(t)\rangle_H + i\hbar\frac{du}{dt}\frac{\partial}{\partial u}|\psi(t)\rangle_H,\qquad (67)$$

where the first term on the r.h.s. denotes of course derivative with respect to the explicit time dependence of the state. The dissipation contribution to the energy is thus described by the "translations" in the u variable. It is then interesting to consider the derivative

$$\frac{\partial S}{\partial U} = \frac{1}{T}.\qquad (68)$$

From (52), by using $S \equiv \frac{2J_2}{\hbar}$ and $U \equiv 2\Omega\mathcal{C}$, we obtain $T = \hbar\Gamma$. Equation (68) is the defining relation for temperature in thermodynamics (with $k_B = 1$) so that one could formally regard $\hbar\Gamma$ (which dimensionally is an energy) as the temperature, provided the dimensionless quantity S is identified with the entropy. In such a case, the "full Hamiltonian" (52) plays the role of the free energy $\mathcal{F}$: $H = 2\Omega\mathcal{C} - (\hbar\Gamma)\frac{2J_2}{\hbar} = U - TS = \mathcal{F}$. Thus $2\Gamma J_2$ represents the heat contribution in H (or $\mathcal{F}$). Of course, consistently, $\frac{\partial \mathcal{F}}{\partial T}\big|_{\Omega} = -\frac{2J_2}{\hbar}$. In conclusion $\frac{2J_2}{\hbar}$ behaves as the entropy, which is not surprising since it controls the dissipative (thus irreversible) part of the dynamics.

We can also take the derivative of $\mathcal{F}$ (keeping T fixed) with respect to Ω. We then have

$$\frac{\partial \mathcal{F}}{\partial \Omega}\Big|_T = \frac{\partial U}{\partial \Omega}\Big|_T = mr^2\Omega\,, \tag{69}$$

which is the angular momentum: this is to be expected since it is the conjugate variable of the angular velocity Ω. It is also suggestive that the temperature $\hbar\Gamma$ is actually given by the background zero-point energy: $\hbar\Gamma = \frac{\hbar\Omega}{2}$.

In the light of the above results, the condition (57) can be then interpreted as a condition for an adiabatic physical system. $\frac{2J_2}{\hbar}$ might be viewed as an analogue of the Kolmogorov–Sinai entropy for chaotic dynamical systems.

6   Conclusions

In this report we have reviewed some aspects of the quantization of the damped harmonic oscillator.

In the framework of the path integral formulation developed by Schwinger and by Feynman and Vernon, we have discussed the doubling of the phase space degrees of freedom for dissipative systems and thermal field theories. We have shown how the doubled variables are related to quantum noise effects.

We have then discussed some algebraic features of the system of damped-antidamped harmonic oscillators which allows for a canonical treatment of quantum dissipation.

We also considered the relation of this model with the 't Hooft proposal, according to which the loss of information due to dissipation in a classical deterministic system manifests itself in the quantum features of the system. We have shown that the quantum spectrum of the harmonic oscillator can be obtained from the dissipative character of the underlying deterministic system.

Finally, we have discussed the thermodynamical features of our system.

Acknowledgments

We would like to thank the organizers of the Piombino workshop "DICE 2002" on "Decoherence, Information, Complexity and Entropy" where some of the results contained in this report have been presented. We also thank the ESF network COSLAB, MIUR, INFN, INFM and EPSRC for partial support.

References

1. J. Schwinger: J. Math. Phys. **2** (1961), 407
2. R.P. Feynman and F.L. Vernon: Annals Phys. **24**, 118 (1963).

3. G. 't Hooft: in *Basics and Highlights of Fundamental Physics*, Erice, (1999); hep-th/0003005
4. G. 't Hooft: hep-th/0104080; hep-th/0105105; quant-ph/0212095
5. Y. Takahashi and H. Umezawa: Collective Phenomena **2**, 55 (1975)
6. H. Umezawa: *Advanced field theory: micro, macro and thermal concepts* (American Institute of Physics, New York 1993);
    H. Umezawa, M. Matsumoto and M. Tachiki: *Thermo Field Dynamics and Condensed States* (North-Holland, Amsterdam 1982)
7. E. Celeghini, M. Rasetti, and G. Vitiello: Ann. Phys. **215**, 156 (1992)
8. M. Blasone, E. Graziano, O.K. Pashaev, and G. Vitiello: Ann. Phys. **252** 115, (1996)
9. M. Blasone and P. Jizba: Can. J. Phys. **80**, 645 (2002)
10. Y.N. Srivastava, G. Vitiello and A. Widom: Ann. Phys. **238**, 200 (1995)
11. M. Blasone, Y.N. Srivastava, G. Vitiello, and A. Widom: Ann. Phys. **267**, 61 (1998)
12. M. Blasone, P. Jizba, and G. Vitiello: Phys. Lett. A **287**, 205 (2001)
13. R. Banerjee: Mod. Phys. Lett. A **17**, 631 (2002); R. Banerjee and P. Mukherjee: J. Phys. A **35**, 5591 (2002)
14. E. Alfinito and G. Vitiello: Class. Quant. Grav. **17**, 93 (2000)
15. G. Vitiello: Int. J. Mod. Phys. B **9**, 973 (1995)
16. E. Alfinito and G. Vitiello: Int. J. Mod. Phys. B **14**, 853 (2000)
17. G. Vitiello: *My Double Unveiled* (John Benjamins, Amsterdam 2001)
18. S. Sivasubramanian, Y.N. Srivastava, G. Vitiello and A. Widom: 'Quantum dissipation induced noncommutative geometry', quant-ph/0301005
19. H. Bateman: Phys. Rev. **38**, 815 (1931);
    P.M. Morse and H. Feshbach: *Methods of Theoretical Physics*, Vol.I (McGraw–Hill, New York 1953), p. 298;
    H. Dekker: Phys. Rep. **80**, 1 (1981);
    H. Feshbach and Y. Tikodinsky: in *Transactions of the N.Y. Academy of Sciences* (1977), volume dedicated to I. Rabi
20. R.P. Feynman: *Statistical Mechanics* (The Benjamin/Cummings Publ. Co., Reading, MA 1972)
21. H. Haken: *Laser Theory* (Springer-Verlag, Berlin 1984)
22. B.G. Wybourne: *Classical Groups in Physics*, (John Wiley & Sons, London 1974)
23. J.S. Bell, *Speakable and unspeakable in Quantum Mechanics* (Cambridge University Press, Cambridge 1987)
24. H.T. Elze and O. Schipper: Phys. Rev. D **66**, 044020 (2002)
25. M. Blasone, E. Celeghini, P. Jizba, and G. Vitiello: 'Quantization, group contraction and zero-point energy', quant-ph/0208012
26. J. Anandan and Y. Aharonov: Phys. Rev. Lett. **65**, 1697 (1990)

Chaotic Quantization:
Maybe the Lord Plays Dice, After All?

Tamas S. Biró[1], Berndt Müller[2], and Sergei G. Matinyan[3]

[1] MTA KFKI RMKI, P.O. Box 49, 1525 Budapest, Hungary
[2] Department of Physics, Duke University, Durham, NC 27708, USA
[3] Yerevan Physics Institute, Yerevan, Armenia

Abstract. We argue that the *quantized* non-Abelian gauge theory can be obtained as the infrared limit of the corresponding *classical* gauge theory in a higher dimension. We show how the transformation from classical to quantum field theory emerges, and relate Planck's constant to quantities defined in the underlying classical gauge theory.

1 Introduction

The question, how gravitation and quantum mechanics can be merged into a consistent unified theory of all fundamental interactions, is still open. Logically, either general relativity (GR), or quantum mechanics (QM), or possibly both, will have to be replaced by a different theory at a more fundamental level. The almost universally accepted notion is that it is GR which needs to be replaced, while QM presumably provides a truly fundamental description of nature. Superstring theory, describing our four-dimensional space-time as the low-energy limit of a ten or eleven-dimensional theory, is widely accepted as the most promising approach, but neither the precise form nor the full content of this theory is entirely understood at the present time.

What about the other option, considering QM as the low-energy limit of a more fundamental theory? This question has been raised by 't Hooft, who conjectured that quantum mechanics can logically arise as the low-energy limit of a microscopically deterministic, but dissipative theory [1, 2]. Explicit, but highly simplified examples for such a mechanism have been constructed [3, 4]. In a recent publication [5] with Biró and Matinyan, we showed how (Euclidean) quantum field theory can emerge in the infrared limit of a higher-dimensional, nonlinear classical field theory (Yang-Mills theory). We called this phenomenon *chaotic quantization* to distinguish it from the formal technique named stochastic quantization [6], not realizing that this term was already introduced by C. Beck several years earlier [7] for essentially the same mechanism. What is special about Yang-Mills fields, however, is that they "quantize themselves", as we shall discuss below. In Sect. 2, we introduce the concept of chaotic quantization of a system with one degree of freedom, which we extend to field theory in Sect. 3. In Sect. 4 we review the chaotic properties of classical Yang-Mills theory, before we analyze their chaotic self-quantization in Sect. 5. In the final section, we enumerate and discuss several open problems.

T.S. Biró, B. Müller, and S.G. Matinyan, Chaotic Quantization: Maybe the Lord Plays Dice, After All?, Lect. Notes Phys. **633**, 164–179 (2004)
http://www.springerlink.com/

2   Chaotic Quantization

A classical physical system encodes much more information than the analogous quantized system. Consider, for instance a point particle in one dimension. The classical system is defined by the pair of coordinates (x, p), implying that every point in the continuous phase space represents a different state. For the quantum system, on the other hand, the uncertainty relation limits the localization of the state in phase space to the finite element $\Delta x \Delta p \sim \hbar$. This observation suggests that it may be useful to consider classical systems, whose internal dynamics results in a self-afflicted loss of information.

Deterministically chaotic systems satisfy this condition. For such a system, the rate of information loss is encoded in the Lyapunov exponent λ, defined as the loarithmic rate of divergence between neighboring trajectories:

$$|x_1(t) - x_2(t)| \sim \mathcal{E}^{\lambda t} \qquad (\lambda > 0) \tag{1}$$

or equivalently in the eigenvalue γ of the Perron-Frobenius operator, defined as the logarithmic rate of convergence of the phase space density to its stationary limit:

$$\rho(x, t) \to \rho_{\lim}(x) + \rho'(x)\mathcal{E}^{-\gamma t} \qquad (\gamma > 0). \tag{2}$$

Before we pursue this idea further, let us recall the method of stochastic quantization [6]. Consider a quantum field $\phi(x)$ in Euclidean space with action $S[\phi]$. The domain of ϕ is formally extended into a fifth dimension denoted by τ. If the field $\phi(x, \tau)$ obeys the Langevin-type equation

$$\tfrac{\partial}{\partial \tau}\phi(x, \tau) = -\tfrac{\delta S}{\delta \phi}(x, \tau) + \xi(x, \tau), \tag{3}$$

where $\xi(x, \tau)$ represents local white noise defined by the moments

$$\langle \xi(x, \tau) \rangle = 0, \qquad \langle \xi(x, \tau)\xi(x', \tau') \rangle = 2\delta(x - x')\delta(\tau - \tau'), \tag{4}$$

then the long-time average of any physical observable converges to the quantum mechanical vacuum expectation value:

$$\lim_{T \to \infty} \tfrac{1}{T} \int_0^T d\tau \, \mathcal{O}[\phi(x, \tau)] = \langle \mathcal{O}[\phi(x)] \rangle_{\mathrm{QM}}. \tag{5}$$

Beck's [8] suggestion was to replace the artificial white noise ξ with the "noise" generated by a deterministic, but chaotic (more precisely: φ-mixing [9]) process.

Following Beck [10, 11], let us start by considering a dynamical system with two variables x, y, which evolves in discrete time steps of length τ. We denote the state of the system at $t_n = n\tau$ as (x_n, y_n). We are interested in the dynamics in the "physical" variable y, if the motion in x is chaotic on short time scales. We define the evolution of the system as follows:

$$(x_{n+1}, y_{n+1}) = f(x_n, y_n) = (T(x_n), \lambda y_n + \tau^{1/2} x_n). \tag{6}$$

Here the map T is assumed to be φ-mixing [9]. Equation (6) can be considered as the stroboscobic map of the differential equation

$$\dot{y} = -\gamma y + \tau^{1/2} \sum_{n=1}^{\infty} x_{n-1} \delta(t - n\tau) , \qquad (7)$$

with $x_{n+1} = T(x_n)$ and $\lambda = \exp(-\gamma \tau)$. Obviously, the variables $\{x_n\}$ take the role of the noise in this equation.

The Langevin equation (7) is equivalent to the Perron-Frobenius equation for the evolution of the phase space density of an ensemble of systems,

$$\rho_{n+1}(x',y') = \sum_{(x,y)\in f^{-1}(x',y')} \frac{\rho_n(x,y)}{\lambda|\partial T/\partial x|} = \sum_{x\in T^{-1}(x')} \frac{\rho_n(x,(y'-\tau^{1/2}x)/\lambda)}{\lambda|\partial T/\partial x|} \qquad (8)$$

Expanding (8) into powers of $(\gamma\tau)^{1/2} \equiv \bar{\tau}^{1/2}$, taking the limit $\tau \to 0$, and interpolating $\rho_n(x,y)$ to a function $\rho(x,y,t)$ which depends continuously upon time,

$$\rho(x,y,t) = \varphi(x,y,t) + \bar{\tau}^{1/2}a(x,y,t) + \bar{\tau}b(x,y,t) + \bar{\tau}^{3/2}c(x,y,t) + \cdots , \qquad (9)$$

yields a set of coupled equations for the coefficient functions,

$$\varphi(x',y,t) = \sum_{x\in T^{-1}(x')} \frac{1}{|\partial T/\partial x|} \varphi(x,y,t)$$

$$a(x',y,t) = \sum_{x\in T^{-1}(x')} \frac{1}{|\partial T/\partial x|} \left(a(x,y,t) - x\frac{\partial}{\partial y}\varphi(x,y,t) \right)$$

$$\text{etc.} \qquad (10)$$

Being interested in the dynamics of the physical variable y only, we define projected functions

$$p_0(y,t) = \int dx\, \varphi(x,y,t) ; \qquad \alpha(y,t) = \int dx\, a(x,y,t) ; \qquad \text{etc.} \qquad (11)$$

For complete maps T, it is possible to show that

$$f(x',y,t) = \sum_{x\in T^{-1}(x')} \frac{g(x,y,t)}{|\partial T/\partial x|} \qquad (12)$$

for all y and t implies

$$\int dx\, f(x,y,t) = \int dx\, g(x,y,t) . \qquad (13)$$

The first equation in the expansion of the Perron-Frobenius equation in powers of $\bar{\tau}^{1/2}$ then becomes a tautology, while the second one takes the form

$$\frac{\partial}{\partial y} \int dx\, x\, \varphi(x,y,t) = 0 . \qquad (14)$$

The desired dynamical equation for the phase space density $p_0(y,t)$ is obtained as the third equation (at order $\bar{\tau}$),

$$\frac{\partial}{\partial y} \int dx\, x\, a(x,y,t) = \frac{\partial}{\partial y}(y\, p_0(y,t)) + \frac{1}{2}\frac{\partial^2}{\partial y^2} \int dx\, x^2 \varphi(x,y,t) - \frac{\partial p_0}{\partial t}. \tag{15}$$

If $h(x)$ is an invariant of the map T, $\varphi(x,y,t) = h(x)p_0(y,t)$ is a solution of the first of the set of equations (10) for any function $p_0(y,t)$, and (14) requires $\langle x \rangle \equiv \int dx\, x h(x) = 0$. Finally, (15) turns into

$$\frac{\partial}{\partial t}p_0(y,t) = \frac{\partial}{\partial y}(y\, p_0(y,t)) + \frac{\langle x^2 \rangle}{2}\frac{\partial^2}{\partial y^2}p_0(y,t) - \frac{\partial}{\partial y}\int dx\, x\, a(x,y,t). \tag{16}$$

The left-hand side and the first two terms on the right-hand side have the form of a Fokker-Planck (FP) equation; the last term on the right-hand side represents a source term. For symmetric maps, $T(x) = T(-x)$, it is easy to see that the source term vanishes, and the projected phase space density $p_0(y,t)$ satisfies a homogeneous FP equation. The first correction in $\bar{\tau}^{1/2}$, $\alpha(y,t)$, also satisfies a FP equation, but in this case, the source term does not vanish. In other words, the deviations in the equation for the exact projected phase space density $p(y,t) = \int dx\, \rho(x,y,t)$ from a FP equation are proportional to $\bar{\tau}^{1/2}$.

An explicit solution can be found for, e. g., the Ulam map $T(x) = 1 - 2x^2$ with $x \in [-1,1]$. The stationary solution of the FP equation is:

$$p_0(y,t) \equiv p_0(y) = \left(\frac{2}{\pi}\right)^{1/2} \mathcal{E}^{-2y^2}; \tag{17}$$

and the full solution written as a power series in $\bar{\tau}^{1/2}$ is:

$$p(y,t) = p_0(y)\left[1 + \bar{\tau}^{1/2}\left(2y - \frac{8}{3}y^3 + O(\bar{\tau})\right)\right]. \tag{18}$$

To obtain the Euclidean Schrödinger equation, instead of the FP equation, we need to introduce a potential $V(y)$ and rescale the auxiliary variable x according to [7]

$$T(x) \longrightarrow T(x)\,\mathcal{E}^{\bar{\tau}V(y)/\hbar}$$

$$y_{n+1} = \lambda y_n + \bar{\tau}^{1/2}x_n \longrightarrow y_{n+1} = \lambda y_n + \left(\frac{\hbar\bar{\tau}}{m\sigma^2}\right)^{1/2}x_n, \tag{19}$$

where $\sigma^2 = \langle x^2 \rangle \equiv \int dx\, x^2 h(x)$. Identifying p_0 with the Euclidean wave function ψ, we find that it satisfies the imaginary-time Schrödinger equation,

$$\hbar\frac{\partial}{\partial t}\psi(y,t) + \left[-\frac{\hbar^2}{2m}\frac{\partial^2}{\partial y^2} + V(y)\right]\psi(y,t) \equiv \mathcal{S}_{\mathrm{E}}\psi(y,t) = 0. \tag{20}$$

The correction linear in $\bar{\tau}^{1/2}$ satisfies the same equation, but with an additional source term,

$$\mathcal{S}_{\mathrm{E}}\alpha(y,t) = \left(\frac{\hbar}{2m}\right)^{3/2}\frac{\partial^3}{\partial y^3}\psi(y,t). \tag{21}$$

The complete wave function is $\tilde{\psi} = \psi + \bar{\tau}^{1/2}\alpha + \cdots$, which approaches the Schrödinger wave function in the limit $\bar{\tau} = \gamma\tau \to 0$.

The important insight to take away from this derivation is that an appropriate chaotic process can serve as the source of the random noise required for the stochastic quantization of a dynamical system if the time scale on which the chaotic process randomizes is sufficiently short, so that the corrections are negligible.

3 Extension to Field Theories

Can this mechanism of quantizing classical systems be generalized to fields $\Phi(x, t)$ with x being a point in three-dimensional space? How this can be done is most easily seen, when the field theory is defined on a lattice, rather than a spatial continuum. Then all one needs to do is introduce a map T together with some internal space $\{\xi^i\}$ at each lattice point x_i. Beck has proposed to define the evolution law including a nearest-neighbor coupling [7]

$$\xi_{n+1}^i = (1 - g)T\left(\xi_n^i\right) + \frac{g}{2d}\sum_{\nu=1}^{d}\left(\xi_n^{i+e_\nu} + \xi_n^{i-e_\nu}\right), \qquad (22)$$

which has the continuum limit

$$\xi_{n+1}^{(x)} = T\left(\xi_n^{(x)}\right) + \frac{g'}{2d}\nabla^2\xi_n^{(x)}, \qquad (23)$$

with appropriate coupling constants g, g'. We will not follow this route further here and refer to Beck's recent monograph [12].

A more natural approach consists in identifying the local internal map space with a compact Lie group $\mathcal{G}_x$. The simplest realization of this idea is the SU(2) gauge theory, i. e., the Yang-Mills field theory. In this case, the internal degrees of freedom (color) of the gauge field provide the local space for the chaotic map, and the nonlinear dynamics of the gauge field uniquely defines the map. As we shall see below, there is no need to introduce new degrees of freedom beyond those provided by the gauge field (in one additional dimension) itself [5]. Before exploring this idea in more detail, however, we need to review what is known about the chaotic dynamics of Yang-Mills fields.

4 Interlude: Chaotic Properties of Yang-Mills Fields

The chaotic nature of classical non-Abelian gauge theories was first recognized twenty years ago [13,14]. Over the past decade, extensive numerical solutions of spatially varying classical non-Abelian gauge fields on the lattice have revealed that the gauge field has positive Lyapunov exponents that grow linearly with the energy density of the field configuration and remain well-defined in the limit of small lattice spacing a or weak-coupling [15,16]. More recently, numerical studies have shown that the $(3+1)$-dimensional classical non-Abelian lattice gauge theory exhibits global hyperbolicity. This conclusion is based on calculations of the complete spectrum of Lyapunov exponents [17] and on the long-time statistical properties of local fluctuations of the Kolmogorov-Sinai (KS) entropy in the classical SU(2) gauge theory [18].

It is useful to note some important relationships between ergodic and periodic orbits for globally hyperbolic dynamical systems. The ergodic Lyapunov exponents $\lambda_{r,i}$ are obtained by numerical integration of a randomly chosen ergodic trajectory, denoted by its origin r. In a Hamiltonian hyperbolic dynamical system with d degrees of freedom the sum of its $d-1$ positive ergodic Lyapunov exponents is obtained as the ergodic mean of the local expansion rate,

$$\lim_{t \to \infty} h_r(t) \equiv \lim_{t \to \infty} \frac{1}{t} \int_0^t \chi(\boldsymbol{x}(t'))\, dt' = \sum_{i=1}^{d-1} \lambda_{r,i} = h_{\mathrm{KS}} \ . \tag{24}$$

Here h_{KS} denotes the Kolomogorov-Sinai entropy and

$$\chi(\boldsymbol{x}(t)) = \frac{d}{dt} \ln \det \left(\frac{\partial \boldsymbol{x}(t)}{\partial \boldsymbol{x}(0)} \right)_{\mathrm{expanding}} \tag{25}$$

is the local rate of expansion along the trajectory $\boldsymbol{x}(t)$. Due to the equidistribution of periodic orbits in phase space it is possible to evaluate the ergodic mean in (24) by weighted sums over periodic orbits p. In fact, for hyperbolic systems the thermodynamic formalism allows to express certain invariant measures on phase space in terms of averages over periodic orbits [19, 20]. Labeling periodic orbits by p (rather than a starting point), and denoting their periods and positive Lyapunov exponents by T_p and $\lambda_{p,i}$, respectively, the connection between the positive ergodic Lyapunov exponents and those of periodic orbits is given by

$$h_{\mathrm{KS}} = \lim_{t \to \infty} \frac{\sum_{t \leq T_p \leq t+\varepsilon} \left(\sum_{i=1}^{d-1} \lambda_{p,i} \right) \exp\left(-\sum_{i=1}^{d-1} \lambda_{p,i} T_p \right)}{\sum_{t \leq T_p \leq t+\varepsilon} \exp\left(-\sum_{i=1}^{d-1} \lambda_{p,i} T_p \right)} \ , \tag{26}$$

where $\varepsilon > 0$ is a small number. The *topological pressure* $P(\beta)$ is a useful tool for analyzing invariant measures on phase space in terms of periodic orbits. This function can be expressed as

$$P(\beta) = \lim_{t \to \infty} \frac{1}{t} \ln \sum_{t \leq T_p \leq t+\varepsilon} \exp\left(-\beta \sum_{i=1}^{d-1} \lambda_{p,i} T_p \right) \ . \tag{27}$$

The relation (26) then follows from (24) and from the identity $-P'(1) = h_{\mathrm{KS}}$.

In order to apply this formalism to the Yang-Mills field, the gauge theory needs to be formulated as a Hamiltonian system on a spatial lattice. How to do this is well known: The Hamiltonian lattice gauge theory was formulated by Kogut and Susskind [21] in order to study the nonperturbative properties of nonabelian gauge theories, such as quark confinement. Denoting the lattice spacing by a and the nonabelian coupling constant by g, the Hamiltonian for the SU(2) gauge theory is given by

$$H = \frac{g^2}{2a} \sum_{x,i} \mathrm{tr}\, E_{x,i}^2 + \frac{2}{g^2 a} \sum_{x,ij} (2 - \mathrm{tr}\, U_{x,ij}) \ . \tag{28}$$

Here the $U_{x,i} \in \mathrm{SU}(2)$ are called the link variables at point x in the coordinate direction i, and the $E_{x,i} \in \mathrm{LSU}(2)$ denote the color-electric field strengths components. $U_{x,ij}$ denotes the plaquette product of four link variables, starting and ending at x and circumscribing the elementary square in the i, j directions. The chaotic nature of this theory was demonstrated [15] by numerical simulations, and Gong [17] obtained the complete Lyapunov spectrum for lattice volumes L^3 with $L = 1, 2, 3$. Bolte et al. [18] extended these calculations to the lattices of size $L = 4, 6$.

For sufficiently long trajectories and fixed energy per lattice site the Lyapunov spectrum shows a unique shape, independent of the lattize size, as shown in Fig. 1. Indeed, for a completely hyperbolic system, physical intuition dictates that the KS entropy h_{KS} is an extensive quantity. For this to be true, the sum over all positive Lyapunov exponents must scale like the lattice volume L^3 and the shape of the distribution of Lyapunov exponents must be independent of L. Figure 1 confirms this expectation.

The KS entropy is a *global* property of the dynamical system. The next step of detail of the ergodic nature of the system is provided by the fluctuations in the quasi-local average of $\chi(\boldsymbol{x}(t))$. These fluctuations are obtained by integrating (25) up to a sampling time t_{s}. For sampling times t_{s} much longer than the largest correlation time one expects that observables sampled along ergodic trajectories exhibit Gaussian fluctuations about their ergodic mean. Waddington [22] has shown that for Anosov systems (i.e., fully hyperbolic systems on compact phase

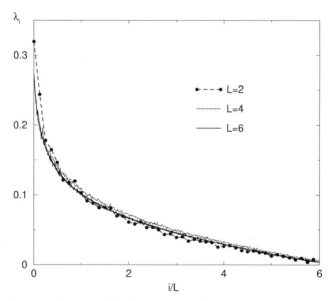

Fig. 1. Distribution of numerically obtained ergodic Lyapunov exponents for a classical SU(2) gauge theory on lattices of size $L = 2, 4, 6$. The index i numbers the Lyapunov exponents and the abscissa is scaled with L^3. The energy per plaquette was chosen as $1.8/g^2 a$

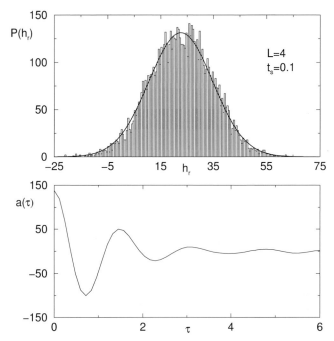

Fig. 2. Top: The distribution of the sum of local expansion rates $h_r(t_s)$ for L=4 and a short sampling time $t_s = 0.1$, together with a Gaussian fit. Bottom: The autocorrelation function $a(\tau)$ for this distribution

spaces) the probability distribution for $h_r(t_s)$ is Gaussian with mean h_{KS},

$$P[h_r(t_s)] \to \exp\left(-\frac{(h_r(t_s)-h_{KS})^2}{2\Delta h_r(t_s)^2}\right) \qquad \text{for} \quad t_s \to \infty . \qquad (29)$$

and square variance proportional to $P''(1)$,

$$\Delta h_r(t) \to \sqrt{P''(1)/t} \qquad \text{for} \quad t \to \infty . \qquad (30)$$

The variance (30) can be related to the autocorrelation function

$$a(\tau) = \langle \chi(\boldsymbol{x}(\tau))\, \chi(\boldsymbol{x}(0))\rangle - h_{KS}^2 \qquad (31)$$

of the local ergodic Lyapunov exponents, through the relation

$$t\,(\Delta h_r(t))^2 = \int_{-t}^{+t} \left(t - \frac{|\tau|}{t}\right) a(\tau)\, d\tau \to P''(1) . \qquad (32)$$

Figure 2 (top) shows the distribution of sampled valued of $h_r(t_s)$, (obtained on a single, very long trajectory) for a short sampling interval $t_s = 0.1$ for a $L = 4$ lattice. The bottom part of Fig. 2 shows the autocorrelation function $a(\tau)$ obtained for the same trajectory. A numerical fit of the function indicates that $a(\tau)$ decays exponentially for large times. The value predicted for Δh_r by

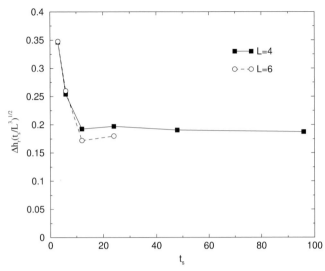

Fig. 3. $\Delta h_r(t_s)$, scaled with $(t_s/L^3)^{1/2}$, as a function of t_s

(30,32) is in excellent agreement with the value obtained by a Gaussian fit to the sampled distribution.

One can also study how the width of the Gaussian scales with the lattice size L. To a very good approximation one finds that it is proportional to $\sqrt{L^3}$. If one includes the dependence on the sampling time, the variance of h_r scales like $\sqrt{L^3/t_s}$ (see Fig. 3). As the mean value h_{KS} scales like L^3, this result confirms the Gaussian nature of the fluctuations: $\Delta h_r/h_{KS} \sim (L^3 t_s)^{-1/2}$.

These numerically obtained results provide strong (although not mathematically conclusive) evidence that the SU(2) lattice gauge theory is a strongly chaotic (Anosov, φ-mixing) system with properties required for the formalism of Sect. 2.

5   Yang-Mills Fields Quantize Themselves

The results discussed in the previous section imply that correlation functions of physical observables decay rapidly, and that long-time averages of observables for a single initial gauge field configuration coincide with their microcanonical phase-space average, up to Gaussian fluctuations which vanish in the long observation time limit as $t_s^{-1/2}$. Since the relative fluctuations of extensive quantities scale as $L^{-3/2}$, the microcanonical (fixed-energy) average can be safely replaced by the canonical average when the spatial volume probed by the observable becomes large. In the following we discuss the hierarchy of time and length scales on which this transformation occurs [5].

According to the cited results, the classical non-Abelian gauge field self-thermalizes on a finite time scale τ_{eq} given by the ratio of the equilibrium entropy and the KS-entropy, which determines the growth rate of the coarse-grained entropy,

$$\tau_{\text{eq}} = S_{\text{eq}}/h_{\text{KS}}.\tag{33}$$

At weak coupling, the KS-entropy for the $(3+1)$-dimensional SU(2) gauge theory scales as

$$h_{\text{KS}} \sim g^2 E \sim g^2 T (L/a)^3\,,\tag{34}$$

where E is the total energy of the field configuration and T is the related temperature related to E by (for SU(N_c))

$$\epsilon \equiv \frac{E}{L^2} = 2(N_c^2 - 1)\frac{T}{a^3} + \mathcal{O}(g^2)\,.\tag{35}$$

The equilibrium entropy of the lattice is independent of the energy and proportional to the number of degrees of freedom of the lattice, $S_{\text{eq}} \sim (L/a)^3$. The time scale for self-equilibration is thus given by

$$\tau_{\text{eq}} \sim \frac{E}{h_{\text{KS}}T} \sim (g^2 T)^{-1}\,.\tag{36}$$

When one is interested only in long-term averages of observables, it is thus sufficient to consider the *thermal* classical gauge theory on a three-dimensional spatial lattice. Furthermore, on time scales $t \gg \tau_{\text{eq}}$, the Yang-Mills field generates a random Gaussian process, which is required for chaotic quantization.

The dynamic properties of thermal non-Abelian gauge fields at such long distances have been studied in much detail [23–25]. While these studies have been made exclusively for the thermal quantum field theory, their results are readily transcribed to the thermal classical gauge field with a lattice cutoff. The real-time dynamics of the gauge field at long distances and times can be described, at leading order in the coupling constant g, by a Langevin equation

$$\sigma_c \frac{\partial A}{\partial t} = -D \times B + \xi\,,\tag{37}$$

where D is the gauge covariant spatial derivative, $B = D \times A$ is the magnetic field strength, and ξ denotes Gaussian distributed (white) noise with zero mean and variance

$$\langle \xi_i(x,t)\xi_j(x',t')\rangle = 2\sigma_c T \delta_{ij}\delta^3(x-x')\delta(t-t')\,.\tag{38}$$

Here σ_c denotes the color conductivity [26] of the thermal gauge field which is determined by the ratio ω^2/γ of the plasma frequency ω and the damping rate γ of a thermal gauge field excitation. In the classical field theory with a lattice cutoff, the color conductivity scales as

$$\sigma_c \sim (a \ln[d_{\text{mag}}/d_{\text{el}}])^{-1}\,.\tag{39}$$

This relation implies that the color conductivity is an ultraviolet sensitive quantity, which depends on the lattice cutoff a.

We now consider observers measuring physical quantities on long time and distance scales $(t, L \gg a, (g^2 T)^{-1})$. The random process defined by the Langevin equation (37) generates three-dimensional field configurations with a probability distribution $P[A]$ determined by the Fokker-Planck equation

$$\sigma \frac{\partial}{\partial t} P[A] = \int d^3x \frac{\delta}{\delta A} \left(T \frac{\delta P}{\delta A} + \frac{\delta W}{\delta A} P[A] \right), \tag{40}$$

where $W[A]$ denotes the magnetic energy functional

$$W[A] = \int d^3x \frac{1}{2} B(x)^2. \tag{41}$$

Any non-static excitations of the magnetic sector of the gauge field, i.e. magnetic fields $B(k)$ not satisfying $k \times B = 0$, die away rapidly on a time scale of order σ_c/k^2, where k denotes the wave vector of the field excitation. Long-term averages are determined by the static magnetic field sector weighted by the stationary solution of the FP equation (40):

$$P_0[A] = e^{-W[A]/T}. \tag{42}$$

The observer measures

$$\langle \mathcal{O}[A] \rangle = \int \mathcal{D}A \mathcal{O}[A] \mathcal{E}^{-W[A]/T}. \tag{43}$$

The magnetic field

$$B_i = \frac{1}{2} \epsilon_{ijk} F^{jk} \equiv \frac{\sqrt{a}}{2} \epsilon_{ijk} f^{jk} \tag{44}$$

defines a three dimensional field strength tensor f^{jk}, and W/T can be identified with the three-dimensional action S_3 measured in units of Planck's constant $\hbar = aT$ [5],

$$\frac{W}{T} \equiv \frac{S_3}{\hbar} = -\frac{1}{4\hbar} \int dx_3 \int d^2x \, f^{ik} f_{ik}. \tag{45}$$

The rescaling of the gauge field strength by the fundamental length scale a is required for dimensional reasons. The same rescaling also determines the three-dimensional coupling constant to be

$$g_3^2 = \frac{g^2}{a} = \frac{g^2 T}{\hbar_3}. \tag{46}$$

The central result of this section is that the highly excited classical $(3+1)$-dimensional Yang-Mills theory reduces to a *vacuum* quantum Yang-Mills theory in three Euclidean dimensions for an observer who is only interested in physics at long distance and time scales, with vacuum expectation values of the standard form,

$$\langle \mathcal{O}[A] \rangle_3 = \int \mathcal{D}A \mathcal{O}[A] \mathcal{E}^{-S_3[A]/\hbar}. \tag{47}$$

Planck's constant is determined by two microscopic quantities of the "fundamental" theory, a and T:

$$\hbar = aT. \tag{48}$$

The existence of both, a fundamental length scale and an energy scale, is critical to the emergence of a constant with the dimensions of an action.

Some noteworthy comments:

- It is important to note that the dimensional reduction is not induced by a compactification of the time coordinate, either in real or imaginary time. We have not asumed a thermal ensemble of gauge fields in the original Minkowski space theory, and the random solution of the $(3 + 1)$-dimensional classical field theory does not satisfy periodic boundary conditions in imaginary time. The effective dimensional reduction is not caused by a discreteness of the excitations with respect to the time-like dimension, but by the dissipative nature of the $(3 + 1)$-dimensional dynamics. Magnetic field configurations satisfying $D \times B = 0$ can be thought of as low-dimensional attractors of the dissipative motion, and the chaotic dynamical fluctuations of the gauge field around the attractor can be consistently interpreted as quantum fluctuations of a vacuum gauge field in 3-dimensional Euclidean space.
- The dimensional reduction by chaotic fluctuations and dissipation does not occur in scalar field theories, because there is no dynamical sector that survives long-time averaging. Quasi-thermal fluctuations generate a dynamical mass for the scalar field and thus eliminate any arbitraily slow field modes. An exception may be the case where the excitation energy of the scalar field is just right to put the quasi-thermal field at the critical temperature of a second-order phase transition, where arbitraily slow modes exist as fluctuations of the order parameter. In the case of gauge fields, the transverse magnetic sector is protected by the gauge symmetry, and it is this sector which survives the time average without any need for fine-tuning of the microscopic theory.
- Generalizing the results of Sect. 2 and Sect. 4, we expect corrections to the vacuum quantum field theory in three dimensions to be of the order of the relative fluctuations of the KS entropy within a spatial region of size Δx,

$$\frac{\Delta h_r}{h_{\mathrm{KS}}} \sim (g^2 T \Delta x)^{-2} \sim \left(\frac{a}{g^2 \hbar \Delta x} \right)^2 . \tag{49}$$

If, e.g., a is of the order of the Planck length and Δx is any physically accessible length scale, the corrections to the dynamics of the quantized Yang-Mills field will be exceedingly small, indeed.

6   Open Problems

Our example for the chaotic quantization of a three-dimensional gauge theory in Euclidean space raises a number of questions:

1. Does the principle of chaotic quantization generalize to higher dimensions, in particular, to quantization in four dimensions?
2. Can the method be extended to describe field quantization in Minkowski space?
3. Can gravity be included in this framework? Does the nonlinearity of classical general relativity provide a source of random noise at short distances, allowing for an effective quantum theory to emerge on long distance and time scales?

The first question is most easily addressed. As long as globally hyperbolic classical field theories can be identified in higher dimensions, our proposed mechanism should apply. Although we do not know of any systematic study of dicretized field theories in higher dimensions, a plausibility argument can be made that Yang-Mills fields exhibit chaos in (4+1) dimensions. For this purpose, we consider the infrared limit of a spatially constant gauge potential [13,14]. For the SU(N_c) gauge field in ($D+1$) dimensions in the $A_0 = 0$ gauge, there are $3(N_c^2-1)$ interacting components of the vector potential and $3(N_c^2-1)$ canonically conjugate momenta, which depend only on the time coordinate. The remaining gauge transformations and Gauss' law allow to eliminate $2(N_c^2-1)$ degrees of freedom. Next, rotational invariance in D dimensions permits to reduce the number of dynamical degrees of freedom by twice the number of generators of the group SO(D), i.e. by $D(D-1)$. This leaves a $(D-1)(2N_c^2-2-D)$-dimensional phase space of the dynamical degrees of freedom and their conjugate momenta. For the dynamics to be chaotic, this number must be at least three. For the simplest gauge group SU(2), this condition permits infrared chaos in $2 \leq D \leq 5$ dimensions, including the interesting case $D = 4$. Higher gauge groups are needed to extend the chaotic quantization scheme to gauge fields in $D > 5$ dimensions. Of course, this reasoning does not establish full chaoticity of the Yang-Mills field in these higher dimensions, it only indicates the possibility. Numerical studies will be required to establish the presence of strong chaos in these classical field theories.

The second question is more difficult. A formal answer could be that the Minkowski-space quantum field theory can (and even must) be obtained by analytic continuation from the Euclidean field theory. Any observable in the Minkowski space theory that can be expressed as a vacuum expectation value of field operators can be obtained in this manner. If this argument appears somewhat unphysical, one might consider a completely different approach, beginning with a chaotic classical field theory defined in (3+2) dimensions. Field theories defined in spaces with two time-like dimensions were first proposed by Dirac in the context of conformal field theory [27] and have recently been considered as generalizations of superstring theory [28]. In this case, one time dimension is effectively eliminated by gauge fixing.

In the absence of similar explicit constraints, field theories with two timelike dimensions, even if the second time direction is compact, exhibit unphysical properties, such as a lack of causality and unitarity [29]. For example, the Coulomb potential of a point charge in the presence of a second, curled up timelike dimension with period L is complex,

$$V(r) = \frac{\alpha}{r} \left(1 + 2 \sum_{n=1}^{\infty} \mathcal{E}^{-2i\pi nr/L} \right). \tag{50}$$

The lack of causality is closely related to the problem of the existence of timelike closed loops, which confuse the distinction between past and future. These problems are avoided, if the second timelike dimension is "thermal", i. e. if it is compact in the imaginary time direction. The factor i then disappears from the exponent of (50), and the corrections to the Coulomb law are real and exponentially suppressed at large distances. In the context of the mechanism discussed in Sect. 5, the physical time dimension may be defined as the coordinate orthogonal to the total 5-momentum vector P^μ of the initial field configuration. Whether this reasoning applies to the case, where the "thermal" field theory is really an ergodic one, remains to be confirmed, but it is quite plausible.

In the presence of two time-like dimensions, the "energy" becomes a two-component vector $\boldsymbol{E}$, which is a part of the $(D+2)$-dimensional energy-momentum vector. If we select an initial field configuration with energy $\boldsymbol{En}$, where $\boldsymbol{n}$ is a two-dimensional unit vector, this choice defines a preferred time-like direction $t\boldsymbol{n}$, in which the field thermalizes. Conservation of the energy-momentum vector ensures that the total energy component orthogonal to $\boldsymbol{n}$ always remains zero. The choice of an initial field configuration corresponds to a spontaneous breaking of the global SO(D,2) symmetry down to a global SO(D,1) symmetry. Whether this process leads to an effective quantum field theory in the $(D+1)$-dimensional Minkowski space, remains to be investigated.

Finally, what about gravity? One reason, why this question is difficult to answer, is that little is known about the properties of general relativity as a dynamical system. Due to its different gauge group structure, gravity has more "capacity to resist" chaos than the Yang-Mills fields. Local invariance against coordinate transformations is incompatible with the concept of Hamiltonian dynamics, which has a preferred time direction. The Hamiltonian version of general relativity [30] used in cosmology (mixmaster universe) reflects this exceptional situation in GR. Even the most basic definitions of deterministic chaos are not directly applicable to general relativity and require appropriate generalizations. Many special, chaotic solutions of Einstein's equations have been found [31, 32]. Most important among these is the eternally oscillating chaotic behavior discovered in [33] for the generic solution of the vacuum Einstein equations in the vicinity of a spacelike singularity, which has the character of deterministic chaos. However, it is not known whether the generic solution exhibits chaos, as in the case of the Yang-Mills theory. It is not even clear what the proper framework for a systematic numerical study of this question would be. For example, the Lyapunov exponents, which were so effectively used in Yang-Mills theory, are coordinate dependent in GR due to the general covariance of Einstein's equations against coordinate transformations.

What is clear, is that general relativity shares many of the properties, which allow nonabelian gauge theories to chaotically quantize themselves: Einstein's equations are strongly nonlinear and have a large set of gauge invariances which could guarantee the survival of a dynamical sector at long distances in the presence of quasi-thermal noise. Under such conditions, GR may not even require a short-distance cutoff, because the thermal Schwarzschild radius $(2GT)$ defines an effective limit to short-distance dynamics which can couple to the dynamics at large distance scales. Applying the relation (48) determining $\hbar$ for the Yang-Mills field, one might conjecture that the analogous relation for gravity has the form $\hbar \sim GT^2$. If the temperature parameter T were of the order of the Planck mass, this relation would yield the observed magnitude of $\hbar$.

However, it is not clear whether the same mechanism – chaos, or exponential growth of sensitivity to initial conditions – which causes information loss in the dynamics of Yang-Mills fields, must also operate in general relativity. 't Hooft has speculated that microscopic black hole formation may be the mechanism that causes the loss of information in the case of gravity [1]. Again, a much better understanding of the structure of generic solutions of Einstein's equations must be a prerequisite to an exploration of these interesting questions.

Acknowledgments

This work was supported in part by a grant from the U.S. Department of Energy (DE-FG02-96ER40495), by the American-Hungarian Joint Fund TÉT (JFNo. 649) and by the Hungarian National Science Fund OTKA (T 019700). One of us (B.M.) acknowledges the support by a U. S. Senior Scientist Award from the Alexander von Humboldt Foundation.

References

1. G. 't Hooft: Class. Quant. Grav. **16**, 3283 (1999); hep-th/0003004
2. G. 't Hooft: hep-th/0104219
3. M. Blasone, P. Jizba, and G. Vitello: Phys.Lett. A **287**, 205 (2001)
4. G. 't Hooft: hep-th/0104080
5. T.S. Biró, S.G. Matinyan, and B. Müller: Found. Phys. Lett. **14**, 471 (2001); see also hep-th/9908031
6. G. Parisi and Y.S. Wu: Sci. Sin. **24**, 483 (1981)
7. C. Beck: Nonlinearity **8**, 423 (1995)
8. C. Beck: Physica D **85**, 459 (1995)
9. P. Billingsley: *Probability and Measure* (Wiley, New York 1978)
10. C. Beck and G. Roepstorff: Physica A **145**, 1 (1987)
11. C. Beck: J. Stat. Phys. **79**, 875 (1995)
12. C. Beck: 'Spatio-temporal Chaos and Vacuum Fluctuations of Quantized Fields', hep-th/0207081
13. S.G. Matinyan, G.K. Savvidy, and N.G. Ter-Arutyunyan-Savvidy: Sov. Phys. JETP **53**, 421 (1981)
14. T.S. Biró, S.G. Matinyan, and B. Müller: *Chaos and Gauge Field Theory* (World Scientific, Singapore 1994)
15. B. Müller and A. Trayanov: Phys. Rev. Lett. **68**, 3387 (1992)
16. C. Gong: Phys. Lett. B **298**, 257 (1993)
17. C. Gong: Phys. Rev. D **49**, 2642 (1994)
18. J. Bolte, B. Müller, and A. Schäfer: Phys. Rev. D **61**, 054506 (2000)
19. W. Parry and M. Pollicott: Astérisque **187–188** (1990)
20. P. Gaspard: *Chaos, Scattering and Statistical Mechanics* (Cambridge University Press, Cambridge 1998)
21. J.B. Kogut and L. Susskind: Phys. Rev. D **10**, 3468 (1974)
22. S. Waddington: Ann. Inst. Henri Poincaré C **13**, 445 (1996)

23. D. Bödeker: Phys. Lett. B **426**, 351 (1998); ibid. **559**, 502 (1999)
24. P. Arnold, D.T. Son and L.G. Yaffe: Phys. Rev. D **59**, 105020 (1999); **60**, 025007 (1999)
25. P. Arnold, D.T. Son, and L.G. Yaffe: Phys. Rev. D **55**, 6264 (1997)
26. A.V. Selikhov and M. Gyulassy: Phys. Lett. B **316**, 373 (1993); Phys. Rev. C **49**, 1726 (1994)
27. P.A.M. Dirac: Ann. Math. **37**, 429 (1936)
28. I. Bars and C. Kounnas: Phys. Lett. B **402**, 25 (1997); I. Bars: Phys. Rev. D **62**, 046007 (2000)
29. F.J. Yndurain: Phys. Lett. B **256**, 15 (1991)
30. C. Misner: Phys. Rev. Lett. **22**, 1071 (1961); Phys. Rev. **186**, 1319 and 1328 (1969)
31. *Deterministic Chaos in General Relativity*, ed. by D. Hobill, A. Burd, and A. Cooley, NATO ASI Series B: Physics, Vol. 332 (Plenum Press, New York, 1994)
32. For a recent review see: S.G. Matinyan: gr-qc/0010054. In: *Proceedings of the Ninth Marcel Grossmann Conference*, Rome, Italy, July 2000, vol. 1
33. V.A. Belinskii, E.M. Lifshitz, and I.M. Khalatnikov: Adv. Phys. **19**, 525 (1970); ibid. **31**, 639 (1982)

Quantum Correlations in Classical Statistics

Christof Wetterich

Institut für Theoretische Physik, Universität Heidelberg, Philosophenweg 16, 69120 Heidelberg, Germany

Abstract. Quantum correlations can be naturally formulated in a classical statistical system of infinitely many degrees of freedom. This realizes the underlying noncommutative structure in a classical statistical setting. We argue that the quantum correlations offer a more robust description with respect to the precise definition of observables.

1   Quantum Structures in Classical Statistics

Classical observables commute, quantum mechanical operators do not - this basic difference reflects itself in a different behavior of classical correlation functions and quantum correlations. We will argue here that these different properties are connected to the formulation of the concept of correlation functions rather than to the "classical" or "quantum" character of the system itself. Quantum correlations can be formulated in classical statistics just as well as classical correlation functions may be defined in a quantum system.

In a quantum system it is well known that a commuting operator product can be defined via the time ordering of operators. A definition of a correlation function based on the time ordered product of two operators has the same commutative properties as the classical correlation function. The reason why practical quantum mechanics uses noncommuting products like $\hat{Q}(t_1)\hat{P}(t_2)$ rather than the time ordered product $T(\hat{Q}(t_1)\hat{P}(t_2))$ is rooted in the subtleties of the definition of the latter when $t_2 = t_1$. Operators for continuous time are defined by a limit process starting from discrete time steps. The expectation value of the quantum product $\langle\hat{Q}(t_1)\hat{P}(t_2)\rangle$ is insensitive to the precise definition of the limiting procedure whereas $\langle T(\hat{Q}(t_1)\hat{P}(t_2))\rangle$ is not. The "quantum correlation" is therefore more "robust" than the "classical correlation" $\langle T(\hat{Q}(t_1)\hat{P}(t_2))\rangle$. We will see that a similar issue of a robust definition of correlation functions is actually present in classical statistics as well. An investigation of the question of relevant information in the classical probability distribution will lead us to the proposal of a robust quantum correlation for a classical statistical system.

The formulation of the basic partition function for classical statistical systems with infinitely many degrees of freedom uses implicitly an assumption of "completeness of the statistical information". This means that we assign a probability to everyone of the infinitely many configurations. The specification

C. Wetterich, Quantum Correlations in Classical Statistics, Lect. Notes Phys. **633**, 180–195 (2004)
http://www.springerlink.com/

of the probability distribution contains therefore an "infinite amount of information". This contrasts with the simple observation that only a finite amount of information is available in practice for the computation of the outcome of any physical measurement. A concentration on measurable quantities suggests that the assumption of completeness of the statistical information may have to be abandoned. In this note we explore consequences of "incomplete statistics" which deals with situations where only partial information about the probability distribution is available. In particular, we consider extended systems for which only local information about the probability distribution is given. From another point of view we ask which part of the information contained in the standard classical probability distribution is actually relevant for the computation of expectation values of local observables. We will see that the quantum mechanical concepts of states, operators and evolution also emerge naturally in this setting [1].

As an example, let us consider a classical statistical system where the infinitely many degrees of freedom φ_n ($n \in Z$) are ordered in an infinite chain. We concentrate on a "local region" $|\tilde{n}| < \bar{n}$ and assume that the probability distribution $p[\varphi]$ has a "locality property" in the sense that the relative probabilities for any two configurations of the "local variables" $\varphi_{\tilde{n}}$ are independent of the values that take the "external variables" φ_m with $|m| > \bar{n}$. Furthermore, we assume that the probability distribution for the $\varphi_{\tilde{n}}$ is known for given values of the variables $\varphi_{\bar{n}}, \varphi_{-\bar{n}}$ at the border of the local interval. As an example, we may consider a probability distribution

$$p[\varphi] = p_>[\varphi_{m \geq \bar{n}}]p_0[\varphi_{-\bar{n} \leq \tilde{n} \leq \bar{n}}]p_<[\varphi_{m \leq -\bar{n}}],$$

$$p_0[\varphi] = \exp\left\{ -\sum_{|\tilde{n}| < \bar{n}} \left[\tfrac{\epsilon}{2}\mu^2\varphi_{\tilde{n}}^2 + \tfrac{\epsilon}{8}\lambda\varphi_{\tilde{n}}^4 + \tfrac{M}{2\epsilon}(\varphi_{\tilde{n}} - \varphi_{\tilde{n}-1})^2 \right] \right.$$

$$\left. -\tfrac{M}{2\epsilon}(\varphi_{\bar{n}} - \varphi_{\bar{n}-1})^2 \right\}, \tag{1}$$

where $p_>$ and $p_<$ are only constrained by the overall normalization of $p[\varphi]$ and we will consider the limit $\epsilon \to 0$. This statistical system cannot be reduced to a system with a finite number of degrees of freedom since the probability for the occurrence of specific values of the "border variables" $\varphi_{\bar{n}}, \varphi_{-\bar{n}}$ depends on the values of the external variables φ_m and their probability distribution. The statistical information about this system is incomplete, if we do not specify the probability distribution $p_>p_<$ for the external variables φ_m completely.

Local observables are constructed from the local variables $\varphi_{\tilde{n}}$. As usual, their expectation values are computed by "functional integrals" where the probability distribution $p[\varphi]$ appears as a weight factor. We will ask the question what is the minimal amount of information about the probability distribution for the external variables φ_m which is necessary for a computation of expectation values of local observables. One finds that this information can be summarized in "states" $|\psi\rangle, \{\psi|$ that can be represented as ordinary functions $\{\psi(\varphi_{\bar{n}})|, |\psi(\varphi_{-\bar{n}})\}$. Since these functions depend each only on one variable, the specification of the states contains much less information than the full probability distribution $p_> p_<$ which depends on infinitely many variables $\varphi_{m \geq \bar{n}}$, $\varphi_{m \leq -\bar{n}}$. The states contain the minimal information for "local questions" and are therefore the appropriate quantities for our formulation of incomplete statistics. We will see in Sect. 6 that any further information about the probability distributions $p_> [\varphi_m], p_< [\varphi_m]$ beyond the one contained in the state vectors is actually irrelevant for the computation of expectation values of local observables.

The expectation values of all local observables can be computed from the knowledge of the local probability distribution and the states $|\psi\}$ and $\{\psi|$. For this computation one associates to every local observable $A[\varphi]$ an appropriate operator $\hat{A}$ and finds the prescription familiar from quantum mechanics

$$\langle A[\varphi] \rangle = \{\psi|\hat{A}|\psi\} \,. \tag{2}$$

There is a unique mapping $A[\varphi] \to \hat{A}$ for every local observable which can be expressed in terms of an appropriate functional integral. We find that for simple observables $A[\varphi]$ the operators $\hat{A}$ correspond precisely to familiar operators in quantum mechanics. For example, the observable $\varphi(\tilde{n})$ can be associated to the operator $\hat{Q}(\tau)$ in the Heisenberg picture where time is analytically continued, $\tau = it$, and $\tilde{n} = \tau/\epsilon$. f Local correlation functions involving derivatives may be ambiguous in the continuum limit $\epsilon \to 0$. This problem is well known in functional integral formulations of quantum field theories. We show how to avoid this problem by defining correlations in terms of equivalence classes of observables. In fact, two observables $A_1[\varphi]$, $A_2[\varphi]$ can sometimes be represented by the same operator $\hat{A}$. In this case $A_1[\varphi]$ and $A_2[\varphi]$ are equivalent since they cannot be distinguished by their expectation values for arbitrary states. They have the same expectation values for all possible probability distributions. We argue that the concept of correlation functions should be based on the equivalence classes of observables rather than on specific implementations. Equivalent observables should lead to equivalent correlations. For this purpose we define a product between equivalence classes of observables which can be associated to the product of operators. For example, we associate a non-commutative product $\varphi(\tilde{n}_1) \circ \varphi(\tilde{n}_2)$ to the operator product $\hat{Q}(\tau_1)\hat{Q}(\tau_2)$. It is striking how the non-commutativity of quantum mechanics arises directly from the question what are meaningful correlation functions. We find that the "quantum correlation" based on $\varphi_1 \circ \varphi_2$ has better "robustness properties" as compared to the usual classical correlation. We hope that these considerations shed new light on the question if quantum mechanics can find a formulation in terms of classical statistics [2] or general statistics [3].

2   States and Operators

Consider a discrete ordered set of continuous variables $\varphi_n \equiv \varphi(\tau)$, $\tau = \epsilon n$, $n \in Z$ and a normalized probability distribution $p(\{\varphi_n\}) \equiv p[\varphi] = \exp(-S[\varphi])$ with $\int D\varphi e^{-S[\varphi]} \equiv \prod_n (\int_{-\infty}^{\infty} d\varphi_n) p[\varphi] = 1$. We will assume that the action S is local in a range $-\bar{\tau} < \tau < \bar{\tau}$, i.e.

$$S = -\ln p = \int_{-\bar{\tau}}^{\bar{\tau}} d\tau' \mathcal{L}(\tau') + S_>(\bar{\tau}) + S_<(-\bar{\tau}),$$

$$\mathcal{L}(\tau') = V(\varphi(\tau'), \tau') + \tfrac{1}{2} Z(\tau')(\partial_{\tau'} \varphi(\tau'))^2. \tag{3}$$

Here we have used a continuum notation ($n_{1,2} = \tau_{1,2}/\epsilon$) which can be translated into a discrete language by

$$\int_{\tau_1}^{\tau_2} d\tau' \mathcal{L}(\tau') = \epsilon \sum_{n=n_1+1}^{n_2-1} \mathcal{L}_n + \tfrac{\epsilon}{2} \left[V_{n_2}(\varphi_{n_2}) + V_{n_1}(\varphi_{n_1}) \right]$$

$$+ \tfrac{\epsilon}{4} \left[Z_{n_2} \left(\tfrac{\varphi_{n_2} - \varphi_{n_2-1}}{\epsilon} \right)^2 + Z_{n_1} \left(\tfrac{\varphi_{n_1+1} - \varphi_{n_1}}{\epsilon} \right)^2 \right] \tag{4}$$

with

$$\mathcal{L}_n = V_n(\varphi_n) + \tfrac{Z_n}{4\epsilon^2} \{ (\varphi_{n+1} - \varphi_n)^2 + (\varphi_n - \varphi_{n-1})^2 \}. \tag{5}$$

This corresponds to a discrete derivative

$$(\partial_\tau \varphi(\tau))^2 = \tfrac{1}{2} \left\{ \left(\tfrac{\varphi(\tau+\epsilon) - \varphi(\tau)}{\epsilon} \right)^2 + \left(\tfrac{\varphi(\tau) - \varphi(\tau-\epsilon)}{\epsilon} \right)^2 \right\}$$

$$= \tfrac{1}{2\epsilon^2} \left\{ (\varphi_{n+1} - \varphi_n)^2 + (\varphi_n - \varphi_{n-1})^2 \right\}. \tag{6}$$

The boundary terms in (4) are chosen such that $S_>(\bar{\tau})$ is independent of all $\varphi(\tau')$ with $\tau' < \bar{\tau}$ whereas $S_<(-\bar{\tau})$ only depends on $\varphi(\tau' \leq -\bar{\tau})$. Except for the overall normalization of p no additional assumptions about the form of $S_>(\bar{\tau})$ and $S_<(-\bar{\tau})$ will be made. In case of S being local also at $\bar{\tau}$ we note that $S_>(\bar{\tau})$ contains a term $\tfrac{\epsilon}{2}[V(\varphi(\bar{\tau}), \bar{\tau}) + V(\varphi(\bar{\tau} + \epsilon), \bar{\tau} + \epsilon)] + \tfrac{\epsilon}{4}(Z(\bar{\tau}) + Z(\bar{\tau} + \epsilon)) \left(\tfrac{\varphi(\bar{\tau}+\epsilon) - \varphi(\bar{\tau})}{\epsilon} \right)^2$, which involves a product $\varphi(\bar{\tau} + \epsilon)\varphi(\bar{\tau})$ and therefore links the variables with $\tau > \bar{\tau}$ to the ones with $\tau \leq \bar{\tau}$.

We are interested in local observables $A[\varphi; \tau]$ which depend only on those $\varphi(\tau')$ where $\tau - \tfrac{\delta}{2} \leq \tau' \leq \tau + \tfrac{\delta}{2}$. (We assume $-\bar{\tau} < \tau - \tfrac{\delta}{2}, \bar{\tau} > \tau + \tfrac{\delta}{2}$.) As usual, the expectation value of A is

$$\langle a(\tau) \rangle = \int D\varphi A[\varphi; \tau] e^{-S[\varphi]}. \tag{7}$$

As mentioned in the introduction, our investigation concerns the question what we can learn about expectation values of local observables and suitable products thereof in a situation where we have no or only partial information about $S_>(\bar{\tau})$ and $S_<(-\bar{\tau})$. It is obvious that the full information contained in S is not needed if only expectation values of local observables of the type (7) are to be computed. On the other hand, $\langle a(\tau) \rangle$ cannot be completely independent of $S_>(\bar{\tau})$ and $S_<(-\bar{\tau})$ since the next-neighbor interactions (4) relate "local variables" $\varphi(\tau - \tfrac{\delta}{2} < \tau' < \tau + \tfrac{\delta}{2})$ to the "exterior variables" $\varphi(\tau' > \bar{\tau})$ and $\varphi(\tau' < -\bar{\tau})$.

In order to establish the necessary amount of information needed from $S_>(\bar{\tau})$ and $S_<(-\bar{\tau})$ we first extend $S_>$ and $S_<$ to values $|\tau| < \bar{\tau}$

$$S_<(\tau_1) = S_<(-\bar{\tau}) + \int_{-\bar{\tau}}^{\tau_1} d\tau' \mathcal{L}(\tau') \ , \ \ S_>(\tau_2) = S_>(\bar{\tau}) + \int_{\tau_2}^{\bar{\tau}} d\tau' \mathcal{L}(\tau') , \quad (8)$$

where we note the general identity

$$S_>(\tau) + S_<(\tau) = S . \quad (9)$$

The expectation value (7) can be written as

$$\langle a(\tau) \rangle = \int d\varphi(\tau + \tfrac{\delta}{2}) \int d\varphi(\tau - \tfrac{\delta}{2}) \int D\varphi_{(\tau' > \tau + \frac{\delta}{2})} e^{-S_>(\tau + \frac{\delta}{2})}$$

$$\times \int D\varphi_{(\tau - \frac{\delta}{2} < \tau' < \tau + \frac{\delta}{2})} A[\varphi; \tau] \exp\{-\int_{\tau - \frac{\delta}{2}}^{\tau + \frac{\delta}{2}} d\tau'' \mathcal{L}(\tau'')\}$$

$$\times \int D\varphi_{(\tau' < \tau - \frac{\delta}{2})} e^{-S_<(\tau - \frac{\delta}{2})} . \quad (10)$$

This suggests the introduction of the "states"

$$|\psi(\varphi(\tau - \tfrac{\delta}{2}); \ \tau - \tfrac{\delta}{2})\} = \int D\varphi_{(\tau' < \tau - \frac{\delta}{2})} e^{-S_<(\tau - \frac{\delta}{2})} ,$$

$$\{\psi(\varphi(\tau + \tfrac{\delta}{2}); \ \tau + \tfrac{\delta}{2})| = \int D\varphi_{(\tau' > \tau + \frac{\delta}{2})} e^{-S_>(\tau + \frac{\delta}{2})} , \quad (11)$$

and the operator

$$\hat{A}_\delta(\varphi(\tau + \tfrac{\delta}{2}), \ \varphi(\tau - \tfrac{\delta}{2}); \ \tau)$$

$$= \int D\varphi_{(\tau - \frac{\delta}{2} < \tau' < \tau + \frac{\delta}{2})} A[\varphi; \tau] \exp\left\{-\int_{\tau - \frac{\delta}{2}}^{\tau + \frac{\delta}{2}} d\tau'' \mathcal{L}(\tau'')\right\} . \quad (12)$$

We note that $|\psi\}$ is a function of $\varphi(\tau - \tfrac{\delta}{2})$ since the latter appears in $S_<(\tau - \tfrac{\delta}{2})$ and is not included in the ("functional") integration (11). Similarly, $\{\psi|$ depends on $\varphi(\tau + \tfrac{\delta}{2})$ whereas $\hat{A}$ is a function of the two variables $\varphi(\tau + \tfrac{\delta}{2})$ and $\varphi(\tau - \tfrac{\delta}{2})$. Using a notation where $|\psi\}$ and $\{\psi|$ are interpreted as (infinite dimensional) vectors and $\hat{A}$ as a matrix, one has

$$\langle a(\tau) \rangle = \{\psi(\tau + \tfrac{\delta}{2}) \hat{A}_\delta(\tau) \psi(\tau - \tfrac{\delta}{2})\} \quad (13)$$

$$\equiv \int d\varphi_2 \int d\varphi_1 \{\psi(\varphi_2; \tau + \tfrac{\delta}{2}) | \hat{A}_\delta(\varphi_2, \varphi_1; \tau) | \psi(\varphi_1; \tau - \tfrac{\delta}{2})\} .$$

This form resembles already the well-known prescription for expectation values of operators in quantum mechanics. In contrast to quantum mechanics (13) still involves, however, two different state vectors.

The mapping $A[\varphi; \tau] \to \hat{A}_\delta(\tau)$ can be computed (cf. (12)) if $\mathcal{L}(\tau')$ is known for $|\tau'| < \bar{\tau}$. The only information needed from $S_>(\bar{\tau})$ and $S_<(-\bar{\tau})$ is therefore contained in the two functions $\{\psi(\varphi)|$ and $|\psi(\varphi)\}$! The specification of these states (wave functions) at $\bar{\tau}$ and $-\bar{\tau}$ and of $\mathcal{L}(|\tau| < \bar{\tau})$ completely determines the expectation values of *all* local observables!

We will see below the close connection to the states in quantum mechanics. In our context we emphasize that for any given S these states can be computed as well defined functional integrals (11). Due to (9) they obey the normalization

$$\{\psi(\tau)\psi(\tau)\} \equiv \int d\varphi\{\psi(\varphi;\tau)||\psi(\varphi;\tau)\} = 1 \,. \tag{14}$$

Incomplete statistics explores statements that can be made for local observables and appropriate products thereof without using information about $S_>$ or $S_<$ beyond the one contained in the states $|\psi\}$ and $\{\psi|$.

3   Evolution in Euclidean Time

For a "locality interval" $\delta > 0$ the expression (13) involves states at different locations or "Euclidean times" $\tau + \frac{\delta}{2}$ and $\tau - \frac{\delta}{2}$. We aim for a formulation where only states at the same τ appear. We therefore need the explicit mapping from $|\psi(\tau - \frac{\delta}{2})\}$ to a reference point $|\psi(\tau)\}$ and similar for $\{\psi(\tau + \frac{\delta}{2})|$. This mapping should also map $\hat{A}_\delta$ to a suitable operator such that the structure (13) remains preserved. The dependence of states and operators on the Euclidean time τ is described by evolution operators ($\tau_2 > \tau_1$, $\tau_2 > \tau_f$, $\tau_i > \tau_1, \tau_f = \tau + \frac{\delta}{2}$, $\tau_i = \tau - \frac{\delta}{2}$),

$$\begin{aligned}
|\psi(\tau_2)\} &= \hat{U}(\tau_2, \tau_1)|\psi(\tau_1)\} \,, \\
\{\psi(\tau_1)| &= \{\psi(\tau_2)|\hat{U}(\tau_2, \tau_1) \,, \\
\hat{A}(\tau_2, \tau_1) &= \hat{U}(\tau_2, \tau_f)\hat{A}(\tau_f, \tau_i)\hat{U}(\tau_i, \tau_1) \,,
\end{aligned} \tag{15}$$

or differential operator equations ($\epsilon \to 0$)

$$\partial_\tau|\psi(\tau)\} = -\hat{H}(\tau)|\psi(\tau)\} \,. \tag{16}$$

The evolution operator has an explicit representation as a functional integral,

$$\hat{U}(\varphi(\tau_2), \varphi(\tau_1); \tau_2, \tau_1) = \int D\varphi_{(\tau_1 < \tau' < \tau_2)} \exp\left\{-\int_{\tau_1}^{\tau_2} d\tau'' \mathcal{L}(\tau'')\right\} \,, \tag{17}$$

and obeys the composition property ($\tau_3 > \tau_2 > \tau_1$)

$$\hat{U}(\tau_3, \tau_2)\hat{U}(\tau_2, \tau_1) = \hat{U}(\tau_3, \tau_1) \tag{18}$$

with

$$\hat{U}(\varphi_2, \varphi_1; \tau, \tau) = \delta(\varphi_2 - \varphi_1) \,. \tag{19}$$

It can therefore be composed as a product of transfer matrices or "infinitesimal" evolution operators

$$\hat{U}(\tau + \epsilon, \tau) = e^{-\epsilon \hat{H}(\tau + \frac{\epsilon}{2})} . \tag{20}$$

In case of translation symmetry for the local part of the probability distribution, i. e. for V and Z independent of τ, we note the symmetry in $\varphi_1 \leftrightarrow \varphi_2$

$$\hat{U}(\tau + \epsilon, \tau) = \hat{U}^T(\tau + \epsilon, \tau) , \ \hat{H}(\tau + \tfrac{\epsilon}{2}) = \hat{H}^T(\tau + \tfrac{\epsilon}{2}) = \hat{H} . \tag{21}$$

In this case the real symmetric matrix $\hat{H}$ has real eigenvalues E_n. Then the general solution of the differential equation (16) may be written in the form

$$|\psi(\tau)\} = \sum_n \psi_0^{(n)} e^{-E_n \tau}, \ \{\psi(\tau)| = \sum_n \bar{\psi}_0^{(n)} e^{E_n \tau} \tag{22}$$

where $\psi_0^{(n)}$ and $\bar{\psi}_0^{(n)}$ are eigenvectors of $\hat{H}$ with eigenvalues E_n. Here we recall that the construction (11) implies that $|\psi\}$ and $\{\psi|$ are real positive functions of φ for every τ. This restricts the allowed values of the coefficients $\psi_0^{(n)}, \bar{\psi}_0^{(n)}$.

We next want to compute the explicit form of the Hamilton operator $\hat{H}$. It is fixed uniquely by the functional integral representation (17) for $\hat{U}$. In order to obey the defining equation (20), the Hamilton operator $\hat{H}$ must fulfill for arbitrary $|\psi(\varphi)\}$ the relation (with $Z = Z(\tau + \frac{\epsilon}{2}) = \frac{1}{2}(Z(\tau + \epsilon) + Z(\tau))$

$$\int d\varphi_1 \hat{H}(\varphi_2, \varphi_1) |\psi(\varphi_1)\} = -\lim_{\epsilon \to 0} \frac{1}{\epsilon} \Big\{ \int d\varphi_1 \tag{23}$$

$$\exp \left[-\tfrac{\epsilon}{2}(V(\varphi_2) + V(\varphi_1)) - \tfrac{Z}{2\epsilon}(\varphi_2 - \varphi_1)^2 \right] |\psi(\varphi_1)\} - |\psi(\varphi_2)\} \Big\} .$$

The solution of this equation can be expressed in terms of the operators

$$\hat{Q}(\varphi_2, \varphi_1) = \varphi_1 \delta(\varphi_2 - \varphi_1) ,$$
$$\hat{P}^2(\varphi_2, \varphi_1) = -\delta(\varphi_2 - \varphi_1) \frac{\partial^2}{\partial \varphi_1^2} , \tag{24}$$

as

$$\hat{H}(\tau) = V(\hat{Q}, \tau) + \frac{1}{2Z(\tau)} \hat{P}^2 . \tag{25}$$

This can be established by using under the φ_1-integral the replacement

$$e^{-\frac{Z}{2\epsilon}(\varphi_2 - \varphi_1)^2} \to \left(\frac{2\pi\epsilon}{Z} \right)^{1/2} \delta(\varphi_2 - \varphi_1) \exp \left(\frac{\epsilon}{2Z} \frac{\partial^2}{\partial \varphi_1^2} \right) , \tag{26}$$

which is valid by partial integration if the integrand decays fast enough for $|\varphi_1| \to \infty$. We note that the operators $\hat{Q}$ and $\hat{P}^2$ do not commute, e.g.

$$[\hat{P}^2, \hat{Q}](\varphi_2, \varphi_1) = -2\delta(\varphi_2 - \varphi_1) \frac{\partial}{\partial \varphi_1} . \tag{27}$$

The Hamilton operator can be used in order to establish the existence of the inverse of the "infinitesimal" evolution operator, $\hat{U}^{-1}(\tau + \epsilon, \tau) = e^{\epsilon \hat{H}(\tau + \frac{\epsilon}{2})}$. Then the inverse $\hat{U}^{-1}(\tau_2, \tau_1)$ is defined by the multiplication of "infinitesimal" inverse evolution operators, and we can extend the composition property (18) to arbitrary τ be defining

$$\hat{U}(\tau_2, \tau_1) = \hat{U}^{-1}(\tau_1, \tau_2) \qquad (28)$$

for $\tau_2 < \tau_1$. (For a given dependence of $\hat{U}$ on the variables τ_2 and τ_1 the matrix $\hat{U}(\tau_1, \tau_2)$ obtains from $\hat{U}(\tau_2, \tau_1)$ by a simple exchange of the arguments τ_1 and τ_2.) Using (15), this allows us to write the expectation value of a local observable in a form involving states at the same τ-variable

$$\langle a(\tau) \rangle = \{\psi(\tau)\hat{U}(\tau, \tau + \tfrac{\delta}{2})\hat{A}_\delta(\tau)\hat{U}(\tau - \tfrac{\delta}{2}, \tau)\psi(\tau)\} . \qquad (29)$$

4   Schrödinger and Heisenberg Operators

In this section we want to exploit further the mapping between incomplete statistics and quantum mechanics for situations where expectation values like $\langle \varphi(\tau) \rangle$ may depend on τ. A typical question one may ask within incomplete classical statistics is the following: Given a large set of measurements of observables with support at a given value $\tau = 0$, like $\langle \varphi^p(0) \rangle$, $\langle (\partial_\tau \varphi(0))^{p'} \rangle$, etc., what can one predict for the expectation values of similar observables at some other location $\tau \neq 0$? It is obvious that the evolution operator $\hat{U}$ is the appropriate tool to tackle this type of questions.

The existence of the inverse evolution operator allows us to associate to an observable $A(\tau)$ the operator $\hat{A}_S(\tau)$ in the Schrödinger representation (cf. (29))

$$\hat{A}_S(\tau) = \hat{U}(\tau, \tau + \tfrac{\delta}{2})\hat{A}_\delta(\tau)\hat{U}(\tau - \tfrac{\delta}{2}, \tau) . \qquad (30)$$

The expectation value of the observable A can be expressed by the expectation value of the operator $\hat{A}_S$ in a way analogous to quantum mechanics

$$< a(\tau) > = \{\psi(\tau)|\hat{A}_S(\tau)|\psi(\tau)\} = \mathrm{Tr}\rho(\tau)\hat{A}_S(\tau) . \qquad (31)$$

For the second identity we have introduced the "density matrix"

$$\rho(\varphi_1, \varphi_2, \tau) = |\psi(\varphi_1, \tau)\}\{\psi(\varphi_2, \tau)| = \int D\varphi_{(\tau' \neq \tau)} e^{-S(\varphi_1, \varphi_2)} ,$$
$$\mathrm{Tr}\rho(\tau) = 1 , \qquad (32)$$

where $S(\varphi_1, \varphi_2)$ obtains from S by replacing $\varphi(\tau) \to \varphi_1$ for all "kinetic" terms involving $\varphi(\tau' < \tau)$ and $\varphi(\tau) \to \varphi_2$ for those involving $\varphi(\tau' > \tau)$, whereas for potential terms $e^{-\epsilon V(\varphi(\tau))} \to e^{-\frac{\epsilon}{2}(V(\varphi_1) + V(\varphi_2))}$.

In order to make the transition to the Heisenberg picture, we may select a reference point $\tau = 0$ and define

$$\hat{U}(\tau) \equiv \hat{U}(\tau, 0) , \; \rho \equiv \rho(\tau = 0) , \; \rho(\tau) = \hat{U}(\tau)\rho\hat{U}^{-1}(\tau) . \qquad (33)$$

This specifies the Heisenberg picture for the τ-dependent operators

$$\hat{A}_H(\tau) = \hat{U}^{-1}(\tau)\hat{A}_S(\tau)\hat{U}(\tau)$$
$$\langle a(\tau) \rangle = \mathrm{Tr}\rho\hat{A}_H(\tau) \qquad (34)$$

We note that for two local observables A_1, A_2 the linear combinations $A = \alpha_1 A_1 + \alpha_2 A_2$ are also local observables. The associated operators obey the same linear relations $\hat{A} = \alpha_1 \hat{A}_1 + \alpha_2 \hat{A}_2$, where $\hat{A}$ stands for $\hat{A}_\delta$, $\hat{A}_S$ or $\hat{A}_H$. The relation (34) is the appropriate formula to answer the question at the beginning of this section. One may use the set of measurements of expectation values at $\tau = 0$ to gather information about ρ. Once ρ is determined with sufficient accuracy, the expectation values $\langle a(\tau) \rangle$ can be computed. Of course, this needs a computation of the explicit form of the Heisenberg operator $\hat{A}_H(\tau)$.

It is instructive to observe that some simple local observables have a τ-independent operator representation in the Schrödinger picture. This is easily seen for observables $A(\tau)$ which depend only on the variable $\varphi(\tau)$. The mapping reads

$$A(\tau) = f(\varphi(\tau)) \to \hat{A}_S(\tau) = f(\hat{Q}) \,. \tag{35}$$

Observables depending only on one variable $\varphi(\tau)$ therefore have the Heisenberg representation (cf. (35))

$$A(\tau) = f(\varphi(\tau)) \to \hat{A}_H(\tau) = f(\hat{Q}(\tau)) \,. \tag{36}$$

Here we have used the definition

$$\hat{Q}(\tau) = \hat{U}^{-1}(\tau)\hat{Q}\hat{U}(\tau) \,. \tag{37}$$

More generally, one finds for products of functions depending on the variables $\varphi(\tau_1), \varphi(\tau_2)...\varphi(\tau_n)$ with $\tau_1 < \tau_2 < ...\tau_n$ the Heisenberg operator

$$\begin{aligned} A(\tau_1, ...\tau_n) &= f_1(\varphi(\tau_1)f_2(\varphi(\tau_2))...f_n(\varphi(\tau_n)) \longrightarrow \\ \hat{A}_H(\tau) &= \hat{U}^{-1}(\tau_n)f_n(\hat{Q})\hat{U}(\tau_n, \tau_{n-1})...\hat{U}(\tau_2, \tau_1)f_1(\hat{Q})\hat{U}(\tau_1) \\ &= f_n(\hat{Q}(\tau_n))...f_2(\hat{Q}(\tau_2))f_1(\hat{Q}(\tau_1)) \,. \end{aligned} \tag{38}$$

This important relation follows directly from the definitions (12), (30), (34). We observe that $\hat{A}_H$ depends on the variables τ_i which are the arguments of A but shows no dependence on the reference point τ. (Only $\hat{A}_\delta$ and $\hat{A}_S$ depend on τ.)

We can use (38) to find easily the Heisenberg operators for observables involving "derivatives", e.g.

$$\begin{aligned} A &= \tilde{\partial}_\tau \varphi(\tau_1) = \tfrac{1}{2\epsilon}(\varphi(\tau_1 + \epsilon) - \varphi(\tau_1 - \epsilon)) \,, \\ \hat{A}_H &= \tfrac{1}{2\epsilon}\{\hat{U}^{-1}(\tau_1 + \epsilon)\hat{Q}\hat{U}(\tau_1 + \epsilon) - \hat{U}^{-1}(\tau_1 - \epsilon)\hat{Q}\hat{U}(\tau_1 - \epsilon)\} \\ &= -\tfrac{1}{Z(\tau_1)}\hat{R}(\tau_1) + O(\epsilon) \,, \end{aligned} \tag{39}$$

where we have assumed that $\hat{H}$ is a smooth function of τ. Here $\hat{R}$ is defined by

$$\hat{R}(\varphi_2, \varphi_1) = \delta(\varphi_2 - \varphi_1)\tfrac{\partial}{\partial \varphi_1} \,, \quad \hat{R}^2 = -\hat{P}^2 \,, \tag{40}$$

and we use, similar to (37), the definitions

$$\hat{R}(\tau) = \hat{U}^{-1}(\tau)\hat{R}\hat{U}(\tau) \,, \quad \hat{P}^2(\tau) = \hat{U}^{-1}(\tau)\hat{P}^2\hat{U}(\tau) \,. \tag{41}$$

Two different definitions of derivatives can lead to the same operator $\hat{A}_H$. An example is the observable

$$A = \partial_\tau^> \varphi(\tau_1) = \tfrac{1}{\epsilon}(\varphi(\tau_1 + \epsilon) - \varphi(\tau_1)).\qquad(42)$$

Up to terms of order ϵ the associated Heisenberg operator is again given by $\hat{A}_H = -Z(\tau_1)^{-1}\hat{U}^{-1}(\tau_1)\hat{R}\hat{U}(\tau_1)$ and therefore the same as for $\tilde{\partial}_\tau \varphi(\tau_1)$ in (39). Applying the same procedure to the squared derivative observable (6) yields

$$A = (\partial_\tau \varphi)^2(\tau_1) \longrightarrow \hat{A}_H = \tfrac{1}{\epsilon Z} - \tfrac{1}{Z^2}\hat{P}^2(\tau_1),\qquad(43)$$

where we have assumed for simplicity a τ-independent Hamiltonian $\hat{H}$. It is remarkable that this operator differs from the square of the Heisenberg operator associated to $\tilde{\partial}_\tau \varphi(\tau_1)$ by a constant which diverges for $\epsilon \to 0$. Indeed, one finds

$$A = (\tilde{\partial}_\tau \varphi(\tau_1))^2 \to \hat{A}_H = \tfrac{1}{2\epsilon Z} - \tfrac{1}{Z^2}\hat{P}^2(\tau_1),$$
$$A = (\partial_\tau^> \varphi(\tau_1))^2 \to \hat{A}_H = \tfrac{1}{\epsilon Z} - \tfrac{1}{Z^2}\hat{P}^2(\tau_1).\qquad(44)$$

Equation (44) teaches us that the product of derivative observables with other observables can be ambiguous in the sense that the associated operator and expectation value depend very sensitively on the precise definition of the derivative. This ambiguity of the derivative observables in the continuum limit is an unpleasant feature for the formulation of correlation functions. It survives when the discussion is extended to observables that are smoothened over a certain interval instead of being strictly local [1].

In the next sections we will see how this problem is connected with the concept of quantum correlations. We will argue that the ambiguity in the classical correlation may be the basic ingredient why a description of our world in terms of quantum statistics is superior to the use of classical correlation functions.

5   Correlation Functions

A basic concept for any statistical description are correlation functions for a number of observables $A_1[\varphi], A_2[\varphi], \ldots$ In particular, a two-point function is given by the expectation value of an associative product of two observables $A_1[\varphi]$ and $A_2[\varphi]$. For local observables A_1, A_2 the product should again be a local observable which must be defined uniquely in terms of the definitions of A_1 and A_2. This requirement, however, does not fix the definition of the correlation uniquely. The standard "classical product", i.e. the simple multiplication of the functionals $A_1[\varphi] \cdot A_2[\varphi]$ (in the same sense as the "pointwise" multiplication of functions) fulfills the general requirements[1] for a correlation function. Other definitions can be conceived as well. In this section we will introduce a quantum correlation which equals the classical ("pointwise") correlation only

---

[1] This holds provided that the product results in a meaningful observable with finite expectation value.

for τ-ordered non-overlapping observables. In contrast, for two local observables with overlapping support we will find important differences between the quantum and classical correlation. In particular, we will discover the effects of the non-commutativity characteristic for quantum mechanics.

Incomplete statistics draws our attention to an important issue in the formulation of meaningful correlation functions. Consider the two versions of the derivative observable $\tilde{\partial}_\tau \varphi$ and $\partial_\tau^> \varphi$ defined by (39) and (40), respectively. In the continuum limit ($\epsilon \to 0$) they are represented by the same operator $\hat{A}_H$. In consequence, both definitions lead to the same expectation value for any state $|\psi\}, \{\psi|$. The two versions of derivative observables cannot be distinguished by any measurement and should therefore be identified. On the other hand, the classical products $\tilde{\partial}_\tau \varphi(\tau_1) \cdot \tilde{\partial}_\tau \varphi(\tau_2)$ and $\partial_\tau^> \varphi(\tau_1) \cdot \partial_\tau^> \varphi(\tau_2)$ are represented by different operators for $\tau_1 = \tau_2$, as can be seen from (44). This means that the two versions of derivative observables lead to different classical correlation functions! Obviously, this situation is unsatisfactory since for $\epsilon \to 0$ no difference between the two versions could be "measured" for the observables themselves. We find this disease unacceptable for a meaningful correlation and require as a criterion for a meaningful correlation function that two observables which have the same expectation values for all (arbitrary) probability distributions should also have identical correlation functions. We have shown that two observables which are represented by the same Heisenberg operator have indeed the same expectation values for all possible probability distributions and should therefore be considered as equivalent. They should therefore lead to indistinguishable correlation functions.

As we have already established, the two derivative observables $A_1 = \tilde{\partial}_\tau \varphi(\tau)$ and $A_2 = \partial_\tau^> \varphi(\tau)$ are indistinguishable in the continuum limit, whereas their classical correlations are not. We may therefore conclude that the classical correlation $A_1 \cdot A_2$ is not a meaningful correlation function. In this section we propose the use of a different correlation based on a quantum product $A_1 \circ A_2$. By construction, this correlation will always obey our criterion of "robustness" with respect to the precise choice of the observables. It should therefore be considered as an interesting alternative to the classical correlation. At this place we only note that the "robustness problem" is not necessarily connected to the continuum limit. The mismatch between indistinguishable observables and distinguishable "classical" correlations can appear quite generally also for $\epsilon > 0$.

Our formulation of a quantum correlation will be based on the concepts of equivalent observables and products defined for equivalence classes. In fact, the mapping $A(\tau) \to \hat{A}_H(\tau)$ is not necessarily invertible on the space of all observables $A(\tau)$. This follows from the simple observation that already the map (12) contains integrations. Two different integrands (observables) could lead to the same value of the integral (operator) for arbitrary fixed boundary values $\varphi(\tau - \frac{\delta}{2})$, $\varphi(\tau + \frac{\delta}{2})$. It is therefore possible that two different observables $A_a(\tau)$ and $A_b(\tau)$ can be mapped into the same Heisenberg operator $\hat{A}_H(\tau)$. Since the expectation values can be computed from $\hat{A}_H(\tau)$ and ρ only, no distinction between $\langle a_a \rangle$ and $\langle a_b \rangle$ can then be made for arbitrary ρ. All local observables $A(\tau)$ which correspond to the same operator $\hat{A}_H(\tau)$ are equivalent.

We are interested in structures that only depend on the equivalence classes of observables. Addition of two observables and multiplication with a scalar can simply be carried over to the operators. This is not necessarily the case, however, for the (pointwise) multiplication of two observables. If $A_a(\tau)$ and $A_b(\tau)$ are both mapped into $\hat{A}_H(\tau)$ and a third observable $B(\tau)$ corresponds to $\hat{B}_H(\tau)$, the products $A_a \cdot B$ and $A_b \cdot B$ may nevertheless be represented by different operators. It is then easy to construct states where $\langle a_a B \rangle \neq \langle a_b B \rangle$ and the pointwise product does not depend only on the equivalence class.

On the other hand, the (matrix) product of two operators $\hat{A}_H \hat{B}_H$ obviously refers only to the equivalence class. It can be implemented on the level of observables by defining a unique "standard representative" of the equivalence class as

$$\bar{A}[\varphi, \tau] = F[\hat{A}_H(\tau)]. \tag{45}$$

Using the mapping $A[\tau] \rightarrow \hat{A}_H(\tau)$ (12), (30), (34), we define the quantum product of two observables as

$$A(\varphi, \tau) \circ B(\varphi, \tau) = F[\hat{A}_H(\tau)\hat{B}_H(\tau)] \equiv (A \circ B)[\varphi, \tau]. \tag{46}$$

This product is associative, but not commutative. (By definition, the operator associated to the observable $(A \circ B)(\varphi, \tau)$ is $\hat{A}_H(\tau)\hat{B}_H(\tau)$ and the product $A \circ B$ is isomorphic to the "matrix multiplication" $\hat{A}\hat{B}$ if restricted to the subspace of operators $\bar{A} = F[\hat{A}], \bar{B} = F[\hat{B}]$.) The correlations (e.g. expectation values of products of observables) formed with the product $\circ$ reflect the non-commutative structure of quantum mechanics. This justifies the name "quantum correlations". Nevertheless, we emphasize that the "quantum product" $\circ$ can also be viewed as just a particular structure among "classical observables".

The definition of the quantum product is unique on the level of operators. On the level of the classical observables, it is, however, not yet fixed uniquely by (46). The precise definition obviously depends on the choice of a standard representation $F[\hat{A}_H(\tau)]$ for the equivalence class of observables represented by $\hat{A}_H$. We will choose a linear map $F[\alpha_1 \hat{A}_{H,1} + \alpha_2 \hat{A}_{H,2}] = \alpha_1 F[\hat{A}_{H,1}] + \alpha_2 F[\hat{A}_{H,2}]$ with the property that it inverses the relation (38). For "time-ordered" $\tau_1 < \tau_2 < ... \tau_n$ the map F should then obey

$$F[f_n(\hat{Q}(\tau_n))...f_2(\hat{Q}(\tau_2))f_1(\hat{Q}(\tau_1))] = f_1(\varphi(\tau_1))f_2(\varphi(\tau_2))...f_n(\varphi(\tau_n)). \tag{47}$$

It is easy to see how this choice exhibits directly the noncommutative property of the quantum product between two observables. As an example let us consider the two observables $\varphi(\tau_1)$ and $\varphi(\tau_2)$ with $\tau_1 < \tau_2$. The quantum product or quantum correlation depends on the ordering

$$\varphi(\tau_2) \circ \varphi(\tau_1) = \varphi(\tau_2)\varphi(\tau_1),$$
$$\varphi(\tau_1) \circ \varphi(\tau_2) = \varphi(\tau_2)\varphi(\tau_1) + F[[\hat{Q}(\tau_1), \hat{Q}(\tau_2)]]. \tag{48}$$

The noncommutative property of the quantum product for these operators is directly related to the commutator

$$[\hat{Q}(\tau_1), \ \hat{Q}(\tau_2)] = \hat{U}^{-1}(\tau_1)\hat{Q}\hat{U}(\tau_1, \tau_2)\hat{Q}\hat{U}(\tau_2)$$
$$-\hat{U}^{-1}(\tau_2)\hat{Q}\hat{U}(\tau_2, \tau_1)\hat{Q}\hat{U}(\tau_1) \,. \tag{49}$$

Only for time-ordered arguments the quantum correlation coincides with the classical correlation.

The map F can easily be extended to operators involving derivatives of φ. We concentrate here for simplicity on a translation invariant probability distribution in the local region with constant $Z(\tau) = Z$. The mappings (with $\tau_2 \geq \tau_1 + \epsilon$)

$$
\begin{aligned}
F(\hat{R}(\tau)) &= -Z\partial_\tau^> \varphi(\tau)\,, \\
F(\hat{R}(\tau)\hat{Q}(\tau)) &= -Z\varphi(\tau)\partial_\tau^> \varphi(\tau)\,, \\
F(\hat{R}(\tau_2)\hat{R}(\tau_1)) &= Z^2 \partial_\tau^> \varphi(\tau_2)\partial_\tau^> \varphi(\tau_1)
\end{aligned}
\tag{50}
$$

are compatible with (47). This can be seen by noting that the τ-evolution of $\hat{Q}(\tau)$ according to (37) implies for $\epsilon \to 0$ the simple relation

$$\partial_\tau \hat{Q}(\tau) = [\hat{H}, \hat{Q}(\tau)] = -Z^{-1}\hat{R}(\tau)\,. \tag{51}$$

A similar construction (note $[\hat{Q}(\tau + \epsilon), \hat{Q}(\tau)] = -\epsilon/Z$) leads to

$$F(\hat{R}^2(\tau)) = Z^2(\partial_\tau^> \varphi(\tau))^2 - Z/\epsilon\,, \tag{52}$$

and we infer that the quantum product of derivative observables at equal sites differs from the pointwise product

$$\partial_\tau^> \varphi(\tau) \circ \partial_\tau^> \varphi(\tau) = (\partial_\tau^> \varphi(\tau))^2 - 1/(\epsilon Z)\,. \tag{53}$$

From the relations (48) and (53) it has become clear that the difference between the quantum product and the "pointwise" classical product of two observables is related to their τ-ordering and "overlap". Let us define that two observables $A_1[\varphi]$ and $A_2[\varphi]$ overlap if they depend on variables $\varphi(\tau)$ lying in two overlapping τ-ranges $\mathcal{R}_1$ and $\mathcal{R}_2$. (Here two ranges do not overlap if all τ in $\mathcal{R}_1$ obey $\tau \leq \tau_0$ whereas for $\mathcal{R}_2$ one has $\tau \geq \tau_0$, or vice versa. This implies that non-overlapping observables can depend on at most one common variable $\varphi(\tau_0)$.) With this definition the quantum product is equal to the classical product if the observables do not overlap and are τ-ordered (in the sense that larger τ are on the left side).

In conclusion, we have established a one-to-one correspondence between classical correlations $\varphi(\tau_2)\varphi(\tau_1)$ and the product of Heisenberg operators $\hat{Q}(\tau_2)\hat{Q}(\tau_1)$ provided that the τ-ordering $\tau_2 \geq \tau_1$ is respected. This extends to observables that can be written as sums or integrals over $\varphi(\tau)$ (as, for example, derivative observables) provided the τ-ordering and non-overlapping properties are respected. For well separated observables no distinction between a quantum and classical τ-ordered correlation function would be needed. In particular, this holds also for "smoothened" observables A_i that involve (weighted) averages over $\varphi(\tau)$ in a range $\mathcal{R}_i$ around τ_i. Decreasing the distance between τ_2 and τ_1, the new features of the quantum product $A_1(\tau_2) \circ A_1(\tau_1)$ show up only once the distance becomes small enough so that the two ranges $\mathcal{R}_1$ and $\mathcal{R}_2$ start to overlap. In an extreme form the difference between quantum and classical correlations becomes apparent for derivative observables at the same location. Quite generally, the difference between the quantum and classical product is seen most easily on the level of the associated operators

$$A_1 \circ A_2 \to \hat{A}_1 \hat{A}_2 \,,$$
$$A_1 \cdot A_2 \to T(\hat{A}_1 \hat{A}_2) \,. \tag{54}$$

Here T denotes the operation of τ-ordering. The τ-ordered operator product is commutative $T(\hat{A}_1 \hat{A}_2) = T(\hat{A}_2 \hat{A}_1)$ and associative $T(T(\hat{A}_1 \hat{A}_2)\hat{A}_3) = T(\hat{A}_1 T(\hat{A}_2 \hat{A}_3)) \equiv T(\hat{A}_1 \hat{A}_2 \hat{A}_3)$. As we have seen in the discussion of the derivative observables, it lacks, however, the general property of robustness with respect to the precise definition of the observables. This contrasts with the noncommutative product $\hat{A}_1 \hat{A}_2$. This discussion opens an interesting perspective: The difference between classical and quantum statistics seems to be a question of the appropriate definition of the correlation function. Simple arguments of robustness favor the choice of the quantum correlation! This remark remains valid if we consider averaged or smoothened observables instead of "pointlike" observables [1]. In a sense, the successful description of nature by quantum-mechanical operators and their products gives an "experimental indication" that quantum correlations should be used!

6   Incomplete Classical Statistics, Irrelevant and Inaccessible Information

Our discussion of incomplete classical statistics may perhaps have led to the impression that the quantum mechanical properties are somehow related to the missing information. This is by no means the case! In fact, our investigation of the consequences of incomplete information about the probability distribution was useful in order to focus the attention on the question which information is really necessary to compute the expectation values of local observables. We can now turn back to standard "complete" classical statistics where the full probability distribution $p[\varphi(\tau)]$ is assumed to be known. We concentrate here on a general class of probability distributions which can be factorized in the form $p = p_> p_0 p_<$ according to (1) – it may be called "factorizable" or "F-statistics". For example, all systems which have only local and next-neighbor interactions are of this form. Within F-statistics the states remain defined according to (11).

We emphasize that any additional information contained in $p[\varphi]$ which goes beyond the local distribution $p_0[\varphi]$ and the states $|\psi\}$ and $\{\psi|$ does not change a iota in the expectation values of local observables and their correlations! The additional information is simply *irrelevant* for the computation of local expectation values. A given probability distribution specifies $p_<$ and $p_>$ uniquely. This determines $|\psi\}$ and $\{\psi|$ and we can then continue with the preceding discussion in order to calculate the expectation values of local observables. The precise form of $p_<$ and $p_>$ which has led to the given states plays no role in this computation.

Since all information contained in $p_<$ and $p_>$ beyond the states $|\psi\}$ and $\{\psi|$ is irrelevant for local expectation values, it is also *inaccessible* by any local measurements. In fact, even the most precise measurements of expectation values and correlation functions for arbitrarily many local observables could at best lead to a reconstruction of the states $|\psi\}$ and $\{\psi|$. This sheds new light on the notion of "incompleteness" of the statistical information discussed in this note. In fact, within F-statistics the "incomplete" information contained in the states $|\psi\}$ and $\{\psi|$ constitutes the most complete information that can possibly be gathered by local measurements! Since any real measurement is local in time and space all assumptions about information beyond the states concern irrelevant and inaccessible information and cannot be verified by observation!

7   Conclusions and Discussion

Within a simple example of classical statistics for coupled unharmonic oscillators on a chain we have formulated a description in terms of states and operators in analogy to quantum mechanics. The state vectors and the operators can be expressed in terms of classical functional integrals. Expectation values of classical observables can be evaluated as "quantum mechanical" expectation values of appropriate operators in appropriate states. Typical quantum mechanical results like the relations between the expectation values in stationary states or the uncertainty relation can be taken over to the classical system [1]. The simple fact that quantum-mechanical information can be used in practice to establish properties of expectation values in a standard classical statistical system demonstrates in a simple way that quantum-mechanical features are indeed genuine properties of classical statistical systems. Our procedure inverts the construction of the Euclidean path integral for a quantum mechanical system in the ground state or thermal state [4,5], with a generalization to a wider class of states.

The introduction of "quantum mechanical" operators associated with every local classical observable allows us to define equivalence classes of observables which cannot be distinguished by any measurement of their expectation values. We argue that the definition of the correlation function should be consistent with this equivalence structure. We require that indistinguishable observables must lead to the same correlation function. This leads to the introduction of a quantum correlation within the classical statistical setting. We point out that the quantum correlation constitutes a more robust definition of the correlation function with respect to the precise details of the definition of observables, both for classical and quantum statistical systems. The basic conceptual distinction between quantum statistics and classical statistics disappears in this respect. The similarity can be extended to the emergence of typical characteristics of quantum statistics like the superposition of states and interference for classical statistical systems [1]. This raises the question [2] if it could be possible to understand the mysteries of the basics of quantum mechanics within a formulation of a classical statistical problem with infinitely many degrees of freedom.

References

1. C. Wetterich: hep-th/0104074.
2. A. Einstein, B. Podolski, and N. Rosen: Phys. Rev. **47**, 777 (1935);
   J.S. Bell: Physics **1**, 195 (1964);
   S. Weinberg: Phys. Rev. Lett. **62**, 485 (1989);
   G. 't Hooft: gr-qc/9903084
3. C. Wetterich: Nucl. Phys. B **314**, 40 (1989); Nucl. Phys. B **397**, 299 (1993)
4. R.P. Feynman: Rev. Mod. Phys. **367**, 20 (1948); Phys. Rev. **80**, 440 (1950)
5. J. Zinn-Justin: *Quantum Field Theory and Critical Phenomena* (Oxford University Press, Oxford 1996);
   S. Weinberg: *The Quantum Theory of Fields* (Cambridge University Press, Cambridge 1995)

Quantum Mechanics and Discrete Time from "Timeless" Classical Dynamics

Hans-Thomas Elze

Instituto de Física, Universidade Federal do Rio de Janeiro, C.P. 68.528,
21941-972 Rio de Janeiro, RJ, Brazil

Abstract. We study classical Hamiltonian systems in which the intrinsic proper time evolution parameter is related through a probability distribution to the physical time, which is assumed to be discrete.

This is motivated by the "timeless" reparametrization invariant model of a relativistic particle with two compactified extra-dimensions. In this example, discrete physical time is constructed based on quasi-local observables.

Generally, employing the path-integral formulation of classical mechanics developed by Gozzi *et al.*, we show that these deterministic classical systems can be naturally described as unitary quantum mechanical models. The emergent quantum Hamiltonian is derived from the underlying classical one. It is closely related to the Liouville operator. We demonstrate in several examples the necessity of regularization, in order to arrive at quantum models with bounded spectrum and stable ground state.

1 Introduction

Recently we have shown that for a particle with time-reparametrization invariant dynamics, be it relativistic or nonrelativistic, one can define quasi-local observables which characterize the evolution in a gauge invariant way [1, 2].

We insist on quasi-local measurements in describing the evolution, which respect reparametrization invariance of the system. Then, as we have argued, the physical time necessarily becomes discrete, its construction being based on a Poincaré section which reflects ergodic dynamics, by assumption. Most interestingly, due to inaccessability of globally complete information on trajectories, the evolution of remaining degrees of freedom appears as in a quantum mechanical model when described in relation to the discrete physical time.

While we pointed out in the explicitly ergodic examples of references [1, 2] that such emergent discrete time leads to what may, for obvious reasons, be called "stroboscopic" quantization, we report here how this occurs quite generally in classical Hamiltonian systems, if time is discrete and related to the proper time of the equations of motion in a statistical way [3]. In our concluding section, we will briefly comment about extensions, where the prescribed probabilistic mapping of physical onto proper time shall be abandoned in favour of a selfconsistent treatment. A closed system has to include its own "clock", if it is not entirely static, reflecting the experience of an observer in the Universe.

Previous related work on the "problem of time" has always assumed that global features of the trajectory of the system are accessible to the observer.

H.-T. Elze, Quantum Mechanics and Discrete Time from "Timeless" Classical Dynamics, Lect. Notes Phys. **633**, 196–220 (2004)
http://www.springerlink.com/

This makes it possible, in principle, to express the evolution of an arbitrarily selected degree of freedom relationally in terms of others [4, 5]. Thereby the Hamiltonian and possibly additional constraints have been eliminated in favour of Rovelli's "evolving constants of motion" [6]. For a recent development aiming at solving the constraints after discretization see [7]. While appealing by its conceptual clarity, incorporating nonlocal observations seems unrealistic to us in any case.

In distinction, we point out that the emergent discrete time in our approach naturally leads to the "stroboscopic" quantization of the system [1–3]. Quantum theory thus appears to originate from "timeless" classical dynamics, due to the lack of globally complete information [3].

Another approach to deterministically induced quantization is proposed in [8], where the consequences of incomplete statistics are analyzed, leading towards Euclidean quantum field theory under very general assumptions. Various other arguments considering quantization as an emergent property of classical systems have recently been proposed, for example, in references [9–13], concerning quantum gravity and dissipation at a fundamental level, "chaotic quantization", and matrix models, respectively.

Our approach tries to illuminate from a different angle how to arrive at quantum models which describe dynamically evolving systems. In particular, we believe that there may be an intimate connection with how the "problem of time" is resolved for a local observer, namely by counting suitably defined and locally measurable incidents.

We remark that the possibility of a fundamentally discrete time (and possibly other discrete coordinates) has been explored before, ranging from an early realization of Lorentz symmetry in such a case [14] to detailed explorations of its consequences and consistency in classical mechanics, quantum field theory, and general relativity [15–17]. However, no detailed models giving rise to such discreteness have been proposed. Quantization, then, is always performed in an additional step, as usual.

In particular, the work by Gambini, Pullin *et al.* aims at a consistent canonical quantization of gravity via discretization [7, 17]. Discretization of time is performed in a *static* fashion, i.e. independently of the evolution. As shown there, the major advance lies in the possibility to satisfy the constraints, in principle, by suitably choosing the Lagrange multipliers. However, the extraneous discretization is reflected by the persistence of a discrete variable n after quantization, which apparently has no physical meaning.

Presently, we shall see a vague resemblance to the persistence of proper time τ in our approach, as long as the discrete physical time is related through a *given* probability distribution to proper time. Our formalism, however, is set up in such a way that the "clock" degrees of freedom can be treated dynamically as part of the system. This should help to pinpoint the role of proper time in the resulting model where, in our case, *quantization emerges* instead of being imposed on the system.

In the following section, we recall the model from [2], in order to motivate the emergence of discrete time from quasi-local observations. This is the starting point of our heuristic derivation of a quantum mechanical picture of what appear fundamentally classical systems.

To put our approach into perspective, we remark that there is clearly no need to follow such construction leading to a discrete physical time in ordinary mechanical systems or field theories, where time is an external classical parameter, commonly called "t". However, assuming for the time being that truly fundamental theories will turn out to be diffeomorphism invariant, adding further the requirement of the observables to be quasi-local (modulo a fundamental length scale), when describing the evolution, then such an approach seems natural, which may lead to quantum mechanics as an emergent description or "effective theory" on the way.

2   Discrete Time of a Relativistic Particle with Extra-dimensions

We consider the (5+1)-dimensional model of a "timeless" relativistic particle (rest mass m) with the action

$$S = \int \mathrm{d}s\, L \ , \tag{1}$$

where the Lagrangian is defined by

$$L \equiv -\tfrac{1}{2}(\lambda^{-1}\dot{x}_\mu \dot{x}^\mu + \lambda m^2) \ . \tag{2}$$

Here λ stands for an arbitrary "lapse" function of the evolution parameter s, $\dot{x}^\mu \equiv \mathrm{d}x^\mu/\mathrm{d}s$ ($\mu = 0, 1, \ldots, 5$), and the metric is $g_{\mu\nu} \equiv \mathrm{diag}(1, -1, \ldots, -1)$. Units are such that $\hbar = c = 1$.

With this form of the Lagrangian, instead of the frequently encountered $L\mathrm{d}s \propto (g_{\mu\nu}\mathrm{d}x^\mu \mathrm{d}x^\nu)^{1/2}$ which emphasizes the geometric (path length) character of the action, the presence of a constraint is immediately obvious, since there is no s-derivative of λ.

Two spatial coordinates, $x^{4,5}$ in (2), are toroidally compactified,

$$x^{4,5} \equiv 2\pi R[\phi^{4,5}] \ , \tag{3}$$

$$[\phi] \equiv \phi - n \ , \quad \phi \in [n, n+1[\ , \tag{4}$$

for any integer n, i.e. the angular variables are periodically continued; henceforth we set $R = 1$, for convenience. Alternatively, we can normalize the angular variables to the square $[0, 1[\times [0, 1[$, of which the opposite boundaries are identified, thus describing the surface of a torus with main radii equal to one.

While full Poincaré invariance is broken, as in other currently investigated models with compactified higher dimensions, the usual one remains in fourdimensional Minkowski space together with discrete rotational invariance in the presently two extra-dimensions; also translational symmetry persists. Furthermore, the internal motion on the torus is ergodic with an uniform asymptotic density for almost all initial conditions, in particular if the ratio of the corresponding initial momenta is an irrational number.

Setting the variations of the action to zero, we obtain

$$\frac{\delta S}{\delta \lambda} = \tfrac{1}{2}(\lambda^{-2}\dot{x}_\mu \dot{x}^\mu - m^2) = 0 \ , \tag{5}$$

$$\frac{\delta S}{\delta x_\mu} = \frac{\mathrm{d}}{\mathrm{d}s}(\lambda^{-1}\dot{x}^\mu) = 0 \ . \tag{6}$$

In terms of the canonical momenta,

$$p_\mu \equiv \frac{\partial L}{\partial \dot{x}^\mu} = -\lambda^{-1}\dot{x}_\mu \ , \tag{7}$$

the equations of motion (6) become simply $\dot{p}^\mu = 0$, while (5) turns into the mass-shell constraint $p^2 - m^2 = 0$.

The equations of motion are solved by

$$x^\mu(s) = x_i^\mu - p^\mu \int_0^s \mathrm{d}s' \lambda(s') \equiv x_i^\mu + p^\mu \tau(s) \ , \tag{8}$$

where the conserved (initial) momentum p^μ is constrained to be on-shell and x_i denotes the initial position. Here we also defined the fictitious proper time (function) τ, which allows us to formally eliminate the lapse function λ from (5)-(6), using $x^\mu(\tau) \equiv x^\mu(s)$ and $\dot{x}^\mu(s) = -\lambda(s)\partial_\tau x^\mu(\tau)$.

In order to arrive at a physical space-time description of the motion, the proper time needs to be determined in terms of observables. In the simplest case, the result should be given by functions $x^{\mu \neq 0}(x^0)$, provided there is a physical clock measuring $x^0 = x_i^0 + p^0\tau$.

Similarly as in the nonrelativistic example studied in [1], the lapse function introduces a gauge degree of freedom into the dynamics, which is related to the reparametrization of the evolution parameter s. In fact, the action, (1)-(2), is invariant under the set of gauge transformations

$$s \equiv f(s') \ , \quad x^\mu(s) \equiv x'^\mu(s') \ , \quad \lambda(s)\frac{\mathrm{d}s}{\mathrm{d}s'} \equiv \lambda'(s') \ . \tag{9}$$

It can be shown that the corresponding infinitesimal tranformations actually generate the evolution of the system. This is the basis of statements that there is no time in systems where dynamics is pure gauge, i.e. of the "problem of time". We refer to [1] for further discussion.

Instead, with an evolution obviously taking place in such systems, we conclude from these remarks that the space-time description of motion requires a gauge invariant construction of a suitable time, replacing the fictitious proper time τ. To this we add the important requirement that such construction should be based on quasi-local measurements, since global information (such as invariant path length) is generally not accessible to an observer in more realistic, typically nonlinear or higher-dimensional theories.

2.1 "Timing" through an Extra-dimensional Window

Our construction of a physical time is based on the assumption that an observer in (3+1)-dimensional Minkowski space can perform measurements on full (5+1)-dimensional trajectories, however, only within a quasi-local window to the two extra-dimensions. In particular, the observer records the incidents ("units of change") when the full trajectory hits an idealized detector which covers a small convex area element on the torus (compactified coordinates $x^{4,5}$).[1]

Thus, our aim is to construct time as an emergent quantity related to the increasing number of incidents measured by the reparametrization invariant incident number

$$I \equiv \int_{s_i}^{s_f} ds' \lambda(s') D(x^4(s'), x^5(s')) \ , \tag{10}$$

where $x^{4,5}$ describe the trajectory of the particle in the extra-dimensions, the integral is taken over the interval which corresponds to a given invariant path $x_i^\mu \to x_f^\mu$, and the function D represents the detector features. Operationally it is not necessary to know the invariant path, in order to count the incidents.

In the following examples we choose for D the characteristic function of a small square of area d^2, $D(x^4, x^5) \equiv C_d(x^4) C_d(x^5)$, with $C_d(x) \equiv \Theta(x)(1 - \Theta(x - d))$, which could be placed arbitrarily. Our results will not depend on the detailed shape of this idealized detector, if it is sufficiently small. More precisely, an incident is recorded only when, for example, the trajectory either leaves or enters the detector, or according to some other analogous restriction which could be incorporated into the definition of D. Furthermore, in order not to undo records, we have to restrict the lapse function λ to be (strictly) positive, which also avoids trajectories which trace themselves backwards (or stall). The records correspond to a uniquely ordered series of events in Minkowski space, which are counted, and only their increasing total number is recorded, which is the Lorentz invariant incident number.

Considering particularly the free motion on the torus, solution (8) yields

$$[\phi(\tau)] = [\phi_0 + \pi\tau] \ , \tag{11}$$

where ϕ is the vector formed of the angles $\phi^{4,5}$, and correspondingly π, with $\pi^{4,5} \equiv p^{4,5}/2\pi R$; the quantities in (11) are periodically continued, as before, see (3)-(4). Without loss of generality we choose $\phi_0 = 0$ and $\pi^5 > \pi^4 > 0$, and place the detector next to the origin with edges aligned to the positive coordinate axes for simplicity.

Since here we are not interested in what happens between the incidents, we reduce the description of the internal motion to coupled maps. For proper time intervals $\Delta\tau$ with $\pi^4 \cdot \Delta\tau = 1$, the ϕ^4-motion is replaced by the map $m \longrightarrow m+1$, where m is a nonnegative integer, while

$$[\phi^5] = [Pm] \ , \tag{12}$$

---

[1] One could invoke a popular distinction between brane and bulk matter as in string theory inspired higher-dimensional cosmology, in order to construct more realistic models involving local interactions.

with $P \equiv \pi^5/\pi^4 > 1$. Then, also the detector response counting incidents can be represented as a map,

$$I(m+1) = I(m) + \Theta(\delta - [\phi^5]) + \Theta([\phi^5] - (1 - P\delta)) \, , \qquad (13)$$

with $I(0) \equiv 1$, and where $\delta \equiv d/2\pi R$ corresponds to the detector edge length d, assumed to be sufficiently small, $P\delta \ll 1$. The two Θ-function contributions account for the two different edges through which the trajectory can enter the detector in the present configuration.

The nonlinear two-parameter map (13) has surpring universal features, some of which we explored in [2]. Here, first of all, following the reparametrization invariant construction up to this point, we identify the *physical time T* in terms of the incident number I from (13),

$$T \equiv \frac{I}{\delta(\pi^4 + \pi^5)} \, . \qquad (14)$$

A statistical argument for the scaling factor $\delta^{-1}(\pi^4 + \pi^5)^{-1}$, based on ergodicity, has been given in [1], which applies here similarly.

We show in Fig. 1 how the physical time T typically is correlated with the fictitious proper time τ. The proper time is extracted as those m-values when incidents happen, $\tau = m + 1 \iff I(m+1) - I(m) = 1$, corresponding to a particularly simple specification of the detector response. For a sufficiently small detector other such specifications yield the same results as described here. This is achieved by always rounding the extracted proper time values to integers, involving negligible errors of order $\delta, \delta/P \ll 1$.

We find that the time T does not run smoothly. This is due to the coarse-grained description of the internal motion: as if we were reading an analog clock

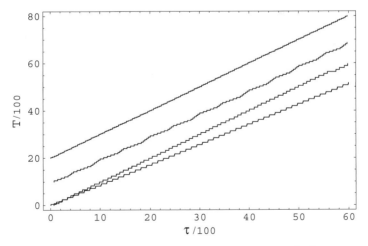

Fig. 1. The physical time T as a function of proper time τ with detector parameter $\delta = 0.005$ and ratio of initial internal momenta $P = \sqrt{31}$ (top), $e, \sqrt{2}, \pi$ (bottom) (see main text); *upper two curves* displaced upwards by $+10$ and $+20$ units, respectively, for better visibility (from [2])

under a stroboscopic light. In our construction, it is caused by the reduction of the full motion to a map (Poincaré section), corresponding to the recording of the physical incidents by the quasi-local detector.

Furthermore, already after a short while, i.e. at low incident numbers, the constructed time approximates well the proper time τ on average.[2] The fluctuations on top of the observed linear dependence result in the *discreteness* of the constructed time.

While in reference [2] we further analyzed the statistical properties of the map considered in this example, we take from here only the result that a physical time can be constructed based on suitable localized observations. Furthermore, after embarking on some useful formal developments in the next section, we will incorporate the resulting probabilistic mapping between discrete physical time and proper time of the equations of motion in Sect. 4.

3 Classical Mechanics in Path-Integral Form

Classical mechanics can be cast into path-integral form, as originally developed by Gozzi, Reuter and Thacker [18], and with recent addenda reported in [19]. While the original motivation has been to provide a better understanding of geometrical aspects of quantization, we presently use it as a convenient tool. We refer the interested reader to the cited references for details, on the originally resulting extended (BRST type) symmetry in particular. We suitably incorporate time-reparametrization invariance, assuming equations of motion written in terms of proper time.

Let us begin with a $(2n)$-dimensional classical phase space $\mathcal{M}$ with coordinates denoted collectively by $\varphi^a \equiv (q^1, \ldots, q^n; p^1, \ldots, p^n)$, $a = 1, \ldots, 2n$, where q, p stand for the usual coordinates and conjugate momenta. Given the proper-time independent Hamiltonian $H(\varphi)$, the equations of motion are

$$\frac{\partial}{\partial \tau} \varphi^a = \omega^{ab} \frac{\partial}{\partial \varphi^b} H(\varphi) \ , \tag{15}$$

where ω^{ab} is the standard symplectic matrix and τ denotes the proper time; summation over indices appearing twice is understood.

To the equation of motion we add the (weak) Hamiltonian constraint, $C_H \equiv H(\varphi) - \epsilon \simeq 0$, with ϵ a suitably chosen parameter. This constraint has to be satisfied by the solutions of the equations of motion. Generally, it arises in reparametrization invariant models, similarly as the mass-shell constraint in the case of the relativistic particle [2]. It is necessary when the Lagrangian time parameter is replaced by the proper time in the equations of motion. In this way, an arbitrary "lapse function" is eliminated, which otherwise acts as a Lagrange multiplier for this constraint.

We remark that field theories can be treated analogously, considering indices a, b, etc. as continuous variables.

---

[2] The apparent excursion for the parameter value $P = \pi$ in Fig. 1, does not persist for longer times, as shown in [2].

Starting point for our following considerations is the *classical* generating functional,

$$Z[J] \equiv \int_H \mathcal{D}\varphi \, \delta[\varphi^a(\tau) - \varphi_{cl}^a(\tau)] \exp(i \int d\tau \, J_a \varphi^a) \ , \qquad (16)$$

where $J \equiv \{J_{a=1,\dots,2n}\}$ is an arbitrary external source, $\delta[\cdot]$ denotes a Dirac δ-functional, and φ_{cl} stands for a solution of the classical equations of motion satisfying the Hamiltonian constraint; its presence is indicated by the subscript "H" on the functional integral. The relevant boundary conditions shall be discussed in the following section. It is important to realize that $Z[0]$ gives weight 1 to a classical path satisfying the constraint and zero otherwise, integrating over all initial conditions.

Using the functional equivalent of $\delta(f(x)) = |df/dx|_{x_0}^{-1} \cdot \delta(x - x_0)$, the δ-functional under the integral for Z can be replaced according to

$$\delta[\varphi^a(\tau) - \varphi_{cl}^a(\tau)] => \delta[\partial_\tau \varphi^a - \omega^{ab}\partial_b H] \det[\delta_b^a \partial_\tau - \omega^{ac}\partial_c \partial_b H] \ , \qquad (17)$$

slightly simplifying the notation, e.g. $\partial_b \equiv \partial/\partial\varphi^b$. Here the modulus of the functional determinant has been dropped [18, 19].

Finally, the δ-functionals and determinant are exponentiated, using the functional Fourier representation and ghost variables, respectively. Thus, we obtain the generating functional in the convenient form,

$$Z[J] = \int_H \mathcal{D}\varphi \mathcal{D}\lambda \mathcal{D}c \mathcal{D}\bar{c} \, \exp\left(i \int d\tau(L + J_a \varphi^a)\right) \ , \qquad (18)$$

which we abbreviate as $Z[J] = \int_H \mathcal{D}\Phi \, \exp(i \int d\tau L_J)$. The enlarged phase space is $(8n)$-dimensional, consisting of points described by the coordinates $(\varphi^a, \lambda_a, c^a, \bar{c}_a)$. The effective Lagrangian is now given by [18, 19]

$$L \equiv \lambda_a \left(\partial_\tau \varphi^a - \omega^{ab}\partial_b H\right) + i\bar{c}_a \left(\delta_b^a \partial_\tau - \omega^{ac}\partial_c \partial_b H\right)c^b \ , \qquad (19)$$

where $c^a, \bar{c}^a$ are anticommuting Grassmann variables. We remark that an entirely bosonic version of the path-integral exists [19].

This completes our brief review of how to put (reparametrization invariant) classical mechanics into path-integral form.

4   From Discrete Time to "States"

We recall from our previous example that the discrete physical time t has been obtained by counting suitably defined incidents, i.e., coincidences of points of the trajectory of the system with appropriate detectors [1, 2]. Thus, it is given by a nonnegative integer multiple of some unit time, $t \equiv nT$. Then, we would like to express the proper time τ which parametrizes the evolution in terms of t.

Here we assume instead that the physical time t is mapped onto a normalized probability distribution P of proper time values τ,

$$P(\tau; t) \equiv \exp\left(-S(\tau; t)\right) \ , \quad \int d\tau \, P(\tau; t) = 1 \ . \qquad (20)$$

For uniqueness, we require that if $S(\tau; t_1)$ and $S(\tau; t_2)$, for $t_1 \neq t_2$, have overlapping support, then they should coincide in this region.

Thus, we describe the idealized case that the system can be separated into degrees of freedom which are employed in the construction of a physical "clock", yielding the values of t, and remaining degrees of freedom evolving in proper time. Neglecting the interaction between both components, and the details of the clock in particular, we describe the relation between physical and proper time by a probability distribution. In this situation, the Hamiltonian constraint only applies to the remaining degrees of freedom, while generally the system will be constrained as a whole. These aspects were exemplified in detail in the simple models of [1,2].

Correspondingly, we introduce the modified generating functional,

$$Z[J] \equiv \int_H d\tau_i d\tau_f \int \mathcal{D}\Phi \, \exp\left(i \int_{\tau_i}^{\tau_f} d\tau \, L_J - S(\tau_i; t_i) - S(\tau_f; t_f)\right) , \qquad (21)$$

instead of (18), using the condensed notation introduced there. In the present case, $Z[0]$ sums over all classical paths satisfying the constraint with weight $P(\tau_i; t_i) \cdot P(\tau_f; t_f)$, depending on their initial and final proper times, while all other paths get weight zero. In this way, the distributions of proper time values $\tau_{i,f}$ associated with the initial and final physical times, t_i and t_f, respectively, are incorporated.

Next, we insert $1 = \int d\tau P(\tau; t)$ into the expression for Z, with an arbitrarily chosen physical time $t > t_0$, and with $t_i = t_f \equiv t_0$. We require the two sets of trajectories created in this way to present branches of forward ("$>$") and backward ("$<$") motion. This leads us to factorize the path-integral into two connected ones,

$$Z[J] = \int d\tau \, P(\tau; t) \cdot \int d\tau_f \int_H \mathcal{D}\Phi_< \, \exp\left(i \int_\tau^{\tau_f} d\tau' \, L_J^< - S(\tau_f; t_0)\right)$$

$$\cdot \int d\tau_i \int_H \mathcal{D}\Phi_> \, \exp\left(i \int_{\tau_i}^\tau d\tau'' \, L_J^> - S(\tau_i; t_0)\right)$$

$$\cdot \prod_a \delta(\varphi_>^a(\tau) - \epsilon(a)\varphi_<^a(\tau)) , \qquad (22)$$

where $\epsilon(a \leq n) \equiv 1$, $\epsilon(a \geq n+1) \equiv -1$, and $J \equiv J_>, J_<$, depending on the branch. The ordinary δ-functions assure continuity of the classical paths in terms of the coordinates q^a, $a = 1, \ldots, n$, and reflect the momenta p^a, $a = 1, \ldots, n$, at proper time τ.

We observe that the generating functional will only be independent of the physical time t, in the absence of an external source, if we assume that the probability distribution P does not explicitly depend on time, $-\log P(\tau; t) = S(\tau; t) \equiv S(\tau - t)$, and if we suitably specify the *boundary conditions*. We set

$$\varphi_>^a(\tau_i) = \epsilon(a)\varphi_<^a(\tau_f) \equiv \phi^a(t_0) , \quad a = 1, \ldots 2n . \qquad (23)$$

Note that the boundary conditions are defined at the physical time t_0, to which correspond the distributed values of the proper times $\tau_{i,f}$. This establishes a one-to-one correspondence between both sets of trajectories. They could be viewed as closed loops with reflecting boundary conditions at both ends, t_0 and t, and fixed initial condition at t_0.

Exponentiating the δ-functions via Fourier transformation, the generating functional can be recognized indeed as a scalar product of a "state" and its adjoint. We define the normalized states by the path-integral,

$$|\tau, \pi_a; t\rangle \equiv Z[J]^{-1/2} \int d\tau_i \int_H \mathcal{D}\Phi \, \exp\left(i \int_{\tau_i}^{\tau+t} d\tau' \, L_J - S(\tau_i; t_0) + i\pi_a \varphi^a(\tau+t)\right),$$
(24)

and, similarly, the adjoint states,

$$\langle\tau, \pi_a; t| \equiv Z[J]^{-1/2} \int d\tau_f \int_H \mathcal{D}\Phi \, \exp\left(i \int_{\tau+t}^{\tau_f} d\tau' \, L_J - S(\tau_f; t_0) - i\pi_a \varphi^a(\tau+t)\right),$$
(25)

where the paths are forward and backward going as indicated by the integral boundaries in the exponent respectively, dropping ">, <"; note that the summation is to be read as $\sum_a \epsilon(a)\pi_a\varphi^a$ in (25).

The redundancy in designating the states, which depend on the sum of proper and physical time, only arises here, since the probability distribution P is assumed not to be explicitly depending on the physical time, for simplicity.

The scalar product of two such states is now defined, and calculated, as follows:

$$\langle t_2|t_1\rangle \equiv \int d\tau d\pi \, P(\tau)\langle\tau, \pi; t_2|\tau, \pi; t_1\rangle = \delta_{t_2, t_1} \, ,$$
(26)

with $d\pi \equiv \prod_a (d\pi_a/2\pi)$. In particular, we have $\langle t|t\rangle = 1$, which corresponds to (22), using definitions (24) and (25). Furthermore, for $t_1 \neq t_2$, we find that states are orthogonal, $\langle t_2|t_1\rangle = 0$, by the symmetry of the motion on the forward and backward branches, for a correspondingly symmetric source J, and by uniqueness of the Hamiltonian flow generating the paths.

A remark is in order here concerning the integration over $d\pi$ above, which originates from exponentiating the δ-functions of (22). Due to the presence of the Hamiltonian constraints on both branches of a trajectory, one of the δ-functions is redundant. We absorb the resulting $\delta(0)$ in the normalization of the states.

Finally, the symmetry between the states and the adjoint states, given the stated assumptions, is perfect. We find

$$\langle\tau, \pi; t| = |\tau, \pi; t\rangle^* \, ,$$
(27)

which is a desirable property of states in a Hilbert space (in "τ, π-representation"). However, from our heuristic discussion this appears as a restriction which could be relaxed, resulting in a less familiar relation between the vector space and its dual.[3]

5 Unitary Evolution

Following the same approach which led to the definition of states in (24),(25), we consider the time evolution of states in the absence of a source, $J = 0$. Suitably inserting "1", as before, and splitting the path-integral, we obtain

$$|\tau', \pi'; t'\rangle = \int d\tau d\pi \, P(\tau) U(\tau', \pi'; t' | \tau, \pi; t) |\tau, \pi; t\rangle \ , \qquad (28)$$

with the kernel

$$U(\tau', \pi'; t' | \tau, \pi; t) \equiv \int \mathcal{D}\Phi \, \exp\left(i \int_{\tau+t}^{\tau'+t'} d\tau'' \, L + i\pi' \cdot \varphi(\tau' + t') - i\pi \cdot \varphi(\tau + t)\right), \qquad (29)$$

where the integral is over *all paths* running between $\tau + t$ and $\tau' + t'$, subject to the constraint; here we abbreviate $\pi \cdot \varphi \equiv \pi_a \varphi^a$. We interpret this as a matrix element of the evolution operator $\widehat{U}(t' | t)$.

Then, it is straightforward to establish the following composition rule:

$$\widehat{U}(t'' | t') \cdot \widehat{U}(t' | t) = \widehat{U}(t'' | t) \ , \qquad (30)$$

where integration over the intermediate variables, say τ', π', with appropriate weight factor $P(\tau')$, is understood. These integrations effectively remove the "1", which is inserted when factorizing path-integrals, and link the endpoint coordinates of one classical path to the initial of another.

Since the Hamiltonian constraint is a constant of motion, there is no need to constrain the path integral representing the evolution operator. Integrating over intermediate variables removes all contributions violating the Hamiltonian constraint, provided we work with properly constrained states. This will be further discussed in the following section.

The physical-time dependence of the evolution operator amounts to translations of proper time variables. Therefore, we may study further properties of $\widehat{U}$ without explicitly keeping it. This simplicity, of course, is related to the analogous property of the states, which we mentioned.

We begin by rewriting the functional integral of (29),

$$U(\tau', \pi'; \tau, \pi) = \int \mathcal{D}\varphi \, \delta[\varphi^a(\tau) - \varphi^a_{cl}(\tau)] \exp\left(i\pi' \cdot \varphi(\tau') - i\pi \cdot \varphi(\tau)\right) \ , \qquad (31)$$

cf. Sect. 3, where the paths run between τ and τ', integrating over all initial conditions. Fixing the initial condition of a classical path, we can pull the exponential factors out of the integral, due to the δ-functional, and integrate over all

---

[3] One might consider only forward going paths, for example, with boundary conditions on the coordinates set at $\pm t_0$. In this case, however, it is not obvious how to obtain the correspondent of (27).

initial conditions in the end:

$$U(\tau',\pi';\tau,\pi) = \int \mathrm{d}\varphi_i(\tau)\ \exp\left(i\pi'\cdot\varphi_f(\tau') - i\pi\cdot\varphi_i(\tau)\right)\int \mathcal{D}\varphi\ \delta[\varphi^a(\tau) - \varphi_{cl}^a(\tau)],$$
(32)

where $\varphi_f(\tau')$ denotes the endpoint of the path singled out by the particular initial condition, $\varphi_i(\tau)$. The functional integral equals one. Then, we obtain the simple but central result

$$U(\tau',\pi';\tau,\pi) = \int \mathrm{d}\varphi\ \exp\left(i\pi'\exp[\widehat{\mathcal{L}}(\tau'-\tau)]\cdot\varphi - i\pi\cdot\varphi\right)$$
(33)

$$\equiv \mathcal{E}(\pi',\pi;\tau'-\tau)\ ,$$
(34)

where $\mathrm{d}\varphi \equiv \prod_a \mathrm{d}\varphi^a$, and with the Liouville operator

$$\widehat{\mathcal{L}} \equiv -\frac{\partial H}{\partial\varphi}\cdot\omega\cdot\frac{\partial}{\partial\varphi}\ ,$$
(35)

which is employed in order to propagate the classical solution from the initial condition at τ to proper time τ'; ω is the symplectic matrix.

Using (33), one readily confirms (30) once again. In particular, then $\widehat{U}(t|t')\cdot\widehat{U}(t'|t) = \widehat{U}(t|t)$, which is not diagonal, in general, in this τ,π-representation. We have: $U(\tau',\pi';t|\tau,\pi;t) = \mathcal{E}(\pi',\pi;\tau'-\tau)$, as defined in (34).

In order to proceed, we consider the time dependence of the evolution kernel $\mathcal{E}$. It is determined by the equation

$$i\partial_\tau\mathcal{E}(\pi',\pi;\tau) = -\int \mathrm{d}\varphi\ \exp\left(i\pi'\cdot\varphi(\tau) - \pi\cdot\varphi\right)\pi'\cdot\omega\cdot\frac{\partial}{\partial\varphi}H(\varphi(\tau))$$

$$= \widehat{\mathcal{H}}(\pi', -i\partial_{\pi'})\mathcal{E}(\pi',\pi;\tau)\ ,$$
(36)

with the effective *Hamilton operator*

$$\widehat{\mathcal{H}}(\pi, -i\partial_\pi) \equiv -\pi\cdot\omega\cdot\frac{\partial}{\partial\varphi}H(\varphi)|_{\varphi=-i\partial_\pi}\ .$$
(37)

Here we used (33)–(34), together with the equation of motion (15). The initial condition is

$$\mathcal{E}(\pi',\pi;0) = (2\pi)^{2n}\delta^{2n}(\pi'-\pi)\ ,$$
(38)

as read off from (33).

Using (36), we finally obtain the *Schrödinger equation* which describes the evolution of the states in physical time,

$$i\partial_t\langle\tau,\pi|\Psi(t)\rangle = \int \mathrm{d}\tau'\mathrm{d}\pi'\ P(\tau')i\partial_t\mathcal{E}(\pi,\pi';\tau+t-\tau')\langle\tau',\pi'|\Psi(0)\rangle$$

$$= \widehat{\mathcal{H}}(\pi, -i\partial_\pi)\langle\tau,\pi|\Psi(t)\rangle\ .$$
(39)

Here we implicitly employed the relation $\langle\tau,\pi|\Psi(t)\rangle = \langle\tau+t,\pi|\Psi(0)\rangle = \langle 0,\pi|\Psi(t+\tau)\rangle$, in order to analytically continue to real values of t and to perform the derivative, despite that the physical time is discrete.

Clearly, the hermitian Hamiltonian must be incorporated in a unitary transfer matrix, in order to describe the evolution through one discrete physical time step. It is plausible that presently the need for regularization of the Hamilton operator, demonstrated in subsequent sections, arises here. It is also conceivable that in a more realistic situation, with clock degrees of freedom forming dynamically part of the system, this particular complication is alleviated.

Furthermore, considering stationary states, we have

$$\langle \tau, \pi | \Psi_E(t) \rangle \equiv \exp(-iEt) \langle \tau, \pi | \Psi(0) \rangle = \exp(-iE(t+\tau)) \langle 0, \pi | \Psi(0) \rangle \quad (40)$$

$$\equiv \exp(-iE(t+\tau)) \langle \pi | \Psi_E \rangle \ , \quad (41)$$

due to the previously discussed additivity of proper and physical time in the present context. Similarly, the Hamiltonian $\widehat{\mathcal{H}}$ is independent of the probability distribution P, mapping physical to proper time, since in the presently idealized situation the clock is decoupled from the system.

Note that there is *no* $\hbar$ in our equations. If introduced, it would merely act as a conversion factor of units. On the other hand, there is an intrinsic scale corresponding to the clock's unit time interval T, which could be analyzed in a more complete treatment where clock and mechanical system are part of the Universe and interact.

Before we will illustrate in some examples the type of quantum Hamiltonians that one obtains, we have to first address the classical observables and their place in the emergent quantum theory, in particular we need to implement the classical Hamiltonian constraint. We recall that in a reparametrization invariant classical theory the Hamiltonian constraint is an essential ingredient related to the gauge symmetry one is dealing with.

6   Observables

It follows from our introduction of states in Sect. 4, see particularly (21)–(25), how the classical observables of the underlying mechanical system can be determined. Considering observables which are function(al)s of the phase space variables φ, the definition of their expectation value at physical time t is obvious:

$$\langle O[\varphi]; t \rangle \equiv \int d\tau \, P(\tau; t) O[-i \tfrac{\delta}{\delta J(\tau)}] \log Z[J]|_{J=0} \quad (42)$$

$$= \int d\tau d\pi \, P(\tau - t) \langle \tau, \pi; 0 | O[\varphi(\tau)] | \tau, \pi; 0 \rangle \quad (43)$$

$$= \int d\tau d\pi \, P(\tau) \langle \tau, \pi; t | O[\varphi(\tau + t)] | \tau, \pi; t \rangle \quad (44)$$

$$= \int d\tau d\pi \, P(\tau) \langle \tau, \pi; t | O[-i\partial_\pi] | \tau, \pi; t \rangle \quad (45)$$

$$= \langle \Psi(t) | \hat{O}[\varphi] | \Psi(t) \rangle \ , \quad (46)$$

where

$$\hat{O}[\varphi] \equiv O[\hat{\varphi}] \ , \quad \hat{\varphi} \equiv -i\partial_\pi \ , \tag{47}$$

in τ, π-representation. In (43)–(44) the notation is symbolical, since the observable should be properly included in the functional integral defining the ket state, for example.

Thus, a classical observable is represented by the corresponding function(al) of a suitably defined *momentum* operator. Furthermore, its expectation value at physical time t is represented by the effective quantum mechanical expectation value of the corresponding operator with respect to the physical-time dependent state under consideration, which incorporates the weighted average over the proper times τ, according to the distribution P. Not quite surprisingly, the evaluation of expectation values involves an integration over the whole τ-parametrized "history" of the states.

Furthermore, making use of the evolution operator $\hat{U}$ of Sect. 5, in order to refer observables at different proper times $\tau_1, \tau_2, \dots$ to a common reference point τ, one can construct *correlation functions* of observables as well, similarly as in [8], for example.

The most important observable for our present purposes is the classical Hamiltonian, $H(\varphi)$, which enters the Hamiltonian constraint of a classical reparametrization invariant system. It is, by assumption, a constant of the classical motion. However, it is easy to see that also its quantum descendant, $\hat{H}(\varphi) \equiv H(\hat{\varphi})$, is conserved, since it commutes with the effective Hamiltonian of (37),

$$[\hat{H}, \hat{\mathcal{H}}] = H(-i\partial_\pi) \, \pi \cdot \omega \cdot \tfrac{\partial}{\partial\varphi} H(\varphi)|_{\varphi=-i\partial_\pi} - \pi \cdot \omega \cdot \tfrac{\partial}{\partial\varphi} H(\varphi)|_{\varphi=-i\partial_\pi} H(-i\partial_\pi)$$

$$= \tfrac{\partial}{\partial\varphi} H(\varphi)|_{\varphi=-i\partial_\pi} \cdot \omega \cdot \tfrac{\partial}{\partial\varphi} H(\varphi)|_{\varphi=-i\partial_\pi} = 0 \ , \tag{48}$$

due to the antisymmetric character of the symplectic matrix. Therefore, it suffices to implement the Hamiltonian constraint at an arbitrary time.

Then, the constraint of the form $C_H \equiv H[\varphi] - \epsilon \simeq 0$ may be incorporated into the definition of the states in (24) by including an extra factor $\delta(C_H)$ into the functional integral, and analogously for the adjoint states. Exponentiating the δ-function, we pull the exponential out of the functional integral, as before. Thus, we find the following operator representing the constraint:

$$\hat{C} \equiv \int d\lambda \, \exp\left(i\lambda(\hat{H}(\varphi) - \epsilon)\right) = \delta(\hat{C}_H) \ , \tag{49}$$

which acts on states as a projector. Of course, a corresponding number of projectors should be included into the definition of the generating functional, see (22), for appropriate normalization of the states.

Supplementing (42)–(46) by the insertion of the Hamiltonian constraint, the properly constrained expection values of observables should be calculated according to

$$\langle O[\varphi]; t\rangle_H \equiv \langle \Psi(t)|\hat{O}[\varphi]\hat{C}|\Psi(t)\rangle \ , \tag{50}$$

which will deviate from the results of the previous definition.

Finally, also the eigenvalue problem of stationary states, see (40)–(41), should be studied in the projected subspace,

$$\widehat{\mathcal{H}}\widehat{C}|\Psi\rangle = E\widehat{C}|\Psi\rangle \ , \tag{51}$$

to which we shall return in the following examples.

7   Examples of Emergent Quantum Systems

7.1   Quantum Harmonic Oscillator from Classical One Beneath

All *integrable models* can be presented as collections of independent harmonic oscillators. Therefore, we begin with the harmonic oscillator of unit mass and of frequency Ω. The action is

$$S \equiv \int dt \ \left(\tfrac{1}{2\lambda}(\partial_t q)^2 - \tfrac{\lambda\Omega^2}{2}(q^2 - 2\epsilon) \right) \ , \tag{52}$$

where λ denotes the arbitrary lapse function, i.e. Lagrange multiplier for the Hamiltonian constraint, and $\epsilon > 0$ is the parameter fixing the energy presented by this constraint.

Introducing the proper time, $\tau \equiv \int dt \ \lambda$, the Hamiltonian equations of motion and Hamiltonian constraint for the oscillator are

$$\partial_\tau q = p \ , \quad \partial_\tau p = -\Omega^2 q \ , \tag{53}$$

$$\tfrac{1}{2}(p^2 + \Omega^2 q^2) - \epsilon = 0 \ , \tag{54}$$

respectively.

Comparing the general structure of the equations of motion (15) with the ones obtained here, we identify the effective Hamilton operator (37), while the constraint operator follows from (49),

$$\widehat{\mathcal{H}} = -(\pi_q\widehat{\varphi}_p - \Omega^2\pi_p\widehat{\varphi}_q) = -\pi_q(-i\partial_{\pi_p}) + \Omega^2\pi_p(-i\partial_{\pi_q}) \ , \tag{55}$$

$$\widehat{C} = \delta(\widehat{\varphi}_p^2 + \Omega^2\widehat{\varphi}_q^2 - 2\epsilon) = \delta(\partial_{\pi_p}^2 + \Omega^2\partial_{\pi_q}^2 + 2\epsilon) \ , \tag{56}$$

respectively. Here we employ the convenient notation $\varphi^a \equiv (\varphi_q; \varphi_p)$, and correspondingly $\pi^a \equiv (\pi_q; \pi_p)$, $\partial_\pi^a \equiv (\partial_{\pi_q}; \partial_{\pi_p})$. Further simplifying this with the help of polar coordinates, $\pi_q \equiv -\Omega\rho\cos\phi$ and $\pi_p \equiv \rho\sin\phi$, we obtain

$$\widehat{\mathcal{H}} = \Omega\widehat{L}_z = -i\Omega\partial_\phi \ , \tag{57}$$

$$\widehat{C} = \delta(\Delta_2 + 2\epsilon) = \delta(\partial_\rho^2 + \rho^{-1}\partial_\rho + \rho^{-2}\partial_\phi^2 + 2\epsilon) \ , \tag{58}$$

where $\widehat{L}_z$ denotes the z-component of the usual angular momentum operator and Δ_2 the Laplacian in two dimensions.

We observe that the eigenfunctions of the eigenvalue problem posed here factorize into a radial and an angular part. The radial eigenfunction, a cylinder function, is important for the calculation of expectation values of certain operators and the overall normalization of the resulting wave functions. However, it does not influence the most interesting spectrum of the Hamiltonian.

In the absence of the full angular momentum algebra, we discretize the angular derivative. Then, the energy eigenvalue problem consists of

$$\widehat{\mathcal{H}}\psi(\phi_n) = -i(\Omega N/2\pi)\Big(\psi(\phi_{n+1}) - \psi(\phi_n)\Big) = E\psi(\phi_n) \ , \tag{59}$$

with $\phi_n \equiv 2\pi n/N$, $1 \leq n \leq N$, and the continuum limit will be considered momentarily.

A complete orthonormal set of eigenfunctions and the eigenvalues are

$$\psi_m(\phi_n) \ = N^{-1/2}\exp[i(m+\delta)\phi_n] \ , \quad 1 \leq m \leq N \ , \tag{60}$$

$$E_m \ = i(\Omega N/2\pi)\Big(1 - \exp[2\pi i(m+\delta)/N]\Big) \tag{61}$$

$$\overset{N\to\infty}{\longrightarrow} \Omega(m+\delta) \ , \quad m \in \mathbf{N} \ , \tag{62}$$

where δ is an arbitrary real constant.

Obviously, the freedom in choosing the constant δ, which arises from the regularization of the Hamilton operator, is very welcome. Choosing $\delta \equiv -1/2$, we arrive at the *quantum harmonic oscillator*, starting from the corresponding classical system. Thus, we recover in a straightforward way 't Hooft's result, derived from an equivalent cellular automaton [9]. See also [2] for the completion of a similar quantum model. In the following example we will encounter one more model of this kind and demonstrate its solution in detail.

Here, and similarly in following examples, the *eigenvalues are complex*, with the real spectrum only obtained in the continuum limit. This is due to the fact that we discretize first-order derivatives most simply, i.e. asymmetrically. It can be avoided easily by employing a symmetric discretization, if necessary.

We find it interesting that our general Hamilton operator (37) does not allow for the direct addition of a constant energy term, while the regularization performed here does.

7.2   Quantum System with Classical Relativistic Particle Beneath

Introducing proper time as in Sect. 2, but leaving the extra-dimensions for now, the equations of motion and the Hamiltonian constraint of the reparametrization invariant kinematics of a classical relativistic particle of mass m are given by

$$\partial_\tau q^\mu = m^{-1}p^\mu \ , \quad \partial_\tau p^\mu = 0 \ , \tag{63}$$

$$p \cdot p - m^2 = 0 \ , \tag{64}$$

respectively. Here we have $\varphi^a \equiv (q^0,\ldots,q^3;p^0,\ldots,p^3)$, $a = 1,\ldots,8$; four-vector products involve the Minkowski metric, $g_{\mu\nu} \equiv \mathrm{diag}(1,-1,-1,-1)$.

Proceeding as before, we identify the effective Hamilton operator,

$$\widehat{\mathcal{H}} = -m^{-1}\pi_q \cdot \widehat{\varphi}_p = -m^{-1}\pi_q \cdot (-i\partial_{\pi_p}) \ , \tag{65}$$

corresponding to (37); the notation is as introduced after (56), however, involving four-vectors. Furthermore, the Hamiltonian constraint is represented by the operator

$$\widehat{C} = \delta(\widehat{\varphi}_p^2 - m^2) = \delta(\partial_{\pi_p}^2 + m^2) \ , \tag{66}$$

following from (49).

After a Fourier transformation, which replaces the variable π_q by a derivative (four-vector) $+i\partial_x$, and with $\pi_p \equiv \bar{x}$, the Hamiltonian and constraint operators become

$$\widehat{\mathcal{H}} = -m^{-1}\partial_x \cdot \partial_{\bar{x}} \ , \quad \widehat{C} = \delta(\partial_{\bar{x}}^2 + m^2) \ , \tag{67}$$

respectively.

Similarly as in the harmonic oscillator case, the eigenvalue problem is properly defined by solved by discretizing the system on a hypercubic lattice of volume L^8 (lattice spacing $l \equiv L/N$) with periodic boundary conditions, for example. Here we obtain the eigenfunctions

$$\psi_{k_x,k_{\bar{x}}}(x_n, \bar{x}_n) = N^{-1} \exp[i(k_x + \delta_x) \cdot x_n + i(k_{\bar{x}} + \delta_{\bar{x}}) \cdot \bar{x}_n] \ , \tag{68}$$

with coordinates $x_n^\mu \equiv ln^\mu$ and momenta $k_x^\mu \equiv 2\pi k^\mu/L$, with $1 \leq n^\mu, k^\mu \leq N$, and where δ_x^μ are arbitrary real constants, for all $\mu = 0, \ldots, 3$ (analogously $\bar{x}_n^\mu$, $k_{\bar{x}}^\mu$, $\delta_{\bar{x}}^\mu$).

The energy eigenvalues are

$$E_{k_x,k_{\bar{x}}} = -m^{-1}l^{-2}\Big((\exp[il(k_x + \delta_x)^0] - 1)(\exp[il(k_{\bar{x}} + \delta_{\bar{x}})^0] - 1) \tag{69}$$

$$- \sum_{j=1}^{3}(\exp[il(k_x + \delta_x)^j] - 1)(\exp[il(k_{\bar{x}} + \delta_{\bar{x}})^j] - 1)\Big)$$

$$= m^{-1}(k_x + \delta_x) \cdot (k_{\bar{x}} + \delta_{\bar{x}}) + O(l) \ , \tag{70}$$

where is L is kept constant in the continuum limit, $l \to 0$. Furthermore, in this limit, one finds that the Hamiltonian constraint requires timelike "on-shell" vectors $k_{\bar{x}}$, obeying $(k_{\bar{x}} + \delta_{\bar{x}})^2 = m^2$, while leaving k_x unconstrained.

Continuing, we perform also the infinite volume limit, $L \to \infty$, which results in a continuous energy spectrum in (70). We observe that no matter how we choose the constants $\delta_x, \delta_{\bar{x}}$, the spectrum will not be positive definite. Thus, the emergent model is not acceptable, since it does not lead to a stable groundstate.

However, let us proceed more carefully with the various limits involved and show that indeed a well-defined quantum model can be obtained. For simplicity, considering (1+1)-dimensional Minkowski space and anticipating the massless limit, we rewrite (70) explicitly:

$$E_{k,\bar{k}} = -(\tfrac{2\pi}{\sqrt{m}L})^2(\bar{k}^1 + \bar{\delta}^1)\Big((k^0 + \delta^0) + (k^1 + \delta^1)\Big) + O(m) \ , \tag{71}$$

where we suitably rescaled and renamed the constants and the momenta, which run in the range $1 \leq \bar{k}^1, k^{0,1} \leq N \equiv 2s + 1$. Furthermore, we incorporated the Hamiltonian (on-shell) constraint, such that only the positive root contributes: $\bar{k}^0 + \bar{\delta}^0 = |\bar{k}^1 + \bar{\delta}^1| + O(m^2) = -(\bar{k}^1 + \bar{\delta}^1) + O(m^2)$. This can be achieved by suitably choosing $\bar{\delta}^{0,1}$.

In fact, just as in the previous harmonic oscillator case, the choice of the constants is crucial in defining the quantum model. Here we set

$$\bar{\delta}^0 \equiv \tfrac{1}{2} , \quad \bar{\delta}^1 \equiv \tfrac{1}{2} - 2s - 3 , \quad \delta^{0,1} \equiv 0 . \tag{72}$$

This results in the manifestly positive definite spectrum,

$$E(\bar{s}_z, s_z^{0,1}) = (\tfrac{2\pi}{\sqrt{m}L})^2 \Big((\bar{s}_z + s + \tfrac{1}{2}) + 1 \Big) \Big((s_z^0 + s + \tfrac{1}{2}) + (s_z^1 + s + \tfrac{1}{2}) + 1 \Big) + O(m) , \tag{73}$$

with (half)integer quantum numbers $\bar{s}_z, s_z^{0,1}$, all in the range $-s \leq s_z \leq s$, replacing $\bar{k}^1, k^{0,1}$.

Recalling the algebra of the $SU(2)$ generators, with $S_z |s_z\rangle = s_z |s_z\rangle$ in particular, we are led to consider the generic operator:

$$h \equiv S_z + s + \tfrac{1}{2} , \tag{74}$$

i.e., diagonal with respect to $|s_z\rangle$-states of the (half)integer representations determined by s. In terms of such operators, we obtain the regularized Hamiltonian corresponding to (73),

$$\widehat{\mathcal{H}} = (\tfrac{2\pi}{\sqrt{m}L})^2 \Big(1 + \bar{h} + h_0 + h_1 + \bar{h}(h_0 + h_1) \Big) + O(m) , \tag{75}$$

which will turn out to be equivalent to three harmonic oscillators, including a coupling term plus an additional contribution to the vacuum energy.

A Hamiltonian of the type of h has been the starting point of 't Hooft's analysis [9], which we adapt for our purposes in the following.

Continuing with standard notation, we have $S^2 \equiv S_x^2 + S_y^2 + S_z^2 = s(s+1)$, which suffices to obtain the following identity:

$$h = \tfrac{1}{2s+1} \Big(S_x^2 + S_y^2 + \tfrac{1}{4} + h^2 \Big) . \tag{76}$$

Furthermore, using $S_\pm \equiv S_x \pm i S_y$, we define coordinate and conjugate momentum operators,

$$\hat{q} \equiv \tfrac{1}{2}(aS_- + a^* S_+) , \quad \hat{p} \equiv \tfrac{1}{2}(bS_- + b^* S_+) , \tag{77}$$

where a and b are complex coefficients. Calculating the basic commutator with the help of $[S_+, S_-] = 2S_z$ and using (74), we obtain

$$[\hat{q}, \hat{p}] = i(1 - \tfrac{2}{2s+1}h) , \tag{78}$$

provided we set $\Im(a^* b) \equiv -2/(2s+1)$. Incorporating this, we calculate

$$S_x^2 + S_y^2 = \tfrac{(2s+1)^2}{4} \Big(|a|^2 \hat{p}^2 + |b|^2 \hat{q}^2 - (\Im a \cdot \Im b + \Re a \cdot \Re b)\{\hat{q}, \hat{p}\} \Big) . \tag{79}$$

In order to obtain a reasonable Hamiltonian in the continuum limit, we set

$$a \equiv i\frac{\Omega^{-1/2}}{\sqrt{s+1/2}} \quad , \quad b \equiv \frac{\Omega^{1/2}}{\sqrt{s+1/2}} \quad , \quad \Omega \equiv (\frac{2\pi}{\sqrt{mL}})^2 \ . \tag{80}$$

Then, the previous (76) becomes

$$\Omega h = \tfrac{1}{2}\hat{p}^2 + \tfrac{1}{2}\Omega^2\hat{q}^2 + \frac{1}{(2s+1)\Omega}\left(\tfrac{1}{4}\Omega^2 + (\Omega h)^2\right) \ , \tag{81}$$

revealing a nonlinearly modified harmonic oscillator Hamiltonian, similarly as in [2,9].

Now it is safe to consider the continuum limit, $2s + 1 = N \to \infty$, keeping $\sqrt{mL}$ and Ω finite. This produces the usual $\hat{q}, \hat{p}$-commutator in (78) for states with limited energy and the standard harmonic oscillator Hamiltonian in (81).

Using these results in (75), the Hamilton operator of the emergent quantum model is obtained,

$$\widehat{\mathcal{H}} = \Omega + \tfrac{1}{2}\sum_{j=\bar{1},0,1}\left(\hat{p}_j^2 + \Omega^2\hat{q}_j^2\right) + \tfrac{1}{4\Omega}(\hat{p}_{\bar{1}}^2 + \Omega^2\hat{q}_{\bar{1}}^2)\sum_{j=0,1}\left(\hat{p}_j^2 + \Omega^2\hat{q}_j^2\right), \tag{82}$$

where the massless limit together with the infinite volume limit is carried out, $m \to 0$, $L \to \infty$, in such a way that Ω remains finite.

The resulting Hamiltonian here is well defined in terms of continuous operators $\hat{q}$ and $\hat{p}$, as usual, and has a positive definite spectrum. The coupling term might appear slightly less unfamiliar, if the oscillator algebra is realized in terms of bosonic creation and annihilation operators.

We previously calculated the matrix elements of operators $\hat{q}, \hat{p}$ with respect to the SU(2) basis of primordial states in a similar case, showing that localization of the quantum oscillator has little to do with localization in the classical model beneath [1,2] .

Finally, we remark that had we chosen $\bar{\delta}^{0,1} = \delta^{0,1} \equiv 1/2$, instead of (72), then a relative sign between terms would remain, originating from the Minkowski metric, and this would yield the Hamiltonian $\widehat{\mathcal{H}} \propto (1+\bar{h})(h_0 - h_1)$, which is not positive definite. Similarly, any symmetric choice, $\bar{\delta}^{0,1} = \delta^{0,1} \equiv \delta$ would suffer from this problem.

This raises the important issue of the role of canonical transformations, and of symmetries in particular. It is conceivable that symmetries will play a role in restricting the present arbitrariness of the regularization defining a quantum model. We will address further aspects of this in the following section.

8   Remarks on (Non)Integrable Interactions

We resume our discussion of general features of the emergent quantum mechanics. Specifically, let us consider a classical system with n degrees of freedom, for example, a chain of particles with harmonic coupling and anharmonic potentials. Denoting the phase space variables by $\varphi^a \equiv (Q, P)$, where Q, P are

n-component vectors, we assume for definiteness a Hamiltonian of the form, $H(\varphi) \equiv (1/2)P^2 + V(Q)$, i.e. with a kinetic term which is simply quadratic in the momenta.

In this case, following (37) and (49), and with

$$\widehat{Q} \equiv -i\partial_{\pi_Q} \quad , \quad \widehat{P} \equiv -i\partial_{\pi_P} \quad , \tag{83}$$

the Hamiltonian and constraint operators, respectively, are given by

$$\widehat{\mathcal{H}} = -\pi_Q \cdot \widehat{P} + \pi_P \cdot V'(\widehat{Q}) \quad , \tag{84}$$

$$\widehat{C} = \delta(\tfrac{1}{2}\widehat{P}^2 + V(\widehat{Q}) - \epsilon) \quad , \tag{85}$$

where, of course, $V'(Q) \equiv \nabla_Q V(Q)$, and the wave function is considered as a function $\psi(\pi_P, \pi_Q)$ of the indicated vectors.

The previous oscillator example suggests to perform a Fourier transformation to variables x, y, such that the eigenvalue problem becomes

$$\widehat{\mathcal{H}}\psi(x, y) = \Big(x \cdot (-i\partial_y) - V'(y) \cdot (-i\partial_x) \Big)\psi(x, y) = E\psi(x, y) \quad , \tag{86}$$

while the constraint operator equation turns into an algebraic constraint,

$$\widehat{C}\psi(x, y) = \delta(\tfrac{1}{2}x^2 + V(y) - \epsilon)\psi(x, y) = 0 \quad . \tag{87}$$

In agreement with the general result (48), we easily confirm here that $\widehat{\mathcal{H}}\widehat{C}\psi = \widehat{C}\widehat{\mathcal{H}}\psi$. The constraint equation then simply states that the phase space variables are constrained to a constant energy surface of the underlying classical system.

The first order quasi-linear partial differential equation (86) can be studied by the method of characteristics [20]. Thus, one finds one equation taking care of the inhomogeneity (right-hand side), which can be trivially integrated. Furthermore, the remaining $2n$ equations for the characteristics present nothing but the classical Hamiltonian equations of motion.

It follows that integrable classical models can (in principle) be decoupled in this context of the characteristic equations by canonical transformations. This assumes that we can apply them freely at the pre-quantum level, which might not be the case. It would lead us essentially to the collection of harmonic oscillators mentioned at the beginning of Sect. 7.1, and corresponding quantum harmonic oscillators as studied there.

Classical crystal-like models with only harmonic forces, or free field theories, respectively, will thus give rise to corresponding free quantum mechanical systems here. These are constructed in a different way in [9]. Presumably, the (fixing of a large class of) gauge transformations invoked there can be related to the existence of integrals of motion implied by integrability here. In any case, we conclude that in the present framework truly interacting quantum (field) theories might be connected with nonintegrable deterministic systems beneath.

Furthermore, we emphasize that the Hamiltonian equations of motion preclude motion into classically forbidden regions of the underlying system. Nevertheless, *quantum mechanical tunneling* is an intrinsic property of the quantum oscillator models that we obtained, as well as of the anharmonic oscillator example considered in the following. Similarly, *spreading of wave packets* is to be expected in the latter case.

In order to demonstrate additional features of the eigenvalue problem of (86), we concentrate on one degree of freedom with phase space coordinates p, q and with a generic anharmonic potential.

Since the potential depends only on q, by locally stretching or squeezing the coordinate, i.e. by an "oscillator transformation" $q \equiv f(\bar{q})$, we can bring it into oscillator form, such that $V(f(\bar{q})) = (1/2)\bar{q}^2$. Implementing this type of transformation, the equations for one degree of freedom are

$$\frac{-i}{f'(\bar{q})}\left(p\partial_{\bar{q}} - \bar{q}\partial_p\right)\psi(p,\bar{q}) = E\psi(p,\bar{q}) \ , \tag{88}$$

$$\delta\left(\tfrac{1}{2}(p^2 + \bar{q}^2) - \epsilon\right)\psi(p,\bar{q}) = 0 \ , \tag{89}$$

with f' denoting the derivative of f. This is very much oscillator-like indeed and, once more employing polar coordinates, we obtain

$$\frac{-i}{f'(\rho\sin\phi)}\partial_\phi\psi(\rho,\phi) = E\psi(\rho,\phi) \ , \tag{90}$$

$$\delta(\rho^2 - \epsilon)\psi(\rho,\phi) = 0 \ . \tag{91}$$

The eigenvalue problem seems underdetermined. As it stands, it would give rise to an unbounded continuous spectrum, with no ground state in particular.

This apparent defect persists for any number of degrees of freedom. However, as we have seen already, sense can be made of the Hamilton operator by a suitable regularization, especially by discretizing the phase space coordinates. The principles of such regularization we still do not know, other than either preserving or intentionally breaking symmetries.

8.1   An Anharmonic Oscillator

It is worth while to consider one more example, a onedimensional system with Hamiltonian

$$H \equiv \tfrac{1}{2}p^2 + V_0|q| \ , \tag{92}$$

in order to demonstrate the subtleties associated with regularization. For the linear potential, the coordinate dependence of the operator on the left-hand side of (90) is mild, since $f'(\rho\sin\phi) = V_0^{-1}\rho\sin\phi$, and the eigenvalues of its discretized counterpart can be found as follows.

Conveniently discretizing the angular variable as $\phi_n \equiv (2\pi n/N) + (3\pi/2)$, $1 \le n \le N$, the eigenvalue equation becomes

$$\prod_{n=1}^{N}\left(1 - \lambda\cos(2\pi n/N)\right) = 1 \ , \quad \lambda \equiv 2\pi i\sqrt{\epsilon}E/NV_0 \ . \tag{93}$$

Setting $\lambda \equiv 2z/(1+z^2)$, and employing a known identity for the finite product arising here, the eigenvalue equation can be transformed into: $(1+z^N)^2 = (1+z^2)^N$. With hindsight, we choose $N \equiv 2(4N'+1)$ and set $z \equiv +\sqrt{u-1}$, to obtain

$$u^{N/2} + (1-u)^{N/2} = 1 \ . \tag{94}$$

This equation has the nice property that, if u is a solution, then so is $1/u$. The location of the solutions $u = 0, 1, \infty$ suggests to look for further solutions in the form of $u \equiv \exp 2i\alpha$. Thus, combining the equations for u and $1/u$, we arrive at the transcendental equation

$$2\sin(\alpha N/2) = (2\sin\alpha)^{N/2} \ . \tag{95}$$

From the multitude of its solutions, due to periodicity, we need to find N solutions of (93).

Closer inspection shows that, in the limit of large N, solutions of (95) consist essentially of those zeros of $\sin(\alpha N/2)$ which lie inside the intervals $[n\pi - \pi/6, n\pi + \pi/6]$, with integer n. Thus, positive energy solutions will be obtained momentarily from

$$E_\pm = NV_0 \exp(i\tfrac{\pi}{4} \pm \tfrac{3}{2}i\alpha)\left((2/\epsilon)\sin(\pm\alpha)\right)^{1/2} , \tag{96}$$

where either "+" or "−" has to be chosen consistently, corresponding to the solutions coming in pairs $\exp(\pm 2i\alpha)$.

A remark is in order here. Considering only the positive root above, $z \equiv +\sqrt{u-1}$, we avoided negative energy solutions. However, there is a price to pay: careful counting reveils that the positive energy spectrum is doubly degenerate. The finite positive part is obtained from (96), incorporating $\alpha \equiv 2\pi m/N$,

$$E = NV_0 \exp(3i\pi m/N)\left((2/\epsilon)\sin(2\pi m/N)\right)^{1/2}, \quad 0 \le m_0 \le m \le N/12 \,, \tag{97}$$

$$\xrightarrow{N\to\infty} \tilde{V}_0(m + m_0 - 1)^{1/2} \ , \quad m \in \mathbf{N} \ , \tag{98}$$

where m_0 is an arbitrary constant, within the allowed range, which defines the zero-point energy of the emergent quantum model. The continuum limit is to be taken such that $\tilde{V}_0 \equiv V_0(4\pi N/\epsilon)^{1/2}$ stays finite. The additional solutions can be chosen in a way that their real parts move to $+\infty$, as $N \to \infty$.[4]

We remark that the spectrum of (98) differs from the one obtained for the same potential in standard quantum mechanics, where WKB yields $E \propto (m - 1/4)^{2/3}$.

Summarizing, the various illustrated features promise to make genuinely interacting models quite difficult to analyze. We hope that more interesting results will be obtained with the help of spectrum generating algebras or some to-be-developed perturbative methods.

---

[4] Corresponding eigenfunctions, i.e. N-component discrete eigenvectors, are obtained by evaluating products of the kind appearing in (93), with $k-1 \le N-1$ factors for the kth component and a constant for $k = 1$.

9   Conclusions

We pursue the view that quantum mechanics is an emergent description of nature, which possibly can be based on classical, pre-quantum concepts.

Our approach is motivated by a construction of a reparametrization-invariant time. In turn, this is based on the observation that "time passes" when there is an observable change, which is localized with the observer. More precisely, necessary are incidents, i.e. observable unit changes, which are recorded, and from which invariant quantities characterizing the change of the evolving system can be derived.

We recall the model of [2], invoking compactified extra-dimensions in which a particle moves in addition to its relativistic motion in Minkowski space. We employ a window to these extra-dimensions, i.e., we consider a quasi-local detector which registers the particle trajectory passing by. Counting such incidents, we construct an invariant measure of time.

A basic ingredient is the assumption of ergodicity, such that the system explores dynamically the whole allowed energy surface in phase space. This assures that there are sufficiently frequent observable incidents. They reflect properties of the dynamics with respect to (subsets of) Poincaré sections. Roughly, the passing time corresponds to the observable change there. Then, the particle's proper time is linearly related to the physical time, however, subject to stochastic fluctuations.

Thus, the reparametrization-invariant time based on quasi-local observables naturally induces stochastic features in the behavior of the external relativistic particle motion. Due to quasi-periodicity (or, generally, more strongly irregular features) of the emerging discrete time, the remaining predictable aspects appear as in unitary quantum mechanical evolution.

In reparametrization-invariant, "timeless" single-particle systems, this idea has been realized in various forms [1,2]. Presently, this has led us to assume the relation between the constructed physical time t and standard proper time τ of the evolving system in the form of a statistical distribution, $P(\tau; t) = P(\tau - t)$, cf. (20). Here we assume that the distribution is not explicitly time-dependent, which means, the physical clock is decoupled from the system under study. We explore the consequences of this situation for the description of the system.

We have shown how to introduce "states", eventually building up a Hilbert space, in terms of certain functional integrals, (24)–(25), which arise from the study of a suitable classical generating functional. The latter was introduced earlier in a different context, studying classical mechanics in functional form [18,19]. We employ this as a convenient tool, and modify it, in order to describe the observables of reparametrization-invariant systems with discrete time (Sect. 6). Studying the evolution of the states in general (Sect. 5), we are led to the Schrödinger equation, (39). However, the Hamilton operator (37) has a non-standard first-order form with respect to phase space coordinates.

The choice of boundary conditions of the classical paths contributing to the generating functional plays a crucial role and deserves better understanding.

Furthermore, illustrating the emergent quantum models in various examples, we demonstrate that proper regularization of the continuum Hamilton operator is indispensable, in order that well-defined quantum mechanical systems emerge, with bounded spectra and a stable groundstate, in particular. Most desirable is a deeper undertanding of this mapping between the continuum Hamilton operator, which is straightforward to write down, given a classical pre-quantum system, and the effective quantum mechanics, which emerges after proper regularization only. Especially, limitations imposed by symmetries and consistency of the procedure need further study.

It is a common experience that the preservation of continuum symmetries through discretization is difficult, for example, see [14–17]. We wonder, whether other regularization schemes are conceivable. The possiblity, mentioned after (39), that the need for regularization is an artefact of decoupled clock degrees of freedom deserves further study.

We find that truly interacting quantum (field) theories might be connnected to nonintegrable classical models beneath, since otherwise the degrees of freedom represented in the stationary Schrödinger equation here, can principally be decoupled by employing classical canonical transformations.

Finally, we come back to the probabilistic relation between physical time and the evolution parameter figuring in the parameterized classical equations of motion, which is the underlying raison d'être of the presented stroboscopic quantization. One would like to include the clock degrees of freedom consistently into the dynamics, in order to address the closed Universe. This can be achieved by introducing suitable projectors into the generating functional. Their task is to replace a simple quasi-local detector which responds to a particle trajectory passing through in Yes/No fashion; by counting such incidents, an invariant measure of time has been obtained before [1, 2]. In a more general setting, this detector/projector has to be defined in terms of observables of the closed system. In this way, typical conditional probabilities can be handled, such as describing "What is the probability of observable X having a value in a range x to $x + \delta x$, *when* observable Y has value y?". Criteria for selecting the to-be-clock degrees of freedom are still unknown, other than simplicity. Most likely the resulting description of evolution and implicit notion of physical time will correspond to our distribution $P(\tau; t)$ of (20), however, now evolving explicitly with the system. We leave this for future study.

The stroboscopic quantization emerging from underlying classical dynamics may be questioned in many respects. It might violate one or the other assumption of existing no-go theorems relating to hidden variables theories. However, we believe it is interesting to learn more about working examples, before discussing this. Unitary evolution, tunneling effects, and spreading of wave packets are recovered in this framework. Interacting theories remain to be explored.

References

1. H.-T. Elze and O. Schipper: Phys. Rev. **D66**, 044020 (2002)
2. H.-T. Elze: Phys. Lett. **A310**, 110 (2003)
3. H.-T. Elze: 'Quantum mechanics emerging from "timeless" classical dynamics', quant-phys/0306096
4. M. Montesinos, C. Rovelli and Th. Thiemann: Phys. Rev. D **60**, 044009 (1999)
5. M. Montesinos, Gen. Rel. Grav. **33**, 1 (2001)
6. C. Rovelli: Phys. Rev. D **42**, 2638 (1990)
7. R. Gambini, R.A. Porto and J. Pullin, 'Consistent Discrete Gravity Solution of the Problem of Time: a Model', gr-qc/0302064
8. C. Wetterich: Lecture in this volume; quant-ph/0212031
9. G. 't Hooft: 'Quantum Mechanics and Determinism'. In: *Particles, Strings and Cosmology*, ed. by P. Frampton and J. Ng (Rinton Press, Princeton, 2001), p. 275; hep-th/0105105; 'Determinism Beneath Quantum Mechanics', quant-ph/0212095
10. M. Blasone, P. Jizba and G. Vitiello: Lecture in this volume; Phys. Lett. A **287**, 205 (2001)
11. T.S. Biró, B. Müller, and S.G. Matinyan: Lecture in this voulme; T.S. Biró, S.G. Matinyan, and B. Müller: Found. Phys. Lett. **14**, 471 (2001)
12. L. Smolin: 'Matrix models as non-local hidden variables theories', hep-th/0201031
13. S.L. Adler: 'Statistical Dynamics of Global Unitary Invariant Matrix Models as Pre-Quantum Mechanics', hep-th/0206120
14. H.S. Snyder: Phys. Rev. **71**, 38 (1947)
15. T.D. Lee: Phys. Lett. B **122**, 217 (1983)
16. G. Jaroszkiewicz and K. Norton: J. Phys. A **30**, 3115 (1997); A **30**, 3145 (1997); A **31**, 977 (1998)
17. R. Gambini and J. Pullin: Phys. Rev. Lett. **90**, 021301 (2003); 'Discrete quantum gravity: applications to cosmology', gr-qc/0212033; C. Di Bartolo, R. Gambini and J. Pullin: Class. Quant. Grav. **19**, 5275 (2002).
18. E. Gozzi, M. Reuter, and W.D. Thacker: Phys. Rev. D **40**, 3363 (1989); D **46**, 757 (1992)
19. E. Gozzi and M. Regini: Phys. Rev. **D62**, 067702 (2000); A.A. Abrikosov (jr.) and E. Gozzi: Nucl. Phys. Proc. Suppl. **66**, 369 (2000)
20. R. Courant and D. Hilbert: *Methods Of Mathematical Physics*, Vol. II (Interscience Publ., New York 1962)
21. D. Levi, P. Tempesta and P. Winternitz: 'Umbral calculus, difference equations and the discrete Schrödinger equation', nlin.SI/0305047

Part IV

Decoherence in Mesoscopic Systems

Introduction:
Experimental and Theoretical Status
of Decoherence

Maria Carolina Nemes

Departamento de Física, Universidade Federal de Minas Gerais, C.P. 702,
30123-970 Belo Horizonte, MG, Brasil

1   Introduction

The physics of open quantum systems encompasses a vast array of phenomena
ranging from subnuclear to cosmological scales. Nevertheless, in this thematic
kaleidoscope, there are a few central problems, whose generality makes them
relevant at almost any scale. In this brief review I shall concentrate on some of
these problems, guided by and concentrating on those more intimately related
to the contributions presented in this conference.

The past decade can celebrate having overcome several barriers, two of the
most important of them in my opinion being the objections commonly levelled
against the idea of large scale quantum phenomena. Firstly rather general arguments (e.g. like Anderson's argument in his discussion of generalized pairing
schemes in superfluidity [1]) have been put forward ruling out any possibility of
macroscopic coherence. The main obstacle is connected with the interaction with
the environment. According to simple estimates, this deleterious effect is inversely proportional to the "degree of classicality" of the system and decoherence
is, for all practical purposes, instantaneous [2].

Secondly, discussions on decoherence have often been associated with rather
ineffable questions about quantum measurement and quantum cosmology, saying
little that might be quantitatively testable in the real world. In the past ten years,
amazing technological developments in several areas of physics allowed for the
experimental verification of essentially quantum effects like tunneling, various
kinds of interference and entanglement phenomena in mesoscopic systems. So we
have reached the point where we need to know rather urgently what mechanisms
cause decoherence, particularly at the mesoscopic scale, since controlling such
mechanism has immediate and profound technological consequences, besides its
value for fundamental Quantum Mechanics and the (unsolved) problem of its
classical limit.

Finally there is the question of how close mesoscopic systems are to the
classical ones. Of course there are pragmatic answers in some simple cases. For
example, take a uniformly charged ellipsoid, let it rotate around its principal
axis with larger moment of inertia and calculate the intensity of the emitted
classical electromagnetic radiation. The same calculation can be performed at the
quantum level considering a rigid deformed nucleus. It has been shown that for
quadrupolar transition in this nucleus only asymptotically reaches the classical

M.C. Nemes, Introduction: Experimental and Theoretical Status of Decoherence, Lect. Notes Phys.
633, 223–238 (2004)
http://www.springerlink.com/

limit [3]. Also at $60\hbar$, the discrepancy between the results is below $10/100$. So in this case, the correspondence principle has been a helpful guide. In other not so simple situations, it has been argued that the observed quantum phenomena are not really macroscopic (for a recent example see Peres [4], refuted by Garg and Leggett [5]) or that how macroscopic they are is largely a question of semantics. The matter was more or less settled for the word mesoscopic.

Of course not everything is flowers. The impressive progress made in the last decade also left some open questions unanswered and brought some new ones. Let us start with the main achievements and then go over to the open questions.

2   Some New Achievements

2.1   Experimental Evidence for Mesoscopic Quantum Phenomena

In 1935, Schrödinger [6], in order to demonstrate the limitations of quantum mechanics used a thought experiment in which a cat is put in a quantum superposition of alive and dead states. The idea remained an academic puzzle until the 1980's when it was proposed [7–10] that a macroscopic object with many degrees of freedom could behave quantum mechanically, provided the coupling with the surroundings be weak enough. We next discuss some of the experimental progress made in this direction.

Flux Tunneling in SQUIDs: Quantum Tunneling and Quantum Coherence. The theoretical results of references [7–9] correctly predicted to striking accuracy the various experiments on SQUIDs [11]. Since then much progress has been made in demonstrating the macroscopic quantum behavior of superconductors [12–16]. There has been however no experimental demonstration of a quantum superposition of truly macroscopically distinct states. Recently, experimental evidence for this phenomenon has been given by two experimental groups [17, 18]. The basic physics is the following: a persistent current may be induced in superconducting loops by application of small magnetic fields. The flux through the loop may be varied by varying the magnetic field. When the magnetic flux is close to half integer values of the elementary flux ϕ_0, the loop develops two stable persistent current states with two opposite polarities. If a Josephson tunnel junction is present, the potential in the SQUID is schematically represented by a double-well potential in a situation when the applied flux corresponds to a value below $1/2\phi_0$ and precisely at $1/2\phi_0$, where ϕ_0 is the quantum flux unit (see Fig. 1A,B).

Note that in a micrometer-sized loop, millions of Cooper pairs are involved. At very low temperatures, excitations of individual charge carriers around the center of mass of the Cooper pair condensate are prohibited by the superconducting gap. Fig. 1C shows the energy levels and persistent currents of the loop as a function of applied flux ϕ_{ext}. The insets at the top plot show again the double-well potential for ϕ_{ext} below $1/2\phi_0$ (left), at $1/2\phi_0$ (middle) and above $1/2\phi_0$ (right). The dashed lines indicate the two localized persistent current states. The solid lines indicate the quantum levels and show an anticrossing near

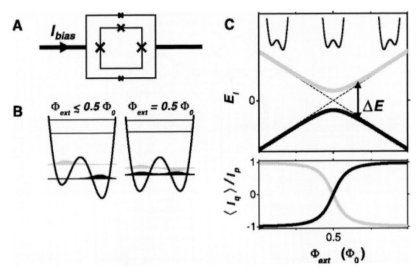

Fig. 1. Schematic representation of the SQUID potential and energy levels as function of the flux. From [18]

$\phi = 1/2\phi_0$ where the energy eigenstates are symmetric and antisymmetric superpositions of the two persistent current states. The coherent superposition would manifest itself in the anticrossing, where the two states would become degenerate without coherent interaction. Coherent tunneling lifts the degeneracy. In references [17, 18] experimental support for the existence of this gap is given.

Of course the potential of such devices for quantum coherent dynamics has stimulated research aimed at applying Josephson junction loops as basic building blocks for quantum computation [19–22].

Dynamics of Mesoscopic Magnets: Environmental Assisted Quantum Tunneling. Recently discovered magnetic molecules [23–26] are of interest due to the very large relaxation time of their magnetization [26] which reaches two months in zero magnetic field at 2K in the $Mn_{12}O_{16}$ molecule (often called "Mn_{12}"). This phenomenon, as we shall see, deals with spin tunneling, not particles. For the Mn_{12} molecule, the total spin is around $10\hbar$. Magnetization measurements [27] and all its magnetic properties are consistent with spin only magnetism. How does this mesoscopic spin come about? Let us understand the $Mn_{12}O_{16}$ molecule a little better: it contains 8 ions Mn^{3+} with spin 2, and 4 ions Mn^{3+} with spin 3/2. The ionic spin are coupled by antiferromagnetic interactions through the oxygen ligands (see Fig. 2). The total spin of the molecule is $10\hbar$. This can be understood as resulting from a Néel ground state in which the eight spins $2\hbar$ are parallel, the four spins $3/2\hbar$ are also parallel, but the spins $2\hbar$ are antiparallel to the spins $3/2\hbar$, so that the total spin is $8 \times 2\hbar - 4 \times 3/2\hbar = 10\hbar$. The experimental evidence that there is quantum tunneling is rather indisputable. The mechanism is the following: in a Mn_{12} acetate crystal, the symmetry

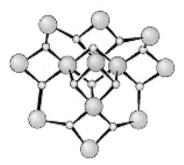

Fig. 2. Structure of the Mn_{12} molecule: Mn ions are represented by large circles and oxygen ions by small circles. The four centered Mn ions have opposite spin orientation with respect to the others. From [30]

axis of all Mn_{12} molecules are aligned in one direction, making the crystal highly anisotropic. Quantum mechanically the spin projection on this anisotropy axis can take only discrete values $m = -10\hbar \cdots 10\hbar$. At zero magnetic field, the two levels of the minimal energy belong to the two opposite spin orientations. The energy barrier in this situation is about 60 K in temperature units. The separation between the levels is about 12 K at the bottom of the spectrum and about 0,6 K at the top.

If a magnetic field is applied at low enough temperatures, it can make, say the $m = -10\hbar$ state, more favorable. Now if the field is removed, most molecules will occupy the level. This negative magnetization of the crystal decays in time due to phonon absorption and subsequent tunneling. Each phonon can change the magnetization by one unit. Consequently, at zero temperature, the molecule must absorb ten phonons before it reaches the top of the barrier. In this situation, however, phonons are scarce and the decay of the magnetic moment may take a long time. This is when the mesoscopic quantum tunneling mechanism becomes effective.

The important issue is: does one have macroscopic coherence too or evidence for that, like the one found in superconductors? The answer seems to be no, at least up to now. Also here, the relevant systems, which prevent mesoscopic spin values (including $Mn_{12}Ac$, $CrNi_6$ and others) may be observed in various quasisteady states, usually separated by a potential barrier (due to the symmetries of the crystals) of a few tens of Kelvin. Sometime ago it was assumed that the transition between these states take place either by tunneling or by thermal activation. Recently it has been shown that in a phonon-type environment, the mesoscopic tunneling rate is hardly affected by the environment. This however does not necessarily mean that the process encompasses mesoscopic correlations: although tunneling is hardly affected, by coupling with the surrounding phonons, there is a very fast coherence loss process, indicating that the system will loose coherence in times of order 10^{-8}s for Mn_{12} molecules. Of course the situation is much more critical for superparamagnetic particles, which for being "more macroscopic" (their spin is of the order $3000\hbar$) will loose coherence at a much faster rate ($\sim 10^{-16}$s [28]). This result is in complete agreement with the decoherence effects described by L. Davidovich [29] for Optical Cavities in which "for macroscopic superpositions quantum coherence decays much faster than for macroscopic observables of the system (in the present case magnetization), its decay time being given by the dissipation time divided by a dimensionless number measuring the 'separation' between the two parts".

Cavity Quantum Electrodynamics: Monitoring Decoherence and the Quantum-Classical Boundary. Modern cavity quantum electrodynamics (cavity QED) sheds light onto the most fundamental aspects of coherence and decoherence in quantum mechanics. One of the main advantages of such experiments is that they can be described by elementary models, being at the same time rich enough to reveal intriguing subtleties of the interplay of coherent dynamics with external coupling. It represents therefore a unique paradigm for matching theory with experiment in the study of quantum coherence. To my opinion the most important contribution of cavity QED to our understanding of mesoscopic quantum phenomena is described in the contribution by L. Davidovich to this conference, and is related to not only creating a quantum superposition of coherent states, but also monitoring its coherence loss to a statistical mixture. The experimental proposal for the experiment [31] has been realized by the Ecole Normale Superieure in Paris by Haroche's group [32]. In the same contribution by L. Davidovich, one finds another major achievement in the area, i.e., measuring the Wigner function by a procedure closely related to the homodyne detection method [33–35], which allows in principle for probing the complete phase space.

Less well known but also very interesting is a complementarity experiment performed with an interferometer showing clear evidence that the quantum-classical boundary is crossed [36]. This of course is not the first experiment designed to illustrate aspects of the complementarity principle. Among these one finds e.g. neutron interferometric double resonance experiment [37] or the complementarity principle probed by dephasing in electron interference [38] and several others using atoms and/or photons [39–45]. But the experiment in [46] is addressed to the quantum-classical limit. They use an interferometer in which the properties of one of the beam splitting elements can be tunned in a continuous manner. An atomic double pulse Ramsey interferometer [47] in which microwave pulses act as beam splitters for the quantum states of the atoms is used. One of the pulses is a coherent field stored in a cavity comprising a small (but important!) adjustable mean photon number. The visibility of the fringes in the final atomic state probability increases with this photon number, illustrating the quantum to classical transition. A theoretical investigation of the experiment is performed in [48], including dissipative and thermal effects which are shown to be small. However it is shown that a simple modification on the experimental arrangement could also allow for the observation of the progressive loss of the capacity of "quantum erasing" as a manifestation of the classical limit of quantum mechanics [48].

It is also important to note the impressive recent experimental and technological developments in this area. An excellent review can be found in [49].

**Observation of Bose-Einstein Condensates: The Question of Quantum
Coherence.** In 1925 Einstein predicted that an ideal quantum gas would undergo a phase transition when the thermal de Broglie wavelength becomes larger
than the mean spacing between particles [50]. For many years the phenomenon,
now known as Bose-Einstein Condensation (BEC), was regarded as an academic
artifact. In 1938, London brought the question up again when superfluidity of
Helium was discovered [51]. However in this case, contrary to Einstein's hypothesis in [50], the interaction is strong.

Bose-Einstein condensates – a low temperature form of matter in which a
macroscopic population of bosons occupies the quantum mechanical ground
state – have been demonstrated for weakly interacting, dilute gases of alkali-
metal [46,52–57]. Dilute Bose-Einstein condensed gases behave as classical matter waves. This remarkable feature has been directly confirmed by several recent
experiments [54–56]. In particular, very clean interference patterns generated
by two overlapping condensates have been observed through absorption imaging
techniques [54]. Theoretical studies have predicted interesting phenomena such
as quantum entanglement of spins [58,59], suppression of quantum state diffusion [60] and interference effects [61]. A comprehensive review of Bose-Einstein
condensation in various systems is given in [62].

Interference phenomena produced by matter waves are key features underlying the quantum mechanical behavior of matter. As discussed in [59] the observation of interference phenomena in condensates does not necessarily need that
the 10^5 atoms are acting coherently. Recently, another major achievement in the
field has been reported. The entanglement of two objects, each consisting of 10^{12}
atoms, has been experimentally demonstrated [63]. In this reference, entanglement is generated via interaction of the two objects (two gas samples of Cesium
atoms) with a pulse of light, which performs a non-local Bell measurement on
collective spins of the samples, experimentally realizing the proposal in [64]. It
is shown that this entangled spin state can be maintained by 0.5 sec.

There are many more exciting and beautiful experiments exhibiting quantum
mechanical properties in other areas such as condensed matter (quantum dot
devices) and the experiments on laser cooled trapped ions [65] and the large C_{60}-
molecules [66]. I apologize for not including them and others, which happened
either for lack of space or knowledge.

2.2   Decoherence Mechanisms and Quantum Computation

The very same mechanism responsible for potential improvements on computation speed using Quantum Mechanics is the one which greatly hinders immediate technical implementation. *Entanglement* between different subsystems is
essential for the production of the states used in information processing; at the

same time, as these q-bits can not be completely isolated from its environment, entanglement with the environmental degrees of freedom is a general feature. The deleterious effect of this coupling is called *decoherence*. Reported theoretical investigation have pointed out that entanglement with the environment can be tuned to some sort of favorable ingredient: in a contribution to the present conference, N.E. Mavromatos has reported on possible quantum mechanical aspects of microtubes. The idea is that mesoscopic quantum coherent states can be formed as a result of the (quantum electromagnetic) interaction of the microtubule dimer with surrounding molecules of ordered water in the interior of the microtubule. There is however a deleterious decoherence effect associated with the dissipation through the walls of the microtubule. Transfer of energy without dissipation, therefore, could only occur if the decoherence time is larger than the average time scale required for energy transfer across the cells. Some order of magnitude estimates indicate the feasibility of observing this mesoscopic coherent states. An interesting formal analogy with Cavity Quantum Electrodynamics is established [67, 68]. Also V. Kendon reported on a study of decoherence effects in the operation of a discrete quantum walk on a line, circle and hypercube [69]. Although high sensitivity to decoherence is reported, which increases with the number of steps in the walk, as the particle is becoming more delocalized at each step, it is shown that the effect of a small amount of decoherence is to entrance the properties of the quantum walk that are desirable for the development of quantum algorithms. Specifically a fast mixing time to uniform distributions. The relevance of such study is the fact that classical random walks are powerful tools to solve hard problems as e.g. estimating the volume of a convex body [70]. It is therefore important to investigate the quantum counterpart of this mechanism and related algorithms [71]. A possible realization of a quantum walk in current cavity quantum electrodynamics [72] and in optical lattices [73] has been advanced.

In a quantum computer, information is stored on quantum variables such as spins, photons or atoms. The elementary unit is a two state quantum system, the q-bit. Quantum computations are performed by the creation of quantum superpositions of q-bits and by the controlled entanglement of the information on the q-bits. Quantum coherence must be conserved to a high degree during these operations. For a quantum computer to be of practical value, the number of q-bits must be at least 10^4. Experimental implementations of q-bits have been done in cavity QED [74], ion traps [75] and nuclear spins of large number of identical molecules [76]. Quantum coherence is high in these systems, but it seems almost impossible to realize the effective number of q-bits.

Let us next discuss the important issue of what has been learnt about decoherence mechanisms in the different physical systems we discussed in the previous section and how to model them.

- Decoherence in superconducting SQUIDs (Sect. 2.1) where large-scale tunneling of magnetization phenomena are at work, is caused principally by coupling to electrons and nuclear spins. This mechanism has been mathematically implemented by the Caldeira-Leggett model [8], where the environment consists of a set of harmonic oscillators linearly coupled to the collective degree of freedom. As discussed before, the model worked amazingly well for SQUID experiments, in situations where the major physical mechanism was the coupling to electrons where one can consider the environment to consist of extended modes [77]. Also the incoherent tunneling of quantum magnets (Sect. 2.1) has been explained within a similar context. A different decoherence mechanism in the same kind of systems can also be responsible for coherence loss, specially at low temperatures. It is the case of localized environmental modes associated with defects, impurities, paramagnetic and the nuclear spin bath. Each of these has a finite Hilbert space in a given energy range, and can thus be mapped to a spin degree of freedom. The most popular mathematical implementation of this scenario is through a set of spin-1/2 systems coupled to the collective coordinate [77]. Recent experiments on crystalline ensembles of molecular magnets [78–85], including direct verification of the role of nuclear spins by isotopic variation [83], also agreed quantitatively with the theoretical prediction for nuclear spin mediated incoherent tunneling in those systems [86,87]. So this model seems to work well too.

Thus, the coherence experiments in SQUIDs and magnets are usually analyzed using the spin-boson and/or central spin models just discussed, because in both cases the collective coordinate dynamics is governed by a "two-well" potential at low energies.

- Decoherence in Bose-Einstein condensates has also been studied. The physical decoherence mechanism in this case is the interaction of the mesoscopic system with the thermal cloud around it. In reference [88], a quantum superposition of two immiscible internal states is considered. A decoherence rate for a such system is calculated and shown to be a significant threat to a macroscopic quantum superposition of Bose-Einstein condensates. An experimental scenario is outlined where the decoherence rate is reduced due to trap engineering of the reservoir. Also environmental effects on the polarization properties of condensates has been investigated [89].

- The predominant component of decoherence in cavity QED systems corresponds simply to the escape or emission of photons, either by absorption into the cavity walls or by scattering or transmission into electromagnetic modes outside the cavity. This decoherence mechanism fits well into the scheme of a coupling to extended modes. The description of decoherence through a linear coupling to a set of harmonic oscillators seems to work well, both qualitatively and quantitatively. Moreover, cavity QED provides a unique paradigm for matching theory with experiment in the study of quantum coherence, entanglement properties and the quantification of the degree of quantum coherence. An example is the well known experiment performed at Ecole Normale in Paris [32] after a simple and ingenious theoretical proposal [31]. Several theoretical investigations followed, which considered a two level atom interacting (either dispersively or resonantly) with the high-Q cavity field [90,91], shedding light on the atom-field entanglement process and the deleterious effect of the coupling to the environment. In fact, a detailed analytical description of the competition between atom-field and field environment couplings (also under the influence of an external source) has been given in the context of the dispersive Jaynes-Cummings model [92]. The purity loss of the global system and of atomic and field subsystems has been studied as a function of time: it is shown that for short times, purity loss is essentially due to the unitary atom-field coupling and in only the next-to-leading order is to be attributed to the coupling with the cavity, or the action of an external source. The role of the external source is subtle: it will tend to compensate for the dissipation of the field intensity *and* to accelerate decoherence of the global and atomic states. Moreover, it is shown that the degree of entanglement of the atom-field system can be, in this case, completely quantified by concurrence [93], even though one has in principle as infinite space of states. For the interesting experimentally available situations, the full system atom-field-cavity can be mapped onto a 2 q-bits system whose quantum correlation content is fully quantified by concurrence.

Cavity QED experiments with strong atom-field coupling has also contributed to the area of quantum computation: the quantum information carried by a two level atom was transferred to the superconducting cavity and, after a delay, to another atom. So a quantum memory has been realized in a superposition of the field in 0 and 1 photon states, and the "holding time" has been measured [94]. Theoretical schemes to inhibit the decoherence of such q-bit have been devised as, e.g., the resonant interaction with a two level atom which temporarily "removes" the photon from the cavity, thereby protecting it from environmental effects [95]. This mechanism can be shown to be very effective in protecting quantum superpositions of 0 and 1 photons. This efficiency strongly depends on the atomic density of the pulses. Could one control the number of atoms sent through the cavity, the efficiency could be twice the typical cavity decay time [96]

2.3   Formal Developments

Quantum mechanics was constructed as a theory of ensembles. However, the latest experimental developments offer the possibility to observe and manipulate single particles. In this case, the observability of quantum jumps, not directly observable in ensembles, leads to conceptual problems of how to describe single realizations of these systems.

Prior to the development of quantum jump methods, all investigation of photon statistics started out from the ensemble description via optical Bloch equations or rate equations. Such equations were used to calculate nonexclusive "probability densities" for the emission of one or several photons at times $t_1, \ldots, t_n$ in the time interval $[0, t]$. Note that only the probability of emission of any photon was asked for. Therefore many more photons might have been emitted in between times t_i. Efforts were made to use nonexclusive "probability densities" to deduce the photon statistics of a single three level ion and the aim was to show that a single ion exhibits bright and dark periods in its resonance fluorescence on the strong transition [97, 98]. This goal was finally realized with the quantum jump approach whose development essentially started with C. Cohen-Tanoudji and J. Dalibard [99] and much at the same time, Zoller, Marte and Walls [100] derived the *exclusive* probability $P_0(t)$ that, after emission at time $t = 0$ no other photon has been emitted in the time interval $[0, t]$ or the exclusive n-photon probability density that in the interval $[0, t]$ n photons are emitted *exactly* at the times $t_1, \ldots, t_n$, without going back to the master equation of the full ensemble. The idea put forward by Zoller *et al.* was not to calculate the complete density operator ρ irrespective of the number of photons that have been emitted, but to *discriminate* between density operators corresponding to different number of emitted photons in the quantized radiation field.

The quantum jump approaches can be formulated somewhat differently in the language of quantum stochastic differential equations [101]. This formulation is quite formal at first sight, and has the advantage that certain operations that use the Markovian approximation become simpler.

Several authors became interested in alternative versions of Quantum Mechanics [102–105] and the investigation of the wave function collapse. In these investigations, stochastic differential equations for the time evolution of the state vector of the system were studied. Stochastic differential equations for the wave function have also been derived by Barchielli [106, 107] from a more abstract mathematical point of view. The approach bu Barchielli also gives a common mathematical basis for bath diffusion and jump processes. A good summary of these approaches can be found in reference [108].

Recently a connection between the quantum jump approach and the decoherent-histories formulation of quantum mechanics has been proven [109]. The decoherent-histories formulation of Quantum Mechanics was introduced by Griffiths [110], Omnès [111, 112] and Gell-Mann and Hartle [113]. In this formalism one describes a quantum system in terms of an exhaustive set of possible histories that must obey a decoherence criterion that prevents them from interfering, so that these histories may be assigned classical probabilities.

As discussed in the contribution by T.A. Brun, over the last years the theory of quantum trajectories has been developed by many authors for various purposes. Applications of the method can be found in [114]. Combining such theoretical efforts with quantum information theory may give insights into the nature of decoherence and quantum trajectories. Simplified models combining these ingredients have been analyzed in the contribution by T.A. Brun.

Another interesting formal development is the introduction of field theoretical tools to study decoherence [115–117]. Recently J.P. Paz and W. Zurek [118] published an interesting review on decoherence and its modelling. The coupling of a Brownian particle to a quantum field is considered. It is shown that some of the results obtained with the usual models of oscillator-type environments are just artifacts of the dipole approximation. This is, e.g., the case with the decoherence rate on distance. Using the master equation of their particle-field model, they show that the decoherence rate does not indefinitely grow with distance, but exhibits saturation.

A field theoretical calculation of the ground state of the electromagnetic field has also been presented in this conference by I. Bialynicki-Birula. It is an interesting, rather unexplored and difficult subject. Will it be possible, in the future, to observe the threshold between Quantum Mechanics and Relativistic Quantum Mechanics, e.g., in cavity QED experiments?

3   Some Open Questions

As discussed in the Introduction, the last decade has witnessed enormous progress in the direction of understanding fundamental problems. This is due to both, the success of experimental and theoretical efforts in several areas. In spite of that, as has been pointed out in contributions to this conference, some loose ends still remain. We briefly discuss a few of them next.

3.1   Mechanisms of Decoherence

So far we have learnt that decoherence is extremely sensitive to the different environmental couplings one finds in different physical situations. Also, so far, the experiments in SQUIDs, nanomagnets and Bose-Einstein condensates have not tested any of available decoherence calculations. We are however confident that they soon will, given the importance of the problem and the advanced technology at disposal today. Such measurements represent an stringent test for theoretical models, which will have to be devised not for their solvability but for their realism of generality.

Another important question in this context is, once we understand the underlying mechanisms, how does decoherence evolve in time. This dynamics of decoherence has so far been tested in cavity QED experiments and discussed in the previous section. However, since experiments are now directly addressing coherence phenomena at the mesoscopic and macroscopic level, the natural question is: how does decoherence affect N-particle entanglement?

The relation between Decoherence and Measurement is a long standing one. One can find papers concentrating on the role of the environment in the measurement problem back in 1958 [119–123]. It is now widely accepted that Decoherence will tend to destroy quantum effects. The measurement problem, however, is far from being considered solved. And now we have another question to add to the puzzle: what are the effects of an external environment on N-particle entanglement?

And last, but not least, what is the role of Decoherence in cosmological scales (This problem has been addressed by I. Byalynick-Birula in this meeting)?

3.2   The Classical Limit of Quantum Mechanics

The dynamical behavior of point particles as described by Newtonian mechanics is considerably altered when one goes over to a nonrelativistic quantum description. The same is valid when one just includes relativistic effects (at the level of special relativity). However one is never mindful about the relativistic-nonrelativistic cross-over and very intrigued by the quantum-classical one! Why? The pragmatic answer is easy: the first cross-over, relativistic-newtonian, is governed by a *scale*, light velocity. The kinematics of both theories is alike. This is not the case in the quantum-classical limit. Of course one also has a typical quantum scale, $\hbar$. But taking the limit $\hbar \to 0$ is far from being a satisfactory answer. The reason is that the kinematical formalism underlying quantum mechanics is completely alien to a classical description and plays a decisive role in quantum phenomena. That is, to my opinion, what makes the problem so difficult. It leads to the fact that quantum phenomena are governed by several time scales [124]

Another aspect related to the same discussion is the quantum counterpart of classically chaotic dynamics. Given the linear structure of Hilbert spaces, there is in principle no room for nonlinear dynamics. Partial answers to this problem have been given, such as the relationship which is almost universal between classically chaotic systems and statistical features of their spectra (for an excellent review on the subject, see [125]). Much less, however is known about the connection between quantum dynamical entanglement and chaos, although work along these lines has been put forward for dissipative regimes [126]. It has also been proposed that the rate of entropy production can be used as an intrinsically quantum test of the chaotic versus regular nature of the evolution [127]. This conjecture has been verified in some specific examples [128], but can not be proven in general. In fact, some counter examples have been found.

So it is fair to say we do not have a sound answer to this puzzle yet.

3.3   Some Experimental Challenges

Experimentalists have played a very decisive role in unveiling several quantum mysteries, in the past years, as I hope has become clear in Sect. 2.1. The success of the experimental research has greatly inspired and stimulated theoreticians. Decoherence has been proven and monitored in [32]. This was not the case in other areas where quantum phenomena have been observed. This will provide for an important test of our knowledge about many body physics.

Also, if decoherence can be monitored in mesoscopic systems where N particle entanglement has been detected, it would teach us a lot about decoherence mechanisms.

There are also interesting achievements in the construction of q-bits, the transferring of information between quantum systems and so on. Controlling decoherence in the big issue in this context and a theoretical as well as an experimental problem. A joint unified effort between theory and experiment will be mandatory for the success of such task, which in itself is a nice feature of the problem.

The great merit of all these puzzles is that they connect quantum mechanics (as a theory conceived for isolated systems) to the rest of the world. And the rest of the world, different areas of physics, share a common, very general open problem: to understand decoherence, whose history has just begun to be written.

References

1. P.W. Anderson: Rev. Mod. Phys. **38**, 298 (1966)
2. W.H. Zurek: Phys. Today **44**, 36 (1991)
3. S.G. Mokarzel, M.C. Nemes, and A.F.R. deToledoPiza: Braz. Journ. of Phys. **26**, 760 (1996)
4. A. Peres: Phys. Rev. Lett. **61**, 2019 (1988)
5. A. Garg and A.J. Leggett: Phys. Rev. Lett. **63**, 2159 (1989)
6. E. Schrödinger: Die Naturwiss. **23**, 807 (1935)
7. A.O. Caldeira and A.J. Leggett: Phys. Rev. Lett. **46**, 211 (1981)
8. A.O. Caldeira and A.J. Leggett: Ann. of Phys. (New York) **149**, 374 (1983)
9. A.J. Leggett, S. Chakravarty, A.T. Dorsey, M.P.A. Fisher, A. Garg and W. Zwerger: Rev. Mod. Phys. **59**, 1 (1987)
10. U. Weiss, H. Grabert, and S. Linkwitz: J. Low Temp. Phys. **68**, 213 (1987)
11. J.M. Martinis, M.H. Devoret, and J. Clarke: Phys. Rev. B **35**, 4682 (1987), and references therein.
12. R. Rouse, S. Han and J.E. Lukens: Phys. Rev. Lett. **75**, 1614 (1995)
13. R. Rouse, S. Han, and J.E. Lukens. In: *Phenomenology of Unification from Present to Future*, ed. by G.D. Patazi, C. Cosmelli, and L. Zanello (World Scientific, Singapore 1998), p. 207
14. J. Clarke, A.N. Cleland, M.H. Devoret, D. Esteve and J.M. Martinis: Science **239**, 992 (1988)
15. P. Silvestrini, V.G. Palmieri, B. Ruggiero, and M. Russo: Phys. Rev. Lett. **79**, 3046 (1997)
16. Y. Nakamura, Y.A. Pashkin, and J.S. Tsai: Nature **398**, 786 (1999)
17. J.R. Friedman, V. Patel, W. Chen, S.K. Tolpygo, and J.E. Lukens: Nature **406**, 43 (2000)
18. C.H. vander Wal, A.C.J. ter Haar, F.K. Wilhelm, R.N. Schouten, C.J.P.M. Harmans, T.P. Orlando, S. Lloyd, and J.E. Mooij: Science **290**, 773 (2000)
19. M.F. Bocko, A.M. Herr, and M.J. Feldman: IEEE Trans. Appl. Supercond. **7**, 3678 (1997)
20. L.B. Ioffe, V.B. Geshkenbein, M.V. Feigel'man, A.L. Fauchère, and G. Blatter: Nature **398**, 679 (1999)
21. J.E. Mooij, T.P. Orlando, L. Levitov, L. Tian, C.H. vander Wal, and S. Lloyd: Science **285**, 1036 (1999)
22. T.P. Orlando, J.E. Mooij, L. Tian, C.H. vander Wal, L.S. Levitov, S. Lloyd, and J.J. Mazo: Phys. Rev.B **60**, 15398 (1999)
23. O. Kahn: *Molecular Magnetism* (VCA, New York 1993)
24. D. Gatteschi, A. Caneshi, L. Pardi, and R. Sessoli: Science **265**, 1054 (1984)
25. A. Caneshi, D. Gatteschi, J. Laugier, P. Rey, R. Sessoli, and C. Zanchini: J. Am. Chem. Soc. **110**, 2795 (1989)
26. R. Sessoli, D. Gatteschi, A. Caneshi, and M. Novak: Nature **365**, 141 (1993)
27. T. Lis: Acta Crystallogr. B **36**, 2042 (1980)
28. E.H. Martins Ferreira, M.C. Nemes, and H.-D. Pfannes, quant-ph/0304185
29. L. Davidovich: Lecture in this volume
30. E.M. Chudnovsky: Science **274**, 938 (1996)

31. L. Davidovich, M. Brune, J.M. Raimond, and S. Haroche: Phys. Rev.A **53**, 1295 (1996)
32. M. Brune, E. Hagley, J. Dreyer, X. Maître, A. Maali, C. Wunderlich, J.M. Raimond, and S. Haroche: Phys. Rev. Lett. **77**, 4887 (1996)
33. B.-G. Englert, N. Sterpi, and H. Walther: Optics Commun. **100**, 576 (1993)
34. L.G. Lutterbach and L. Davidovich: Phys. Rev. Lett. **78**, 2547 (1997)
35. L.G. Lutterbach and L. Davidovich: Optics Express **3**, 147 (1998)
36. P. Bertet, S. Osnaghi, A. Rauschenbeutel, G. Nogues, A. Auffeves, M. Brune, J.M. Raimond and S. Haroche: Nature **411**, 166 (2001)
37. G. Badurek, H. Rauch, and D. Tuppinger: Phys. Rev. A **34**, 2600 (1995)
38. E. Buks, R. Schuster, M. Heilblum, D. Mahalu, and V. Umansky: Nature **391**, 871 (1998)
39. W.M. Wiseman, F.E. Harrison, M.J. Collett, S.M. Tan, D.F. Walls, and R.B. Killip: Phys. Rev. A **56**, 55 (1997)
40. P. Grangier, G. Roger, and A. Aspect: Europhys. Lett. **1**, 173 (1986)
41. M.S. Chapman, T.D. Hammond, A. Lenef, J. Schmiedmayer, R.A. Rubenstein, E. Smith, and D.E. Pritchard: Phys. Rev. Lett. **75**, 3783 (1995)
42. U. Eichman, J.C. Bergquist, J.J. Bollinger, J.M. Gilligan, W.M. Itano, D.J. Wineland, and M.G. Raizen: Phys. Rev. Lett. **70**, 2359 (1993)
43. S. Dürr, T. Nonnand, and G.Rempe: Phys. Rev. Lett. **81**, 5705 (1998)
44. T.J. Herzog, P.G. Kwiat, H. Weinfurter, and A. Zeilinger: Phys. Rev. Lett. **75**, 3034 (1995)
45. P.D.D. Schwindt, P.G. Kwiat and B.-G. Englert: Phys. Ref. A **60**, 4285 (1999)
46. M.H. Anderson, J.R. Ensher, M.R. Mathews, C.E. Wieman, and E.A. Cornell: Science **269**, 198 (1995)
47. N. Ramsey: *Molecular Beams* (Oxford Univ. Press, Oxford, UK 1985)
48. M.O. Terra Cunha and M.C. Nemes, Phys. Lett. A **305**, 313 (2002)
49. S. Haroche. In: *Fundamental Systems in Quantum Optics*, Proc. Les Houches Summer School, Session LIII, ed. by J. Dalibard, J.M. Raimond, and J. Zinn-Justin (Elsevier, Amsterdam 1992)
50. A. Einstein: Sitz. Preuss. Akad. Wiss., p. 3 (1925)
51. F. London: Nature **141**, 643 (1938)
52. K.B. Davis, M.-O. Mewes, M.R. Andrews, N.J. van Druten, D.S. Durfee, D.M. Kurnand, and W.Ketterle: Phys. Rev. Lett. **75**, 3969 (1995)
53. C.C. Bradley, C.A. Sackett, and R.G. Hulet: Phys. Rev. Lett. **78**, 985 (1997)
54. M.R. Andrews, C.G. Townsend, H.-J. Miesner, D.S. Durfee, D.M. Kurnand, and W.Ketterle: Science **275**, 637 (1997)
55. D.S. Hall, M.R. Mathews, C.E. Wieman, and E.A. Cornell: Phys. Rev. Lett. **81**, 1543 (1998)
56. B.P. Anderson and M.A. Kasevich: Science **282**, 1686 (1998)
57. L.Deng, E.W. Hagley, J. Wen, M. Trippenbach, Y. Band, P.S. Julienne, J.E. Samsarian, K. Helmerson, S.L. Rolston et al.: Nature **398**, 218 (1999)
58. J.I. Cirac, M. Lewenstein, K. Mølmer, and P. Zoller: Phys. Rev. A **57**, 1208 (1998)
59. S. Cruz-Barrios, M.C. Nemes, and A.F.R. de Toledo Piza: Europhys. Lett. **61**, 148 (2003)
60. C.K. Law, H. Pu, N.P. Bigelow, and J.H. Eberly: Phys. Rev. A **58**, 531 (1998)
61. C.M. Savage, J. Ruostekoski, and D.F. Walls: Phys. Rev. A **57**, 3805 (1998)
62. A. Griffin, D.W. Smoke, and S. Stringari (eds.): *Bose-Einstein Condensation* (Cambridge University Press, Cambridge, GB 1995)
63. B. Julsgaard, A. Kozhebin and E.S. Polzik: to appear in Nature

64. L.-M. Duan, G. Giedke, J.I. Cirac, and P. Zoller: Phys. Rev. Lett. **84**, 2722 (2000)
65. C. Monroe, D.M. Meekhof, B.E. King, and D.J. Wineland: Science **272**, 1131 (1996)
66. M. Arndt, O. Nairz, J. Vos-Andreae, C. Keller, G. van der Zouw, and A. Zellinger: Nature **401**, 680 (1999)
67. N.E. Mavromatos and A.K. Powell: Lecture in this volume
68. N.E. Mavromatos and N.D. Nanopoulos: Int. J. Mod. Phys. B **12**, 517 (1998)
69. V. Kendon and B. Tregenna: Lecture in this volume
70. M. Dyer, A. Frieze, and R. Kannan: J. of the ACM **38**, 1 (1991)
71. V. Kendon and B. Tregenna: quant-ph/0209005
72. B.C. Sanders and S.D. Barlett: quant-ph/0207028
73. W. Dür, R. Raussendorf, V.M. Kendon, and H.-J. Briegel: quant-ph/0207137
74. M. Dubé and P.C.E. Stamp: to appear in special issue of Chemical Physics on "Quantum Physics of Open Systems", cond-mat/0102156
75. C. Monroe, D.M. Meekhof, B.E. King, W.M. Itano, and D.J. Wineland: Phys. Rev. Lett. **75**, 4714 (1995)
76. Q.A. Turchette, C.J. Hood, W. Lange, H. Mabuchi, and H.J. Kimble: Phys. Rev. Lett. **75**, 4710 (1995)
77. N.V. Prokof'ev and P.C.E. Stamp: Rep. Prog. Phys. **63**, 669 (2000)
78. T. Ohm, C. Sangregorio, and C. Paulsen, Eur. Phys. J. B **6**, 195 (1998)
79. T. Ohm, C. Sangregorio and C. Paulsen: J. Low Temp. Phys. **113**, 1141 (1998)
80. W. Wernsdorfer, T. Ohm, C. Sangregorio, R. Sessoli, D. Mailly, and C. Paulsen: Phys. Rev. Lett. **82**, 3903 (1999)
81. W. Wernsdorfer, R. Sessoli, and D. Gatteschi: Europhys. Lett. **47**, 254 (1999)
82. W. Wernsdorfer and R. Sessoli: Science **284**, 133 (1998)
83. W. Wernsdorfer, A. Caneschi, R. Sessoli, D. Gatteschi, A. Cornia, V. Villar, and C. Paulsen: Phys. Rev. Lett. **84**, 2965 (2000)
84. L. Thomas and B. Barbara: J. Low Temp. Phys. **113**, 1055 (1998)
85. L. Thomas, A. Caneschi, and B. Barbara: Phys. Rev. Lett. **83**, 2398 (1999)
86. N.V. Prokof'ev and P.C.E. Stamp: J. Low Temp. Phys. **104**, 143 (1996)
87. N.V. Prokof'ev and P.C.E. Stamp: J. Phys.: Cond. Matt. **5**, L663 (1993)
88. D.A.R. Dalvit, J. Dziarmaga, and W.H. Zurek: Phys. Rev. A **62**, 013607 (2000)
89. L.M. Kuang, J.H. Li, and B.B. Hu: J. Opt. B - Quantum S O **4**, 295 (2002)
90. H.P. Breuer, U. Dorner, and F. Petruccione: Eur. Phys. J. D **14**, 377 (2001)
91. J.G. Peixoto de Faria and M.C. Nemes: J. Opt. B - Quantum S O **4**, 265 (2002)
92. J.G. Peixoto de Faria and M.C. Nemes: quant-ph/0212063
93. R.F. Werner: Phys. Rev. A **40**, 4277 (1989)
94. X. Maître, E. Hagley, G. Nogues, C. Wunderlich, P. Goy, M. Brune, J.M. Raimond, and S. Haroche: Phys. Rev. Lett. **79**, 769 (1997)
95. J.G. Peixoto de Faria, P.Nussenzveig, M.C. Nemes, and A.F.R. de Toledo Piza: quant-ph/0012037
96. J.G. Peixoto de Faria, P. Nussenzveig, and M.C. Nemes: in progress
97. A. Schenzle and R.G. Brewer: Phys. Rev. A **34**, 3127 (1986)
98. D.T. Pegg and P.L. Knight: Phys. Rev. A **37**, 4303 (1988)
99. C. Cohen-Tanoudji and J. Dalibard: Europhys. Lett. **1**, 441 (1986)
100. P. Zoller, M. Marte, and D.F. Walls, Phys. Rev. A **35**, 198 (1987)
101. C.W. Gardiner, A.S. Parkings, and P.Zoller: Phys. Rev. A **46**, 4363 (1992)
102. P. Pearle: Phys. Rev. D **13**, 857 (1976)
103. G.C. Ghirardi, A. Rimini, and T. Weber: Phys. Rev. D **34**, 470 (1986)
104. L. Diósi: J. of Phys. A **21**, 2885 (1988)

105. L. Diósi: Phys. Rev. A **40**, 1165 (1989)
106. A. Barchielli: Phys. Rev. A **34**, 1642 (1986)
107. A. Barchielli: Quantum Opt. **2**, 423 (1990)
108. P. Zoller and C.A. Gardiner: *Lecture Notes on the Les Houches Summer School on Quantum Fluctuations* (Elsevier, New York 1995)
109. T.A. Brun: Phys. Rev. Lett. **78**, 1833 (1997)
110. R. Griffiths: J. Stat. Phys. **36**, 219 (1984)
111. R. Omnès: J. Stat. Phys. **53**, 893 (1988)
112. R. Omnès: J. Stat. Phys. **57**, 893 (1989)
113. M. Gell-Mann and J.B. Hartle: Phys. Rev. D **47**, 3345 (1993)
114. T.A. Brun: Am. J. of Phys. **70**, 719 (2002)
115. H.-T. Elze: Nucl. Phys. B **436**, 213 (1995)
116. H.-T. Elze: Phys. Lett. B **369**, 295 (1996)
117. H.-T. Elze: Mod. Phys. Lett. A **14**, 2259 (1999)
118. J.P. Paz and W.H. Zurek: quant-ph/0010011
119. H.S. Green: Nuovo Cimento **9**, 880 (1958)
120. A. Daneri, A. Loinger, and G.M. Prosperi: Nucl. Phys. **33**, 297 (1962)
121. A. Daneri, A. Loinger, and G.M. Prosperi: Nuovo Cimento B **44**, 119 (1966)
122. H.D. Zeh: Found. Phys. **1**, 69 (1970)
123. M. Simonius: Phys. Rev. Lett. **40**, 980 (1978)
124. A.C. Oliveira and M.C. Nemes: submitted to Phys. Rev. A
125. M. Gutzwiller: *Chaos in Classical and Quantum Mechanics* (Springer, New York 1950)
126. W.H. Zurek and J.P. Paz: Phys. Rev. Lett. **72**, 2508 (1994)
127. W.H. Zurek and J.P. Paz: Physica D **83**, 300 (1995)
128. K. Furuya, M.C. Nemes, and G.Q. Pellegrino: Phys. Rev. Lett. **80**, 5524 (1998)

Decoherence and Quantum Trajectories

Todd A. Brun

Institute for Advanced Study, Princeton, NJ 08540, USA

Abstract. Decoherence is the process by which quantum systems interact and become correlated with their external environments; quantum trajectories are a powerful technique by which decohering systems can be resolved into stochastic evolutions, conditioned on different possible "measurements" of the environment. By calling on recently-developed tools from quantum information theory, we can analyze simplified models of decoherence, explicitly quantifying the flow of information and randomness between the system, the environment, and potential observers.

1 Introduction

In the last twenty years, the concept of *decoherence* has gradually grown to wide acceptance in the description of open quantum systems: systems which interact with an external environment [1, 2]. Such open systems are ubiquitous in nature. Almost no systems can be considered to be truly isolated, with the possible exception of the universe as a whole. One of the most difficult tasks of experimenters is to insulate the systems they study from the noisy effects of the environment, in order to see quantum effects (such as interference and entanglement) which would otherwise be masked from us. This concealing effect of decoherence is the main reason quantum mechanics was discovered only recently in history: only microscopic systems can be isolated sufficiently well to exhibit quantum effects.

Over the last ten years the theory of *quantum trajectories* has been developed by a wide variety of authors [3–10] for a variety of purposes, including the ability to model continuously monitored open systems [3,5,6], improved numerical calculation [4,10], and insight into the problem of quantum measurement [7–9]. One of the most important benefits of quantum trajectories is that they give a wide range of different descriptions for decoherent systems.

Even more recently, there has been an explosion of interest in *quantum information theory*. This has been largely stimulated by interest in quantum computers, and their potential to solve otherwise intractable problems; but the field has quickly been seen to give a new paradigm for the study of quantum systems, by abstracting their quantum properties from the details of their physical embodiments.

An obvious possibility then suggests itself: to use these new tools of quantum information theory to analyze open quantum systems, giving insights into the

T.A. Brun, Decoherence and Quantum Trajectories, Lect. Notes Phys. **633**, 239–252 (2004)
http://www.springerlink.com/ © Springer-Verlag Berlin Heidelberg 2004

nature of decoherence and quantum trajectories. First, let us review some simple ideas from quantum information, which will suffice to construct simple models of systems and environments. With these models, we can explicitly track the flow of information and randomness in quantum open systems. (For a good general source on quantum computation and information, see [11] and references therein; for a more complete application to quantum trajectories, see [12].)

1.1   Q-bits and Gates

The simplest possible quantum mechanical system is a two-level atom or *q-bit*, which has a two-dimensional Hilbert space $\mathcal{H}_2$. There are many physical embodiments of such a system: the spin of a spin-1/2 particle, the polarization states of a photon, two hyperfine states of a trapped atom or ion, two neighboring levels of a Rydberg atom, the presence or absence of a photon in a microcavity, etc. For our purposes, the particular physical embodiment is irrelevant. Q-bits were introduced in quantum information theory by analogy with *classical* bits, which can take two values, 0 or 1. Just as a q-bit is the simplest imaginable quantum system, a classical bit (or c-bit) is the simplest system which can contain any information.

By convention, we choose a particular basis and label its basis states $|0\rangle$ and $|1\rangle$, which we define to be the eigenstates of the Pauli spin matrix $\hat{\sigma}_z$ with eigenvalues $+1$ and -1, respectively. We similarly define the other Pauli operators $\hat{\sigma}_x, \hat{\sigma}_y$; linear combinations of these, together with the identity $\hat{1}$, are sufficient to produce any operator on a single q-bit.

The most general pure state of a q-bit is

$$|\psi\rangle = \alpha|0\rangle + \beta|1\rangle, \quad |\alpha|^2 + |\beta|^2 = 1 . \tag{1}$$

A global phase may be assigned arbitrarily, so all physically distinct pure states of a single q-bit form a two-parameter space.

If we allow states to be *mixed*, we represent a q-bit by a density matrix ρ; the most general density matrix can be written

$$\rho = p|\psi\rangle\langle\psi| + (1 - p)|\bar{\psi}\rangle\langle\bar{\psi}| , \tag{2}$$

where $|\psi\rangle$ and $|\bar{\psi}\rangle$ are two orthogonal pure states, $\langle\psi|\bar{\psi}\rangle = 0$. The mixed states of a q-bit form a three parameter family.

For *two* q-bits, the Hilbert space $\mathcal{H}_2 \otimes \mathcal{H}_2$ has a tensor-product basis

$$|i\rangle_A \otimes |j\rangle_B \equiv |ij\rangle_{AB} , \quad i, j \in \{0, 1\} ; \tag{3}$$

similarly, for N q-bits we can define a basis $\{|i_{N-1}i_{N-2}\cdots i_0\rangle\}$, $i_k = 0, 1$.

All states evolve according to the Schrödinger equation with some Hamiltonian $\hat{H}(t)$,

$$\frac{d|\psi\rangle}{dt} = -\frac{i}{\hbar}\hat{H}(t)|\psi\rangle. \tag{4}$$

(Henceforth, I will assume $\hbar = 1$.) Over a finite time this is equivalent to applying a unitary operator $\hat{U}$ to the state $|\psi\rangle$,

$$\hat{U} = \mathrm{T} : \exp\left\{-i \int_{t_0}^{t_f} dt\, \hat{H}(t)\right\} : , \qquad (5)$$

where $\mathrm{T} : :$ indicates that the integral should be taken in a time-ordered sense, with early to late times being composed from right to left. For the models I consider in this paper I will treat all time evolution at the level of unitary transformations rather than explicitly solving the Schrödinger equation, so time can be treated as a discrete variable

$$|\psi_n\rangle = \hat{U}_n \hat{U}_{n-1} \cdots \hat{U}_1 |\psi_0\rangle . \qquad (6)$$

For a mixed state ρ, Schrödinger time evolution is equivalent to $\rho \to \hat{U}\rho\hat{U}^\dagger$.

Because these unitary transformations are discrete, a transformation on only one or a few q-bits is analogous to a *logic gate* in classical information theory. Typical classical gates are the NOT (which affects only a single bit), and the AND and the OR (which affect two). Such gates are defined by a truth table, which gives the output for given values of the input bits.

In the quantum case, there is a continuum of possible unitary transformations. I will consider only a limited set of two-bit transformations in this paper, and no transformations involving more than two q-bits; but these models are readily generalized to more complex situations. Let us examine a couple of examples of quantum two-bit transformations. The controlled-NOT gate (or CNOT) is widely used in quantum computation; it can be defined by its action on the tensor-product basis vectors,

$$\hat{U}_{\mathrm{CNOT}} |ij\rangle = |i(i \oplus j)\rangle , \qquad (7)$$

where $\oplus$ denotes addition modulo 2. If the first bit is in state $|0\rangle$ this gate leaves the second bit unchanged; if the first bit is in state $|1\rangle$ the second bit is flipped $|0\rangle \leftrightarrow |1\rangle$. Hence the name: whether a NOT gate is performed on the second bit is *controlled* by the first bit.

Another important gate in quantum computation is the SWAP; applied to the tensor-product basis vectors it gives

$$\hat{U}_{\mathrm{SWAP}} |ij\rangle = |ji\rangle . \qquad (8)$$

As the name suggests, the SWAP gate just exchanges the states of the two bits: $\hat{U}_{\mathrm{SWAP}}(|\psi\rangle \otimes |\phi\rangle) = |\phi\rangle \otimes |\psi\rangle$.

CNOT and SWAP are examples of two-bit *quantum gates*. Such gates are of tremendous importance in the theory of quantum computation. More general unitary transformations can be built up by applying a succession of such quantum gates to the q-bits which make up the system. Such a succession of quantum gates is called a *quantum circuit*.

1.2   Projective Measurements

In the standard description of quantum mechanics, observables are identified with Hermitian operators $\hat{O} = \hat{O}^{\dagger}$. A measurement of a system initially in state $|\psi\rangle$ returns an eigenvalue o_n of $\hat{O}$, and leaves the system in the eigenstate $\hat{\mathcal{P}}_n|\psi\rangle/\sqrt{p_n}$, where $\hat{\mathcal{P}}_n$ is the projector onto the eigenspace corresponding to eigenvalue o_n; the probability of the measurement outcome is $p_n = \langle\psi|\hat{\mathcal{P}}_n|\psi\rangle$. For a mixed state ρ the probability of outcome n is $p_n = \mathrm{Tr}\{\hat{\mathcal{P}}_n\rho\}$ and the state after the measurement is $\hat{\mathcal{P}}_n\rho\hat{\mathcal{P}}_n/p_n$.

Because two observables with the same eigenspaces are completely equivalent to each other (as far as measurement probabilities and outcomes are concerned), we will not worry about the exact choice of Hermitian operator $\hat{O}$; instead, we will choose a complete set of orthogonal projections $\{\hat{\mathcal{P}}_n\}$ which represent the possible measurement outcomes. These satisfy

$$\hat{\mathcal{P}}_n\hat{\mathcal{P}}_{n'} = \hat{\mathcal{P}}_n\delta_{nn'} , \quad \sum_n \hat{\mathcal{P}}_n = \hat{1} . \tag{9}$$

A set of projection operators which obey (9) is often referred to as an *orthogonal decomposition of the identity*. For a single q-bit, the only nontrivial measurements have exactly two outcomes, which we label $+$ and $-$, with probabilities p_+ and p_- and associated projectors of the form

$$\hat{\mathcal{P}}_{\pm} = \tfrac{\hat{1}\pm\boldsymbol{n}\cdot\hat{\boldsymbol{\sigma}}}{2} = |\psi_{\pm}\rangle\langle\psi_{\pm}| , \tag{10}$$

where $\boldsymbol{n}$ is a unit 3-vector and $\hat{\boldsymbol{\sigma}} = (\hat{\sigma}_x, \hat{\sigma}_y, \hat{\sigma}_z)$. The two projectors sum to the identity operator, $\hat{\mathcal{P}}_+ + \hat{\mathcal{P}}_- = \hat{1}$. The average information obtained from a projective measurement on a q-bit is the *Shannon entropy* for the two measurement outcomes,

$$S_{\mathrm{meas}} = -p_+ \log_2 p_+ - p_- \log_2 p_- . \tag{11}$$

The maximum information gain is precisely one bit, when $p_+ = p_- = 1/2$, and the minimum is zero bits when either p_+ or p_- is 0. After the measurement, the state is left in an eigenstate of $\hat{\mathcal{P}}_+$ or $\hat{\mathcal{P}}_-$, so repeating the measurement will result in the same outcome. This repeatability is one of the most important features of projective measurements.

2   Quantifying Quantum Information

2.1   How Much Information in a Q-bit?

The Shannon entropy is the standard measure used in ordinary classical information theory to quantify the information gain from a random source. It is interesting to see how far we can get in quantifying *quantum* information, using only the tools that I've described so far.

To begin with, let's ask a question that's been around from the beginning of quantum information theory: *how much information is contained in a quantum state?* We can consider two possible answers to this question. In the first place, we could measure the quantum system in question. The information gained is quantified by the Shannon entropy of the outcome. Suppose that our system is a q-bit. Then the measurement has at most two possible outcomes, for a maximum information gain of one bit.

On the other hand, there are an infinite number of possible states (1) for a q-bit, forming a continuum of states parametrized by the complex numbers α and β. To completely specify α and β (for instance if we wanted to prepare the system in a particular state) would take an *infinite* number of bits. Thus, it seems that a quantum system can contain far more information than it is possible to extract.

This seeming paradox can be resolved by distinguishing between the physical system and the state, which is a *description* of that system. It is not necessarily surprising that it might take far more information to give a complete description of a system than it is possible to extract from that system. For instance, consider a classical bit x, which can take the values 0 or 1. This bit could be chosen randomly according to a probability distribution $p(x)$, which requires specifying the values $p(0)$ or $p(1)$. Since these are real numbers, to describe them in this case too would require an infinite number of bits.

It is good to bear this distinction between system and state in mind, since it is not always completely clear in quantum information whether one is manipulating the system or the state. For instance, in the well-known protocol of quantum teleportation, it is not the physical system which is transfered, but rather its state.

2.2   Shannon and von Neumann Entropy

The Shannon entropy, or information gain, depends strongly on both the state and the choice of measurement. For a q-bit, this ranges from 0 bits (representing a determined outcome) and 1 bit (representing a maximally uncertain outcome). Because the probabilities depend on the choice of measurement, we cannot associate a definite value of the Shannon entropy with the state. This is unlike the case of a classical probability distribution, where the Shannon entropy is unique.

We can, however, ask the minimum and maximum values of the Shannon entropy for a given state. We consider all possible measurements which are *maximally fine-grained*, i.e., which have D distinct outcomes for a D-dimensional system – two, in the case of a q-bit. For any state of a system with a Hilbert space of dimension D, the *maximum* Shannon entropy is $\log_2 D$. That means that for any state, we can find a measurement which is maximally uncertain.

The minimum, however, is quite different. For a pure state, it is always possible to find a measurement with Shannon entropy 0. For a mixed state this is not true. For any mixed state ρ, the minimum Shannon entropy of a fine-grained measurement is greater than 0.

What is the interpretation of this minimum entropy? We associate it with our *ignorance* of a system. For a pure state, this minimum entropy is zero, which we take to mean that we know *as much as possible* about this system – we have *maximal knowledge*, or *minimal ignorance*. For a mixed state, however, our ignorance is not minimal – we could learn more.

This minimum value of the Shannon entropy has a relatively simple formula,

$$S(\rho) = -\text{Tr}\{\rho \log_2 \rho\} \, . \tag{12}$$

This is essentially the *von Neumann entropy*. It vanishes for pure states, and takes a maximum value of $\log_2 D$ for the *maximally mixed* state $\hat{1}/D$.

2.3   Randomness

Unlike the classical case, having maximal information about a system does *not* imply that we can predict the outcome of any measurement. It means only that there is *some* measurement which we can predict with certainty. For most measurements, the Shannon entropy will not vanish.

Suppose we have a q-bit in the state $\alpha|0\rangle + \beta|1\rangle$, and we carry out a measurement in the $|0\rangle, |1\rangle$ basis. The Shannon entropy for this measurement is $-|\alpha|^2 \log_2 |\alpha|^2 - |\beta|^2 \log_2 |\beta|^2$, which will vanish only if either α or β is zero. I described this before as the information gained from the measurement.

One might logically ask at this point: information about what? The q-bit was initially in a pure state, which I have just stated to represent maximum knowledge. After the measurement, the q-bit is still in a pure state. I cannot, therefore, have gained any further information about the system. We can only conclude that the information I acquired by carrying out the measurement represents pure *randomness*.

We can illustrate this rather spectacularly by considering a q-bit initially in the state $(|0\rangle + |1\rangle)/\sqrt{2}$. We repeatedly and alternately measure the system in the bases $|0\rangle, |1\rangle$ and $(|0\rangle \pm |1\rangle)/\sqrt{2}$. Each time we measure the system, the outcome is completely indeterminate; we gain exactly one bit of information. By continuing this procedure as long as we like, we can gain as many bits of information as we wish. But none of these bits actually represent information about the system. The system starts, and remains for all time, in a pure state. All of the bits we acquire are pure randomness, and the q-bit and measurements form a *randomness pump*.

3   A Simple Plan

Let us now consider a simple model of a quantum process, which forms the basis of the rest of this talk. Consider a very simple quantum system: a single q-bit, which begins in a pure state $|\psi_0\rangle$. We then send in a second q-bit, the *probe*, in state $|\phi_0\rangle$. The two q-bits interact briefly, for a period δt, before flying apart again; this interaction causes them to undergo a joint unitary transformation $\hat{U}$. After they have interacted, we may intercept the probe and measure it, with projection operators $\hat{P}_+$ and $\hat{P}_-$ representing the two possible outcomes of the measurement. A schematic picture of this process is given in Fig. 1.

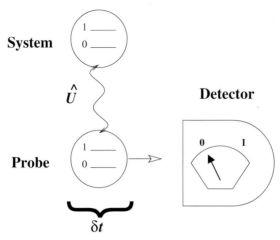

Fig. 1. This is a schematic diagram of the type of model used in this paper. The system is a single two-level system, or q-bit. It interacts briefly with a probe q-bit over a time interval δt, and the probe bit is subsequently measured

The initial state of the system and probe can be written as a simple tensor product $|\psi_0\rangle_s \otimes |\phi_0\rangle_p$. After the system and probe have interacted, however, the new joint state $|\Psi\rangle$ will generally no longer be a product. A state of this type, which cannot be written as a product, is said to be *entangled*. Such states have many curious properties, without classical analogues.

Even though the joint state $|\Psi\rangle$ is pure, we cannot describe either the system or the probe alone by a pure state. It *is* possible, however, to describe them by a *mixed* state, by finding the *reduced density operator*. The reduced density operator ρ_s for the system is found by taking a partial trace of the joint state over the probe degree of freedom,

$$\rho_s = \mathrm{Tr}_p\{|\Psi\rangle\langle\Psi|\} . \qquad (13)$$

This mixed state ρ_s gives exactly the same predictions as the joint state $|\Psi\rangle$ for any measurement which is restricted to the system alone. We can, of course, find a similar reduced density matrix ρ_p for the probe by taking a partial trace over the system.

Provided that the joint state $|\Psi\rangle$ is pure, the mixed states ρ_s and ρ_p must have the same von Neumann entropy: $S(\rho_s) = S(\rho_p)$. (This is true in general, not just for q-bits.) Because this quantity is the same whichever subsystem we trace out, and because it vanishes for product states, it is widely used as a measure of entanglement for pure states: the *entropy of entanglement*, $S_E(|\Psi\rangle)$.

Suppose now that our system and probe have interacted and are in an entangled state $|\Psi\rangle$. What happens if we measure the probe? As mentioned above, the measurement is represented by the two projection operators $\hat{\mathcal{P}}_\pm$ which sum to the identity $\hat{\mathcal{P}}_+ + \hat{\mathcal{P}}_- = \hat{1}$. Because these are projectors onto a two-dimensional Hilbert space, each of them projects onto a one-dimensional subspace. We can therefore write them as $\hat{\mathcal{P}}_+ = |+\rangle\langle+|$, $\hat{\mathcal{P}}_- = |-\rangle\langle-|$. After allowing the system and probe to interact, and then measuring the probe, the system and probe will be in one of two possible joint states,

$$(\hat{1} \otimes \hat{\mathcal{P}}_\pm)|\Psi\rangle = (\hat{1} \otimes \hat{\mathcal{P}}_\pm)\hat{U}|\psi_0\rangle \otimes |\phi_0\rangle$$
$$= \hat{A}_\pm|\psi_0\rangle \otimes |\pm\rangle , \qquad (14)$$

(where we have not renormalized the final state). The system and probe are once more in a product state. The operators $\hat{A}_\pm$ are determined by the unitary transformation $\hat{U}$ and the initial state of the probe $|\phi_0\rangle$. The probabilities of the two outcomes are

$$p_\pm = (\langle\psi_0| \otimes \langle\phi_0|)\hat{U}^\dagger(\hat{1} \otimes \hat{\mathcal{P}}_\pm)\hat{U}(|\psi_0\rangle \otimes |\phi_0\rangle)$$
$$= \langle\psi_0|\hat{A}_\pm^\dagger\hat{A}_\pm|\psi_0\rangle . \qquad (15)$$

The fact that these two probabilities must add to 1 for any state $|\psi_0\rangle$ implies that $\hat{A}_+^\dagger\hat{A}_+ + \hat{A}_-^\dagger\hat{A}_- = \hat{1}$.

If we discard the probe after the measurement and renormalize the state, the system is left in the new state $\hat{A}_\pm|\psi_0\rangle/\sqrt{p_\pm}$. This is quite similar to the effects of a projective measurement; indeed, if $\hat{A}_\pm$ are projectors, this reduces to the usual formula for a projective measurement. Because of this, and because we are indirectly acquiring information about the system by measuring the probe, this is commonly referred to as a *generalized measurement*.

How much information is gained in such a generalized measurement? We can calculate this in exactly the same way as for a projective measurement. The Shannon entropy of the generalized measurement is $-p_+ \log_2 p_+ - p_- \log_2 p_-$. This must obviously be greater than the entropy of entanglement $S_E(|\Psi\rangle)$. We can choose projectors $\hat{\mathcal{P}}_\pm$ to *minimize* the Shannon entropy by writing the state $|\Psi\rangle$ in *Schmidt form*:

$$|\Psi\rangle = \sqrt{p_+}|+\rangle_s \otimes |+\rangle_p + \sqrt{p_-}|-\rangle_s \otimes |-\rangle_p . \qquad (16)$$

Choosing the right Schmidt bases $|\pm\rangle_{s,p}$ requires us to know the initial states $|\psi_0\rangle$ and $|\phi_0\rangle$ and the unitary transformation $\hat{U}$.

Of course, in the case described above, the system starts and ends in a pure state; so this generalized measurement also generates randomness. However, it is certainly capable of giving information about the system. Suppose that the initial state of the system is maximally mixed: $\rho_s = \hat{1}/2$, with $S(\rho_s) = 1$. If we have it interact with the probe and carry out the measurement, then with probabilities

$$p_\pm = \text{Tr}\{\hat{A}_\pm\rho_s\hat{A}_\pm^\dagger\} \qquad (17)$$

the system will be left in one of the states

$$\rho_\pm = \hat{A}_\pm\rho_s\hat{A}_\pm^\dagger/p_\pm = \hat{A}_\pm\hat{A}_\pm^\dagger/2p_\pm . \qquad (18)$$

In general, neither of these states will be maximally mixed. The entropy of the system will be diminished by an average amount

$$\Delta S = 1 - p_+ S(\rho_+) - p_- S(\rho_-) . \qquad (19)$$

This represents the actual information gained about the system. This must always be less than or equal to the Shannon entropy of the generalized measurement, $-p_+ \log_2 p_+ - p_- \log_2 p_-$.

Let's look at a couple of examples to see how this works. Suppose the system is initially in the maximally mixed state $\rho_s = \hat{1}/2$; the probe is in the pure state $|\phi_0\rangle = |0\rangle$; and we measure the probe using projectors onto the states $\{|0\rangle, |1\rangle\}$. We let the interaction $\hat{U}$ be the CNOT gate from (7). In this case, we gain exactly 1 bit of information about the system, equal to the Shannon entropy of the measurement, and leaving the system in a pure state $|0\rangle$ or $|1\rangle$.

Suppose that we keep the same initial states and interaction, but instead make the measurement given by $\hat{\mathcal{P}}_\pm = |\pm\rangle\langle\pm|$ where $|\pm\rangle = (|0\rangle \pm |1\rangle)/\sqrt{2}$. In this case, the Shannon entropy is still 1 bit, but we now gain no information about the system; it is left in exactly the same state as it started, the maximally mixed state $\rho_s = \hat{1}/2$.

If we further generalize this scheme and allow the initial state of the probe to be mixed, then it is actually possible to *lose* information about the system; ΔS can be negative. For instance, an initial pure state for the system can become mixed, due to noise from interacting with a mixed environment.

With this very simple model of a quantum process, we can build up everything we need to understand both decoherence and quantum trajectories. We examine them both in the next two sections.

4   Decoherence

In discussing quantum evolution it is usually assumed that the quantum system is very well isolated from the rest of the world. This is a useful idealization, but it is rarely realized in practice, even in the laboratory. In fact, most systems interact at least weakly with external degrees of freedom [1,2]. This is the process of *decoherence*.

One way of taking this into account is to include a model of these external degrees of freedom in our description. Let us assume that in addition to the system in state $|\psi\rangle \in \mathcal{H}_S$ there is an external *environment* in state $|E\rangle \in \mathcal{H}_E$.

Systems which interact with their environments are said to be *open*. Most real physical environments are extremely complicated, and the interactions between systems and environments are often poorly understood. In analyzing open systems, one often makes the approximation of assuming a simple, analytically solvable form for the environment degrees of freedom.

For this paper, I will assume that both the system and the environment consist solely of q-bits. I will also assume a simple form of interaction, namely that the system q-bit interacts with one environment q-bit at a time, and that after interacting they never come into contact again; and that the environment q-bits have no Hamiltonian of their own. This may seem ridiculously oversimplified, but in fact it suffices to demonstrate most of the physics exhibited by much more realistic descriptions.

For this type of model, the Hilbert space of the system is $\mathcal{H}_2$ and the Hilbert space of the environment is $\mathcal{H}_E = \mathcal{H}_2 \otimes \mathcal{H}_2 \otimes \cdots$. I will assume that all the environment q-bits start in some pure initial state $|\phi_0\rangle$, usually $|0\rangle$, though further elaborations of the model could include other pure-state and mixed-state environments. These environment q-bits play a role exactly like the probe in Sect. 3, except that they are never measured.

We can describe this as an effective process on the system alone. After each interaction, the system's reduced state ρ undergoes the evolution

$$\rho \rightarrow \sum_{k=\pm} \hat{A}_k \rho \hat{A}_k^\dagger , \tag{20}$$

so that after t steps the state becomes

$$\rho(t) = \sum_{k_1,\ldots,k_t=\pm} \hat{A}_{k_t} \cdots \hat{A}_{k_1} \rho(0) \hat{A}_{k_1}^\dagger \cdots \hat{A}_{k_t}^\dagger . \tag{21}$$

The operators $\hat{A}_\pm$ depend on the choice of $\hat{U}$ and $|\phi_0\rangle$; they are determined just as in the generalized measurement case described in Sect. 3. However, this determination is not unique; many choices of $\hat{A}_k$ yield the same process (20).

An interesting case is when the interaction $\hat{U}$ is *weak*, that is, close to the identity:

$$\hat{U} = \exp(-i\epsilon\hat{H}) \approx \hat{1} - i\epsilon\hat{H} - (\epsilon\hat{H})^2/2 , \tag{22}$$

where $|\hat{H}| \sim O(1), |\epsilon| \ll 1, \hat{H} = \hat{H}^\dagger$. If the environment q-bits arrive with an average separation of δt, and ϵ is sufficiently small, then the system state ρ will approximately obey a continuous evolution equation

$$\frac{\partial \rho}{\partial t} = -i[\hat{H}_{\text{eff}}, \rho] + \hat{L}\rho\hat{L}^\dagger - (1/2)\hat{L}^\dagger\hat{L}\rho - (1/2)\rho\hat{L}^\dagger\hat{L} . \tag{23}$$

This is a *Lindblad master equation* [14]. In terms of $\hat{H}$, $\hat{H}_{\text{eff}} \propto \langle\phi_0|\hat{H}|\phi_0\rangle$, and $\hat{L} \propto \langle\bar{\phi}_0|\hat{H}|\phi_0\rangle$.

Let's consider a concrete example. Suppose the interaction is $\hat{U} = \exp(-i\epsilon\hat{H})$ with $\hat{H} = \hat{\sigma}_z \otimes \hat{\sigma}_x$, and the environment bits are initially in state $|0\rangle$. Then the reduced density matrix for the system alone will obey (23) with $\hat{H}_{\text{eff}} = 0$ and $\hat{L} = (\epsilon/\sqrt{\delta t})\hat{\sigma}_z$. This master equation will cause a system initially in the pure state $\alpha|0\rangle + \beta|1\rangle$ to evolve in the long time limit to the mixed state $\rho = |\alpha|^2|0\rangle\langle0| + |\beta|^2|1\rangle\langle1|$. As the system state becomes mixed, its von Neumann entropy grows, reflecting a gradual loss of information about the system, or (alternatively) a growing entanglement of the system with the environment. Different interactions or environment states of course will lead to different master equations.

5   Quantum Trajectories

We can readily generalize our model of decoherence by supposing that we have experimental access to the q-bits of the environment. After each bit has interacted with the system, we measure it using some predefined projective measurement; based on the outcome of this measurement, we update our knowledge of the system state. The series of decohering interactions then becomes a series of generalized measurements, as described in Sect. 3.

The evolution of the system state is no longer deterministic; it becomes stochastic due to the randomness of the measurement outcomes. We also now acquire information about the system as it is lost to the environment. Given perfect measurements of the environment, the system will remain always in a pure state.

Rather than the evolution (20), the system now undergoes

$$|\psi\rangle \to \hat{A}_\pm |\psi\rangle / \sqrt{p_\pm} \tag{24}$$

with probabilities $p_\pm = \langle \psi | \hat{A}_\pm^\dagger \hat{A}_\pm | \psi \rangle$. After t steps, the state will have become

$$|\psi(t)\rangle = \hat{A}_{k_t} \cdots \hat{A}_{k_1} |\psi(0)\rangle / \sqrt{p_{k_1,\ldots,k_t}} \,, \tag{25}$$

where the k_i take the values $\pm$. Note that, while there was an ambiguity in $\hat{A}_\pm$ for our decoherence model, by choosing a particular measurement we fix a particular choice of $\hat{A}_\pm$.

This evolution becomes more interesting when the interaction is *weak*, as described in the previous section. In this case, our series of generalized measurements are *weak measurements* [13] – on average, they disturb the system state very little, but also give very little information. The effective evolution of the system becomes approximately continuous, but rather than a master equation, it is a *stochastic Schrödinger equation*.

Let us consider the same system described at the end of Sect. 4, but now including a measurement of each environment q-bit after it has interacted with the system. After the interaction, the joint state of the system and environment bit is

$$\hat{U}|\psi\rangle \otimes |0\rangle \approx (1 - \epsilon^2/2)|\psi\rangle \otimes |0\rangle - i\epsilon\hat{\sigma}_z|\psi\rangle \otimes |1\rangle \,. \tag{26}$$

Suppose that we measure the environment bit in the $\{|0\rangle, |1\rangle\}$ basis. Then with probability $p_0 \approx 1 - \epsilon^2$ the bit will be found in state $|0\rangle$ and the system state will remain unchanged. With a small probability $p_1 \approx \epsilon^2$, however, the environment bit will be found in state $|1\rangle$ and the state of the system will change to $\hat{\sigma}_z|\psi\rangle$. This type of evolution can be approximated as a *quantum jump equation*:

$$|d\psi\rangle = (\hat{\sigma}_z - \hat{1})|\psi\rangle dN \,, \tag{27}$$

where dN is a stochastic differential variable which is 0 most of the time, but occasionally (with probability of ϵ^2 in each interval δt) becomes 1. Writing this in terms of the statistical mean $M[\cdot]$,

$$dN^2 = dN \,, \quad M[dN] = (\epsilon^2/\delta t)dt \,. \tag{28}$$

Instead of the measurement above, we might instead measure the environment bits in the basis $|\pm\rangle = (|0\rangle \pm |1\rangle)/\sqrt{2}$. In terms of this basis, the joint state becomes

$$\hat{U}|\psi\rangle \otimes |0\rangle = \frac{1}{\sqrt{2}} \exp(-i\epsilon\hat{\sigma}_z)|\psi\rangle \otimes |+\rangle + \frac{1}{\sqrt{2}} \exp(+i\epsilon\hat{\sigma}_z)|\psi\rangle \otimes |-\rangle \,. \tag{29}$$

The system state undergoes one of two weak unitary transformations, based on the outcome of the measurement; the two outcomes have equal probability. This evolution is approximated by the continuous *quantum state diffusion* equation

$$|d\psi\rangle = -i\hat{L}|\psi\rangle dW , \qquad (30)$$

where $\hat{L} = (\epsilon/\sqrt{\delta t})\hat{\sigma}_z$ and dW is a real differential stochastic variable obeying

$$M[dW] = 0 , M[dW^2] = dt . \qquad (31)$$

We see how exactly the same physical system can exhibit two very different-looking evolutions based on the choice of environmental measurement. In both cases, if we average over all possible trajectories we recover the solution of the master equation (23),

$$M[|\psi(t)\rangle\langle\psi(t)|] = \rho(t) . \qquad (32)$$

Because of this, these different stochastic Schrödinger equations are often refer-red to as different *unravelings* of the master equation.

6   Quantum Trajectories and Decoherent Histories

As is clear from the previous sections, the formalism of quantum trajectories calls on nothing more than standard quantum mechanics, and as framed above is in no way an alternative theory or interpretation. Everything can be described solely in terms of measurements and unitary transformations, the building blocks of the usual Copenhagen interpretation.

However, many people have expressed dissatisfaction with the standard in-terpretation over the years, usually due to the role of measurement as a funda-mental building block of the theory. Measuring devices are large, complicated things, very far from elementary objects; what exactly constitutes a measure-ment is never defined; and the use of classical mechanics to describe the states of measurement devices is not justified. Presumably the individual atoms, elec-trons, photons, etc., which make up a detector can themselves be described by quantum mechanics. If this is carried to its logical conclusion, however, and a Schrödinger equation is constructed for the measurement process, one obtains not classical behavior, but rather giant macroscopic superpositions such as the famous Schrödinger's cat paradox [15].

One approach to this problem is to retain the usual quantum theory, but to eliminate measurement as a fundamental concept, finding some other inter-pretation for the predicted probabilities. While many interpretations follow this approach, the one that is most closely tied to quantum trajectories is the *de-coherent* (or *consistent*) *histories formalism* of Griffiths, Omnés, Gell-Mann and Hartle [16–18]. In this formalism, probabilities are assigned to *histories* of events rather than measurement outcomes at a single time. These can be grouped into sets of mutually exclusive, exhaustive histories whose probabilities sum to 1. However, not all histories can be assigned probabilities under this interpreta-tion; only histories which lie in sets which are *consistent*, that is, whose histories do not exhibit interference with each other, and hence obey the usual classical probability sum rules.

Each set is basically a choice of description for the quantum system. For the models considered in this paper, the quantum trajectories correspond to histories in such a consistent set. The probabilities of the histories in the set exactly equal the probabilities of the measurement outcomes corresponding to a given trajectory. This equivalence has been shown between quantum trajectories and consistent sets for certain more realistic systems, as well [19–21].

For a given quantum system, there can be multiple consistent descriptions which are *incompatible* with each other; that is, unlike in classical physics, these descriptions cannot be combined into a single, more finely-grained description. In quantum trajectories, different unravelings of the same evolution correspond to such incompatible descriptions. In both cases, this is an example of the complementarity of quantum mechanics.

We see, then, that while quantum trajectories can be straightforwardly defined in terms of standard quantum theory when the environment is subjected to repeated measurements, even in the absence of such measurements there is an interpretation of the trajectories in terms of decoherent histories. Because the consistency conditions guarantee that the probability sum rules are obeyed, one can therefore use quantum trajectories as a calculational tool even in cases where no actual measurements take place.

7 Conclusions

In this paper I have presented a simple model of a system and environment consisting solely of quantum bits, using no more than single-bit measurements and two-bit unitary transformations. The simplicity of this model makes it particularly suitable for demonstrating the properties of decoherence and quantum trajectories. We can quantify the transfer of information from system to environment, the amount of entanglement, and the randomness produced by particular choices of measurement.

Quantum trajectories can often simplify the description of an open quantum system in terms of a stochastically evolving pure state rather than a density matrix. While for the q-bit models of this paper there is no great advantage in doing so, for more complicated systems this can often make a tremendous practical difference [10].

The ideas behind decoherence and quantum trajectories developed largely separately from the ideas which have led to the recent explosion of interest in quantum information; but I would argue that both areas can contribute much to the understanding of the other. I hope that this paper has given support to this view.

Acknowledgments

I would like to thank Steve Adler, Howard Carmichael, Lajos Diósi, Bob Griffiths, Jim Hartle, Rüdiger Schack, Artur Scherer and Andrei Soklakov for their comments, feedback, and criticisms of the ideas behind this paper; and Hans-Thomas Elze for his kind invitation to speak at the DICE conference in Piombino, and for including this work in the resulting volume of lectures. This work was supported in part by NSF Grant No. PHY-9900755, by DOE Grant No. DE-FG02-90ER40542, and by the Martin A. and Helen Chooljian Membership in Natural Science, IAS.

References

1. W.H. Zurek: Phys. Rev. D **24**, 1516 (1981); Phys. Rev. D **26**, 1862 (1982); Physics Today **44**, 36 (1991)
2. E. Joos and H.D. Zeh: Z. Phys. B **59**, 229 (1985)
3. H.J. Carmichael: *An Open Systems Approach to Quantum Optics* (Springer, Berlin 1993)
4. J. Dalibard, Y. Castin, and K. Mølmer: Phys. Rev. Lett. **68**, 580 (1992)
5. R. Dum, P. Zoller, and H. Ritsch: Phys. Rev. A **45**, 4879 (1992)
6. C.W. Gardiner, A.S. Parkins, and P. Zoller: Phys. Rev. A **46**, 4363 (1992)
7. N. Gisin: Phys. Rev. Lett. **52**, 1657 (1984); Helv. Phys. Acta. **62**, 363 (1989)
8. L. Diósi: J. Phys. A **21**, 2885 (1988); Phys. Lett. **129**A, 419 (1988); Phys. Lett. A **132**, 233 (1988)
9. N. Gisin and I.C. Percival: J. Phys. A **25**, 5677 (1992); J. Phys. A **26**, 2233, 2245 (1993)
10. R. Schack, T.A. Brun, and I.C. Percival: J. Phys. A **28**, 5401 (1995)
11. M.A. Nielsen and I.L. Chuang: *Quantum Computation and Quantum Information* (Cambridge University Press, Cambridge 2000)
12. T.A. Brun: Am. J. Phys. **70**, 719–737 (2002).
13. L. Vaidman. In: *Advances in quantum phenomena*, ed. by E.G. Beltrametti and J.-M. Levy-Leblond (Plenum Press, New York 1995), p. 357.
14. G. Lindblad: Commun. Math. Phys. **48**, 119 (1976)
15. E. Schrödinger: Naturwissenschaften **23**, 807, 823, 844 (1935)
16. R.B. Griffiths: J. Stat. Phys. 36, 219 (1984); Phys. Rev. A **54**, 2759 (1996); *ibid.* **57**, 1604 (1998)
17. R. Omnès: *The Interpretation of Quantum Mechanics* (Princeton University Press, Princeton 1994); *Understanding Quantum Mechanics* (Princeton University Press, Princeton 1999)
18. M. Gell-Mann, J.B. Hartle. In: *Complexity, Entropy, and the Physics of Information*, SFI Studies in the Science of Complexity v. VIII, ed. by W. Zurek (Addison-Wesley, Reading, MA, 1990) p. 425; in: *Proceedings of the Third International Symposium on the Foundations of Quantum Mechanics in the Light of New Technology*, ed. by S. Kobayashi, H. Ezawa, Y. Murayama, and S. Nomura (Physical Society of Japan, Tokyo 1990) p. 321; in: *Proceedings of the 25th International Conference on High Energy Physics*, Singapore, 2–8 August 1990, ed. by K.K. Phua and Y. Yamaguchi (World Scientific, Singapore 1990)
19. L. Diósi, N. Gisin, J.J. Halliwell, and I.C. Percival: Phys. Rev. Lett. **21**, 203 (1995)
20. T.A. Brun: Phys. Rev. Lett. **78**, 1833 (1997)
21. T.A. Brun: Phys. Rev. A **61**, 042107 (2000)

Decoherence in Discrete Quantum Walks

Viv Kendon and Ben Tregenna

QOLS, Blackett Laboratory, Imperial College London, London, SW7 2BW, UK

Abstract. We present an introduction to coined quantum walks on regular graphs, which have been developed in the past few years as an alternative to quantum Fourier transforms for underpinning algorithms for quantum computation. We then describe our results on the effects of decoherence on these quantum walks on a line, cycle and hypercube. We find high sensitivity to decoherence, increasing with the number of steps in the walk, as the particle is becoming more delocalised with each step. However, the effect of a small amount of decoherence can be to enhance the properties of the quantum walk that are desirable for the development of quantum algorithms, such as fast mixing times to uniform distributions.

1 Introduction to Quantum Walks

Quantum walks are based on a generalisation of classical random walks, which have found many applications in the field of computing. Examples of the power of classical random walks to solve hard problems include algorithms for solving k-SAT [1], estimating the volume of a convex body [2], and approximation of the permanent [3]. They are a subset of a wider model of computation, cellular automata, which have been proved universal for classical computation. The utility of classical walks suggests that extending the formalism to the quantum regime may assist the new field of quantum information processing in generating further quantum algorithms. Similarly to the classical case, it is also possible to define the notion of quantum cellular automata, whose equivalence to quantum Turing machines has been shown [4].

Most known quantum algorithms are based on the *quantum Fourier transform*, for an introduction to quantum computing and algorithms see, e. g., [5,6]. Quantum versions of random walks provide a distinctly different paradigm in which to develop quantum algorithms. Very recently, two such algorithms have been presented. Shenvi *et al.* [7] proved that a quantum walk can perform the same task as Grover's search algorithm [8], with the same quadratic speed up. Childs *et al.* [9] describe a quantum algorithm for transversing a particular graph exponentially faster than can be done classically. This exponential speed up is very promising, though the problem presented is somewhat contrived.

In fact, several possible extensions of classical random walks to the quantum regime have been proposed [10–12], however, here we will only treat the *discrete time, coined quantum walks* [13], subsequently these are referred to simply as

V. Kendon and B. Tregenna, Decoherence in Discrete Quantum Walks, Lect. Notes Phys. **633**, 253–267 (2004)
http://www.springerlink.com/

"quantum walks". Before introducing quantum walks, it is helpful to review the properties of classical random walks. This is followed by an overview of quantum walks on a line, N-cycle, and hypercube. Section 2 presents our results on the effects of decoherence in these quantum walks.

1.1 Classical Random Walks on Graphs

The discrete space on which a random walk takes place can most generally be described as a graph $G(V, E)$ with two components, a set of vertices V, and a set of edges E. An edge may be specified by the pair of vertices that it connects, $e = (v_i, v_o)$. The graph is *undirected* when $(v_i, v_o) \in E$ iff $(v_o, v_i) \in E$. The second essential feature of a classical walk is the (time-independent) transition matrix M, whose elements M_{ij} provide the probability for a transition from vertex v_i to vertex v_j. These probabilities are non-zero only for a pair of vertices connected by an edge,

$$M_{ij} \neq 0 \quad \text{iff} \quad e = (v_i, v_j) \in E . \tag{1}$$

The walk is *unbiased* if the non-zero elements of M are given by $M_{ij} = \frac{1}{d_i}$, where d_i is the degree of the vertex v_i (the number of edges connected to v_i). A graph is called *regular (d-regular)* if all vertices have equal degree (d). The state of a classical random walk at a given time t is described by a probability distribution $P(v, t)$ over the vertices $v \in V$. This distribution evolves at each time step by application of the transition matrix M,

$$P(v, t) = M^t P(v, 0) . \tag{2}$$

A number of features of these classical walks are worthy of note for later comparisons. If G is *connected* (every pair of vertices have a path linking them via a sequence of edges), then the walk tends to a steady state distribution π which is independent of the initial state $P(v, 0)$. Further, if the graph is regular then the limiting distribution is uniform on all the vertices. An exception arises in the case of periodic random walks, however, a "resting probability" for the walk to remain at the current vertex may be added which breaks the periodicity and restores the usual convergence. The rate of convergence to this limiting distribution may described in a number of ways, here we choose the *mixing time*,

$$M_\varepsilon^C = \min \{T | \forall t > T : ||P(v, t) - \pi||_{\mathrm{tv}} < \varepsilon\} . \tag{3}$$

The measure used here is the *total variational distance*,

$$||p_1 - p_2||_{\mathrm{tv}} = \sum_{v_i \in V} |p_1(v_i) - p_2(v_i)| . \tag{4}$$

The mixing time thus gives a measure of the time after which the distribution is within a distance ε of the limiting distribution and remains at least this close. An alternative measure that is useful in classical algorithms, for example 3-SAT [1], is the *hitting time*. This is defined for a pair of vertices v_0, v_1 as the expected time at which a walk starting at v_0 reaches v_1 for the first time.

1.2   Criteria for a Quantum Walk

Given a d-regular graph, $G(V, E)$, the associated Hilbert space may be defined as

$$\mathcal{H}_V = \text{span}\{|v_i\rangle\}_{i=1}^{|V|} . \tag{5}$$

As already mentioned, several extensions of classical random walks to the quantum regime have been proposed, there is no unique way to do this. To guide us, we note there are several properties of classical random walks and quantum systems that we would like such quantum walks to have, if possible. The three desirable properties of the classical transition matrix M are,

- locality: $M_{ij} \neq 0$ iff $e = (v_i, v_j) \in E$, i. e., transitions are only between vertices directly connected by an edge
- homogeneity: if $e = (v_i, v_j) \in E$, $|M_{ij}| = \frac{1}{d_i}$, i. e., equal probability of transition to any neighbouring vertex
- time independence: the current step does not depend in any way on previous steps of the random walk (Markovian)

The quantum transition matrix U, equivalent to M, should have similar properties to M. In addition, it should be

- unitary

because all pure quantum processes are unitary. As proved by Meyer [14,15], this requirement of unitarity is incompatible with the three prior properties, except in very special circumstances that don't produce interesting quantum dynamics. In order to generate a non-trivial quantum evolution, one or more of these constraints must be relaxed. The continuous time quantum walks proposed by Farhi and Gutmann [10], is non-local, in the sense that there is a small probability of the particle moving arbitrarily far away in a given unit time interval. The quantum cellular automata of Meyer [15, 16] are not homogeneous, in that the full dynamics is specified over two time steps rather than one. One may make the quantum walk non-unitary by measuring the particle at each step, but this simply reproduces the classical random walk. Finally, one may consider relaxing the time independence condition slightly, and this is what the coined quantum walks effectively do. For completeness, we note that there is an equivalent formulation of the coined quantum walk in terms of a simple quantum process on a directed graph, derived from the original undirected graph, that is due to Watrous [17].

1.3   Coined Quantum Walks

By analogy with the classical walk, in which one pictures flipping a coin at each time-step to determine which edge to leave the current vertex by, interesting quantum results may be obtained by the introduction of an explicit (quantum) coin. In addition to the particle, with associated Hilbert space $\mathcal{H}_V$ as given in (5), there is an auxiliary quantum system, the coin, with a Hilbert space of dimension d (the degree of the graph) $\mathcal{H}_C = \mathbf{C}^d$. The transition matrix is then defined as a unitary matrix U acting on the tensor product of these two spaces. This unitary is constructed from two separate operators, a "coin flip" and a conditional translation. For an unbiased walk, the coin flip operator, $\mathbf{C}$ is a unitary matrix whose elements all have equal modulus in the computational basis of the coin. This adds significant new degrees of freedom to the system, as the relative phases of these elements may be chosen arbitrarily. The conditional translation $\mathbf{T}$ moves the particle along an edge to an adjacent vertex determined by the state of the coin. The evolution of the system from an initial state $|\psi(0)\rangle_{\mathrm{CV}}$ to $|\psi(t)\rangle_{\mathrm{CV}}$ after t steps is thus given by,

$$|\psi(t)\rangle_{\mathrm{CV}} = [\mathbf{T} \cdot (\mathbf{C} \otimes \mathbf{I})]^t \, |\psi(0)\rangle_{\mathrm{CV}} \,. \tag{6}$$

The coin can also be thought of as forming a d-state quantum memory from one step to the next, allowing a much wider range of dynamics to be fully reversible (a necessary property of unitary evolution).

Unitary evolution is also completely deterministic, so the choice of initial condition is never "washed out", rather, it plays a key role in determining the outcome of the quantum walk. Unitarity also means that the joint state of the coin and particle can never reach a steady state. Even the induced probability distribution over the nodes obtained by tracing out the coin,

$$P(v,t) = \,_{\mathrm{V}}\langle v|\mathrm{Tr}_{\mathrm{C}}\left[|\psi(t)\rangle_{\mathrm{CV}}\langle\psi(t)|\right]|v\rangle_{\mathrm{V}} \,, \tag{7}$$

does not converge to a long-time limit. However, it is possible to define a time-averaged probability distribution that does reach a steady state,

$$\overline{P(v,T)} = \sum_{t=0}^{T-1} \frac{P(v,t)}{T} \,. \tag{8}$$

Operationally, this is easy to produce, it is the distribution obtained by sampling the particle location at some time t uniformly selected at random from $0 \leq t < T$. Using this distribution it is possible to define a quantum mixing time similar to that in (3),

$$M_\varepsilon^{\mathrm{Q}} = \min\left\{T | \forall t > T : ||\overline{P(v,t)} - \pi||_{\mathrm{tv}} < \varepsilon\right\} \,. \tag{9}$$

The second measure discussed for classical random walks was the hitting time. This can also be extended to quantum walks, but as before will require some modification. In the classical case it is possible to measure the location of the particle at each time step to ascertain whether or not it has reached the desired vertex. The equivalent quantum measurement disturbs the system; if a complete projection onto the vertex Hilbert space is performed at each step then all coherences are lost and a classical distribution results. There are two approaches to this problem that have been used in the literature [18]. It would be possible to wait until a chosen time, T, and then perform a full measurement over all the vertices. If the probability of being at the desired vertex is greater than some value p which is bounded below by an inverse polynomial in the size of the graph, it is said that the walk has a (T, p) one-shot hitting time. (This lowerr bound ensures that standard amplification procedures can efficiently raise the success probability arbitrarily close to unity.) The second alternative is to perform a partial measurement at each time step. Projecting onto the subspaces given by the desired vertex v, $I\!\!P_v = |v\rangle\langle v|$, and its orthogonal complement, $I\!\!P_{\bar{v}} = \mathbf{I} - |v\rangle\langle v|$ will halt the walk once the vertex v is reached. A walk has a (T, p) concurrent hitting time if $I\!\!P_v$ is measured with probability greater or equal to p, which again must have as a lower bound an inverse polynomial in the graph size, in a time T. Both these quantities have been shown to exhibit an exponential speed up over the classical case for a walk between opposite corners of a hypercube [18].

Explicit solution of a quantum walk has proved to be a difficult problem. In certain cases standard techniques from classical graph theory have been applied with some success, namely solution in the Fourier space of the problem [13, 19], and also the technique of generating functions [20]. Fourier solutions are possible when the graph is of a particular form, known as a Cayley graph. Any discrete group has an associated Cayley graph, in which the elements of the group form the vertices and the edges are placed by choosing a complete set of generators, g_i and placing an edge between the vertices a and b iff, $a = g_i b$ for one of the g_i. This produces an undirected graph results if the group is Abelian. We will next consider three specific cases of such graphs for which analytical solution has proved possible.

1.4 Coined Quantum Walk on a Line

Consider the case when the graph consists of an lattice of points at integer positions on an infinite line. The Hilbert space of the particle is represented by the integers,

$$\mathcal{H}_V = \text{span}\{|x\rangle : x \in \mathbf{Z}\} . \tag{10}$$

At each time interval the particle can move one step to either the left or the right, the coin thus requires only a two dimensional Hilbert space which can conveniently be written as,

$$\mathcal{H}_C = \{|R\rangle, |L\rangle\} . \tag{11}$$

The conditional translation $\mathbf{T}$ thus has the following action on the basis states,

$$\mathbf{T}|R, x\rangle = |R, x + 1\rangle , \qquad \mathbf{T}|L, x\rangle = |L, x - 1\rangle . \tag{12}$$

The evolution of the probability distribution induced on the lattice may now be compared with that of the analogous classical walk on a line. It is sufficient to us the Hadamard operator as the coin flip,

$$\mathbf{H} = \tfrac{1}{\sqrt{2}} \begin{pmatrix} 1 & 1 \\ 1 & -1 \end{pmatrix} , \tag{13}$$

all others are essentially equivalent [19, 20]. We also choose the initial state to be

$$|\psi(0)\rangle = \tfrac{1}{\sqrt{2}} (|R\rangle + i|L\rangle) \otimes |0\rangle . \tag{14}$$

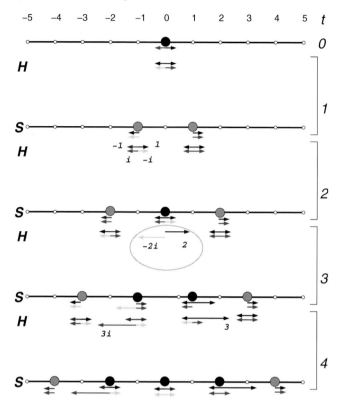

Fig. 1. A schematic representation of the first four steps of the evolution of the 1-D walk with a Hadamard coin. Circles show the particle positions and arrows show the state of the coin, with relative magnitudes indicating the amplitude of the components of the wavefunction

This choice produces a symmetric distribution, as in the classical case. Since **H** and **T** contain only real elements, there is no interference between the part of the walk due to the initial state $|R\rangle$ and that due to $i|L\rangle$. Thus the natural (in terms of basis states) outcome of a quantum walk is actually biased, unlike a classical random walk, a basic demonstration of the initial condition affecting the entire outcome of the walk. Four steps of this evolution are shown schematically in Fig. 1 and the respective probability distributions are shown in Fig. 2 after 100 time steps. The classical walk forms a binomial distribution, with a mean of 0 and standard deviation $\sigma_C = \sqrt{T}$. Figure 2 displays several of the interesting features of a quantum walk. The central interval $[-2\sqrt{T}, 2\sqrt{T}]$ is essentially uniform in the quantum case, with oscillating peaks outside this region up to $[-T/\sqrt{2}, T/\sqrt{2}]$. This is due to interference between the large number of possible paths to each point in this range.

The quantum walk on a line has been solved exactly [19, 21] using both real space (path counting) and Fourier space methods. The solutions are complicated, mainly due to the "parity" property, namely, that the solutions must have

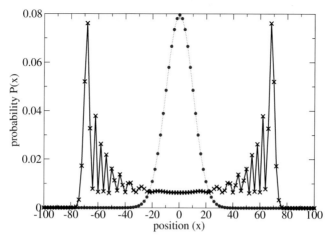

Fig. 2. A comparison of the probability distribution of a classical and coined quantum walk on a line, after 100 time steps. Only even points are plotted as both functions are zero on the odd grid points

support only on even(odd)-numbered lattice sites at even(odd) times. The moments can be calculated for asymptotically large times, for a walk starting at the origin, the standard deviation is

$$\lim_{T \to \infty} \sigma_Q = \left(1 - \frac{1}{\sqrt{2}}\right)^{1/2} T \ . \tag{15}$$

The standard deviation σ_Q is thus linear in T, in contrast to $\sqrt{T}$ for the classical walk.

Quantum walks on a line have now been studied in considerable detail. Discussions of absorbing boundaries have been given [19,20,22], with applications to halting problems in mind. Extensions to multiple coins have been made by Brun *et al.* [23,24]. However, though useful for understanding the basic properties of quantum walks, the walk on a line is too simple to yield interesting quantum problems for significant algorithms.

1.5   Coined Quantum Walk on a *N*-Cycle

If periodic boundary conditions are applied, instead of a walk on an infinite line, a closed walk on a *N*-cycle is obtained. For this system, which wraps round on itself, the standard deviation is inappropriate and mixing times must be considered. The classical result is well known,

$$\overline{M}_\varepsilon^C = O(N^2/\varepsilon) \ . \tag{16}$$

Here the time average of the probability distribution, $\overline{P(v,T)}$ has been used in (3) for ease of comparison with the quantum case. The usual mixing time defined by (3) using $P(v,T)$ scales as $M_\varepsilon^C = O(N^2 \log(1/\varepsilon))$. The scaling of ε

is not important for the quantum–classical comparisons we will do here, more details can be found in [13]. The equivalent quantity for a quantum walk was bounded above by Aharonov *et al.* [13], and shows a quadratic speed-up,

$$M_\varepsilon^Q \leq O\left(\tfrac{N \log N}{\varepsilon^3}\right) . \tag{17}$$

Bounds have also been established in the case of more general graphs and it is conjectured that mixing times can be improved at most polynomially by quantum walks [13]. Numerical studies, see [25], suggest the form of the quantum mixing time is actually $M_\varepsilon^Q = O\left(N/\varepsilon\right)$, and explain why tighter analytical bounds are hard to obtain.

1.6   Coined Quantum Walk on a Hypercube

A hypercube of dimension N is a Cayley graph with 2^N vertices labelled by the bit-strings of length N. The Hilbert space for this graph is spanned by the basis states $|x\rangle$, $x \in [0, 2^N]$. Each vertex has degree N, so the state space of the coin is $\mathcal{H}_C = \mathbf{C}^N$. The basis vectors of this space are denoted by $|a\rangle$, $a \in [0, N]$. These states correspond to the N vectors $|e_a\rangle$ where e_a is the vector with all zeroes except for a single one in the ath position. The translation operator for the walk on a hypercube can then be defined as,

$$\mathbf{T}|a, x\rangle = |a, x \oplus e_a\rangle . \tag{18}$$

Any $N \times N$ unitary matrix with all elements of unit modulus may be used as an unbiased coin, however, this is not the most natural choice given the symmetries of the hypercube. Instead, a biased coin has been selected, which distinguishes the edge along which the particle arrived at the vertex from all the others [26]. This "Grover" coin acts thus,

$$\mathbf{G}|a\rangle = \tfrac{2}{N} \sum_b |b\rangle - |a\rangle . \tag{19}$$

The full evolution for a single time step of this walk is then given by $\mathbf{U} = \mathbf{T} \cdot (\mathbf{G} \otimes \mathbf{I})$.

The quantum walk on the hypercube can be solved analytically by mapping it to a walk on a line with a variable coin operator, provided the symmetry of the hypercube is maintained throughout. To preserve the symmetry between particle states with equal Hamming weights, defined to be the number of 1's in their bit-string, the initial state must be chosen to be localised on the hypercube at $|0\rangle$ with the coin in an equal superposition of all its states, $|\psi(0)\rangle = (1/\sqrt{N}) \sum_a |a\rangle \otimes |0\rangle$. With this ansatz it has been shown by Moore and Russell [26] that the walk has an exact *instantaneous* mixing time, when the distribution transiently becomes exactly uniform, at $T = \pi N/4$. More importantly, Kempe [18] found the first exponential gap between classical and quantum walks for hitting times on the hypercube. Both the concurrent and one-shot hitting times for a walk to travel from $|0\rangle$ to $|2^N\rangle$ are found to be $\pi N/2$, with small error probability. This is exponentially faster than the classical hitting time, $T = 2^{N-1}$. This is not a true quantum speed up, since there are more sophisticated classical algorithms than a random walk to reach the opposite corner that exploit the symmetry of the hypercube. Nonetheless, it is important as the first indication that quantum walks have the potential for driving quantum algorithms with exponentially faster, later confirmed by Childs $et\ al.$ [9] on a more random graph.

2   Decoherence in Quantum Walks

We will now study each of the three systems just introduced in the presence of decoherence. A simple model of decoherence is chosen here, at each time step of the quantum walk a measurement is made in the computational basis with a probability p. We consider three cases, where the measurement is over only the coin degrees of freedom, only over the particle states, or is a complete measurement of both. Given a unitary transform $\mathbf{U} = \mathbf{T} \cdot (\mathbf{C} \otimes \mathbf{I})$ for each step of the walk as described by (6), the effect of this decoherence model can be described as a discrete master equation,

$$\varrho(t+1) = (1-p)U\varrho(t)U^{\dagger} + p\sum_i \mathbb{P}_i U \varrho(t) U^{\dagger} \mathbb{P}_i \ . \tag{20}$$

The summation runs over the dimensions of the Hilbert space on which the decoherence occurs, either the coin $\mathcal{H}_C$, the particle $\mathcal{H}_V$ or both $\mathcal{H}_C \otimes \mathcal{H}_V$. The projectors $\mathbb{P}_i$ are defined to act in the computational basis. When $p = 0$ the ideal quantum walk is obtained, and for $p = 1$ when a measurement is made at each time step it produces the classical random walk.

The numerical work summarised in the remainder of this paper has been presented more fully in [25].

2.1   Decoherence in a Quantum Walk on a Line

Since there is an experimental proposal to implement a quantum walk on a line in an optical lattice [27], as well as the three examples for $\mathbb{P}_i$ given above, we considered the likely form of experimental errors, and also modeled the effect of an imperfect Hadamard on the coin. The Hadamard operation may be considered to be a "rotation" about the computational basis by $\pi/4$, (actually a rotation and reflection since $\det(\mathbf{Rot}(\theta)) = -1$),

$$\mathbf{Rot}(\theta) = \begin{pmatrix} \sin(\theta) & \cos(\theta) \\ \cos(\theta) & -\sin(\theta) \end{pmatrix} \ . \tag{21}$$

The error model used in this case consisted of a Gaussian spread of standard deviation $\sqrt{p}\,\pi/4$ about the ideal value of $\theta = \pi/4$.

All types of decoherence model produce the same general form for the decay of $\sigma_p(T)$ from the quantum to the classical value, with small differences in the

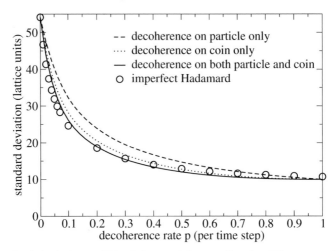

Fig. 3. Standard deviation $\sigma_p(T)$ of the particle on a line for different models of decoherence, for $T = 100$ time steps

rates, as shown in Fig. 3. The slope of $\sigma_p(T)$ is finite as $p \to 0$ and zero at $p = 1$. We have calculated $\sigma_p(T)$ analytically for $pT \ll 1$ and $T \gg 1$ for the case where $I\!\!P_i$ is the projector onto the preferred basis $\{|a, x\rangle\}$, i. e., the decoherence affecting both particle and coin. Details are given in [28]), the result is

$$\sigma_p(T) \leq \sigma(T) \left[1 - \tfrac{pT}{6\sqrt{2}} + \tfrac{p}{\sqrt{2}}(1 - 1/\sqrt{2}) + O(p^2, 1/T) \right] . \qquad (22)$$

The first order dependence is thus proportional to pT, so the sensitivity to decoherence grows linearly in T for a given decoherence rate p.

There are interesting differences in the shape of the distribution of the particle position for each of the types of decoherence. The decoherence rate that gives the closest to uniform distribution has been selected and plotted in Fig. 4, along with the pure quantum and classical distributions for comparison. When the particle position is subject to decoherence that tends to localise the particle in the standard basis, this produces a highly uniform distribution between $\pm T/\sqrt{2}$ for a particular choice of p. The optimal decoherence rate p_u can be obtained by minimising the total variational distance between the actual and uniform distributions,

$$\nu(p, T) \equiv ||P(x, p, T) - P_u(T)||_{\text{tv}} \equiv \sum_x |P(x, p, T) - P_u(T)| , \qquad (23)$$

where $P(x, p, T)$ is the probability of finding the particle at position x after T time steps, regardless of coin state [compare (7)], and $P_u(T) = \sqrt{2}/T$ for $-T/\sqrt{2} \leq x \leq T/\sqrt{2}$ and zero otherwise. The optimum decoherence rate depends on the number of steps in the walk, it can be determined numerically that $p_u T \simeq 2.6$ for decoherence on both the particle and the coin, and $p_u T \simeq 5$ for decoherence on the particle only. These differences in the quality of the uniform

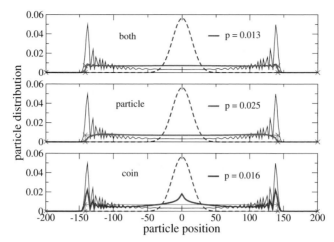

Fig. 4. Distribution of the particle position for a quantum walk on a line after $T = 200$ time steps. Pure quantum (dotted), fully classical (dashed), and decoherence at rate shown on part of system indicated by key (solid). Uniform distribution between $-T/\sqrt{2} \le x \le T/\sqrt{2}$ (crosses) also shown

distribution are independent of p and T, and provide an order of magnitude (0.6 down to 0.06) improvement in ν over the pure quantum value. Decoherence just on the coin does not enhance the uniformity of the distribution, as Fig. 4 shows, there is a cusp at $x = 0$.

2.2   Decoherence in a Quantum Walk on a N-Cycle

We now consider a walk on a N-cycle subjected to decoherence. There is an experimental proposal for implementation of a quantum walk on a cycle in the phase of a cavity field [29], in which further aspects of decoherence in such quantum walks are considered. Recall from Sect. 1.5 that the pure quantum walk on a cycle with N odd, is known [13] to mix in time $\le O(N \log N)$ if a Hadamard coin is used. The quantum walk on a cycle with N even does not mix to the uniform distribution with a Hadamard coin, but can be made to do so by appropriate choice of coin flip operator [30]. Under the action of a small amount of decoherence, the mixing time becomes shorter for all cases, typical results are shown in Fig. 5. In particular, decoherence causes the even-N cycle to mix to the uniform distribution even when a Hadamard coin is used. The asymptotes in Fig. 5 for N even and decoherence on the coin only, for $p < 2/N$, are well fitted by $\varepsilon p M_\varepsilon \simeq N/4$ for N divisible by 2, and $\varepsilon p M_\varepsilon \simeq N/16$ for N divisible by 4. For larger p, the mixing time tends to the classical value of $N^2/16\varepsilon$. Although for N divisible by 4, the (coin-decohered) mixing time shows a minimum below the classical value at $p \simeq 2/N$, this mixing time is still quadratic in N. Thus although noise on the coin causes the even-N cycle to mix to the uniform distribution, it does not produce a significant speed up over the classical random walk.

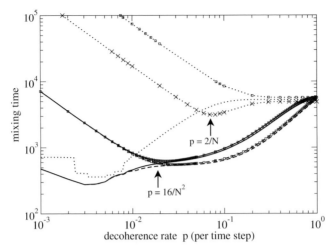

Fig. 5. Numerical data for mixing times on cycles of size $N = 29$ and $N = 30$ ($\square$), for coin (dotted), particle (dashed) and both (solid) subject to decoherence, using $\varepsilon = 0.01$. Also $N = 28$ ($\times$) for coin. Both axes logarithmic

For decoherence on the particle position, with $p < 16/N^2$, $\varepsilon p M_\varepsilon \simeq 1/(N/2 - 1)$ for N divisible by 2, and $\varepsilon p M_\varepsilon \simeq 1/(N/4 + 3)$ for N divisible by 4. At $p \simeq 16/N^2$, there is a minimum in the mixing time at a value roughly equal to the $(N + 1)$-cycle pure quantum mixing time, $M_\varepsilon^{(\min)} \sim \alpha N/\varepsilon$ (with α a constant of order unity). Decoherence on the particle position thus causes the even-N cycle to mix in linear time for a suitable choice of decoherence rate $p^{(\min)} \sim 16/N^2$, independent of ε so long as $\varepsilon < 1/N$.

For all types of decoherence, the odd-N cycle shows a minimum mixing time at a position somewhat earlier than the even-N cycle, roughly $p = 2/N^2$, but because of the oscillatory nature of $\overline{P(x, p, T)}$, the exact behaviour [25] is not a smooth function of p or ε. As decoherence on the particle (or both) increases, at $p \simeq 16/N^2$, the mixing time passes through an inflexion and from then on behaves in a quantitatively similar manner to the adjacent-sized even-N cycles, including scaling as $M_\varepsilon^{(\min)} \sim \alpha N/\varepsilon$ at the inflexion. Thus for at least $0 \leq p \lesssim 16/N^2$ the mixing time stays linear in N, and exhibits the quantum speed up over the classical N^2.

2.3   Decoherence in a Quantum Walk on a Hypercube

Recalling from Sect. 1.6, we are interested in the hitting time to the opposite corner, which was shown by Kempe [18] to be polynomial, an exponential speed up over a classical random walk. Kempe discusses two types of hitting times, one-shot, where a measurement is made after a pre-determined number of steps, and concurrent, where the desired location is monitored continuously to see if the particle has arrived. In each case, the key parameter is the probability P_h of finding the particle at the chosen location. Here we consider only target locations

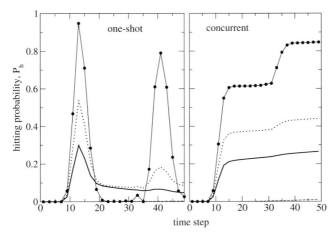

Fig. 6. Hitting probability on a 9–dimensional hypercube for one-shot (left) and concurrent (right), perfect walk (circles),with $p = 0.05$ (dotted), $p = 0.1 \simeq 1/9$ (solid). Classical hitting probability barely visible (dashed)

exactly opposite the starting vertex. We have calculated P_h numerically following the scheme of [18, 26] with a N-dimensional coin and the Grover coin operator, defined in (19). Figure 6 shows how P_h is affected by decoherence. All forms of decoherence have a similar effect on P_h, reducing the peaks and smoothing out the troughs. For the one-shot hitting time this is useful, raising P_h in the trough to well above the classical value, so it is no longer necessary to know exactly when to measure. For $p \lesssim 1/N$, the height of the first peak scales as $P_h(p) = P_h(0) \exp\{-(N + \alpha)p\}$, where $0 \lesssim \alpha \lesssim 2$ depending on whether coin, particle or both are subject to decoherence. Thus P_h decreases exponentially in p, but $p \simeq 1/N$ only lowers P_h by a factor of $1/e$, still exponentially better than classical. Continuous monitoring of the target location as in the concurrent hitting time is already a type of controlled decoherence, no new features are produced by the addition of unselective decoherence, but there is still a range of $0 < p \lesssim 1/N$ within which the quantum speed up is preserved.

2.4   Summary of Decoherence Effects in Coined Quantum Walks

For the walk on a line, whilst the effect of decoherence is to reduce the standard deviation rapidly towards the classical value, it is possible to generate highly uniform distributions over a range proportional to T for small values of the decoherence parameter, p. Uniform sampling is one of the basic tasks for which classical random walks are used, so being able to do this over a quadratically larger range with a quantum walk is certainly promising, though no quantum algorithms using this have been described to date. For a walk on a cycle the effects of decoherence are more beneficial. An optimum rate of decoherence exists for which the rate of mixing is enhanced beyond the pure quantum bound. Further, any amount of decoherence removes the effect of the coin flip operator on the steady state, allowing all such walks to converge to a uniform distribution. Fast mixing to a uniform distribution is again an important basic property required for efficient random sampling, and is the limiting factor in many classical algorithms. For a hypercube, decoherence can still be tolerated so long its rate is kept smaller than $O(1/N)$, which is logarithmic in the system size (2^N).

Thus, for both experiments and algorithms, decoherence need only be controlled down to finite low levels, rather than negligible levels, in order to observe the intriguing quantum effects displayed by coined quantum walks. Many open problems remain concerning the best ways to do this, how to exploit the full power of the extra degrees of freedom provided by the quantum coin, and how to make quantum walks perform useful tasks over more general graphs that provide real quantum computational advantages over classical algorithms.

References

1. U. Schöning: 'A probabilistic algorithm for k-SAT and constraint satisfaction problems'. In: *40th Annual Symposium on Found. of Comp. Sci.* (IEEE Computer Society Press, Los Alamitos, CA, 1999), p. 17.
2. M. Dyer, A. Frieze and R. Kannan: J. of the ACM, **38**(1), 1 (1991).
3. M. Jerrum, A. Sinclair and E. Vigoda: 'A polynomial-time approximation algorithm for the permanent of a matrix with non-negative entries'. In: *Proc. 33rd STOC* (Assoc. for Comp. Machinery, New York 2001), p. 712.
4. J. Watrous. In: *Proc. 36th Ann. Symp. on Foundations of Computer Science* (Assoc. for Comp. Machinery, New York 1995), p. 528.
5. P.W. Shor: 'Introduction to quantum algorithms', quant-ph/0005003.
6. M.A. Nielsen and I.J. Chuang: *Quantum Computation and Quantum Information* (Cambridge University Press, Cambridge, UK, 2000).
7. N. Shenvi, J. Kempe and K.B. Whaley: Phys. Rev. A, **67**, 052307 (2003).
8. L.K. Grover: 'A fast quantum mechanical algorithm for database search'. In: *Proc of the 28th Annual STOC* (Assoc. for Comp. Machinery, New York 1996), p. 212.
9. A.M. Childs, R. Cleve, E. Deotto, E. Farhi, S. Gutmann and D.A. Spielman: 'Exponential algorithmic speedup by quantum walk', quant-ph/0209131.
10. E. Farhi and S. Gutmann: Phys. Rev. A, **58**, 915 (1998).
11. A.M. Childs, E. Farhi and S. Gutmann: Quantum Information Processing, **1**, 35 (2002).
12. G. Grossing and A. Zeilinger: Complex Systems, **2**, 197 (1988).
13. D. Aharanov, A. Ambainis, J. Kempe and U. Vazirani: 'Quantum walks on graphs'. In: *Proc. 33rd STOC* (Assoc. for Comp. Machinery, New York 2001), p. 50.
14. D.A. Meyer: Phys. Lett. A, **225**, 337 (1996).
15. D.A. Meyer: 'Unitarity in one dimensional nonlinear quantum cellular automata', quant-ph/9605023.
16. D.A. Meyer: J. Stat. Phys., **85**, 551 (1996).
17. J. Watrous: J. Comp. System Sciences, **62**, 376 (2001).
18. J. Kempe: 'Quantum random walks hit exponentially faster', quant-ph/0205083.
19. A. Ambainis, E. Bach, A. Nayak, A. Vishwanath and J. Watrous: 'One-dimensional quantum walks'. In: *Proc. 33rd STOC* (Assoc. for Comp. Machinery, New York, 2001), p. 60.
20. E. Bach, S. Coppersmith, M.P. Goldschen, R. Joynt and J. Watrous: 'One-dimensional quantum walks with absorbing boundaries', quant-ph/0207008.

21. A. Nayak and A. Vishwanath: 'Quantum walk on the line', quant-ph/0010117.
22. T. Yamasaki, H. Kobayashi and H. Imai: 'An analysis of absorbing times of quantum walks'. In: *Unconventional Models of Computation, Third International Conference, UMC 2002, Kobe, Japan, October 15-19, 2002, Proceedings*, volume 2509 of *Lecture Notes in Computer Science* (ISBN 3-540-44311-8), ed. by C. Calude, M. J. Dinneen and F. Peper (Springer Verlag, Berlin 2002), p. 315.
23. T.A. Brun, H.A. Carteret and A. Ambainis: Phys. Rev. A, **67**, 052317 (2003).
24. T.A. Brun, H.A. Carteret and A. Ambainis: Phys. Rev. A, **67**, 032304 (2003).
25. V. Kendon and B. Tregenna: 'Decoherence in a quantum walk on a line'. In: *Quantum Communication, Measurement & Computing (QCMC'02)*, ed. by Jeffrey H. Shapiro and Osamu Hirota (Rinton Press, 2002).
26. C. Moore and A. Russell: 'Quantum walks on the hypercube'. In: *Proc. 6th Intl. Workshop on Randomization and Approximation Techniques in Computer Science (RANDOM '02)*, ed. by J.D.P. Rolim and S. Vadhan (Springer Verlag, Berlin 2002), p. 164.
27. W. Dür, R. Raussendorf, H.-J. Briegel and V.M. Kendon: Phys. Rev. A, **66**, 05231 (2002).
28. V. Kendon and B. Tregenna: Phys. Rev. A, **67**, 042315 (2003).
29. B. Sanders, Si. Bartlett, B. Tregenna and P.L. Knight: Phys. Rev. A, **67**, 042305 (2003).
30. B. Tregenna, W. Flanagan, R. Maile and V. Kendon: 'Controlling quantum walks: coins and initial states', *New J. Phys.*, (2003), to appear.

Decoherence and Quantum-State Measurement in Quantum Optics

Luiz Davidovich

Instituto de Física, Universidade Federal do Rio de Janeiro, C.P. 68.528
Rio de Janeiro, RJ 21941-972, Brazil

Abstract. This paper discusses work developed in recent years, in the domain of quantum optics, which has led to a better understanding of the classical limit of quantum mechanics. New techniques have been proposed, and experimentally demonstrated, for characterizing and monitoring in real time the quantum state of an electromagnetic field in a cavity. They allow the investigation of the dynamics of the decoherence process by which a quantum-mechanical superposition of coherent states of the field becomes a statistical mixture.

1 Introduction

One of the most subtle problems in contemporary physics is the relation between the macroscopic world, described by classical physics, and the microscopic world, ruled by the laws of quantum mechanics. Among the several questions involved in the quantum-classical transition, one stands out in a striking way. As pointed out by Einstein in a letter to Max Born in 1954 [1], it concerns "the inexistence at the classical level of the majority of states allowed by quantum mechanics," namely coherent superpositions of classically distinct states. Indeed, while in the quantum world one frequently comes across coherent superpositions of states (like in Young's two-slit interference experiment, in which each photon is considered to be in a coherent superposition of two wave packets, centered around the classical paths which stem out of each slit), one does not see macroscopic objects in coherent superpositions of two distinct classical states, like a stone which could be at two places at the same time. There is an important difference between a state of this kind and one which would involve just a classical alternative: the existence of quantum coherence between the two localized states would allow in principle the realization of an interference experiment, complementary to the simple observation of the position of the object. We know all this already from Young's experiment: the observation of the photon path (that is, a measurement able to distinguish through which slit the photon has passed) unavoidably destroys the interference fringes.

If one assumes that the usual rules of quantum dynamics are valid up to the macroscopic level, then the existence of quantum interference at the microscopic level necessarily implies that the same phenomenon should occur between distinguishable macroscopic states. This was emphasized by Schrödinger in his famous "cat paradox" [2]. An important role is played by this fact in quantum

L. Davidovich, Decoherence and Quantum-State Measurement in Quantum Optics, Lect. Notes Phys.
633, 268–286 (2004)
http://www.springerlink.com/

measurement theory, as pointed out by von Neumann [3]. Indeed, let us assume for instance that a microscopic two-level system (states $|+\rangle$ and $|-\rangle$) interacts with a macroscopic measuring apparatus, in such a way that the pointer of the apparatus points to a different (and classically distinguishable!) position for each of the two states, that is, the interaction transforms the joint atom-apparatus initial state into

$$|+\rangle|\uparrow\rangle \to |+\rangle'|\nearrow\rangle\,,$$
$$|-\rangle|\uparrow\rangle \to |-\rangle'|\nwarrow\rangle\,,$$

where one has allowed for a change in the state of the two-level system, due to its interaction with the measurement apparatus.

The linearity of quantum mechanics implies that, if the quantum system is prepared in a coherent superposition of the two states, say $|\psi\rangle = (|+\rangle + |-\rangle)/\sqrt{2}$, the final state of the complete system should be a coherent superposition of two product states, each of which corresponding to a different position of the pointer:

$$(1/\sqrt{2})(|+\rangle + |-\rangle)|\uparrow\rangle$$
$$\to (1/\sqrt{2})(|+\rangle'|\nearrow\rangle + |-\rangle'|\nwarrow\rangle) = (1/\sqrt{2})(|\nearrow\rangle' + |\nwarrow\rangle')\,,$$

where in the last step it was assumed that the two-level system is incorporated into the measurement apparatus after their interaction (for instance, an atom that gets stuck to the detector). One gets, therefore, as a result of the interaction between the microscopic and the macroscopic system, a coherent superposition of two classically distinct states of the macroscopic apparatus. This is actually the situation in Schrödinger's cat paradox: the cat can be viewed as a measuring apparatus of the state of a decaying atom, the state of life or death of the cat being equivalent to the two positions of the pointer. This would imply that one should be able in principle to get interference between the two states of the pointer: it is precisely the lack of evidence of such phenomena in the macroscopic world that motivated Einstein's concern.

Faced with this problem, von Neumann introduced through his collapse postulate [3] two distinct types of evolution in quantum mechanics: the deterministic and unitary evolution associated to the Schrödinger equation, which describes the establishment of a correlation between states of the microscopic system being measured and distinguishable classical states (for instance, distinct positions of a pointer) of the macroscopic measurement apparatus; and the probabilistic and irreversible process associated with measurement, which transforms coherent superpositions of distinguishable classical states into statistical mixtures. This separation of the whole process into two steps has been the object of much debate [4–6]; indeed, it would not only imply an intrinsic limitation of quantum mechanics to deal with classical objects, but it would also pose the problem of drawing the line between the microscopic and the macroscopic world.

Several possibilities have been explored as solutions to this paradox, including the proposal that a small non-linear term in the Schrödinger equation, although unnoticeable for microscopic phenomena, could eliminate the coherence between distinguishable macroscopic states, thus transforming the quantum superpositions into statistical mixtures [4]. The non-observability of the coherence between the two positions of the pointer has been attributed both to the lack of non-local observables with matrix elements between the two corresponding states [7] as well as to the fast decoherence due to interaction with the environment [8–10]. This last approach has been emphasized in recent years: decoherence follows from the irreversible coupling of the observed system to a reservoir [8,9]. In this process, the quantum superposition is turned into a statistical mixture, for which all the information on the system can be described in classical terms, so our usual perception of the world is recovered. Furthermore, for macroscopic superpositions quantum coherence decays much faster than the macroscopic observables of the system, its decay time being given by the dissipation time divided by a dimensionless number measuring the "separation" between the two parts. The statement that these two parts are macroscopically separated implies that this separation is an extremely large number. Such is the case for biological systems like "cats" made of huge number of molecules. In the simple case mentioned by Einstein [1], of a particle split into two spatially separated wave packets by a distance d, the dimensionless measure of the separation is $(d/\lambda_{dB})^2$, where λ_{dB} is the particle de Broglie wavelength. For a particle with mass equal to 1 g at a temperature of 300 K, and $d = 1$ cm, this number is about 10^{40}, and the decoherence is for all purposes instantaneous. This would provide an answer to Einstein's concern: the decoherence of macroscopic states would be too fast to be observed.

In this paper, it will be shown that the study of the interaction between atoms and electromagnetic fields in cavities can help us understand some aspects of this problem. In fact, many recent contributions in the field of quantum optics have led not only to the investigation of the subtle frontier between the quantum and the classical world, but also of hitherto unsuspected quantum mechanical processes like teleportation. Research on quantum optics is therefore intimately entangled with fundamental problems of quantum mechanics.

The whole area of "cavity quantum electrodynamics" is a very recent one. It concerns the interactions between atoms and discrete modes of the electromagnetic field in a cavity, under conditions such that losses due to dissipation and atomic spontaneous emission are very small. Usually, one deals with atomic beams crossing cavities with a high quality factor Q (defined as the product of the angular frequency of the mode and its lifetime, $Q = \omega\tau$). The atoms, prepared in special states and detected after interacting with the field, serve two purposes: they are used to manipulate the field in the cavity, so as to produce the desired states, and also to measure the field.

Several factors contributed to the development of this area. The production of superconducting Niobium cavities, with extremely high quality factors, up to the order of 10^{10}, allows one to keep a photon in the cavity for a time of the order of a fraction of a second. New techniques of atomic excitation (alkaline atoms, like Rubidium and Cesium, are frequently used for this purpose) to highly excited levels (principal quantum numbers of the order of 50), and with maximum angular momentum ($\ell = n - 1$) – the so-called planetary Rydberg atoms – have led to the production of atomic beams that interact strongly even with very weak fields, of the order of one photon, due to the large magnitude of the relevant electric dipoles. Besides, the lifetime of these states is large – of the order of the millisecond – which may be understood semiclassically, from the correspondence principle (which should be valid for $n \sim 50$): the electron is always very far away from the nucleus, and therefore its acceleration is small, implying weak radiation and a long lifetime. One should also mention the new techniques of atomic velocity control, which allow the production of approximately monokinetic atomic beams, leading to a precise control of the interaction time between atom and field. For a review of some of the main problems and results in this field, see [11].

2 Coherent Superpositions of Mesoscopic States in Cavity QED

2.1 Building the Coherent Superposition

We show now how, by carefully tailoring the interactions between two-level atoms and one mode of the electromagnetic field in a cavity, one can produce quantum superpositions of distinguishable coherent states of the field, thus mimicking the superposition of two classically distinct states of a pointer.

For a harmonic oscillator, a coherent state [12] is obtained by displacing the ground state in phase space. In general, the position will be displaced by x and the momentum by p, so that the displacement can be characterized by the complex amplitude $\alpha = \sqrt{m\omega/2\hbar}(x + ip/m\omega)$, where m and ω are respectively the mass and the angular frequency of the oscillator. The state is thus denoted by $|\alpha\rangle$, and it can be physically realized by applying a classical force to the oscillator. Coherent states are "quasi-classical" states: they evolve in time without changing their shape, the corresponding wave packet oscillating around the equilibrium position like a classical particle. Furthermore, they are minimum-uncertainty states: the product of the uncertainties in position and momentum is equal to the minimum value allowed by the Heisenberg uncertainty principle.

A one-mode electromagnetic field can be described by a harmonic oscillator Hamiltonian, in which the position and momentum are replaced by the *quadratures* of the field. These are defined as the amplitudes q_1 and q_2 of the cosine and sine terms in the time-dependent expression of the field,

$$E = E_0 \left[q_1 \cos(\boldsymbol{k} \cdot \boldsymbol{r} - \omega t) + q_2 \sin(\boldsymbol{k} \cdot \boldsymbol{r} - \omega t) \right] . \tag{1}$$

This expression is analogous to the one that yields the position of a harmonic oscillator at time t, as a function of the initial position and momentum,

$$x(t) = x(0) \cos \omega t + [p(0)/m\omega] \sin \omega t , \tag{2}$$

so that the quadratures q_1 and q_2 play a role analogous to the position $x(0)$ and momentum $p(0)$ (conveniently normalized). In the same way that, in quantum

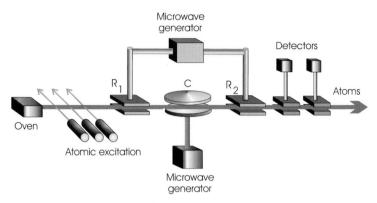

Fig. 1. Experimental arrangement for producing and measuring a coherent superposition of two coherent states of the field in cavity **C**

mechanics, the position and momentum are non-commuting operators, quantization of the electromagnetic field is achieved by requiring that the operators corresponding to the quadratures, $\hat{q}_1$ and $\hat{q}_2$, satisfy $[\hat{q}_1, \hat{q}_2] = i$. They are related to the photon annihilation operator $\hat{a}$ by

$$\hat{a} = (\hat{q}_1 + i\hat{q}_2)/\sqrt{2}. \tag{3}$$

The complex amplitude of the field is defined as $\alpha = (q_1 + iq_2)/\sqrt{2}$.

For an electromagnetic field, a coherent state also corresponds to a displaced ground state (in this case the vacuum state of the electromagnetic field). It can be explicitly written as (for a one-mode field)

$$|\alpha\rangle = \hat{D}(\alpha, \alpha^*)|0\rangle = \exp\left(\alpha\hat{a}^\dagger - \alpha^*\hat{a}\right)|0\rangle, \tag{4}$$

where $|0\rangle$ is the vacuum state of the electromagnetic field and $\hat{D}(\alpha, \alpha^*)$ is the *displacement operator*. The average number of photons in the coherent state $|\alpha\rangle$ is $|\alpha|^2$ [12]. This state can be physically realized by turning on a classical current (for instance, a microwave generator) when the field is in the vacuum state. The corresponding evolution operator in the interaction picture is then closely related to the displacement operator.

The method for generating the quantum superposition of two coherent states, proposed in [13], and sketched in Fig. 1, involves a beam of circular Rydberg atoms [14] crossing a high-Q cavity **C** in which a coherent state is previously injected (this is accomplished by coupling the cavity to a classical source – a microwave generator – through a wave guide). The utilization of circular levels is due to their strong coupling to microwaves and their very long radiative decay times, which makes them ideally suited for preparing and detecting long-lived correlations between atom and field states [15]. On either side of the high-Q cavity there are two low-Q cavities ($\mathbf{R_1}$ and $\mathbf{R_2}$), which remain coupled to a microwave generator. The fields in these two cavities can be considered as classical. As a matter of fact, it can be shown that, for the experiments realized so far, the average number of photons in these cavities is of the order of one. How come then this field behaves classically? This is due to the highly dissipative character of these cavities: again, dissipation helps to turn the quantum-mechanical world into a classical one. The classical behavior of the fields in these low-Q cavities was demonstrated in [16].

This set of two low-Q cavities constitutes the usual experimental arrangement in the Ramsey method of interferometry [15, 17]. Two of the (highly excited) atomic levels, which we denote by $|e\rangle$ (the upper level) and $|g\rangle$ (the lower one), are resonant with the microwave fields in cavities $\mathbf{R_1}$ and $\mathbf{R_2}$. The interaction of a two-level atom with a resonant electromagnetic field is analogous to the interaction between a spin and a magnetic field: it amounts to a rotation transformation applied to the two states. The intensity of the fields in $\mathbf{R_1}$ and $\mathbf{R_2}$ is chosen so that, for the selected atomic velocity, effectively a $\pi/2$ pulse is applied to the atom as it crosses each cavity. For a properly chosen phase of the microwave field, this pulse transforms the state $|e\rangle$ into the linear combination $(|e\rangle + |g\rangle)/\sqrt{2}$, and the state $|g\rangle$ into $(-|e\rangle + |g\rangle)/\sqrt{2}$.

Therefore, if each atom is prepared in the state $|e\rangle$ just prior to crossing the system, after leaving $\mathbf{R_1}$ the atom is in a superposition of two circular Rydberg states $|e\rangle$ and $|g\rangle$,

$$|\psi_{\text{atom}}\rangle = \frac{1}{\sqrt{2}}(|e\rangle + |g\rangle). \tag{5}$$

On the other hand, the superconducting cavity is assumed not to be in resonance with any of the transitions originating from those two atomic states. This means that the atom does not suffer a transition, and does not emit or absorb photons from the field. This property is further enhanced by the fact that the cavity mode is such that the field slowly rises and decreases along the atomic trajectory, so that, for sufficiently slow atoms, the atom-field coupling is adiabatic. However, the cavity is tuned in such a way that it is much closer to resonance with respect to one of those transitions, say the one connecting $|e\rangle$ to some intermediate state $|i\rangle$. The relevant level scheme is illustrated in Fig. 2. This implies that, if the atom crosses the cavity in state $|e\rangle$, dispersive effects can induce an appreciable phase shift on the field in the cavity. That is, the atom acts like a refraction index, changing the frequency of the field while the interaction is on – the corresponding energy change is just the AC-Stark shift, which for a Fock state of the electromagnetic field is proportional to the number of photons in the cavity. This frequency shift, multiplied by the interaction time between the atom and the mode, leads to a phase shift of the field in the cavity, if the atom is in state $|e\rangle$. The phase shift is negligible, however, if the atom is in state $|g\rangle$. For a principal quantum number equal to 50 in the state e, and the cavity tuned close to the $50 \rightarrow 51$ circular to circular transition (around 50 GHz), a phase shift of the order of π per photon is produced by an atom crossing the centimeter size cavity with a velocity of about 100 m/s [13].

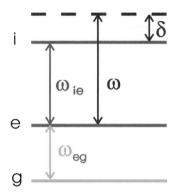

Fig. 2. Atomic level scheme: The transition $i \leftrightarrow e$ is detuned by δ from the frequency ω of a mode of cavity **C**, while the transition $e \leftrightarrow g$ is resonant with the fields in $\mathbf{R_1}$ and $\mathbf{R_2}$. State $|g\rangle$ is not affected by the field in **C**

Note that, if there is a coherent state in the cavity, a phase shift of ϕ per photon if the atom is in state $|e\rangle$ implies that

$$|e\rangle|\alpha\rangle \to |e\rangle \exp(-|\alpha|^2/2) \sum_{n=0}^{\infty} \frac{\alpha^n}{\sqrt{n!}} e^{in\phi}|n\rangle = |e\rangle|\alpha e^{i\phi}\rangle, \qquad (6)$$

that is, the phase of the coherent state is shifted by ϕ. The above equation makes use of the expansion of a coherent state in terms of Fock states.

After the atom has crossed the cavity, in a time short compared to the field relaxation time and also to the atomic radiative damping time, the state of the combined atom-field system can be written as

$$|\psi_{\text{atom+field}}\rangle = \frac{1}{\sqrt{2}} (|e; -\alpha\rangle + |g; \alpha\rangle), \qquad (7)$$

assuming that the phase shift is π per photon if the atom is in the excited state. The entanglement between the field and atomic states is analogous to the correlated two-particle states in the Einstein-Podolski-Rosen (EPR) paradox [18–20]. The two possible atomic states e and g are here correlated to the two field states $|-\alpha\rangle$ and $|\alpha\rangle$, respectively. After the atoms leave the superconducting cavity, one can detect them in the e or g states, by sending them through two ionization chambers, the first one having a field smaller than the second, so that it ionizes the atom in the e state, but not in the g state, while the second ionizes the atoms that remain in state g (Fig. 1). In the actual experiment, this detection system is replaced by a single chamber, with a static electric field that increases linearly along the direction of atomic motion. This measurement projects the field in the cavity either onto the state $|\alpha\rangle$ (if the atom is detected in state g), or onto the state $|-\alpha\rangle$ (if the atom is detected in state e). However, as in an EPR experiment [20], one may choose to make another kind of measurement, letting the atom cross, after it leaves the superconducting cavity, a second classical microwave field ($\mathbf{R_2}$ in Fig. 1), which amounts to applying to the atom another

$\pi/2$ pulse. The state (7) gets transformed then into

$$|\psi'_{\text{atom+field}}\rangle = \tfrac{1}{2}\left(|e; -\alpha\rangle - |e; \alpha\rangle + |g; \alpha\rangle + |g; -\alpha\rangle\right).\qquad(8)$$

If one detects now the atom in the state $|g\rangle$ or $|e\rangle$, the field is projected onto the state

$$|\psi_{\text{cat}}\rangle = \frac{1}{N_1}\left(|\alpha\rangle + e^{i\psi_1}|-\alpha\rangle\right),\qquad(9)$$

where $N_1 = \sqrt{2\left[1 + \cos\psi_1 \exp(-2|\alpha|^2|)\right]}$ and $\psi_1 = 0$ or π, according to whether the detected state is g or e, respectively. One produces therefore a coherent superposition of two coherent states, with phases differing by π. For $|\alpha|^2 \gg 1$, this is a "Schrödinger cat-like" state.

Superpositions of coherent states of the field were produced in the experiment reported in [21], and were detected by a procedure proposed in [22, 23].

2.2   Measuring the Coherent Superposition

Once the quantum superposition is produced, how could one tell the difference between such a superposition and a statistical mixture of the two coherent states? This can be done by simply sending another atom, in the same initial state as the first one. It can be shown then [23] that, for the state (9), with $|\alpha| \gg 1$, there is a perfect correlation between the measurements of the first and the second atom: both are always detected in the same state. On the other hand, for the corresponding statistical mixture the probability of detecting the second atom in state $|e\rangle$ is 50%, independently of which state was detected for the first atom. By delaying the sending of the second atom, one may thus explore the dynamical process by which the quantum superposition is transformed into a statistical mixture, due to the always present dissipation in a non-perfect cavity.

The time-dependent behavior of the conditional probability for measuring the second atom in the upper state, knowing that the first atom was also measured in the upper state, is displayed in Fig. 3. The sharp decay of this conditional probability from the perfectly coherent situation to the plateau associated with an incoherent superposition defines the *decoherence time*. This time can be shown to be equal to the dissipation time for the field in the cavity divided by twice the average number of photons in the field. Thus, it becomes shorter as the field becomes more macroscopic. Note also that the plateau eventually disappears, and the probability for measuring the second atom in the state $|e\rangle$ goes to zero. This can be easily understood: the field in the cavity **C** leaks out, and therefore the sole effect on the atom initially prepared in the state $|e\rangle$ is the sum of two $\pi/2$ pulses in the cavities **R₁** and **R₂**, that is a π pulse, which takes the atom into the state $|g\rangle$.

An experimental realization of this proposal was made in 1996 by Haroche's group at Ecole Normale Supérieure, in Paris [21]. The dynamical measurement of the decoherence process, as proposed above, was in agreement with the theoretical predictions.

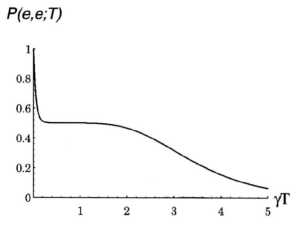

$P(e,e;T)$

Fig. 3. Conditional probability for finding the second atom in state $|e\rangle$ if the first atom was detected in state $|e\rangle$, as a function of time (in units of the field damping time)

3   The Wigner Distribution

One might wonder if it could be possible to get, from the above experimental setup, a more complete information on the field in the cavity. This was shown to be indeed possible in [24, 25]: a slight modification of the above experiment leads to the reconstruction of the so-called Wigner distribution of the field in the cavity, which provides a complete description of the quantum state of the field in phase space.

Phase space probability distributions are very useful in classical statistical physics. Averages of relevant functions of the positions and momenta of the particles can be obtained by integrating these functions with those probability weights.

In quantum mechanics, similar averages are calculated by taking the trace of the product of the density operator that describes the system with the observable of interest. Heisenberg's inequality forbids the existence in phase space of bonafied probability distributions, since one cannot determine simultaneously the position and the momentum of a particle. In spite of this, phase space distributions may still play a useful role in quantum mechanics, allowing the calculation of the average of operator-valued functions of the position and momentum operators as classical-like integrals of c-number functions. These functions are associated to those operators through correspondence rules, which depend on a previously defined operator ordering.

From all phase space representations, the Wigner distribution [30] is the most natural one, when one looks for a quantum-mechanical analog of a classical probability distribution in phase space. It is in fact the only distribution that leads to the correct marginal distributions, for any direction of integration in phase space [27,28]. Let us consider for simplicity a one-dimensional problem, for a particle with position q and momentum p. We take these to be dimensionless variables, measured in terms of some typical position and momentum of the system, which play the role of natural units (for a harmonic oscillator, the natural units would be the uncertainties in position and momentum of the ground state). If the state of the particle is characterized by the density operator $\hat{\varrho}$, then we should have not only

$$\int \mathcal{D}p\, W(q,p) = \langle q|\hat{\varrho}|q\rangle \,, \int \mathcal{D}q W(q,p) = \langle p|\hat{\varrho}|p\rangle \,, \tag{10}$$

where $|q\rangle$ and $|p\rangle$ are eigenstates of the operators $\hat{q}$ and $\hat{p}$, respectively, but also

$$P(q_\theta) = \int W(q_\theta \cos\theta - p_\theta \sin\theta, q_\theta \sin\theta + p_\theta \cos\theta)\mathcal{D}p_\theta \,. \tag{11}$$

where now

$$P(q_\theta) = \langle q_\theta|\hat{\varrho}|q_\theta\rangle \,, \tag{12}$$

the rotated coordinate q_θ being defined as

$$q_\theta = q\cos\theta + p\sin\theta \,. \tag{13}$$

One should note that, for a pure state, $\langle q|\hat{\varrho}|q\rangle = |\psi(q)|^2$, $\langle p|\hat{\varrho}|p\rangle = |\tilde{\psi}(p)|^2$. One should also note that from (10) it follows immediately the normalization property,

$$\int \mathcal{D}p\, \mathcal{D}q\, W(q,p) = 1 \,. \tag{14}$$

Expression (11), which yields the probability distribution for q_θ in terms of the function $W(q,p)$, is called a *Radon transform*. Note that this transform may be defined independently of quantum mechanics, and in fact it was investigated in 1917 by the mathematician Johan Radon [29]. He showed that, if one knows $P(q_\theta)$ for all angles θ, then one can uniquely recover the function $W(q,p)$, through the so-called *Radon inverse transform*. Quantum mechanics comes into play if one now identifies $P(q_\theta)$, given by the Radon transform (11), with the quantum expression (12). It follows then that (11) and (12) uniquely determine the function $W(q,p)$, in terms of the density operator $\hat{\varrho}$ of the system. The function $W(q,p)$ is in this case precisely the Wigner function of the system, expressed in terms of the density matrix in the position representation by

$$W(q,p) = \frac{1}{2\pi} \int_{-\infty}^{+\infty} e^{ipx} \left\langle q - \frac{x}{2} \middle| \hat{\varrho} \middle| q + \frac{x}{2} \right\rangle \mathcal{D}x \,, \tag{15}$$

which, except for a normalization constant, is the famous expression written down by Wigner [30] in his article *"On the Quantum Correction for Thermodynamic Equilibrium,"* published in 1932.

The demonstration of this result can be found in [27, 28]. Let us note that Radon's result is the mathematical basis of tomography. In fact, application of this procedure to medicine (see Fig. 4) has brought the Nobel prize in Medicine to Cormack and Hounsfield in 1979.

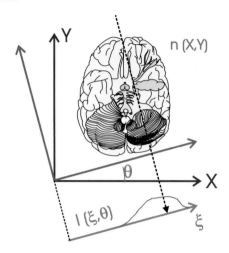

Fig. 4. Medical tomography: Measurement of the X-ray absorption for all angles along a plane allows one to reconstruct the absorptive part of the refraction index for a slice of the organ under investigation

The tomographic procedure has a simple interpretation for a harmonic oscillator. From (2), it is clear that in this case measuring the quadratures for all angles is equivalent to measuring the position of the harmonic oscillator for all times from 0 to $2\pi/\omega$. This implies that the measurement of $|\psi(x,t)|^2$ for $0 < t \leq 2\pi/\omega$ allows one to reconstruct the state $\psi(x,t)$ of the harmonic oscillator.

The question about what is the minimum set of measurements needed to reconstruct the state of a system is actually a very old problem in quantum mechanics. In his article on quantum mechanics in the *Handbuch der Physik* in 1933 [31], Pauli stated that "the mathematical problem, as to whether for given functions $W(x)$ and $\tilde{W}(p)$ [probability distributions in position and momentum space], the wave function ψ, if such a function exists, is always uniquely determined has still not been investigated in all its generality." One knows now the answer to this question: the probability distributions $W(x)$ and $\tilde{W}(p)$ do not form a complete set in the tomographic sense, and therefore are not sufficient to determine uniquely the quantum state of the system.

In 1949, it was shown by Moyal [32] that the Wigner distribution can be used to calculate averages of symmetric operator functions of q and p, as classical-like integrals in phase space. Thus, for instance,

$$\mathrm{Tr}\left(\hat{\varrho}\left\{\hat{q}^2\hat{p}\right\}_{\mathrm{sim}}\right) = \mathrm{Tr}\left[\hat{\varrho}\left(\hat{q}^2\hat{p} + \hat{q}\hat{p}\hat{q} + \hat{p}\hat{q}^2\right)/3\right] = \int \mathcal{D}q\mathcal{D}p\, W(q,p)q^2p, \quad (16)$$

where $W(q,p)$ is the Wigner function corresponding to the density operator $\hat{\varrho}$.

For a harmonic oscillator, an alternative expression for the Wigner function may be obtained by expressing the position and momentum operators $\hat{q}$ and $\hat{p}$

(or, alternatively, the quadrature operators $\hat{q}_1$ and $\hat{q}_2$) in terms of the annihilation and creation operators $\hat{a}$ and $\hat{a}^\dagger$, defined by (3).

One gets then [33]

$$W(\alpha, \alpha^*) = 2\mathrm{Tr}\left[\hat{\varrho}\hat{D}(\alpha, \alpha^*)e^{i\pi\hat{a}^\dagger\hat{a}}\hat{D}^{-1}(\alpha, \alpha^*)\right], \tag{17}$$

where the displacement operator is defined by (4). Since $\hat{\mathcal{P}} = \exp(i\pi\hat{a}^\dagger\hat{a})$ is the parity operator (note that $\hat{\mathcal{P}}\hat{q}\hat{\mathcal{P}} = -\hat{q}$, $\hat{\mathcal{P}}\hat{p}\hat{\mathcal{P}} = -\hat{p}$), this expression shows that the Wigner function is proportional to the average of the displaced parity operator.

The Wigner function given by (17) involves actually a different normalization with respect to the one defined by (15): one must set $W \to 2\pi W$, so that

$$\int (\mathcal{D}^2\alpha/\pi)\,W(\alpha, \alpha^*) = 1. \tag{18}$$

It is easy to check that the Wigner function is real and bounded. With the normalization (18), it satisfies the bound

$$|W((\alpha, \alpha^*)| \leq 2. \tag{19}$$

However, it may become negative: this is related to the fact that a bonafied phase space distribution cannot exist in quantum mechanics.

3.1 Measuring the Wigner Function

It was only in 1989 that Risken and Vogel suggested that the technique of homodyne detection could be used to reconstruct the Wigner function of a running electromagnetic wave [34]. Indeed, this technique allows the measurement of the probability distribution of an arbitrary quadrature of the electromagnetic field $q_\theta = q_1\cos\theta + q_2\sin\theta$, and one is then able to reconstruct the Wigner function through the inverse Radon transform.

The first experimental demonstration of this procedure was achieved in 1993 by Smithey et al. [35]. In view of the low detection efficiency in those experiments, the detected distribution was actually a smoothed version of the Wigner function, closely related to the so-called Husimi distribution. A much better result was achieved by Mlynek's group in 1995 [36], clearly displaying a highly compressed Gaussian, corresponding to the experimentally obtained Wigner function of a squeezed state of light emerging from an optical parametric oscillator (squeezed states are minimum uncertainty states such that the variance of one of the quadratures is smaller than the one corresponding to the vacuum state of the field). A procedure closely related to the homodyne detection method was used to reconstruct the vibrational state of a molecule by T. J. Dunn et al. [37].

Using a different (but also indirect) method, the Wigner function of the center-of-mass state of an ion trapped in a harmonic trap, and placed in the first excited state of the harmonic potential, was measured by Wineland's group at NIST [38].

3.2   Examples of Wigner Functions

Some examples of Wigner functions are shown in Fig. 5. The Wigner function corresponding to the ground state of a harmonic oscillator (or the vacuum of the electromagnetic field) is a Gaussian, centered around the origin of phase space. For a squeezed state, one gets a compressed Gaussian. On the other hand, for eigenstates of the harmonic oscillator – corresponding, for an electromagnetic field, to states with well-defined number of photons – the Wigner function is negative in some regions of phase space, as shown in Fig. 5b. As mentioned before, this is an evidence that it cannot be considered a bonafied probability distribution. Note also that, while the statistical mixture of two coherent states (which are displaced ground states) corresponds to a sum of two Gaussians, the Wigner function corresponding to the quantum superposition of two coherent states exhibits interference fringes, a clear signature of coherence. Decoherence leads to the disappearance of these fringes. Therefore, the measurement of the Wigner function of the electromagnetic field would be a clear-cut way of distinguishing between a coherent superposition and a mixture of the two coherent states. Furthermore, if one could make this measurement fast enough, one would be able to follow the decoherence process in real time.

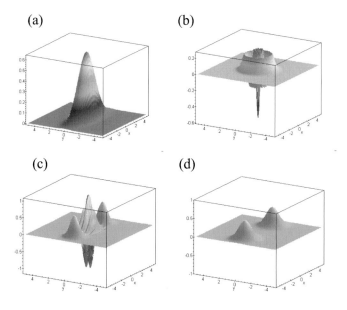

Fig. 5. Examples of Wigner distributions: (a) Squeezed state; (b) Harmonic oscillator eigenstate with $n = 3$; (c) Superposition of two coherent states, $|\psi\rangle \propto |\alpha_0\rangle + |-\alpha_0\rangle$, with $\alpha_0 = 3$; (d) Statistical mixture $\frac{1}{2}(|\alpha_0\rangle\langle\alpha_0| + |-\alpha_0\rangle\langle-\alpha_0|)$, also with $\alpha_0 = 3$

4   Direct Measurement of the Wigner Function

Once the proper state of the field is produced in the cavity, how would one be able to measure it? As shown in [25, 26], it is actually possible to measure the Wigner function of the field by a relatively simple scheme, which provides directly the value of the Wigner function at any point of phase space. This is in contrast with the tomographic procedure, or the method adopted at NIST, which yield the Wigner function only after some integration or summation. Furthermore, and also in contrast with those methods, the scheme proposed in [25, 26] is not sensitive to detection efficiency, as long as one atom is detected within a time shorter than the decoherence time. A similar procedure can be applied to the reconstruction of the vibrational state of a trapped ion [25], and also in some cases to molecules [39]. We will discuss here only the application to the electromagnetic field.

The basic experimental scheme for measuring the Wigner function [25] coincides with the one used to produce the "Schrödinger cat"-like state, illustrated in Fig. 1. A high-Q superconducting cavity $\mathbf{C}$ is placed between two low-Q cavities ($\mathbf{R_1}$ and $\mathbf{R_2}$ in Fig. 1). The cavities $\mathbf{R_1}$ and $\mathbf{R_2}$ are connected to the same microwave generator. Another microwave source is connected to $\mathbf{C}$, allowing the injection of a coherent state into this cavity. This system is crossed by a velocity-selected atomic beam, such that an atomic transition $e \leftrightarrow g$ is resonant with the fields in $\mathbf{R_1}$ and $\mathbf{R_2}$, while another transition $e \leftrightarrow i$ is quasi-resonant (detuning δ) with the field in $\mathbf{C}$, so that the atom interacts dispersively with this field if it is in state e, while no interaction takes place in $\mathbf{C}$ if the atom is in state g. The relevant level scheme is shown in Fig. 2. Just before $\mathbf{R_1}$, the atoms are promoted to the highly excited circular Rydberg state $|e\rangle$ (typical principal quantum numbers of the order of 50, corresponding to lifetimes of the order of some milliseconds). As each atom crosses the low-Q cavities, it sees a $\pi/2$ pulse, so that $||e\rangle \to [|e\rangle + |g\rangle]/\sqrt{2}$, and $|g\rangle \to [-|e\rangle + |g\rangle]/\sqrt{2}$. If the atom is in state e when crossing $\mathbf{C}$, there is an energy shift of the atom-field system (Stark shift), which dephases the field, after an effective interaction time between the atom and the cavity mode. We assume that the one-photon phase shift is equal to π. We call this a conditional phase shift, since it depends on the atomic state.

The atom is detected and the experiment is repeated many times, for each amplitude and phase of the injected field α, starting from the same initial state of the field. In this way, the probabilities P_e and P_g of detecting the probe atom in states e or g are determined. It was shown in [26] that

$$P_g - P_e = W(-\alpha, -\alpha^*)/2 , \tag{20}$$

where the Wigner function in this expression is defined in (17), with the normalization (18). Therefore, the difference between the two probabilities yields a direct measurement of the Wigner function!

The derivation of (20), developed in [25, 26], is based on expression (17) for the Wigner function. Indeed, one may notice that the experimental procedure discussed above amounts to implementing experimentally on the state to be reconstructed the two operations explicitly represented in (17): the displacement operation (implemented through the injection of the coherent microwave field) and the parity operation (implemented through the conditional π-phase shift). In particular, the distribution in (20) clearly satisfies (19), since $|P_g - P_e| \leq 2$.

An important feature of this scheme is the insensitivity to the detection efficiency of the atomic counters, of the order of $40\pm15\%$ in recent experiments [13].

One should note that this method allows the measurement of the Wigner function at each time t, allowing therefore the monitoring of the decoherence process "in real time." It is interesting, in this respect, to compare the procedure described above with the one described before in this article, as proposed in [23], with the objective of observing the decoherence of a Schrödinger cat-like state. As we have seen, it was proposed in that reference that the decoherence of the state $|\pm\rangle = (|\alpha\rangle \pm |-\alpha\rangle)/N_\pm$ could be observed by measuring the joint probability of detecting in states $|e\rangle$ or $|g\rangle$ a pair of atoms, sent through the system depicted in Fig. 1, both atoms being prepared initially in the same state. Detection of the first atom prepares the coherent superposition of coherent states. Detection of the second atom probes the state produced in **C**. Since no field was injected into the cavity between the two atoms, it is clear now that the experiment proposed in [23] amounts to a measurement of the Wigner function at the origin of phase space, which is non zero for the pure state $|\pm\rangle$, vanishes after the decoherence time, and increases again as dissipation takes place, bringing the field to the vacuum state. In the experiment realized by Brune *et al.* [13], both $|e\rangle$ and $|g\rangle$ lead to dephasings (in opposite directions) of the field in **C**. In this case, it is easy to show that the Wigner function is again recovered, as long as the one-photon phase shift is $\phi = \pi/2$ (with opposite signs for e and g), and a dephasing $\eta = \pi/2$ is applied to the second Ramsey zone [25].

Getting a π phase shift per photon imposes stringent conditions on the experiment. The interaction time between the atom and the cavity field should be large enough, which implies using slow atoms, with a precisely controlled speed. Furthermore, the interaction time between the atom and the cavity field should be much smaller than the damping time of the field in the cavity, and therefore a very good cavity is required.

An easier task consists in measuring the value of the Wigner function at the origin of phase space when one knows beforehand that the field in the cavity contains at most one photon. In this case, one does not need to inject a field into the cavity, and the dispersive interaction leading to the phase shift of the field can be replaced by a resonant 2π interaction between levels e and i (see Fig. 2). This interaction takes the atom from state e to state i and then back to state e (thus the name "2π-rotation", in view of the analogy with the full rotation of a spin $1/2$), if there is one photon in the field. Exactly as it would happen with a spin $1/2$ object, the state changes sign under this transformation: $|e\rangle|1\rangle_{\text{field}} \rightarrow -|e\rangle|1\rangle_{\text{field}}$. On the other hand, nothing happens if the atom is in state g or if there is no photon in the field. The conditional one-photon π phase change is thus accomplished in this case with a resonant interaction, which requires an interaction time much shorter than the dispersive case. This idea was

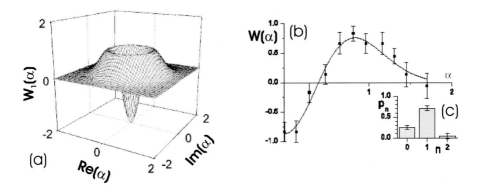

Fig. 6. Wigner function for a one-photon state: (a) Distribution in phase space; (b) Distribution measured in [41]; (c) Corresponding photon-number distribution, showing that one does not have a pure one-photon state, due to imperfections in the preparation process, the possible decay of the photon in the cavity, and the contamination with thermal photons

implemented in an experiment at Ecole Normale Supérieure, in Paris [40]. The one-photon state was produced by sending an excited atom through the empty cavity, where the atom suffers a π transition, leaving one photon in the cavity, from which it exits in the state g. This was the first time a negative value was measured for the Wigner function of an electromagnetic field, namely the value at the origin of the Wigner function corresponding to a one-photon state [this distribution is shown in Fig. 6a].

More recently [41], the Paris group was able to measure the full Wigner function for a one-photon state in the cavity, using the technique proposed in [25]. The result is displayed in Fig. 6b, which exhibits a slice of the cylindrically symmetric distribution. From the Wigner function, it is possible to get the photon-number distribution, which is displayed in Fig. 6c. This distribution shows that the state produced in the cavity is not a perfect one-photon state, which explains the fact that the value of the Wigner function at the origin of phase space is larger than -2, the value it should have for a one-photon state. An interesting feature of this measurement is that it probes a region of the phase space with area smaller than $\hbar$, which corresponds to the negative region of the Wigner function displayed in Fig. 6. It is thus an explicit demonstration of the fact that it is possible in principle to probe regions of phase space as small as one wants!

5   Conclusion

Since the invention of the laser, the field of quantum optics has been a very active field. Its discoveries have had not only an important technological impact, but have also led to experiments and proposals that probe fundamental questions of quantum mechanics. Some of these questions were discussed here: experiments in the field of cavity quantum electrodynamics have helped us to probe the subtle boundary between the classical and the quantum world, and have allowed the monitoring of the decoherence process, which is at the heart of quantum theory of measurement. The development of new techniques for probing the quantum state of the electromagnetic field in a cavity have led to the experimental unveiling of the Wigner function of a one-photon field, thus demonstrating the feasibility of probing regions of phase space with area smaller than $\hbar$. These methods may lead to a new generation of experiments, which will probe the dynamics of the quantum state of the electromagnetic field.

New challenges involve the demonstration of the teleportation process between two-level atoms [42], as well as trying to control the decoherence process, which is the main villain of quantum computers. Several proposals for fighting decoherence have been made in the last years, ranging from quantum error correction schemes [43] to feedback implementations [44,45], from the realization of q-bits in symmetric subspaces decoupled from the environment [46] to dynamical decoupling techniques [47] and reservoir engineering [48].

On a fundamental level, difficult problems still persist, related to the classical limit of non-linear systems, where chaotic behavior may play an important role [49].

Even though fundamental problems related to the classical limit of quantum mechanics and the quantum theory of measurement remain to be solved, I think it is fair to say that quantum optics has helped us to understand and observe an important piece of this puzzle.

Acknowledgments

This work was partially supported by PRONEX (Programa de Apoio a Núcleos de Excelência), CNPq (Conselho Nacional de Desenvolvimento Científico e Tecnológico), FAPERJ (Fundação de Amparo à Pesquisa do Estado do Rio de Janeiro), FUJB (Fundação Universitária José Bonifácio), and the Millennium Institute on Quantum Information. It is a pleasure to acknowledge the collaboration on the subjects covered by this paper with my students A.R.R. Carvalho, M. França Santos, T.B.L. Kist, L.G. Lutterbach, and P. Milman, and with my colleagues M. Brune, S. Haroche, R.L. de Matos Filho, M. Orszag, J.M. Raimond, and N. Zagury.

References

1. Letter from Albert Einstein to Max Born in 1954, cited by E. Joos. In: *New Techniques and Ideas in Quantum Measurement Theory*, ed. by D.M. Greenberger (New York Academy of Science, New York 1986)
2. E. Schrödinger: Naturwissenschaften **23**, 807 (1935); **23**, 823 (1935); **23**, 844 (1935). English translation by J.D. Trimmer: Proc. Am. Phys. Soc. **124**, 3235 (1980)
3. J. von Neumann: *Die Mathematische Grundlagen der Quantenmechanik* (Springer-Verlag, Berlin 1932). English translation by R.T. Beyer: *Mathematical Foundations of Quantum Mechanics*, (Princeton University Press, Princeton, NJ, 1955)

4. E. Wigner: in *The Scientist Speculates*, ed. by I.J. Good (William Heinemann, London 1962), p. 284, and also in: *Symmetries and Reflections* (Indiana University Press, Bloomington 1967), p. 171. See also E. Wigner: Am. J. of Phys. **31**, No. 1 (1963)

5. *Quantum Theory and Measurement*, ed. by J.A. Wheeler and W.H. Zurek (Princeton Univ. Press, Princeton 1983); W. Zurek: Phys. Today **44**, No. 10, 36 (1991); R. Omnès: *The Interpretation of Quantum Mechanics* (Princeton University Press, Princeton, NJ 1994); D. Giulini, E. Joos, C. Kiefer, J. Kupsch, I.-O. Stamatescu, and H.D. Zeh: *Decoherence and the Appearance of a Classical World in Quantum Theory* (Springer-Verlag, Berlin 1996)

6. K. Hepp: Helv. Phys. Acta **45**, 237 (1972); J.S. Bell: Helv. Phys. Acta **48**, 93 (1975)

7. K. Gottfried: *Quantum Mechanics* (Benjamim, Reading, MA , 1966), Sect. IV

8. H.D. Zeh: Found. Phys. **1**, 69 (1970); W.H. Zurek: Phys. Rev. D **24**, 1516 (1981); **26**, 1862 (1982); W.G. Unruh and W.H. Zurek: Phys. Rev. D **40**, 1071 (1989); W.H. Zurek: Phys. Today **44**, No. 10, 36 (1991); B.L. Hu, J.P. Paz, and Y. Zhang: Phys. Rev. D **45**, 2843 (1992)

9. H. Dekker: Phys. Rev. A **16**, 2116 (1977); A.O. Caldeira and A.J. Leggett: Physica (Amsterdam) **121A**, 587 (1983); Phys. Rev. A **31**, 1059 (1985)

10. E. Joos and H.D. Zeh: Z. Phys. B **59**, 223 (1985); G.J. Milburn and C.A. Holmes: Phys. Rev. Lett. **56**, 2237 (1986); F. Haake and D. Walls: Phys. Rev. A **36**, 730 (1987)

11. S. Haroche: 'Cavity Quantum Electrodynamics'. In: *Fundamental Systems in Quantum Optics, Proc. Les Houches Summer School, Session LIII*, ed. by J. Dalibard, J.M. Raimond and J. Zinn-Justin (Elsevier, Amsterdam 1992); see also *Cavity Quantum Electrodynamics*, ed. by P. Berman (Academic Press, New York 1994)

12. R.J. Glauber: Phys. Rev. **131**, 2766-2788, 1963

13. M. Brune, S. Haroche, J.M. Raimond, L. Davidovich, and N. Zagury: Phys. Rev. A **45**, 5193 (1992)

14. R.G. Hulet and D. Kleppner: Phys. Rev. Lett. **51**, 1430 (1983); A. Nussenzveig *et al*: Euro. Phys. Lett. **14**, 755 (1991)

15. M. Brune, P. Nussenzveig, F. Schmidt-Kaler, F. Bernardot, A. Maali, J.M. Raimond, and S. Haroche: Phys. Rev. Lett. **76**, 1800 (1996)

16. J.I. Kim, K.M. Fonseca Romero, A.M. Horiguti, L. Davidovich, M.C. Nemes, and A.F.R. de Toledo Piza: Phys. Rev. Lett. **82**, 4737 (1999)

17. N.F. Ramsey: *Molecular Beams* (Oxford Univ. Press, New York 1985)

18. A. Einstein, B. Podolski, and N. Rosen: Phys. Rev. **47**, 777 (1935)

19. J.S. Bell: Physics (Long Island City, N.Y.) **1**, 195 (1964)

20. S.J. Freedman and J.S. Clauser: Phys. Rev. Lett. **28**, 938 (1972); A. Aspect, J. Dalibard, and G. Roger, Phys. Rev. Lett. **49**, 1804 (1982)

21. M. Brune, E. Hagley, J. Dreyer, X. Maître, A. Maali, C. Wunderlich, J.M. Raimond, and S. Haroche: Phys. Rev. Lett. **77**, 4887 (1996)

22. L. Davidovich, A. Maali, M. Brune, J.M. Raimond, and S. Haroche: Phys. Rev. Lett. **71**, 2360 (1993)

23. L. Davidovich, M. Brune, J.M. Raimond, and S. Haroche: Phys. Rev. A **53**, 1295 (1996)

24. B.-G. Englert, N. Sterpi, and H. Walther: Optics Commun. **100**, 526 (1993)

25. L.G. Lutterbach and L. Davidovich: Phys. Rev. Lett. **78**, 2547 (1997)

26. L.G. Lutterbach and L. Davidovich: Optics Express **3**, 147 (1998)

27. J. Bertrand and P. Bertrand: Found. Phys. **17**, 397 (1987)

28. U. Leonhardt: *Measuring the Quantum State of Light* (Cambridge University Press, Cambridge 1997)
29. J. Radon: *Berichte über die Verhandlungen der Königlich-Sächsischen Gesellschaft der Wissenschaften zu Leipzig, Mathematisch-Physikalische Klasse* **69**, 262 (1917)
30. E. Wigner: Phys. Rev. **40**, 749 (1932)
31. W. Pauli: 'Die allgemeinen Prinzipien des Wellenmechanik'. In: *Handbuch der Physik*, ed. by H. Geiger ad K. Scheel (Springer-Verlag, Berlin 1933). English translation: W. Pauli: *General Principles of Quantum Mechanics* (Springer-Verlag, Berlin 1980)
32. J.E. Moyal: Proc. Cambridge Phil. Soc. **45**, 99 (1949)
33. K.E. Cahill and R.J. Glauber: Phys. Rev. **177**, 1857 (1969); *ibid.* **177**, 1882 (1969).
34. K. Vogel and H. Risken: Phys. Rev. A **40**, 2847 (1989)
35. D.T. Smithey, M. Beck, M.G. Raymer, and A. Faridani: Phys. Rev. Lett. **70**, 1244 (1993)
36. G. Breitenbach, T. Müller, S.F. Pereira, J.-Ph. Poizat, S. Schiller, and J. Mlynek: J. Opt. Soc. Am. B **12**, 2304 (1995)
37. T.J. Dunn, I.A. Walmsley, and S. Mukamel: Phys. Rev. Lett. **74**, 884 (1995)
38. D. Leibfried, D.M. Meekhof, B.E. King, C. Monroe, W.M. Itano, and D.J. Wineland: Phys. Rev. Lett. **77**, 4281 (1996); see also Physics Today **51**, no. 4, 22 (1998)
39. L. Davidovich, M. Orszag, and N. Zagury: Phys. Rev. Lett. **57**, 2544 (1998)
40. G. Nogues, A. Rauschenbeutel, S. Osnaghi, P. Bertet, M. Brune, J.M. Raimond, S. Haroche, L.G. Lutterbach, and L. Davidovich: Phys. Rev. A **62**, 054101 (2000)
41. P. Bertet, A. Auffeves, P. Maioli, S. Osnaghi, T. Meunier, M. Brune, J.M. Raimond, and S. Haroche, Phys. Rev. Lett. **89**, 200402 (2002)
42. L. Davidovich, N. Zagury, M. Brune, J.M. Raimond, and S. Haroche: Phys. Rev. A **50**, R895 (1994)
43. P.W. Shor: Phys. Rev. A **52**, 2493 (1995); D. Gottesman: *ibid.* **54**, 1862 (1996); A. Ekert and C. Macchiavello: Phys. Rev. Lett. **77**, 2585 (1996); A.R. Calderband *et al*: *ibid.* **78**, 405 (1997)
44. H. Mabuchi and P. Zoller: Phys. Rev. Lett. **76**, 3108 (1996)
45. D. Vitali, P. Tombesi and G.J. Milburn: Phys. Rev. Lett. **79**, 2442 (1997); Phys. Rev. A **57**, 4930 (1998)
46. P. Zanardi and M. Rasetti: Phys. Rev. Lett. **79**, 3306 (1997); D.A. Lidar, I.L. Chuang, and K.B. Whaley: Phys. Rev. Lett. **81**, 2594 (1998); D. Braun, P.A. Braun and F. Haake: Opt. Comm. **179**, 195 (2000)
47. L. Viola, E. Knill and S. Lloyd: Phys. Rev. Lett. **82**, 2417 (1999); L. Viola and S. Lloyd: Phys. Rev. A **58**, 2733 (1998)
48. A.R.R. Carvalho, P. Milman, R.L. de Matos Filho, and L. Davidovich: Phys. Rev. Lett. **86**, 4988 (2001)
49. W.H. Zurek and J.P. Paz: Phys. Rev. Lett. **72**, 2508 (1994); S. Habib, K. Shizume and W.H. Zurek: Phys. Rev. Lett. **80**, 4361 (1998)

The Structure of the Vacuum
and the Photon Number

Iwo Białynicki-Birula

Center for Theoretical Physics, Polish Academy of Sciences,
Lotnikow 32/46, 02-668 Warsaw, Poland
and Institute of Theoretical Physics, Warsaw University

Abstract. It is shown that the universal expression describing the number of photons in terms of the electromagnetic field vectors, discovered long time ago by Zeldovich, makes its appearance in the quantum description of the ground state of the electromagnetic field. It is also argued that this rather unorthodox way of counting the photons may have consequences for astrophysics and cosmology.

1   Introduction

The standard definition of the vacuum state of the electromagnetic field as the state with no photons is very simple and extremely useful in most applications but it does not carry any information about many features of this fundamental state. In particular, it does not tell us anything about the electromagnetic structure of the vacuum state. A similar situation occurs in standard wave mechanics. We may define the ground state of the harmonic oscillator as the state that is annihilated by the annihilation operator but this does not tell us what is the probability distribution of the position or momentum. This additional information is contained in the wave function and to find the wave function one must go through the procedure of solving the eigenvalue problem for the energy operator – the Hamiltonian. For a harmonic oscillator this procedure is routinely carried out in (almost) every textbook of quantum mechanics but in quantum electrodynamics this is (almost) never done. Fortunately, the analogy between the quantum mechanical harmonic oscillator and the electromagnetic field is so close that one may reproduce the quantum mechanical results by pure analogy, without going through detailed calculations. This method will be adopted here and it will lead in a straightforward way to final results. Instead of the wave function I shall use the Wigner function which carries the same information as the wave function but in the case of the electromagnetic field exhibits more symmetry. In particular, we shall find that the formula for Wigner function of the ground state in quantum electrodynamics involves the formula for the number of photons discovered long time ago by Zeldovich [1]. Thus, the structure of the ground state and the number of photons are intimately related. One may wonder whether such a relationship also holds for non-Abelian gauge theories.

I. Białynicki-Birula, The Structure of the Vacuum and the Photon Number, Lect. Notes Phys. **633**,
287–295 (2004)
http://www.springerlink.com/

2   Wigner Function in Quantum Mechanics

My aim is to introduce the Wigner function of the electromagnetic field and to do this I shall proceed by analogy with nonrelativistic quantum mechanics. The standard definition of the Wigner function for an n-dimensional configuration space is [2,3]

$$W(\boldsymbol{p}, \boldsymbol{r}) = \int \frac{d^n \eta}{(2\pi\hbar)^n} e^{i\boldsymbol{\eta}\cdot\boldsymbol{p}/\hbar} \psi(\boldsymbol{r} - \boldsymbol{\eta}/2)\psi^*(\boldsymbol{r} + \boldsymbol{\eta}/2). \tag{1}$$

The ground state of the harmonic oscillator in one dimension is described by the wave function

$$\psi(q) = \left(\frac{m\omega}{\pi\hbar}\right)^{1/4} \exp\left(-\frac{m\omega q^2}{2\hbar}\right). \tag{2}$$

The corresponding Wigner function $W_G(p, q)$ is

$$W_G(p, q) = \frac{1}{\pi\hbar}\exp\left(-\frac{2E(p,q)}{\hbar\omega}\right), \tag{3}$$

where $E(p, q)$ is the classical expression for the energy of the oscillator

$$E(p, q) = \frac{p^2}{2m} + \frac{m\omega^2 q^2}{2}. \tag{4}$$

The expression in the exponent may be viewed as twice the number of excitations: the total energy $E(p, q)$ divided by the energy of one excitation $\hbar\omega$. This result is most easily extended to many dimensions by employing complex notation. Let me define classical analogs of annihilation and creation operators A and A^*,

$$A = \frac{p}{\sqrt{2m}} + i\sqrt{\frac{m}{2}}\omega q, \tag{5}$$

to rewrite the energy in the form

$$E(p, q) = A^* A. \tag{6}$$

In n dimensions the complex variables A and A^* become n-dimensional vectors $\boldsymbol{A}$ and $\boldsymbol{A}^*$ and the energy is

$$E(p_i, q^i) = \boldsymbol{A}^* \cdot \boldsymbol{A}. \tag{7}$$

In general the frequencies do not form a diagonal matrix so that the components A_i,

$$A_i = \frac{p_i}{\sqrt{2m}} + i\sqrt{\frac{m}{2}}\omega_{ij} q^j, \tag{8}$$

are not to be identified with normal modes. Taking this into account, one may extend the formula (3) to n dimensions as follows

$$W_G(p_i, q^i) = \frac{1}{(\pi\hbar)^n} \exp\left(-2\boldsymbol{A}^* \frac{1}{\hbar\omega} \boldsymbol{A}\right). \tag{9}$$

Therefore, one has to calculate the inverse of the frequency matrix to find the Wigner function of the ground state.

The Wigner function for mixed states is defined as a weighted sum of the pure state components. In particular, the Wigner function for the thermal state of the harmonic oscillator in one dimension is (see, for example, [3])

$$W_T(p,q) = \frac{\tanh(\hbar\omega/2kT)}{\pi\hbar} \exp\left(-2\frac{\tanh(\hbar\omega/2kT)\, E(p,q)}{\hbar\omega}\right). \tag{10}$$

Generalization from one dimension to many dimensions consists in replacing q and p by n dimensional vectors and ω by the matrix $\hat{\omega}$.

3   Wigner Functional for the Electromagnetic Field

Since the electromagnetic field may be viewed as one huge, infinitely dimensional harmonic oscillator, it is possible to write down the expressions for the Wigner function (or rather Wigner functional) of the electromagnetic field by an extrapolation from n to infinitely many dimensions. The starting point is the standard expression for the energy of the electromagnetic field

$$E = \int d^3r \left[\frac{\boldsymbol{D}(\boldsymbol{r},t)\boldsymbol{D}(\boldsymbol{r},t)}{2\epsilon_0} + \frac{\boldsymbol{B}(\boldsymbol{r},t)\boldsymbol{B}(\boldsymbol{r},t)}{2\mu_0}\right]. \tag{11}$$

With the use of the complex Riemann-Silberstein vector [4]

$$\boldsymbol{F} = \frac{\boldsymbol{D}}{\sqrt{2\epsilon_0}} + i\frac{\boldsymbol{B}}{\sqrt{2\mu_0}}, \tag{12}$$

the field energy can be cast into the form analogous to (7)

$$E = \int d^3r\, \boldsymbol{F}^*(\boldsymbol{r},t)\cdot\boldsymbol{F}(\boldsymbol{r},t). \tag{13}$$

In order to write down the Wigner functional W_G of the ground state, one has to evaluate the inverse of the frequency matrix in the coordinate representation. This can easily be done using the Fourier transformation

$$\left[\frac{1}{\hbar\hat{\omega}}\right](\boldsymbol{r},\boldsymbol{r}') = \int \frac{d^3k}{(2\pi)^3} \frac{e^{i\boldsymbol{k}\cdot(\boldsymbol{r}-\boldsymbol{r}')}}{\hbar c|\boldsymbol{k}|} = \frac{1}{2\pi^2\hbar c|\boldsymbol{r}-\boldsymbol{r}'|^2}, \tag{14}$$

and at the end one arrives at the formula

$$W_G[\boldsymbol{D},\boldsymbol{B}] = \exp(-2N[\boldsymbol{D},\boldsymbol{B}]), \tag{15}$$

where

$$N[\boldsymbol{D},\boldsymbol{B}] = \frac{1}{2\pi^2\hbar c}\int d^3r \int d^3r' \frac{\boldsymbol{F}^*(\boldsymbol{r},t)\boldsymbol{F}(\boldsymbol{r}',t)}{|\boldsymbol{r}-\boldsymbol{r}'|^2}$$

$$= \frac{1}{4\pi^2\hbar}\int d^3r \int d^3r' \left[\sqrt{\frac{\mu_0}{\epsilon_0}}\frac{\boldsymbol{D}(\boldsymbol{r},t)\boldsymbol{D}(\boldsymbol{r}',t)}{|\boldsymbol{r}-\boldsymbol{r}'|^2} + \sqrt{\frac{\epsilon_0}{\mu_0}}\frac{\boldsymbol{B}(\boldsymbol{r},t)\boldsymbol{B}(\boldsymbol{r}',t)}{|\boldsymbol{r}-\boldsymbol{r}'|^2}\right]. \tag{16}$$

I have explicitly indicated the arguments of the Wigner functional. The magnetic induction $\boldsymbol{B}$ (strictly speaking the vector potential $\boldsymbol{A}$) is the counterpart of the position q and the electric displacement vector $\boldsymbol{D}$ is the counterpart of the momentum p. There is only one element missing in (15): the overall normalization. One cannot extend the normalization used in the finite-dimensional case to the infinite number of dimensions because it would produce an infinite prefactor. Therefore, the Wigner functional of the full electromagnetic field can only be used to determine *relative* probabilities of different field configurations in a given state. The same expression (15) has been obtained before [5], starting from the field functional of the ground state of the electromagnetic field.

The Wigner functional of the thermal state of the electromagnetic field can be found by the same method as in the case of the ground state by integrating the single oscillator formula (10) over all wave vectors. The integration over k, after integration by parts (the oscillating boundary value at ∞ should be killed by an appropriate regularization), reduces to the following integral

$$\int \frac{d^3k}{(2\pi)^3} e^{i\boldsymbol{k}\cdot\boldsymbol{r}} \frac{\tanh(\hbar\omega_k/2kT)}{\hbar\omega_k} = \int_0^\infty \frac{dk\, k^2}{2\pi^2} \frac{\sin(kr)}{kr} \frac{\tanh(\hbar ck/2kT)}{\hbar ck}$$

$$= \frac{1}{4\pi^2\hbar kTr^2} \int_0^\infty dk\, \frac{\cos(kr)}{\cosh^2(\hbar ck/2kT)} \tag{17}$$

that can be found in Gradshteyn and Ryzhik [6]

$$\int_0^\infty dx\, \frac{\cos(ax)}{\cosh^2(bx)} = \frac{a\pi}{2b^2 \sinh(a\pi/2b)}. \tag{18}$$

Finally, one obtains

$$\int \frac{d^3k}{(2\pi)^3} e^{i\boldsymbol{k}\cdot\boldsymbol{r}} \frac{\tanh(\hbar\omega_k/2kT)}{\hbar\omega_k} = \frac{kT}{2\pi\hbar^2 c^2 |\boldsymbol{r}| \sinh(\pi kT|\boldsymbol{r}|/\hbar c)}. \tag{19}$$

The Wigner functional (up to the normalization factor) for the thermal state of the full electromagnetic field is

$$W_T[\boldsymbol{D}, \boldsymbol{B}] = \exp\left(-2\Theta[\boldsymbol{D}, \boldsymbol{B}]\right), \tag{20}$$

where the exponent $\Theta[\boldsymbol{B}, \boldsymbol{D}]$ is (the factor of 2 has been misplaced in [5])

$$\Theta[\boldsymbol{D}, \boldsymbol{B}] = \frac{kT}{4\pi\hbar^2 c^2} \int d^3r \int d^3r'$$

$$\times \left[\frac{\sqrt{\mu_0/\epsilon_0}\, \boldsymbol{D}(\boldsymbol{r},t)\cdot\boldsymbol{D}(\boldsymbol{r}',t) + \sqrt{\epsilon_0/\mu_0}\, \boldsymbol{B}(\boldsymbol{r},t)\cdot\boldsymbol{B}(\boldsymbol{r}',t)}{|\boldsymbol{r}-\boldsymbol{r}'| \sinh(\pi kT|\boldsymbol{r}-\boldsymbol{r}'|/\hbar c)} \right]. \tag{21}$$

In the limit, when $T \to 0$, this expression tends to the ground state form (15) while for $T \to \infty$ it goes to zero. Both these limits are in full agreement with our expectations. At zero temperature, excited states do not contribute and at very high temperatures all configurations become equally probable: the distribution is becoming flat, every configuration carries zero weight. The Wigner functional of a Gaussian form, such as the one for the ground state or the thermal state, is positive definite. Therefore, it has a direct probabilistic interpretation. These Wigner functionals determine the statistical distribution of electric and magnetic fields. In the ground state, the probability to find a particular configuration of the electromagnetic field depends only on the number of photons in this configuration. There is a significant difference between the ground state and the thermal state. In the ground state the correlations between the field values at different points are long range but in the thermal state they are short range. The range is determined by the thermal length $\lambda_T = \hbar c/kT$. At the temperature of cosmic microwave background radiation (CMB) this length is $\lambda_T = 8.4 \cdot 10^{-4}$m while at the room temperature $\lambda_T = 7.6 \cdot 10^{-6}$m. Thermal fluctuations wipe out correlations. At macroscopic distances the fields become statistically independent.

4   The Number of Photons

The dimensionless number $N[\boldsymbol{B}, \boldsymbol{D}]$ appearing in the Wigner functional of the electromagnetic field for the ground state must possess some special physical significance. This number has the full symmetry of free Maxwell theory. It is a perfect scalar, invariant under Poincaré transformations and under conformal transformations [7]. As has been observed long time ago by Zeldovich [1], this quantity after field quantization becomes the photon number operator. In order to see this one may use the decomposition of the field operator $\hat{\boldsymbol{F}}$ into plane waves

$$\hat{\boldsymbol{F}} = \sum_{\boldsymbol{k}} \sqrt{\tfrac{\hbar \omega_{\boldsymbol{k}}}{\epsilon_0 V}} \boldsymbol{e}(\boldsymbol{k}) \left(e^{i\boldsymbol{k}\cdot\boldsymbol{r} - i\omega_{\boldsymbol{k}}t} \hat{a}_{\boldsymbol{k}} + e^{-i\boldsymbol{k}\cdot\boldsymbol{r} + i\omega_{\boldsymbol{k}}t} \hat{b}_{\boldsymbol{k}}^{\dagger} \right), \tag{22}$$

where $\hat{a}_{\boldsymbol{k}}$ and $\hat{b}_{\boldsymbol{k}}$ are the annihilation operators of photons with opposite circular polarization. The substitution of this decomposition into the quantum counterpart of the formula (16) leads (after all the integrations and normal ordering) to the well known expression for the photon number operator

$$\hat{N} = \tfrac{1}{2\pi^2\hbar} \int d^3r \int d^3r' \frac{\hat{\boldsymbol{F}}^{*}(\boldsymbol{r},t)\hat{\boldsymbol{F}}(\boldsymbol{r}',t)}{|\boldsymbol{r}-\boldsymbol{r}'|^2} = \sum_{\boldsymbol{k}} \left(\hat{a}_{\boldsymbol{k}}^{\dagger}\hat{a}_{\boldsymbol{k}} + \hat{b}_{\boldsymbol{k}}^{\dagger}\hat{b}_{\boldsymbol{k}} \right). \tag{23}$$

Thus, the expression (16) is the classical limit of the photon number operator. In this limit the creation and annihilation operators should be replaced by complex amplitudes,

$$\left(\hat{a}_{\boldsymbol{k}}, \hat{b}_{\boldsymbol{k}}^{\dagger} \right) \rightarrow \left(\alpha_{\boldsymbol{k}}, \beta_{\boldsymbol{k}}^{*} \right), \tag{24}$$

and the number of photons operator becomes a pure, dimensionless number.

The total number of photons may be written as the space integral of the following "photon density" $\rho(\boldsymbol{r}, t)$

$$\rho(\boldsymbol{r}, t) = \tfrac{1}{2\pi^2\hbar c} \int d^3\eta \, \frac{\boldsymbol{F}^{*}(\boldsymbol{r}+\boldsymbol{\eta}/2)\boldsymbol{F}(\boldsymbol{r}-\boldsymbol{\eta}/2)}{|\boldsymbol{\eta}|^2}. \tag{25}$$

Together with the "photon current" $\boldsymbol{j}(\boldsymbol{r}, t)$,

$$\boldsymbol{j}(\boldsymbol{r}, t) = \frac{1}{2i\pi^2\hbar} \int d^3\eta \, \frac{\boldsymbol{F}^*(\boldsymbol{r}+\boldsymbol{\eta}/2) \times \boldsymbol{F}(\boldsymbol{r}-\boldsymbol{\eta}/2)}{|\boldsymbol{\eta}|^2}, \tag{26}$$

$\rho(\boldsymbol{r}, t)$ satisfies the continuity equation $\partial_t\rho + \nabla \cdot \boldsymbol{j} = 0$ but it cannot be treated as a genuine photon density since it is not positive everywhere. This was to be expected because photons are not strictly localizable.

5   Number of Photons for Various Field Configurations

It is worth stressing that the formula (16) is universal. One may use it even when the decomposition of the field into modes is unknown. Given any configuration of the classical electromagnetic field at a time t, one can calculate this number. It may even be applied to static fields, for which, according to traditional wisdom, the notion of the number of photons seems meaningless. I shall give several estimates of the number of photons in such unusual situations[1].

First, let us consider the number of photons in the Coulomb field of an elementary charge,

$$N_C = \frac{1}{4\pi^2\epsilon_0\hbar c} \frac{e^2}{(4\pi)^2} \int d^3r \int d^3r' \frac{\boldsymbol{r}\boldsymbol{r}'}{r^3 r'^3 |\boldsymbol{r}-\boldsymbol{r}'|^2} \, . \tag{27}$$

This expression is ultraviolet and infrared divergent but both these infinities are removed by physical cutoffs. The ultraviolet cutoff l for electrons should be placed around the electron Compton wave length, $l = 10^{-13}$m because of the QED vacuum polarization and for protons around the proton radius $l = 10^{-15}$m. The infrared cutoff R, or the screening length, is due to the polarization of the surrounding medium; all charges in reality are eventually screened. With these provisions the number of photons in the Coulomb field is

$$\begin{aligned} N_C &= \frac{\alpha}{2\pi} \int_l^R dr \int_l^R dr' \int_{-1}^1 du \frac{u}{r^2+r'^2-2rr'u} \\ &= \frac{\alpha}{2\pi} \left(2\ln(\tfrac{R}{l}) + 2\ln(1 - \tfrac{l^2}{R^2}) + (\tfrac{l}{R} + \tfrac{R}{l}) \ln\left(\tfrac{1+l/R}{1-l/R}\right) - 4\ln 2 \right) \\ &\approx \frac{\alpha}{\pi} \left(\ln(\tfrac{R}{l}) + 1 - 2\ln 2 \right), \end{aligned} \tag{28}$$

where α is the fine structure constant. This number is tiny; it amounts only to 0.08 Coulomb photons per one isolated proton in the intergalactic space where the screening radius is about 1 m and only 0.024 photons in the Coulomb field of the hydrogen atom where the screening radius is given by the Bohr radius.

The situation is quite different for static magnetic fields because they are not screened and may extend over vast regions. In the simplest case, for a constant

---

[1] Note that while formula (16) can be applied to time independent fields, the decomposition (22) into propagating photon modes and the ensuing number operator (23) strictly should exclude the static contribution. Thus, result (16) appears more generally applicable than its interpretation related to (23) [note added by editor].

field occupying a sphere of radius R, the number of photons is given by the formula

$$N = \frac{B^2}{4\pi^2\,\mu_0\hbar c}\int_{r<R} d^3r \int_{r'<R} d^3r'\Big(\frac{1}{|\boldsymbol{r}-\boldsymbol{r}'|^2}\Big) = \frac{B^2 R^4}{\mu_0\hbar c}\,I, \tag{29}$$

where the remaining dimensionless triple integral happens to be equal to 1,

$$I = 2\int_0^1 d\xi\,\xi^2 \int_0^1 d\eta\,\eta^2 \int_{-1}^1 du\,\frac{1}{\xi^2+\eta^2-2\xi\eta u} = 1. \tag{30}$$

A constant field is not a realistic magnetic field configuration since magnetic field lines must close. However, for every field configuration that occupies a region of volume V the number of photons will be given by the formula

$$N = 2.5\cdot 10^{31}\, B^2 V^{4/3}\,\kappa, \tag{31}$$

where κ is a numerical coefficient depending on the field configuration and the shape of the region and B and V are measured in teslas and cubic meters, respectively. For all static field configurations varying in space on the scale of R, there will be no dramatic cancellations and κ will be of the order of 1. Therefore, the number of photons in a static magnetic field scales with the four-thirds power of the volume V and may become huge for large objects. The galactic magnetic field is rather weak [8], about $5\cdot 10^{-10}$ T, but the large volume of the Milky Way makes the difference. Taking the diameter as 60 000 ly and the thickness as 1000 ly one obtains for the volume about $10^{12}\,\mathrm{ly}^3$ cube and that amounts to $V = 10^{60}\,\mathrm{m}^3$. Upon inserting these numbers into (31) one obtains for the Milky Way about 10^{92} magnetic photons. This result is to be compared with the number of CMB photons. At $T = 2.728$ K there are $4.12\cdot 10^8$ photons in one cubic meter. This gives only about 10^{68} CMB photons in the Milky Way. The ratio of these two numbers is 10^{24}! One may also estimate the number of Coulomb photons in the Milky Way. Assuming (very roughly!) that the whole mass of the Milky Way, equal to about 10^{12} solar masses, is composed of hydrogen atoms, with our estimate of 0.024 Coulomb photons for one hydrogen atom, we obtain again 10^{68}, the same number as for the CMB photons.

Are all these estimates of any value? The ratio of the number of photons to the number of baryons in the Universe plays an important role in cosmology. I have shown that the traditional counting of photons may have to be modified when the evaluation of the photon number takes into account also static fields. Of course, the number of photons inside the galaxies is very small as compared to the number of photons outside the galaxies. However, there is some evidence [8] that intergalactic magnetic fields exist and are only a few orders of magnitude weaker than the galactic ones. Should this be true and indeed there is some large scale magnetic field in the intergalactic space, even a much weaker one than the galactic field, our counting of the photons in the Universe would be quite different.

6   Photons Trapped by Gravitational Field

Very large number of photons stored in the galactic magnetic field raises the question, how they were captured in this state at the first place. Of course, one may pursue the standard line of reasoning and try to explain the magnetic field by some kind of dynamo mechanism [8], but I am tempted to speculate that the galactic magnetic field may be due to the BEC mechanism. To make this idea more plausible, one would have to show that there might exist a photon mode in the galaxy that would produce the galactic magnetic field when filled with a huge number of photons. This mode cannot be strictly stationary because (excluding black holes) gravity cannot bind massless particles. However, it would be sufficient to show that there is a very long lived state playing the role of the ground state of radiation. In order to see whether this is possible, let me analyze the Maxwell equations in curved space from that point of view. Maxwell equations in curved space expressed in terms of complex field vectors $\boldsymbol{F}$ and $\boldsymbol{G}$ have the form [4]

$$i\partial_t \boldsymbol{F}(\boldsymbol{r},t) = c\boldsymbol{\nabla} \times \boldsymbol{G}(\boldsymbol{r},t)\,, \tag{32}$$

$$\boldsymbol{\nabla}\cdot\boldsymbol{F}(\boldsymbol{r},t) = 0\,, \tag{33}$$

where

$$\boldsymbol{G} = \frac{\boldsymbol{E}}{\sqrt{2\mu_0}} + i\frac{\boldsymbol{H}}{\sqrt{2\epsilon_0}}\,. \tag{34}$$

The metric tensor is hidden in the "gravitational constitutive relations" that connect the vectors $\boldsymbol{F}$ and $\boldsymbol{G}$,

$$F^i = -\frac{1}{g_{00}}\left(\frac{g^{ij}}{\sqrt{-g}} + i\,g_{0k}\varepsilon^{ikj}\right)G_j. \tag{35}$$

The Maxwell equations for the photon wave function in a rotationally symmetric gravitational field described by the metric tensor

$$\{g_{\mu_0\nu}\} = \mathrm{Diag}\{g(r), -h(r), -h(r), -h(r)\}, \tag{36}$$

expressed in terms of the $\boldsymbol{G}$ vector, read

$$i\partial_t f(r)\boldsymbol{G}(\boldsymbol{r},t) = c\nabla \times \boldsymbol{G}(\boldsymbol{r},t), \tag{37}$$

where

$$f(r) = \sqrt{h(r)/g(r)}. \tag{38}$$

Stationary solutions of Maxwell equations after the separation of the time variable obey the eigenvalue equation

$$\nabla \times \boldsymbol{G}(\boldsymbol{r}) = \kappa f(r)\boldsymbol{G}(\boldsymbol{r}), \tag{39}$$

where $\kappa = \omega/c$ is the length of the wave vector. Owing to the spherical symmetry, the solution can be sought as a sum of angular momentum eigenstates. For each pair of eigenvalues J and M of the total angular momentum and its projection on the z axis, the solution is a sum of three terms

$$G = \frac{A(r)}{r} \mathbf{Y}_{JM}^{(e)} + \frac{B(r)}{r} \mathbf{Y}_{JM}^{(m)} + \frac{C(r)}{r} \mathbf{Y}_{JM}^{(l)}. \tag{40}$$

Upon substituting this sum into (39) one obtains a set of three coupled equations for A, B and C. The elimination of the functions $A(r)$ and $C(r)$ leads to the following equation for the function $B(r)$,

$$\frac{d^2 B}{dr^2} - \frac{1}{f} \frac{df}{dr} \frac{dB}{dr} - \frac{J(J+1)}{r^2} B + \kappa^2 f^2 B = 0. \tag{41}$$

After the substitution $B(r) = \sqrt{f(r)}\psi(r)$, this equation takes on the form of the time-independent Schrödinger equation for the radial wave function $\psi(r)$

$$-\frac{d^2 \psi(r)}{dr^2} + \frac{J(J+1)}{r^2} \psi(r) + U(r)\psi(r) = \kappa^2 \psi(r), \tag{42}$$

where the potential function $U(r)$ is

$$U(r) = \frac{3}{4f^2} \left(\frac{df}{dr} \right)^2 - \frac{1}{2f} \frac{d^2 f}{d\rho^2} - \kappa^2 (f^2 - 1). \tag{43}$$

For all asymptotically flat metrics, $U(r)$ vanishes at infinity. Therefore, bound states with positive energy ($\kappa^2 > 0$) are ruled out. However, this argument does not exclude a quasi-bound state. A quasi-bound state might be formed when the potential $U(r)$ has a potential well for smaller r which is separated from large values of r by a potential barrier. If a long-lived quasi-bound state would exist photons might condense into such a state. Large number of photons in the Bose-Einstein condensate may produce the magnetic field observed in the Milky Way.

Acknowledgments

I would like to that Zofia Białynicka-Birula for many constructive comments and advice. This work was supported by the KBN Grant 5PO3B 14920.

References

1. Ya.B. Zeldovich: Dokl. Acad. Sci. USSR, **163**, 1359 (1965), (in Russian)
2. E.P. Wigner: Phys. Rev. **40**, 749 (1932)
3. M. Hillery, R.F. O'Connell, M.O. Scully, and E.P. Wigner: Phys. Rep. **106**, 121 (1984)
4. I. Białynicki-Birula: 'Photon wave function'. In: *Progress in Optics*, Vol. XXXVI, ed. by E. Wolf (Elsevier, Amsterdam 1996), p. 245
5. I. Białynicki-Birula: Opt. Commun. **179**, 237 (2000)
6. I.S. Gradshteyn and Ryzhik: *Tables of Integrals, Sums and Products* (Academic, New York 1980)
7. L. Gross: J. Math. Phys. **5**, 687 (1964)
8. L.M. Widrow: Rev. Mod. Phys. **74**, 775 (2002)

On the Possibility of Quantum Coherence in Biological Systems with Application to Quantum Computing

Nicolaos E. Mavromatos and A. Keith Powell

Department of Physics, King's College London, WC2R 2LS, UK

1 Introduction

The recent rapid advances in silicon based electronics and the information processing technology it has generated cannot be expected to continue. Of the obstacles which lie in the way, perhaps the most fundamental is that of the quantum mechanical restrictions imposed by the discrete nature of matter and the way it interacts over the short length scales envisaged for future generations of electronic components.

By simple extrapolation of the well known Moore's Law for computer hardware we can expect to reach the limits of classical electronics, for the purposes of computing, within the next 10-15 years. Although this will correspond to computer processors being clocked at speeds of approaching 100 GHz and transistor gate sizes of a few nanometres, it is for certain that the market for ever more powerful computers will not saturate.

Many groups around the world are confident that the classical concept of computing will be surpassed by new quantum computers. Such computers, exploiting the properties of entangled quantum state functions ought, in principle, to be capable of manipulating data far more rapidly than a conventional computer. Although simple demonstrations have illustrated the principles involved, there are many issues to be addressed with regards to viable hardware for more exotic applications.

It is for the above reasons that attention is currently being focused at alternatives to silicon. One such possibility would be carbon nanotubes, which have attracted much interest recently. These are thin tubular structures of carbon (see Fig. 1), a few nanometres in diameter, which have been shown to have high current carrying capacities, semiconductor properties, quantised conductivity due to quantum coherent phenomena [1], and possible superconductivity [2] and high tensile strength.

However, more exotic and more cost effective possibilities may arise in nature Itself. For instance, it is well established that tubular structures a few nanometres constitute they basic building blocks of all biological cells. Such structures are protein made cylinders called microtubules and they are similar geometrically to carbon nanotubes, although quite different chemically.

Apart from their biological and biochemical significance, one might develop a more exotic, technologically orientated, interest in these biological systems in the context of signal processing for three main reasons:

N.E. Mavromatos and A.K. Powell, On the Possibility of Quantum Coherence in Biological Systems with Application to Quantum Computing, Lect. Notes Phys. **633**, 296–318 (2004)
http://www.springerlink.com/

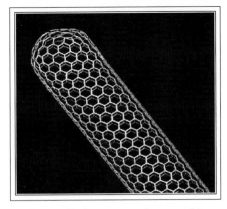

Fig. 1. Carbon nanotube assembly

- i) The length scales and complex environment associated with biological systems does not exclude the possibility that quantum-mechanical phenomena play an important role in their functionality allowing for effects as dissipationless and non-dispersive signal transconduction.
- ii) The possibility of self-assembly of complex systems tailored to fulfill particular signal processing operations.
- iii) The prospect of having available biological sources of technologically important materials

This article deals with the prospects for information processing by developing physical models for biological systems and predicting what might become possible.

2   Quantum Mechanics in Biological Systems

The rôle of Quantum Mechanics in Biological Matter is naively expected to be strongly suppressed, mainly due to the 'macroscopic nature' of most biological entities. The 'open system' nature of biological systems combined with the requirement for them to function at or near room temperature suggests a very fast *collapse* of the pertinent wave-functions, to a classical state. Until recently it was thought that a fundamental disadvantage of exploiting quantum coherent phenomena in macroscopic/mesoscopic systems was the collapse of the pertinent quantum states involved by thermalisation and other dissipative effects associated with operation at finite temperatures. It has now been demonstrated, outside of the biological world, that a macroscopic population of atoms can be quantum entangled for periods of up to milliseconds, at room temperature [3] From the point of view of planned devices, such times are extremely long, allowing the manipulation of data on much faster time scales and its subsequent transfer via macroscopic circuits. Prompted by such discoveries, systems which admit macroscopic quantum entanglement phenomena, at or near room temperature, and their exploitation in potential applications form a fundamentally important area of research.

Hence, decoherence arguments cannot prevent the possibility, in certain special biological matter systems, of having a quantum-mechanical coherent states becoming maintained for some extended interval, which is long enough for certain physical processes to occur as a result of quantum physics. We should remember that coherent states in such systems would be very complicated superpositions of many-body interactions and that small changes to these state functions could be masked from the environment, extending their effective lifetimes. Indeed, biological molecules, such as chlorophyll, appear to have developed complex, regularly organised structures to form molecular environments which modify decoherence effects. The biological systems with which we are also interested, such as tubulin protein dimers and their microtubular assemblies, have regular structures which too must modify the way in which a coherent state collapses. This also believed to be the case in the way that photo-receptive molecules rhodopsin and chlorophyll register and store single photon events.

Fundamental to the structure of all cells and now being realised as important for cell function and signaling are *Cell Microtubules* (see Figs. 2, 3).

We shall argue in this article that it is possible for parts of Cell Microtubules (MT), to operate as quantum-mechanical isolated cavities (see Fig. 4), exhibiting properties analogous to those of electromagnetic cavities used in Quantum Optics [4]. Here we shall mainly review the principle results of [5]. Details are included in [5]. The main idea is that dissipationless energy transfer in such entities proceeds via the formation of solitonic excitations, which could be the

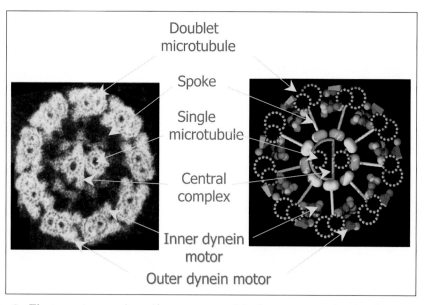

Fig. 2. Electronmicrograph and computer model of axoneme molecule, showing the position and geometrical arrangement of microtubules

Fig. 3. Crithidia flagellum motion brought about by the sliding motions of Cell Micro-tubules in axoneme molecules

results of quantum coherence, notwithstanding a classical interpretation. Until the work of [5] no specific microscopic models had been considered [6] in this respect and the physical origins of the conjectured solitons were unclear. Howe-ver, after the identification of the MT as one of the most important structures of the cell, both functionally and structurally, a model for their dynamics has been presented in [7]. In this model the formation of solitonic structures, and their rôle in energy transfer across the MT, is discussed in terms of classical physics. In [8] the *quantum aspects* of this one-dimensional model have been considered, and argued on the consistent quantisation of the soliton solutions, as well as the fact that such semiclassical solutions may emerge as a result of 'decoherence' due to environmental entanglement, according to recent ideas [9].

The basic assumption of the model used in [8] was that the fundamental structures in the MT (more specifically of the brain MT) are Ising spin chains (one-space-dimensional structures). The interaction of each chain with the neig-hbouring chains and the surrounding water environment had been mimicked by suitable potential terms in the one-dimensional Hamiltonian. The model descri-bing the dynamics of such one-dimensional sub-structures was the ferroelectric distortive spin chain model of [7].

Ferroelecricity is an essential ingredient for the quantum-mechanical mecha-nism for energy transfer we discuss here. It is conjectured [10] that the fer-roelectricity, which might occur in MT arrangements, will be that of hydrated ferroelectrics, i.e. the ordering of the electric dipoles will be due to the interaction of the dimers with the ordered-water molecules in the interior of the microtu-bular cavities. The importance of ferroelectricity lies on the fact that it induces a dynamical dielectric 'constant' $\epsilon(\omega)$ which is dependent on the frequency ω of the excitations in the medium, which in principle can be directly measured. Below a certain frequency, such materials are characterized by almost vanis-hing dynamical dielectric 'constants', which in turns implies that electrostatic interactions, which are inversely proportional to ϵ might be enhanced and thus become dominant against thermal losses. In the case of microtubules, the perti-nent interactions are of electric dipole type, scaling with the distance r as $1/\epsilon r^3$. For ordinary water media, the dielectric constant is of order 80. If ferroelectri-

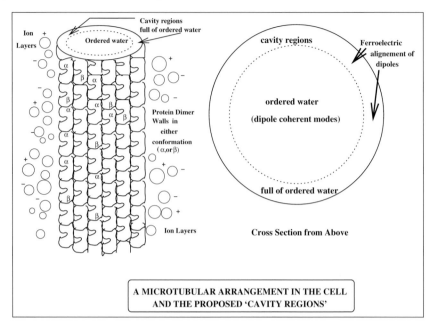

Fig. 4. Conjectural cavity regions in MT sustaining quantum coherence and the rôle of water in MT environment

city occurs, however, this is diminished significantly, with the result that the dipole-dipole interactions may overcome the thermal losses k_BT at room temperatures inside an interior cylindrical region of MT bounded by the dimer walls of thickness a few Angstroms [8].

Once such an isolation is provided, it is possible for these thin interior regions to act as cavities in a way similar to that of quantum optical electromagnetic cavities. The latter are characterized by coherent models of the electromagnetic radiation. In a similar spirit one expects such coherent cavity modes in the biological case as well. Indeed, as discussed in [8], such modes are provided by the interaction of the electric dipole moments of the ordered-water molecules in the interior of MT with the quantised electromagnetic radiation [11, 12]. Such coherent modes are termed dipole quanta. In our case, such modes inside the thin cavity regions will play the role of the cavity modes in quantum optics.

3   Review of the Physical Model for Microtubule Dynamics

MicroTubules (MT) are hollow cylinders comprised of an exterior surface (of cross-section diameter 25 nm)with 13 arrays (protofilaments) of protein dimers called tubuline (see Fig. 5).

The interior of the cylinder (of cross-section diameter 14 nm) contains *ordered water* molecules, which implies the existence of an electric dipole moment and an electric field. The arrangement of the dimers is such that, if one ignores

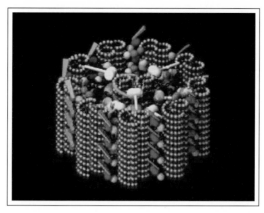

Fig. 5. Axoneme molecule built from protofilament assemblies of MTs. It is a universal feature of cells that there are 13 protofilaments in each MT

their size, they resemble triangular lattices on the MT surface. Each dimer consists of two hydrophobic protein pockets, and has an unpaired electron. There are two possible positions of the electron, called α and β *conformations*. When the electron is in the β-conformation there is a distortion of the electric dipole moment, whose vector is inclined with an angle θ as compared to the α conformation [7]. Experimentally the situation has not been verified yet, given that a measurement of the electric dipole moment is still pending. This is a field of much current interest.

In standard models for the simulation of the MT dynamics [7], the 'physical' degree of freedom - relevant for the description of the energy transfer - is the projection of the electric dipole moment on the longitudinal symmetry axis (x-axis) of the MT cylinder. The θ distortion of the β-conformation leads to a displacement u_n along the x-axis, which is thus the relevant physical degree of freedom. This way, the effective system is one-dimensional (spatial), and one has the possibility of a quantum integrable system [8].

Information processing occurs via interactions among the MT protofilament chains. The system may be considered as similar to a model of interacting Ising chains on a triangular lattice, the latter being defined on the plane stemming from filleting open and flattening the cylindrical surface of MT. Classically, the various dimers can occur in either α or β conformations. Each dimer is influenced by the neighbouring dimers resulting in the possibility of a transition. This is the basis for classical information processing, which constitutes the picture of a (classical) cellular automata.

The *quantum computer character* of the MT network results from the *assumption* that each dimer finds itself in a superposition of α and β conformations [13]. Viewed as a *two-state* quantum mechanical system, the MT tubulin dimers couple to conformational changes with $10^{-9} - 10^{-11}$sec transitions, corresponding to an angular frequency

$$\omega \sim \mathcal{O}(10^{10}) - \mathcal{O}(10^{12}) \text{ Hz} . \tag{1}$$

The scenario for quantum computation in MT presupposes that there exists a macroscopic coherent state among the various chains. Let us now try to understand its emergence. Let u_n be the displacement field of the n-th dimer in a MT chain. The continuous approximation proves sufficient for the study of phenomena associated with energy transfer in biological cells, and this implies that one can make the replacement

$$u_n \to u(x, t) \tag{2}$$

with x a spatial coordinate along the longitudinal symmetry axis of the MT. There is a time variable t due to fluctuations of the displacements $u(x)$ as a result of the dipole oscillations in the dimers. At this stage, t is viewed as a reversible variable.

The effects of the neighbouring dimers (including neighbouring chains) can be phenomenologically accounted for by an effective potential $V(u)$. In the model of [7] a double-well potential was used, leading to a classical kink solution for the $u(x, t)$ field. More complicated interactions are allowed in the string picture, as explained in [8]. More generic polynomial potential have also been considered in [8].

The effects of the surrounding water molecules can be summarized by a viscous force term that damps out the dimer oscillations,

$$F = -\gamma \partial_t u \tag{3}$$

with γ determined phenomenologically at this stage. This friction should be viewed as an environmental effect, which, however, does not lead to energy dissipation, as a result of the non-trivial solitonic structure of the ground-state and the non-zero constant force due to the electric field. This is a well known result, directly relevant to energy transfer in biological systems [6].

In mathematical terms the effective equation of motion for the relevant field degree of freedom $u(x, t)$ reads:

$$u''(\xi) + \rho u'(\xi) = P(u) \tag{4}$$

where $\xi = x - vt$, v is the velocity of the soliton, $\rho \propto \gamma$ [7], and $P(u)$ is a polynomial in u, of a certain degree, stemming from the variations of the potential $V(u)$ describing interactions among the MT chains [8]. In the mathematical literature [14] there has been a classification of solutions of equations of this form. For certain forms of the potential, the solutions include *kink solitons* that may be responsible for dissipation-free energy transfer in biological cells. A typical propagation velocity of the kink solitons of [7] is $v \sim 2$ m/sec [1]. This implies that for moderately long microtubules, of length $L \sim 10^{-6}$ m, such kinks transport energy in

$$t_F \sim 5 \times 10^{-7} \text{ sec}. \tag{5}$$

This scale is larger than the time scale that Fröhlich had conjectured as corresponding to the frequency of the coherent phonons in biological matter ($t \sim$

---

[1] Although, models with $v \sim 30$ m/sec have also been considered.

$10^{-11} - 10^{-12}$ s). In this article, however, we shall keep calling the time (5) the Fröhlich scale, since, upon quantisation, the kink soliton solutions of [7] yield quantum coherent states [8] that are very similar to Fröhlich's phonons.

Therefore, in the above 'phenomenological' approach to the MT physics, the importance of the water environment can be seen formally as follows: were it not for the friction term (3) there would be no stable solitonic structures in the ferroelectric distortive model of [7].

Let us now attempt a microscopic analysis of the physics underlying the interaction of the water molecules with the dimers of the MT. Our investigation points towards the fact that, as a result of the ordered structure of the water environment, there appear *collective* coherent modes, which in turn interact with the dimer structures (mainly through the unpaired electrons of the dimers)leading to the formation of a quantum coherent solitonic state that extends over the whole network of MT. According to the idea put forward in [8], following [9], such coherent states should be viewed as the result of *decoherence* of the dimer system due to its interaction/coupling with the water environment.

As we shall argue below, such a coupling could be detected by a phenomenon analogous to what is happening in atoms interacting with coherent modes of the electromagnetic radiation in *Cavities*, namely the *Vacuum-Field Rabi Splitting* (VFRS) [15]. Our *conjecture* is that the interior of MT, full of ordered water molecules, can be viewed as a *cavity* rather than a *wave guide*; the cavity structure can be formed by 'closing' the ends of the MT, and is a way of providing a fairly isolated system which can sustain coherent modes.

We now remark that in the approach of viewing the MT as cavities, the main source of dissipation will be the leakage of photons or other coherent modes from the cavity. Its rate can be assumed *small* for the MT of the brain, otherwise the incoherent mode will dominate. This dissipation constitutes an ordinary environment, which cannot lead to the formation of pointer states, but instead induces eventual collapse of the solitonic states into completely classical ground states. Hence, according to the above discussion, there are two stages when decoherence plays an important rôle.

- (i) The first stage concerns the coupling of the system of dimers with the coherent modes formed in the ordered water. This coupling decoherence in the dimer sub-system. producing quantum coherent states ('pointer states') according to [8, 9]. The time scale of this decoherence corresponds to the time scale necessary for the formation of the solitonic coherent states of [8].
- (ii) The second stage refers to the decoherence due to ordinary dissipation through the walls of the (imperfect) MT cavities. Such an ordinary environment does not admit a pointer basis [16], but causes collapse of the quantum coherent state of stage (i) down to a classical ground state.

In [5] we estimated the decoherence time scales of both stages (i) ($\sim 10^{-4}$s) and (ii) and compared them with the Fröhlich scale ($\sim 10^{-6}$ s). Moreover, we propose specific experimental tests for the detection of the quantum-mechanical behaviour of the MT arrangements inside the cell. An important feature of all these tests is the Vacuum-Field Rabi Splitting phenomenon, to a brief description of which we now turn for instructive purposes. In this article we will review first the Rabi splitting phenomenon and then expand upon the experimental techniques to falsify the rôle of quantum mechanical behaviour in MT systems.

4   On the Rabi Splitting in Atomic Physics

In this section we shall review briefly the VFRS in atomic physics [15, 17]. The basic phenomenon involves an interaction between two oscillators in resonance, where the frequency degeneracy is *removed* by an amount proportional to the strength of their coupling. In the cavity QED case of [15], one oscillator consists of a small collection of N atoms, whilst the other is a resonant mode of a high-Q(uality) cavity[2]. Immediately after the suggestion of [15], a similar phenomenon has been predicted for absorption spectra of atoms in cavities [18].

Experimentally one excites the coupled atom-cavity system by a tunable field probe [19]. The excitation is then found resonant *not* at the 'bare' atom or cavity frequencies but at the *split* frequencies of the 'dressed' atom-field system. The splitting is enhanced for collections of atoms. For instance, for a system of N atoms, the split is predicted to be [18]

$$\text{Rabi splitting} = 2\lambda\sqrt{N},$$
$$2\lambda = \text{Rabi splitting of a single atom}. \tag{6}$$

Despite its theoretical prediction by means of quantum mechanical oscillator systems coupled with a *quantised* radiation field mode in a cavity, at present there seems to be still a *debate* on the nature of the phenomenon: (i) the dominant opinion is that the Rabi splitting is a manifestation of the *quantum nature* of the electromagnetic radiation (cavity field), and is caused as a result of an *entanglement* between the atom and the cavity coherent modes of radiation. It is a sort of Stark effect, but here it occurs in the absence of an external field [15]. This 'dynamical Stark effect' is responsible for a splitting of the resonant lines of the atoms by an amount proportional to the collective atomic-dipole amplitude. (ii) there is however a dual interpretation [20], which claims that the splitting can be observed in optical cavities as well, and is simply a result of *classical* wave mechanics inside the cavity, where the atomic sample behaves as a *refractive medium* with a *complex* index, which splits the cavity mode into two components.

Irrespectively of this second classical interpretation, one *cannot disprove* the presence of the phenomenon in entangled atom-quantum-coherent mode systems, relevant to our picture of viewing MT filled with ordered water as cavities. We shall try to make specific experimental predictions that could shed light in the formation of quantum coherent states, and their eventual decoherence. As mentioned above, the latter could be due to the interaction of the dimer unpaired

---

[2] The quantity Q is defined as the ratio of the stored-energy to the energy-loss per period [4], by analogy with a damped harmonic oscillator.

spins (playing the rôle of the atoms in the Rabi experiments) with the ordered-water coherent modes (playing the rôle of the cavity fields). Possible scenarios for the origin of such cavity coherent modes in the case of Biological MT will be described below.

The expression effect of Rabi-vacuum splitting in absorption spectra of atoms may be formally summarized as follows [18]: Consider the application of an external field, of frequency Ω. Then, there is a doublet structure (splitting) of the absorption spectrum of the atom-cavity system with peaks at

$$\Omega = \omega_0 - \Delta/2 \pm \tfrac{1}{2}(\Delta^2 + 4N\lambda^2)^{1/2}, \tag{7}$$

where Δ is the detuning of the cavity mode, compared to the atom frequency. For resonant cavities the splitting occurs with equal weights,

$$\Omega = \omega_0 \pm \lambda\sqrt{N}. \tag{8}$$

Notice here the *enhancement* in the effect for multi-atom systems $N \gg 1$. This is the 'Vacuum Field Rabi Splitting phenomenon', predicted originally in emission spectra in [15].

The quantity $2\lambda\sqrt{N}$ is called the 'Rabi frequency' [15]. From the emission-spectrum theoretical analysis an estimate of λ may be inferred which involves the matrix element, $\underline{d}$, of atomic electric dipole between the energy states of the two-level atom [15],

$$\lambda = \frac{E_{vac}\underline{d}.\underline{\epsilon}}{\hbar}, \tag{9}$$

where $\underline{\epsilon}$ is the cavity (radiation) mode polarization, and

$$E_{vac} \sim \left(\frac{2\pi\hbar\omega_c}{\varepsilon_0 V}\right)^{1/2} \tag{10}$$

is the r.m.s. vacuum field amplitude at the centre of the cavity of volume V, and of frequency ω_c, with ε_0 the dielectric constant of the vacuum [3]: $\varepsilon_0 c^2 = \frac{10^7}{4\pi}$, in M.K.S. units. There are simple experiments which confirmed this effect [19], involving beams of Rydberg atoms resonantly coupled to superconducting cavities.

The situation which is of interest to us involves atoms that are *near resonance* with the cavity. In this case $\Delta \ll \omega_0$, but such that $\lambda^2 N/|\Delta|^2 \ll 1$; in such a case, formula (7) yields two peaks that are characterized by dispersive frequency shifts $\propto \frac{1}{\Delta}$,

$$\Omega \simeq \omega_0 \pm \frac{N\lambda^2}{|\Delta|} + \mathcal{O}(\Delta), \tag{11}$$

whilst no energy exchange takes place between atom and cavity mode. This is also the case of interest in experiments using such Rabi couplings to construct Schrödinger's cats in the laboratory, i.e. macro(meso)scopic combinations measuring apparatus + atoms' to verify decoherence experimentally. The first experiment of this sort, which confirms theoretical expectations, is described in [21].

---

[3] For cavities containing other dielectric media, e.g. water in the case of the MT, ε_0 should be replaced by the dielectric constant ε of the medium.

5   Microscopic Mechanisms for the Formation of Quantum-Coherent States in MT and Dissipationless Energy Transfer

Above we have sketched the experimental construction of a mesoscopic quantum coherent state (a 'Schrödinger's cat' (SC)). The entanglement of the atom with the coherent cavity mode, manifested experimentally by the 'vacuum Rabi splitting', leads to a quantum-coherent state for the combined atom-cavity system (SC), comprising of the superposition of the states of the two-level Rydberg atom. Dissipation induced by the leakage of photons in the cavity leads to decoherence of the coherent atom-cavity state within ascertain time scale, which depends crucially on the nature of the coupled system, and the nature of the 'environment'. It is the point of this session to attempt to discuss a similar situation that conjecturally occurs in the systems of MT. The first issue concerns the nature of the 'cavity-field modes', where we argue that the presence of *ordered water*, which seems to occupy the interior of the microtubules [22], plays an important rôle in producing coherent modes, which resemble those of the ordinary electromagnetic field in superconducting cavities, discussed above.

Let us first review briefly some suggestions about the rôle of the electric dipole moment of water molecules in producing coherent modes after coupling with the electromagnetic radiation field [12]. Such a coupling implies a 'laser-like' behaviour. Although it is not clear to us whether such a behaviour characterizes ordinary water, in fact we believe it does not due to the strong suppression of such couplings in the disordered ordinary water, however it is quite plausible that such a behaviour characterizes the ordered water molecules that exist in the interior of MT [22]. If true, then this electric dipole-quantum radiation coupling will be responsible, according to the analysis of [12], for the appearance of *collective* quantum coherent modes. The Hamiltonian used in the theoretical model of [12] is

$$H_{ow} = \sum_{j=1}^{M} [\tfrac{1}{2I} L_j^2 + \underline{A}.\underline{d}_{ej}] \,, \tag{12}$$

where A is the quantised electromagnetic field in the radiation gauge [12], M is the number of water molecules, L_j is the total angular momentum vector of a single molecule, I is the associated (average) moment of inertia, and d_{ej} is the electric dipole vector of a single molecule, $|d_{ej}| \sim 2e \otimes d_e$, with $d_e \sim 0.2$ Angström. As a result of the dipole-radiation interaction in (12) coherent modes emerge, which in [11] have been interpreted as arising from the quantisation of the Goldstone modes responsible for the *spontaneous breaking* of the electric dipole (rotational) symmetry. Such modes are termed 'dipole quanta' in [11].

This kind of mechanism has been applied to microtubules [23], with the conclusion that such coherent modes cause 'super-radiance', i.e. create a specific quantum-mechanical ordering in the water molecules with characteristic times much shorter than those of thermal interaction. In addition, the optical medium inside the internal hollow core of the microtubule is made transparent by the coherent photons themselves [23]. Such phenomena, if observed, could verify the coherent-mode emission from living matter, thereby verifying Fröhlich's ideas.

In our picture of viewing the MT arrangements as cavities, these coherent modes are the quantum coherent 'oscillator' modes of Sect. 3, represented by annihilation and creation operators $a_c, a_c^\dagger$, which play the rôle of the cavity modes, if the ordered-water interior of the MT is viewed as an isolated cavity [4]. The rôle of the small collection of atoms, described in the atomic physics analogue above, is played in this picture by the protein dimers of the MT chains. The latter constitute a two-state system due to the α and β conformations, defined by the position on the unpaired spin in the dimer pockets. The presence of unpaired electrons is crucial to such an analogy. The interaction of the dipole-quanta coherent modes with the protein dimers results in an entanglement which we claim is responsible for the emergence of *soliton quantum coherent states*, extending over large scale, e.g. the MT or even the entire MT network.

The issue, we are concerned with here, is whether such coherent states are responsible for *energy-loss-free transport*, as well as for *quantum computations* due to their eventual collapse, as a result of 'environmental' entanglement of the entire 'MT dimers + ordered water' system. An explicit construction of such solitonic states has been made in the field-theoretic model for MT dynamics of [8], based on classical ferroelectric models for the displacement field $u(x,t)$ discussed in Sect. 2 [7]. The quantum-mechanical picture described here should be viewed as a simplification of the field-theoretic formalism, which, however, is sufficient for qualitative estimates of the induced decoherence.

To study quantitatively the effects of decoherence in MT systems we make the plausible *assumption* that the environmental entanglement of the 'ordered-water cavity' (OWC), which is responsible for dissipation, is attributed *entirely* to the leakage of photons (electromagnetic radiation quanta) from the MT interior of volume V ('cavity'). This leakage may occur from the *nodes* of the MT network, if one assumes fairly isolated interia. This leakage will cause decoherence of the coherent state of the 'MT dimer-OWC' system. The leakage determines the damping time scale T_r [5].

The dimers with their two conformational states α,β play the rôle of the collection of N *two-level* Rydberg atoms in the atomic physics analogue described in Sect. 3. If we now make the assumption that the ordered-water dipole-quantum coherent modes couple to the dimers of the MT chains in a way similar to the

---

[4] This was not the picture envisaged in [12]. However, S. Hameroff, as early as 1974, had conjectured the rôle of MT as 'dielectric waveguides' for photons [24], and in [23] some detailed mathematical construction of the emergence of coherent modes out of the ordered water are presented. In our work in [5], reviewed here, we consider the implications of such coherent modes for the system of dimers, in particular for the formation of kink solitons of [7]. Thus, our approach is different from that in [12,23], where attention has been concentrated only on the properties of the water molecules. We should emphasize that the phenomenon of optical transparency due to superradiance may co-exist with the formation of kink soliton coherent states along the dimer chains, relevant to the dissipation-free energy transfer along the MT, discussed in [5] and here.

one leading to a Rabi splitting, described above, then one may assume a coupling λ_0 of order:

$$\lambda_0 \sim \frac{d_{\mathrm{dimer}} E_{ow}}{\hbar} \tag{13}$$

where d_{dimer} is the single-dimer electric dipole matrix element, associated with the transition from the α to the β conformation, and E_{ow} is a r.m.s. typical value of the amplitude of a coherent dipole-quantum field mode.

Given that each dimer has a mobile charge [25]: $q = 18 \times 2e$, e the electron charge, one may *estimate*

$$d_{\mathrm{dimer}} \sim 36 \times \frac{\varepsilon_0}{\varepsilon} \times 1.6. \times 10^{-19} \times 4.10^{-9}$$
$$\sim 3 \times 10^{-18} \ \mathrm{Cb} \times \mathrm{Angstrom} \tag{14}$$

where we used the fact that a typical distance for the estimate of the electric dipole moment for the 'atomic' transition between the α, β conformations is of $\mathcal{O}(4 \ \mathrm{nm})$, i.e. of order of the distance between the two hydrophobic dimer pockets. We also took account of the fact that, as a result of the water environment, the electric charge of the dimers appears to be screened by the relative dielectric constant of the water, $\varepsilon/\varepsilon_0 \sim 80$. We note, however, that the biological environment of the unpaired electric charges in the dimer may lead to further suppression of d_{dimer} (14). It should be stressed that at present we have no experimental input about this value and this is expected to be an essential step in any experimental programme.

The frequency ω_c and the r.m.s. amplitude E_{ow} of the collective modes of the dipole-quanta has been estimated in [5],

$$\omega_c \sim \epsilon/\hbar \sim 6 \times 10^{12} s^{-1} \ , \qquad E_{ow} \sim 10^4 \ \mathrm{V/m} \,. \tag{15}$$

Notice that the electric fields of such order of magnitude can be provided by the electromagnetic interactions of the MT dimer chains, the latter viewed as giant electric dipoles [7]. This may be seen to suggest that the super-radiance coherent modes ω_c, which in our scenario interact with the unpaired electric charges of the dimers and produce the kink solitons along the chains, owe their existence to the (quantized) electromagnetic interactions of the dimers themselves. This suggests a 'Basic Cycle' for dissipationless energy transfer in MT quantum physics (see Fig. 6).

We assume that the system of $\mathcal{N}$ MT dimers interacts with a *single* dipole-quantum mode of the ordered water and we ignored interactions among the dimer spins[5]. In our work here we concentrate our attention on the formation of a coherent soliton along a single dimer chain, the interactions of the remaining 12 chains in a protofilament MT cylinder being represented by appropriate interaction terms in the effective potential of the chain MT model of [7]. In a moderately long microtubule of length $L \simeq 10^{-6} \ m$ there are $\mathcal{N} = L/8 \simeq 10^2$

---

[5] More complicated situations, including interactions among the dimers, as well as of the dimers with more than one radiation quanta, which might undoubtedly occurring nature, complicate the above estimate.

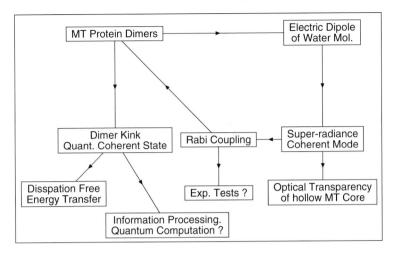

Fig. 6. Suggested 'Basic Cycle' for dissipationless energy transfer in cell MT due to Quantum Physics. The dominant interactions between dimers and water molecules are assumed to be electromagnetic in origin

tubulin dimers of average length 8 nm in each chain. One can easily estimate [5] the Rabi coupling for the entire MT:

$$\text{Rabi coupling for MT} \equiv \lambda_{MT} = \sqrt{\mathcal{N}}\lambda_0 \sim 3 \times 10^{11} s^{-1}, \tag{16}$$

which is, on average, an order of magnitude smaller than the characteristic frequency of the dimers (1). In this way, the perturbative analysis of the previous section, for small Rabi splittings $\lambda \ll \omega_0$, is valid.

Having estimated the Rabi coupling between the dimers and the ordered-water coherent modes one can estimate the average time scale necessary for the formation of the pointer coherent states which arise due to this entanglement [8, 9]. This is the same as the decoherence time due to the water-dimer coupling. In such a case one may use the integrable model of MT [8], which is known to possess pointer coherent states [26].

In [5] we have estimated that the time scale over which such solitonic quantum coherent pointer states in the tubulin dimers are formed (pumped): $t_{\text{pump}} > 10^{-10}$ s. This is not far from the Frölich scale [27] $10^{-11} - 10^{-12}$ s.

To answer the question whether quantum coherent pointer states are responsible for loss-free energy transport across the MT one should examine the time scale of the decoherence induced by the coupling of the MT to their biological environment as a consequence of dissipation through the walls of the MT cylinders. It is this environment that will induce decoherence and eventual collapse of the pointer states formed in the ordered water. Using typical numbers of MT networks, we can estimate this decoherence time in a way similar to the corresponding situation in atomic physics [21],

$$t_{\text{collapse}} = \frac{T_r}{2n\mathcal{N}\sin^2\left(\frac{\mathcal{N}n\lambda_0^2 t}{\Delta}\right)}, \tag{17}$$

where T_r denotes the time scale over which the cavity damps its energy, which in our case is identified with the life-time of the super-radiance coherent mode in the ordered water. Above we took into account that the dominant (dimer)-(dipole quanta) coupling occurs for ordered-water 'cavity' modes which are *almost at resonance* with the dimer oscillators

We also assume that a typical coherent mode of dipole/quanta contains an average of $n = \mathcal{O}(1) - \mathcal{O}(10)$ oscillator quanta. The *macroscopic* character of the Schrödinger's cat dimer-dipole-quanta system comes from the N dimers in a MT (or $\mathcal{N}_{mt}\mathcal{N}$ in MT networks) [8].

The time t appearing in (17) represents the 'time' of interaction of the dimer system with the dipole quanta. A reasonable estimate of this time scale in our MT case can be obtained by equating it with the average life-time of a coherent dipole-quantum state, which, in the super-radiance model of [23] can be estimated as

$$t \sim \frac{c\hbar^2 V}{4\pi d_{ej}^2 \epsilon N_w L} \tag{18}$$

with d_{ej} the electric dipole moment of a water molecule, L the length of the MT, and N_w the number of water molecules in the volume V of the MT. For typical values of the parameters for moderately long MT, $L \sim 10^6$ m, $N_w \sim 10^8$, a typical value of t is

$$t \sim 10^{-4} \text{ sec}. \tag{19}$$

We remark at this point that this is considerably larger than the average life-time of a coherent dipole quantum state in the water model of [12]. The time scale T_r, over which a cavity MT dissipates its energy, can be identified in our model with the average life-time (18) of a coherent-dipole quantum state,

$$T_r \sim t \sim 10^{-4} \text{ sec}, \tag{20}$$

which leads to a naive estimate of the quality factor for the MT cavities, $Q_{MT} \sim \omega_c T_r \sim \mathcal{O}(10^8)$. We note, for comparison, that high-quality cavities encountered in Rydberg atom experiments dissipate energy in time scales of $\mathcal{O}(10^{-3}) - \mathcal{O}(10^{-4})$ sec, and have Q's which are comparable to Q_{MT} above. Thus, it is not unreasonable to expect that conditions like (20), characterizing MT cavities, are met in Nature.

From (17), (16), and (20), one then obtains the following estimate for the collapse time of the kink coherent state of the MT dimers due to dissipation:

$$t_{\text{collapse}} \sim \mathcal{O}(10^{-7}) - \mathcal{O}(10^{-6}) \text{ sec} \tag{21}$$

which is larger or equal than the Fröhlich scale (5) required for energy transport across the MT by an average kink soliton in the model of [7]. The result (21), then, implies that Quantum Physics may be responsible for dissipationless energy transfer across the MT.

Before closing this section we would like to make some important remarks concerning the precise location of the thermally-isolated cavity regions in the MT arrangements. Above, we have made the simplifying assumption that the entire water-full interior of the MT cylinder operates as a cavity region. However, such a space of cross-section diameter 16 nm is probably too big for this to be true. What we envisage as actually happening is that the cavity regions are *thin* regions near the dimer walls, of thickness up to a few Angströms, within which the dipole-dipole interactions between the electric dipole moments of the tubulin dimers and the water molecules dominate over thermal losses (see Fig. 4). In connection to the above issue of determining the exact location of the cavity region, we cannot resist in pointing out the recent experimental findings of D. Sackett [28], who demonstrated the existence of *thin* layers of ions, of thickness of order $\mathcal{O}(7-8)$ Angströms, *outside* the dimer walls of MT cylinders, in which the electrostatic interactions dominate again over thermal losses. All of the above results on the order of magnitude of decoherence time scales are the same in this geometry as can be easily seen. It might well be, therefore, that the cavity regions relevant for the existence of quantum-coherent phenomena in MT lie in thin thermally isolated layers on *both* the interior and the exterior of the tubulin walls. It would be very interesting to explore experimentally such a possibility and this is described next.

6   Key Experimental Work on Microtubules

The experimental work to date on MT has been complicated by the the interaction of the molecular system with environmental factors. It is known that measurements of dielectric properties can be extensively modified by sample preparation techniques and the materials used as sample substrates. There is, however, much progress being made in these areas.

6.1   Dielectric Measurements and Ferroelectric Properties

Conceptually the simplest measurements to be made are those of the dielectric constant measurements of MT across a range of temperatures. The conjectures properties of ferroelectricity will appear as characteristic drops in dielectric values brought about by ferroelectric screening of dipole charges.

Essentially, these measurements can be undertaken by prepared samples being placed in an impedance measuring cell. Measurements of the dielectric properties are then made across a broad range of frequencies. The characteristic dips in values are looked for. Suitable inert substrates identified for this work include sample holders constructed from inert silicon nitride membranes.

6.2   Tests of Rabi Splitting Phenomenon

As reviewed above [8] an indirect verification of the existence of cavity quantum coherent modes in regions of MTs (see Fig. 4) would be the experimental detection of the aforementioned Vacuum field Rabi coupling, λ_{MT}, between the MT dimers and the ordered water quantum coherent modes. This coupling, if present, could be tested experimentally by the same methods used to measure VFRS in atomic physics [19], i.e. by using the MT themselves as *cavity environments*, and considering tunable probes to excite the coupled dimer-water system. Such probes could be pulses of (monochromatic) light, for example, passing through the hollow cylinders of the MT. This would be the analogue of an external field in the atomic experiments described above, which would then resonate, not at the bare frequencies of the coherent dipole quanta or dimers, but at the *Rabi split* ones, and this would have been exhibited by a double pick in the absorption spectra of the dimers [19]. By using MT of different sizes one could thus check on the characteristic $\sqrt{N}$-enhancement of the Rabi coupling for MT systems with N dimers.

The technical complications that might arise in such experiments are associated with the absence of completely resonant cavities in practice. In fact, from our discussion in this article, one should expect a slight detuning Δ between the cavity mode and the dimer, of frequency ω_0. As discussed in Sect. 3, the detuning produces a split of the vacuum-Rabi doublet into a cavity line $\omega_0 + \lambda^2/\Delta$ and an 'atomic line' $\omega_0 - \lambda^2/\Delta$. In atomic physics there are well established experiments [19, 29] to detect such splittings. In fact, detections of such lines is considered a very efficient way of 'quantum non-demolition' measurement [30] for small microwave photon numbers. Such atomic physics experiments, therefore, may be used as a guide in performing the corresponding biological experiments involving MT as cavities.

6.3   Optical Measurements

Optical probing techniques require the use of tunable, highly frequency pulsed lasers (see Fig. 7). Quantum cavity modes will be identifiable as Rabi splits in the excitation spectra. Additionally, such lasers can also be used to induce modulations in the dielectric properties of the materials. When these modulations take the form of interference patterns, laser radiation may be coherently scattered from these structures. The nature of the scattered field, along with delayed excitation pulses can be used to give detailed knowledge of the time-resolved dielectric properties of the sample.

The most common form of this interaction is that of degenerate four-wave mixing (DFWM) where two incident waves, with varying delay, write a holographic grating into the material. A third wave applied is then scattered into a coherent fourth wave, the magnitude of which contains the important details of dipole moments. The spatial orientation of the induced grating can discern anisotropic effects and its spatial frequency can determine the dependence of the phenomenon upon underlying material structure. The interaction can be

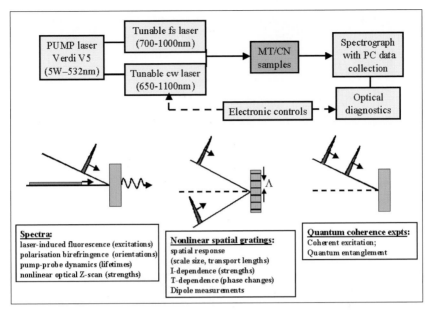

Fig. 7. Optical techniques for probing dielectric properties of MT

thought of as *dynamic holography* where the hologram reading and writing beams are simultaneously supplied to the test medium. The induced modifications to the dielectric properties bring about a dynamically generated scattered wave.

7 Self-Assembly Concepts

It is remarkable that molecules such as tubulin can be polymerised into long chains (see Fig. 8) and then just as straight forwardly be broken down back to monomer elements by catalysed reactions in the laboratory. This in itself suggests that an important aspect to biological systems is not just self-assembly but also reconfiguration.

The axoneme molecule, also formed from MT filaments is responsible for driving the motions of many cells including flagella. The structure and control of such molecules is now becoming understood aided by sophisticated computer models (see Fig. 9).

Underlying these molecules are the structures known as molecular motors such as dynein. Activated by molecules of ATP, these molecular molecules drive by ratcheted molecular reconfigurations provide the motive force behind cell motions. It is relatively easy to attach loads to these molecules, such as carbon nanotubes and then position them with high precision by the chemical activation and deactivation of the molecular motors. Substrates formed of orientated dynein molecules can then transport load molecules to target areas (see Fig. 10).

In this way, sophisticated assemblies may be built up, to support the formation and manipulation of solitons in a paradigm of a quantum computer. One

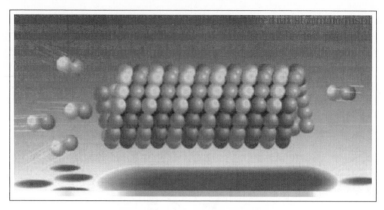

Fig. 8. Polymerisation of MTs

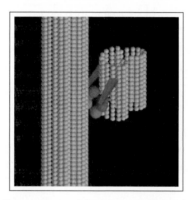

Fig. 9. Computer model of dynein molecular motor

Fig. 10. Microtubules can be arranged on a bed of activated dynein

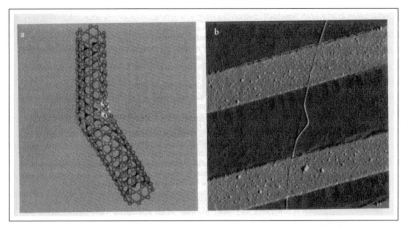

Fig. 11. Laying out carbon nanotubes between electrodes

thing which is striking about this technology is its scalability. The dissipationless nature of quantum based signal processing and the near arbitrary complexity by which structures can be formed by self-assembly gives rise to the prospect of extremely sophisticated information processing architectures (see Fig. 11).

8   Conclusions and Outlook

In [5] and here we have put forward a conjecture concerning the representation of the MT arrangements inside the cell as *isolated* high-Q(uality) *cavities*. We presented a scenario according to which the presence of the ordered water in the interior of the cylindrical arrangements results in the appearance of electric dipole quantum coherent modes, which couple to the unpaired electrons of the MT dimers via Rabi vacuum field couplings, familiar from the physics of Rydberg atoms in electromagnetic cavities [15]. In quantum optics, such couplings are considered as experimental proof of the quantised nature of the electromagnetic radiation. In our case, therefore, if present, such couplings could indicate the existence of the coherent quantum modes of electric dipole quanta in the ordered water environment of MT, conjectured in [11, 12], and used in [5] and here. The suggested mechanism for the production of quantum-coherent states in MT arrangements may be summarised by the 'basic cycle' shown in Fig. 6.

Some generic decoherence time estimates, due to environmental entanglement of the MT cavities, have been given. The conclusion was that only in *fairly isolated* cavities, which we conjecture exist inside the biological cells, decoherence occurs in time scales which are in agreement with the Fröhlich scale ($\sim 5 \times 10^{-7}$ sec) for energy transfer across the MT via the formation of kink solitonic structures. In such a case, *dissipationless energy transfer* might occur in biological systems, an in particular in MT which are the substratum of cells, in much the same way as frictionless electric current transport occurs in superconductors, i.e. via *quantum coherent modes* that extend over relatively large spatial regions. A phenomenological analysis indicated that for moderately long MT networks such a situation could be met if the MT cavities dissipate energy in time scales of order $T_r \sim 10^{-4} - 10^{-5}$sec. This lower bound is comparable to the corresponding scales of atomic cavities [19], $\sim 10^{-4}$sec, which, in turn, implies that the above scenarios, on dissipationless energy transport as a result of the formation of quantum coherent states, have a good chance of being realised at the scales of MT, which are comparable to those in Atomic Physics.

We have discussed briefly experimental methods for elucidating the rôle of quantum mechanics in MTs. These methods are similar in spirit to those applied in quantum optical cavities, to which the MTs present striking similarities. We have been prompted to the conjectures about the existence of such isolated cavity modes in MTs by the geometric analogies of MTs with carbon nanotube assemblies. Such analogies may go beyond the geometrical characteristics, for instance, the rôle of ordered water in both carbon nanotubes and MTs may be important in ensuring non trivial conductivity and related transport properties. It should be stressed that we are only doing *in vitro* studies of MTs. However, we believe that, modulo possible difficulties in extrapolating from *in vitro* (Laboratory) experiments to *in vivo* situations, such experimental set-ups will help in shedding some light on the issue of the quantum mechanical nature of the MT arrangements, and even on the processes of the transmission of electric signals (stimuli) by the neuronal systems.

We close by remarking that, if the quantum-mechanical scenario for MT dynamics discussed above turns out to be right, then, as suggested in [5] and here, this may imply a quantum-hologram picture for information processing of brain MT networks. In this respect, we also mention that the existence of solitonic quantum-coherent states along the MT dimer walls may imply a rôle for these biological entities as 'information gates'. Consider, for instance, a node of three MT. In case two of the branches happen to have a soliton moving, then this soliton will continue to propagate in the out coming branch,and this would correspond to a "1" signal. In contrast, if only one of the branches has a soliton, then, the soliton movement will terminate at the node, and this would correspond to a "0" signal.

Clearly much more work needs to be done before even tentative conclusions are reached, concerning the nature of the MT arrangements. However, we believe that the present work constitutes a useful addition to the programme of understanding *in vitro* the nature of MT, and the associated processes of energy and signal transfer, with a view to applying such findings in order to facilitate the use of biological MTs (or hybrids thereof with carbon nanotubes) as basic building blocks in future quantum computation, if at all possible.

Acknowledgements

We wish to thank a number of people who have been involved with this work, in particular Deeph Chana, Mike Damzen, Mike Holwill, Alan Michette, Dimitri Nanopoulos and Eberhard Unger. It is also a pleasure to thank Hans-

Thomas Elze and the other organizers of the Piombino workshop for providing a thought stimulating meeting. This work is partially supported by EPSRC grant GR/S06080/01.

References

1. S. Roth and V. Krstic: 'GLJA Rikken', 2002 Current Applied Physics, in press; D. Tománek: 2002 Current Applied Physics, in press; C. Papdopoulos *et al.* Phys. Rev. Lett. **85**, 3476 (2000)
2. Z.K. Tang *et al.*: Science **292**, 2462 (2001)
3. B. Julsgaard, A. Kozhekin and E. Polzik: Nature **413**, 400 (2001)
4. S. Haroche and J.M. Raimond: in *Cavity Quantum Electrodynamics*, ed. by P. Berman (Academic Press, New York 1994), p. 123, and references therein
5. N.E. Mavromatos and D.V. Nanopoulos: Int. J. Mod. Phys. B **12**, 517 (1998); Adv. Str. Biol. **5**, 283 (1998). N.E. Mavromatos: Bioelectroch. Bioener. **48**, 273 (1999); N.E. Mavromatos, A. Mershin, and D.V. Nanopoulos: Int. J. Mod. Phys. B **16**, 3623 (2002)
6. P. Lal: Phys. Lett. A **111**, 389 (1985)
7. M.V. Satarić, J.A. Tuszyński, and R.B. Zakula: Phys. Rev. E **48**, 589 (1993); this model of MT dynamics is based on the ferroelectric-ferrodistortive model of: M.A. Collins, A. Blumen, J.F. Currie, and J. Ross: Phys. Rev. B **19**, 3630 (1978)
8. N.E. Mavromatos and D.V. Nanopoulos: Int. J. Mod. Phys. B **11**, 851 (1997)
9. For a comprehensive review see: W.H. Zurek: Phys. Today **44**, No. 10, 36 (1991); see also: W.H. Zurek: Phys. Rev. D **24**, 1516 (1981); A.O. Caldeira and A.J. Leggett: Physica A (Amsterdam) **121**, 587 (1983); Ann. Phys. **149**, 374 (1983)
10. N.E. Mavromatos, D.V. Nanopoulos, I. Samaras, and K. Zioutas: Adv. Struct. Biol. **5**, 127 (1998); quant-ph/9803005
11. E. Del Giudice, S. Doglia, M. Milani, and G. Vitiello: Nucl. Phys. B **251** (FS 13), 375 (1985); *ibid* B **275** (FS 17), 185 (1986)
12. E. Del Giudice, G. Preparata, and G. Vitiello: Phys. Rev. Lett. **61**, 1085 (1988)
13. S. Hameroff and R. Penrose: 'Orchestrated Reduction of Quantum Coherence in Brain Microtubules: a Model of Consciousness'. In: *Towards a science of Consciousness*, The First Tucson Discussions and Debates, ed. by S. Hameroff *et al.* (MIT Press, Cambridge, MA, 1996), p. 507
14. M. Otwinowski, R. Paul, and W.G. Laidlaw: Phys. Lett. A **128**, 483 (1988)
15. J.J. Sanchez-Mondragon, N.B. Narozhny, and J.H. Eberly: Phys. Rev. Lett. **51**, 550 (1983)
16. A. Albrecht: Phys. Rev. D **46**, 5504 (1992)
17. S. Haroche, lecture at 'DICE 2002'
18. G.S. Agarwal: Phys. Rev. Lett. **53**, 1732 (1984)
19. F. Bernardot *et al.*: Europhys. Lett. **17**, 34 (1992)
20. Yifu Zhu *et al.*: Phys. Rev. Lett. **64**, 2499 (1990)
21. M. Brune *et al.*: Phys. Rev. Lett. **77**, 4887 (1996)
22. S. Hameroff: *Ultimate Computing: Biomolecular Consciousness and Nano Technology* (Elsevier North-Holland, Amsterdam 1987); S. Hameroff, S. A. Smith and R.C. Watt: Ann. N.Y. Acad. Sci. **466**, 949 (1986)
23. M. Jibu, S. Hagan, S. Hameroff, K. Pribram, and K. Yasue: Biosystems **32**, 195 (1994)

24. S.R. Hameroff: Am. J. Clin. Med. **2**, 163 (1974)
25. P. Dustin: *MicroTubules* (Springer-Verlag, Berlin 1984);
    Y. Engleborghs: Nanobiology **1**, 97 (1992)
26. J. Ellis, N.E. Mavromatos, and D.V. Nanopoulos: 'Some Physical aspects of Liouville String Dynamics', in *1st International Workshop on Phenomenology of Unification from Present to Future*, 23-26 March 1994, Roma, ed. by G. Diambrini-Palazzi *et al.* (World Scientific, Singapore 1994), p. 187
27. H. Fröhlich. In: *Bioelectrochemistry*, ed. by F. Guttman and H. Keyzer (Plenum, New York 1986)
28. D. Sackett. In: *Subcellular Biochemistry*, Vol. 24 ed. by B.B. Biswas and S. Roy (Plenum Press, New York 1995)
29. M. Brune *et al.*: Phys. Rev. Lett. **65**, 976 (1990)
30. M. Brune *et al.*: Phys. Rev. A **45**, 5193 (1991)

Part V

Entropy, Chaos, and Complexity

Introduction:
From Efficient Quantum Computation
to Nonextensive Statistical Mechanics

Tomaz Prosen

Physics Department, FMF, University of Ljubljana, Ljubljana, Slovenia

These few pages will attempt to make a short comprehensive overview of several contributions to this volume which concern rather diverse topics. I shall review the following works, essentially reversing the sequence indicated in my title:

- First, by C. Tsallis on the relation of nonextensive statistics to the stability of quantum motion "on the edge of quantum chaos".
- Second, the contribution by P. Jizba on information theoretic foundations of generalized (nonextensive) statistics.
- Third, the contribution by J. Rafelski on a possible generalization of Boltzmann kinetics, again, formulated in terms of nonextensive statistics.
- Fourth, the contribution by D.L. Stein on the state-of-the-art open problems in spin glasses and on the notion of complexity there.
- Fifth, the contribution by F.T. Arecchi on the quantum-like uncertainty relations and decoherence appearing in the description of perceptual tasks of the brain.
- Sixth, the contribution by G. Casati on the measurement and information extraction in the simulation of complex dynamics by a quantum computer.

Immediately, the following question arises: What do the topics of these talks have in common? Apart from the variety of questions they address, it is quite obvious that the common denominator of these contributions is an approach to describe and control "the complexity" by simple means. One of the very useful tools to handle such problems, also often used or at least referred to in several of the works presented here, is the concept of Tsallis entropy and nonextensive statistics.

Weinstein, Tsallis and Lloyd [1] study the stability of quantum motion, as characterized by the overlap between a quantum state

$$|\psi_u(t)\rangle = e^{-\mathrm{i}Ht/\hbar}|\psi_i\rangle \ ,$$

evolved by a certain Hamiltonian H, and a state

$$|\psi_p(t)\rangle = e^{-\mathrm{i}H't/\hbar}|\psi_i\rangle \ ,$$

evolved by a slightly perturbed Hamiltonian

$$H' = H + \delta V \ ,$$

T. Prosen, Introduction: From Efficient Quantum Computation to Nonextensive Statistical Mechanics, Lect. Notes Phys. **633**, 321–326 (2004)
http://www.springerlink.com/   © Springer-Verlag Berlin Heidelberg 2004

where δ is a small perturbation strength parameter, namely

$$f(t) = \langle \psi_u(t) | \psi_p(t) \rangle \ .$$

The overlap $f(t)$, or its square $F(t) = |f(t)|^2$, proposed in [2], is also often called *fidelity* or *quantum Loschmidt echo*. The fidelity is equivalent to the probability of return to the initial quantum state after performing a quantum echo – composition of forward evolution generated by the unperturbed Hamiltonian H and backward time-inverted dynamics generated by the perturbed Hamiltonian H'. It has been established [3] that the fidelity for a classically chaotic Hamiltonian, and for initial coherent state $|\psi_i\rangle$, exhibits the so-called Lyapunov decay $F(t) = \exp(-\lambda t)$ with the local finite-time Lyapunov exponent λ, for times that are shorter than the so called log-time $t_{\log} = -\log \hbar / \lambda$. However, for initial states other than the coherent states, e.g. random states, or for times longer than $t_{\log}$ the physics of quantum fidelity is entirely different. The decay of fidelity $F(t)$ depends on the strength of perturbation δ, and essentially on the time autocorrelation function of the perturbation observable $C(t', t'') = \langle V(t'')V(t')\rangle - \langle V(t'')\rangle\langle V(t')\rangle$. – We have denoted $\langle . \rangle := \langle \psi_i | . | \psi_i \rangle$, and $V(t) := e^{iHt/\hbar} V e^{-iHt/\hbar}$. – Namely, using elementary time-dependent quantum perturbation theory one finds [4] the following linear response formula

$$F(t) = 1 - \frac{\delta^2}{\hbar^2} \int_0^t dt' \int_0^t dt'' C(t', t'') + \dots \ . \tag{1}$$

This result can be interpreted in terms of a fluctuation-dissipation relationship: The one minus fidelity, which measures the amount of *dissipation* of quantum information, is equal to the integrated time correlation function of the perturbation (the *fluctuation*). Quite surprisingly, this means that faster decay of correlations, or even more — negative correlation, improves the fidelity.

Using a bit more sophisticated ideas along the same lines one finds that for the case of classically chaotic (and mixing) dynamics, the decay of fidelity is exponential

$$F(t) = \exp(-t/\tau_{\mathrm{ch}}), \quad \tau_{\mathrm{ch}} = \frac{\hbar^2}{2\sigma\delta^2} \ . \tag{2}$$

The constant σ plays the role of diffusion coefficient — an integrated (classical limit of) autocorrelation function $C(t' - t'') = C(t', t'')$,

$$\sigma = \int_{-\infty}^{\infty} C(t) dt \ .$$

It is important to stress that this result, for times beyond the log-time scale $t > t_{\log}$, does not depend on the structure of the initial state.

In the opposite case of regular, *i.e.* integrable, classical dynamics, or for the initial state starting inside KAM islands, the fidelity behaves differently. For coherent initial states, and for perturbation which perturbs the classical frequency of the underlying torus, the decay is Gaussian

$$F(t) = \exp(-(t/\tau_{\mathrm{reg}})^2) \ , \quad \tau_{\mathrm{reg}} = \frac{\hbar^{1/2}}{\kappa\delta} \ . \tag{3}$$

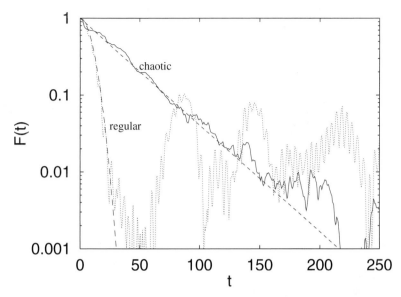

Fig. 1. Fidelity decay of an initial coherent state in a generalized Jaynes-Cummings model for regular and chaotic classical dynamics. Solid and dotted curves indicate numerical results for the chaotic and regular regime, while the dashed curves indicate the corresponding semi-classical exponential and Gaussian

The constant κ can be determined purely from the classical data (see [4] for explicit results), namely the classical limit of the perturbation observable and the phase-space position of the center of the initial wave-packet, and does not depend on $\hbar$ and δ. It is important to stress that the time scale of fidelity decay for classically chaotic dynamics scales as $\tau_{ch} \propto \delta^{-2}$ and is generally longer than the time scale for classically regular dynamics $\tau_{reg} \propto \delta^{-1}$ for sufficiently small perturbation δ and fixed value of effective Planck constant $\hbar$. Interestingly, the behaviour is qualitatively different if the order of the limits $\hbar \to 0$ and $\delta \to 0$ is reversed. In the opposite case we recover the classical intuition which says that the integrable evolution is more stable than the chaotic one. In the Fig. 1, we show the comparison between exponential and Gaussian decay of fidelity (all the parameters except the parameters controlling the nature of classical dynamics are the same in both cases) for the numerical example (see [5] for details) of the generalized Jaynes-Cummings model.

In the linear response regime (1), fidelity decays linearly or quadratically for classically chaotic or regular motion, respectively. However, on the 'edge of quantum chaos', i.e. for initial packets placed on the border between the KAM island and the chaotic sea, one observes the intermediate power-law decay of the fidelity reported by Weinstein et al. [1], which is explained in terms of nonextensive statistics in the critical regime between regular and chaotic hamiltonian dynamics. This result is also consistent with a slow power-law decay of correlations in this critical regime using the correlation-integral formula for fidelity (1). In this regime, Weinstein et al. show that numerical fidelity data are well captured by the formula

$$F_q(t) = [1 - (1 - q)(t/\tau_q)^2]^{1/(1-q)} = \exp_q(-(t/\tau_q)^2) \ ,$$

where $\exp_q$ is a q-exponential function. The authors also give numerical tables for the values of q and τ_q for various values of the parameters in their numerical example of a quantized kicked top.

Jizba [6] discusses information theoretic foundations of generalized statistics, in particular the q−statistics and nonextensive statistics based on the quantitative measures of information due to Renyi and Tsallis. The paper approaches axiomatic definition of generalized statistics. It is shown how Renyi's entropy naturally emerges by solving for the most general information measure preserving the additivity of statistically independent systems while still being compatible with Kolmogorov's probability axioms. Tsallis' entropy, on the contrary, breaks the additivity axiom and satisfies generalized (nonextensive) addition rule. The paper discusses an intimate connection between Rényi's q−parameter and the fractal and multifractal properties of the probability distribution and its generalized dimensions.

As a concrete demonstration of the application of nonextensive statistics in traditional non-equilibrium statistical mechanics, Sherman and Rafelski [7] discuss the generalized Boltzmann equation in the nonideal case of dense systems with long-range interactions. The most important technical assumption is the relaxation of energy and momentum conservation in two-particle collisions. The only remaining condition is a strict conservation of the total energy and momentum of the entire system. As a consequence, the authors work with the fluctuation distributions of the energy and momentum discrepancy in two-particle interactions. Based on certain reasonable assumptions, the authors derive the Tsallis (nonextensive) equilibrium distribution of the Boltzmann momentum density.

Quite remotely from the lines above, Stein [8] introduces us into modern problems of classical statistical physics of spin glasses. In particular, the motivation to study such systems is very general and can go outside the scope of magnetic materials. The general and abstract defining property of a spin glass is the competition between frustration and quenched disorder. The theoretically most interesting question is related to the existence of phase transition points at low but non-vanishing temperature and this property may sensitively depend on the range of interaction and dimensionality of the system. The author reviews the famous approach of Parisi to the solution of the infinite-range Sherrington-Kirkpatrick spin glass model. He raises fundamental questions on interesting open issues such as nonunique equilibrium states, non-self-averaging of the overlap distribution, and even nonexistence of the usual thermodynamic limit for a spin glass.

Arecchi [9] presents a remarkable excursion of a theoretical physicist into neurophysiology, and neurophysics, in particular, of the brain and into cognitive science. The central idea of this quite speculative and intriguing but very interesting contribution is the quantum-like uncertainty relation between the information content of a perception ΔP and a minimal time interval ΔT for the perception to take place

$$\Delta P \Delta T \geq C \ .$$

The analogue of the Planck constant, C, depends on the physiological properties of the brain and may, according to the author, depend on the individual and his or hers motivation to solve certain task. The authors speculates even further and develops analogies for the entanglement and decoherence in the quantum-like perception-time phase space.

Casati and Montangero [10] outline certain recent ideas on the connection between quantum chaos, quantum computation [11], and quantum measurement. In particular, the paper discusses how certain important observable features of quantum dynamical systems, such as the localization length due to quantum dynamical localization, can be determined efficiently on a quantum computer. As an example, the authors outline an efficient quantum algorithm for the simulation of a generic quantum chaotic system, namely the sawtooth map, and describe how accurate estimation of the localization length can be determined by a small amount of measurements in the momentum space. The efficiency of the quantum computation is characterized by a square-root speed-up with respect to the optimal classical algorithm. The quantum computation of a state with the localization length ℓ would require $O(\ell(\log \ell)^2)$ quantum gates, whereas fastest classical computation requires $O(\ell^2 \log \ell)$ steps.

Closing the circle and coming back to the contribution by Weinstein et al. [1], we wish to stress that understanding of the dynamical properties of quantum fidelity is also important for controlling the stability of quantum computation against static imperfections. Using the generalization of the correlation formula for fidelity (1), one can design more robust and stable quantum algorithms by making a dynamical sequence of quantum gates 'more chaotic'. This has been nicely demonstrated in the example of the quantum Fourier transformation [12].

References

1. Y.S. Weinstein, C. Tsallis, and S. Lloyd: Lecture in this volume; quant-ph/0305086; see also: Y.S. Weinstein, C. Tsallis, and S. Lloyd: Phys. Rev. Lett. **89**, 214101 (2002)
2. A. Peres: Phys. Rev. A **30**, 1610 (1984); see also:
   A. Peres: *Quantum Theory: Concepts and Methods* (Kluwer, Dordrecht 1995)
3. R.A. Jalabert and H.M. Pastawski: Phys. Rev. Lett. **86**, 2490 (2001)
4. T. Prosen: Phys. Rev. E **65**, 036208 (2002); T. Prosen and M. Znidaric: J. Phys. A **35**, 1455 (2002)
5. T. Prosen, T.H. Seligman, and M. Znidaric: Phys. Rev. A **67**, 042112 (2003)
6. P. Jizba: Lecture in this volume; cond-mat/0301343
7. T. Sherman and J. Rafelski: Lecture in this volume; physics/0204011
8. D.L. Stein: Lecture in this volume; cond-mat/0301104
9. F.T. Arecchi: Lecture in this volume

10. G. Casati and S. Montangero: Lecture in this volume;
    G. Benenti, G. Casati, S. Montangero, and D.L. Shepelyanski: Phys. Rev. Lett.
    87, 227901 (2001); Eur. Phys. J. D **20**, 293 (2002); Eur. Phys. J. D **22**, 285 (2003)
11. M.A. Nielsen and I.L. Chuang: *Quantum Computation and Quantum Information*
    (Cambridge University Press, Cambridge 2000)
12. T. Prosen and M. Znidaric: J. Phys. A **34**, L681 (2001)

Uncertainty Domains Associated with Time Limited Perceptual Tasks: Fuzzy Overlaps or Quantum Entanglement?

F. Tito Arecchi[1,2]

[1] Department of Physics, University of Firenze
[2] Istituto Nazionale di Ottica Applicata, Largo E. Fermi 6, 50125 Firenze, Italy

Abstract. The current understanding of complexity in computer science is that it corresponds to how the amount of computational resources invested in solving a problem scales with the system size. On the contrary, in natural sciences, a complex system is such that its knowledge cannot be confined to a unique model, but it is stratified over different and mutually irreducible hierarchical levels, each one with its own rules and language [1].

In dealing with cognitive processes complexity arises already in the physical description of how external stimuli (light, sound, pressure, chemicals) are transformed into sensorial perceptions.

I call "neurophysics" the combination of neurodynamical events, whereby neurons are treated as nonlinear dynamical systems, and the peculiar spike synchronization strategy selected in course of the natural evolution as the optimal strategy to elaborate information into relevant cognitive processes. Already at this fundamental level, we come across a possible quantum limitation, here presented explicitly for the first time, which would forbid the brain operations to be fully simulated by a universal computing machine.

1 Neuron Synchronization

It is by now established that a holistic perception emerges, out of separate stimuli entering different receptive fields, by synchronizing the corresponding spike trains of neural action potentials [2, 3]. We recall that action potentials play a crucial role for communication between neurons [4]. They are steep variations in the electric potential across a cell's membrane, and they propagate in essentially constant shape from the soma (neuron's body) along axons toward synaptic connections with other neurons. At the synapses they release an amount of neurotransmitter molecules depending upon the temporal sequences of spikes, thus transforming the electrical into a chemical carrier. As a fact, neural communication is based on a temporal code whereby different cortical areas which have to contribute to the same percept P synchronize their spikes. Limiting for convenience the discussion to the visual system, spike emission in a single neuron of the higher cortical regions results as a trade off between bottom-up stimuli arriving through the LGN (lateral geniculate nucleus) from the retinal detectors and threshold modulation due to top-down signals sent as conjectures by the

F.T. Arecchi, Uncertainty Domains Associated with Time Limited Perceptual Tasks: Fuzzy Overlaps or Quantum Entanglement?, Lect. Notes Phys. **633**, 327–340 (2004)
http://www.springerlink.com/

semantic memory. This is the core of ART (adaptive resonance theory [5]) or other computational models of perception [6] which assume that a stable cortical pattern is the result of a Darwinian competition among different percepts with different strength. The winning pattern must be confirmed by some matching procedure between bottom-up and top-down signals.

It is the aim of this paper to present a fundamental aspect of percept formation, namely, a quantum limitation in information encoding/decoding through spike trains, whenever the processing session is interrupted. In fact, the temporal coding requires a sufficiently long sequence of synchronized spikes, in order to realize a specific percept. If the sequence is interrupted by the arrival of new uncorrelated stimuli, then an uncertainty ΔP emerges in the percept space P. This is related to the finite duration ΔT allotted for the code processing by the uncertainty relation

$$\Delta P \cdot \Delta T \geq C$$

where C is a positive dimensional quantity whose non zero value represents a quantum constraint on the coding. This constraint implies that the percepts are not set-theoretical objects, that is, objects belonging to separate domains, but there are overlap regions where it is impossible to discriminate one percept from another.

In order to establish the above uncertainty relation we must introduce a metric and hence a distance in percept space; furthermore we must consider whether in the case of two different percepts separated by ΔP and observed for a duration ΔT the above relation is just the trivial overlap of two fuzzy sets or it discloses a fundamental quantum entanglement of the type discussed in formulating the Bell inequalities [7]. For this purpose it is necessary to briefly review the neurodynamics of spike formation As for the dynamics of the single neuron, a saddle point instability separates in parameter space an excitable region, where axons are silent, from a periodic region, where the spike train is periodic (equal interspike intervals). If a control parameter is tuned at the saddle point, the corresponding dynamical behavior (homoclinic chaos) consists of a frequent return to the instability [8]. This manifests as a train of geometrically identical spikes, which however occur at erratic times (chaotic interspike intervals). Around the saddle point the system displays a large susceptibility to an external stimulus, hence it is easily adjustable and prone to respond to an input, provided this is at sufficiently low frequencies; this means that such a system is robust against high frequency noise as discussed later.

Such a type of dynamics has been recently dealt with in a series of reports that here I recapitulate as the following chain of linked facts:

- A single spike in a 3D dynamics corresponds to a quasi-homoclinic trajectory around a saddle focus SF (fixed point with 1 (2) stable direction and 2 (1) unstable ones); the trajectory leaves the saddle and returns to it (Fig. 1). We say "quasi-homoclinic" because, in order to stabilize the trajectory away from SF, a second fixed point, namely a saddle node SN, is necessary to assure a heteroclinic connection.

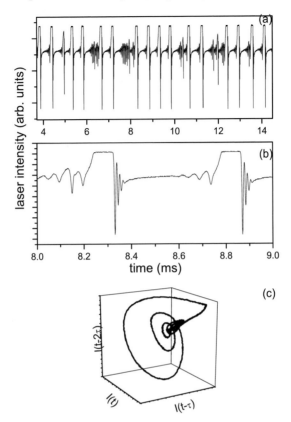

Fig. 1. (a) Experimental time series of the laser intensity for a CO2 laser with feedback in the regime of homoclinic chaos. (b) Time expansion of a single orbit. (c) Phase space trajectory built by an embedding technique with appropriate delays (from [8])

- A train of spikes corresponds to the sequential return to, and escape from, the SF. A control parameter can be set at a value B_C for which this return is erratic (chaotic interspike interval) even though there is a finite average frequency. As the control parameter is set above or below B_C, the system moves from *excitable* (single spike triggered by an input signal) to periodic (yielding a regular sequence of spikes without need for an input), with a frequency monotonically increasing with the separation ΔB from B_C (Fig. 2) [9].

- Around SF, any tiny disturbance provides a large response. Thus the homoclinic spike trains can be synchronized by a periodic sequence of small disturbances (Fig. 3). However each disturbance has to be applied for a minimal time, below which it is no longer effective; this means that the system is insensitive to broadband noise, which is a random collection of fast positive and negative signals [10].

- The above considerations lay the floor for the use of mutual synchronization as the most convenient way to let different neurons respond coherently to

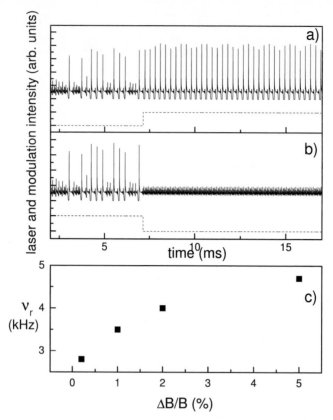

Fig. 2. Stepwise increase, a), and decrease, b), of control parameter B0 by +/- 1/100 brings the system from homoclinic to periodic or excitable behavior. c) In case a) the frequency ν_r of the spikes increases monotonically with ΔB_0 (from [9])

the same stimulus, organizing as a space pattern. In the case of a single dynamical system, it can be fed back by its own delayed signal. As the delay is long enough the system is decorrelated with itself and this is equivalent to feeding an independent system. This process allows to store meaningful sequences of spikes as necessary for a short term memory [11].

- In presence of localized stimuli over a few neurons, the corresponding disturbances propagate by inter-neuron coupling (either excitatory or inhibitory); a synchronized pattern is uniquely associated with each stimulus; different patterns compete and we conjecture that the resulting sensory response, which then triggers motor actions, corresponds by a majority rule to that pattern which has extended over the largest cortical domain ("winner takes all" dynamics). An example is discussed for two inputs to a one-dimensional array of coupled HC systems [12].

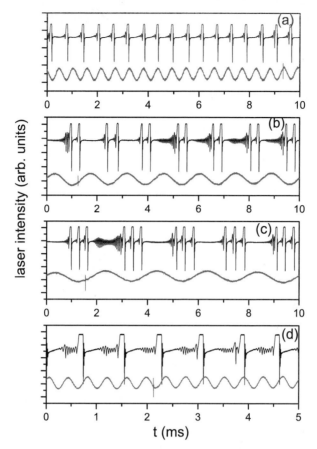

Fig. 3. Experimental time series for different synchronization induced by periodic changes of the control parameter. (a) 1:1 locking, (b) 1:2, (c) 1:3, (d) 2:1 (from [8])

These facts have been established experimentally and confirmed by a convenient model in the case of a class B laser[1] with a feedback loop which readjusts the amount of losses depending on the value of the light intensity output [14].

The above listed facts hold in general for any dynamical system which has a 3-dimensional sub-manifold separating a region of excitability from a region of periodic oscillations: indeed, this separatrix has to be a saddle focus.

---

[1] I recall the classification widely accepted in laser physics. Class A lasers are ruled by a single order parameter, the amplitude of the laser field, which obeys a closed dynamical equation; all the other variables having much faster decay rate, thus adjusting almost instantly to the local field value. Class B lasers are ruled by two order parameters, the laser field and the material energy storage providing gain; the two degrees of freedom having comparable characteristic times and behaving as activator and inhibitor in chemical dynamics [13]

2   Perceptions, Feature Binding, and Qualia

Let us return to the visual system; the role of elementary feature detectors has been extensively studied in the past decades [15]. By now we know that some neurons are specialized in detecting exclusively vertical or horizontal bars, or a specific luminance contrast, etc. However the problem arises: how elementary detectors contribute to a holistic (Gestalt) perception? A hint is provided by [3]. Suppose we are exposed to a visual field containing two separate objects. Both objects are made of the same visual elements, horizontal and vertical contour bars, different degrees of luminance, etc. What are then the neural correlates of the identification of the two objects? We have one million fibers connecting the retina to the visual cortex, through the LGN. Each fiber results from the merging of approximately 100 retinal detectors (rods and cones) and as a result it has its own receptive field. Each receptive field isolates a specific detail of an object (e.g. a vertical bar). We thus split an image into a mosaic of adjacent receptive fields.

Now the "feature binding" hypothesis consists of assuming that all the cortical neurons whose receptive fields are pointing to a specific object synchronize the corresponding spikes, and as a consequence the visual cortex organizes into separate neuron groups oscillating on two distinct spike trains for the two objects.

Direct experimental evidence of this synchronization is obtained by insertion of microelectrodes in the cortical tissue of animals just sensing the single neuron [3]. Indirect evidence of synchronization has been reached for human beings as well, by processing the EEG (electro-encephalo-gram) data [16].

The advantage of such a temporal coding scheme, as compared to traditional rate based codes, which are sensitive to the average pulse rate over a time interval and which have been exploited in communication engineering, has been discussed in a recent paper [17].

Based on the neurodynamical facts reported above, we can understand how this occurs [5]. The higher cortical stages where synchronization takes place have two inputs. One (bottom-up) comes from the sensory detectors via the early stages which classify elementary features. This single input is insufficient, because it would provide the same signal for e.g. horizontal bars belonging indifferently to either one of the two objects. However, as we said already, each neuron is a nonlinear system passing close to a saddle point, and the application of a suitable perturbation can stretch or shrink the interval of time spent around the saddle, and thus lengthen or shorten the interspike interval. The perturbation consists of top-down signals corresponding to conjectures made by the semantic memory.

In other words, the perception process is not like the passive imprinting of a camera film, but it is an active process whereby the external stimuli are interpreted in terms of past memories. A focal attention mechanism assures that a matching is eventually reached. This matching consists of resonant or coherent behavior between bottom-up and top-down signals. If matching does not occur, different memories are tried, until the matching is realized. In presence of a fully new image without memorized correlates, then the brain has to accept the fact that it is exposed to a new experience.

Notice the advantage of this time dependent use of neurons, which become available to be active in different perceptions at different times, as compared to the computer paradigm of fixed memory elements which store a specific object and are not available for others (the so called "grandmother neuron" hypothesis).

We have above presented qualitative reasons why the degree of synchronization represents the perceptual salience of an object. Synchronization of neurons located even far away from each other yields a space pattern on the sensory cortex, which can be as wide as a few square millimeters, involving millions of neurons. The winning pattern is determined by dynamic competition (the so-called "winner takes all" dynamics).

This model has an early formulation in ART and has been later substantiated by the synchronization mechanisms. Perceptual knowledge appears as a complex self-organizing process.

Naively, one might expect that a given "qualia", that is, a private sensation as e.g. the red of a Titian painting, is always coded by the same sequence of spikes. If so, in a near future the corresponding information could be retrieved by a high resolution detector, and hence a Rosetta stone could be established between the spike sequences and the qualia. Such a naive expectation which would lead to a world without privacy, is altogether wrong for the following reasons. After the initial experience of that qualia, the first time one has seen that Titian, any further repetition of that experience, either by memory recollection or by re-watching the painting occurs in presence of new experiential elements (one has become older, his/her store of memories has drastically mutated) and these novelties contribute to feature binding by a modified synchronization pattern. Evidence of such a fact has been established by Freeman [18] reporting the synchronization pattern of the olfactory bulb of a rabbit, recorded by a large number of electrodes; as the same odor is presented twice, with an intermediate odor in between, the two patterns are all together different, even though the animal behavior hints at the same reaction. Freeman's experiment is contrasted by the fact that some olfactory neurons of the locust yield the same bursts of spikes for the same odor [19]. Presumably, lower animals as locusts have a much smaller semantic repertoire than rabbits or humans, and hence for them the dream of the Rosetta stone has some validity.

3   A Metric in Percept Space

A convenient metric space for percepts must be introduced, in order to give a meaning to the measure of ΔP and establish the uncertainty relation.

We discuss two proposals of metrics of spike trains. The first one [20] considers each spike as very short, and each coincidence as an instantaneous event with no time uncertainty. The metric spans a large, yet discrete, space and it can be programmed on a standard computer. A more recent proposal [21] accounts for the physical fact that each spike is spread in time by a filtering process, hence the overlap takes a time breadth t_c and any coincidence is a smooth continuous process. Indeed, as discussed above, in neurodynamics the high sensitivity region around the saddle focus has a finite time width.

The discrete metric satisfies a set-theoretical expectation and is programmable in a computer. The continuous metric fits better neurodynamical facts.

Victor and Purpura [20] have introduced several families of metrics between spike trains as a tool to study the nature and precision of temporal coding. Each metric defines the distance between two spike trains as the minimal "cost" required to transform one spike train into the other via a sequence of allowed elementary steps, such as inserting or deleting a spike, shifting a spike in time, or changing an interspike interval length.

The geometries corresponding to these metrics are in general not Euclidean, and distinct families of cost-based metrics typically correspond to distinct topologies on the space of spike trains. Each metric, in essence, represents a candidate temporal code in which similar stimuli produce responses which are close and dissimilar stimuli produce responses which are more distant.

Spike trains are considered to be points in an abstract topological space. A spike train metric is a rule which assigns a non-negative number $D(S_a, S_b)$ to pairs of spike trains S_a and S_b which expresses how dissimilar they are.

A metric D is essentially an abstract distance. By definition, metrics have the following properties:

- $D(S_a, S_a) = 0$;
- Symmetry: $D(S_a, S_b) = D(S_b, S_a)$;
- Triangle inequality: $D(S_a, S_c) \leq D(S_a, S_b) + D(S_b, S_c)$;
- Non-negativity: $D(S_a, S_b) > 0$ unless $S_a = S_b$.

The metric may be used in a variety of ways – for example, one can construct a neural response space via multidimensional scaling of the pairwise distances, and one can assess coding characteristics via comparison of stimulus-dependent clustering across a range of metrics.

This distance can be calculated efficiently by a dynamic programming algorithm. Although this method has been applied successfully [22], the calculation of the full cost function is quite involved. The reason is that it is not always clear where a displaced spike came from, and if the number of spikes in the trains is unequal, it can be difficult to determine which spike was inserted/deleted.

Rossum has introduced an Euclidean distance measure that computes the dissimilarity between two spike trains [21]. First of all, one filters both spikes trains giving to each spike a duration t_c. In our language, this would be the time extension of the saddle point region of high susceptibility, where perturbations are effective. To calculate the distance, evaluate the integrated squared difference of the two trains. The simplicity of the distance allows for an analytical treatment of simple cases. Numerical implementation is straightforward and fast.

The distance interpolates between, on the one hand, counting non-coincident spikes and, on the other hand, counting the squared difference in total spike count. In order to compare spike trains with different rates, total spike count can be used (large t_c). However, for spike trains with similar rates, the difference in total spike number is not useful and coincidence detection is sensitive to noise. The distance uses a convolution with the exponential function. As an alternative measure, one could convolute the spikes with a square window. In that case the situation becomes somewhat similar to binning followed by calculating the Euclidean distance between the number of spikes in the bins. But in standard binning the bins are fixed on the time axis, therefore two different spike trains yield identical binning patterns as long as spikes fall in the same bin. However, with the proposed distance (convoluting with either a square or an exponential) this does not happen; the distance is zero only if the two spike trains are fully identical (assuming t_c is finite).

The continous distance introduced here is explicitly embedded in Euclidean space, which makes it less general but easier to analyze than the distance introduced by Victor and Purpura.

Interestingly, the distance is related to stimulus reconstruction techniques, where convoluting the spike train with the spike triggered average yields a first order reconstruction of the stimulus [23]. Here the exponential corresponds roughly to the spike triggered average and the filtered spike trains correspond to the stimulus. The distance thus approximately measures the difference in the reconstructed stimuli. This might well explain the linearity of the measure for intermediate t_c.

The above reports on attempts to introduce a metric in percept space are limited in their relevance by our critical consideration on qualia. Indeed, due to the subjective introduction of new elements of experience, two spike trains referring to the same qualia could be far away from each other in the percept space, that is, orthogonal to each other. The simplistic stimulus-response paradigm of behaviorism which works for lower animals as locusts, fails for higher animals where qualia represent a richness which is sensitive to environmental tiny features. We say that environment induces decoherence between two perceptions corresponding to similar stimuli.

For the rest of the paper, we take the over-simplifying assumption (perhaps applicable only to invertebrates) that a percept space is in one-to-one correspondence with a stimulus space, in order to develop persuasive quantum considerations.

4   Role of Duration T in Perceptual Definition: A Quantum Aspect

How does a synchronized pattern of neuronal action potentials become a relevant perception? This is an active area of investigation which may be split into many hierarchical levels. Notice however that, due to the above consideration on qualia, the experiments suggested in this section hold for purely a-semantic perceptions; by this we mean the task of detecting rather elementary features which would not trigger our categorization skills.

Not only the different receptive fields of the visual system, but also other sensory channels as auditory, olfactory, etc. integrate via feature binding into a holistic perception. Its meaning is "decided" in the PFC (pre-frontal cortex) which is a kind of arrival station from the sensory areas and departure for signals going to the motor areas. On the basis of the perceived information, motor actions are started, including linguistic utterances.

Sticking to the neurodynamical level, and leaving to psychophysics the investigation of what goes on at higher levels of organization, we stress here a fundamental temporal limitation.

Taking into account that each spike lasts about 1 ms, that the minimal interspike separation is 4 ms, and that the average decision time at the PCF level is about $T = 200\ ms$, we can split T into $200/4 = 50$ bins of 4 ms duration, which are designated by 1 or 0 depending on whether they have a spike or not. Thus the total number of different messages which can be transmitted is $2^{50} \approx 10^{17}$ that is, well beyond the information capacity of present computers. Even though this number is large, we are still within a finitistic realm. Provided we have time enough to ascertain which one of the 10^{17} different messages we are dealing with, we can classify it with the accuracy of a digital processor, without residual error.

But suppose we expose the cognitive agent to fast changing scenes, for instance by presenting in sequence unrelated video frames with a time separation less than 200 ms. While small gradual changes induce the sense of motion as in movies, big differences imply completely different subsequent spike trains. Here any spike train gets interrupted after a duration ΔT less than the canonical T. This means that the PFC cannot decide among *all* coded perceptions having the same structure up to ΔT, but different afterwards. How many are they: the remaining time is $\tau = T - \Delta T$. To make a numerical example, take a time separation of the video frames $\Delta T = T/2$, then $\tau = T/2$. Thus, in spike space an interval ΔP comprising

$$2^{\tau} \approx 2^{25} \approx 10^8$$

different perceptual patterns is uncertain. The uncertainty can be expressed as

$$\Delta P \approx 2^{\tau} = 2^T 2^{-\Delta T} = P_M 2^{-\Delta T} = P_M e^{-(\ln 2)\Delta T}$$

where we have called $P_M = 2^T$ the maximum information basket over the time T. The exponential relation between ΔP and ΔT can be approximated over a wide range by an hyperbolic relation. Precisely, as we increase ΔT, ΔP shrinks with the uncertainty principle

$$\Delta P \cdot \Delta T \geq C \ .$$

If we adopt the Victor and Purpura metric, with the distances measured by costs in nondimensional units, than C has the dimensions of a time.

If we adopt the Rossum metric, then the distance ΔP is the square of a voltage and therefore can be taken as an energy, so that C has the dimensions of an action, as Planck's constant in the quantum energy-time uncertainty relation.

The problem faced thus far in the scientific literature, of an abstract comparison of two spike trains without accounting for the available time for such

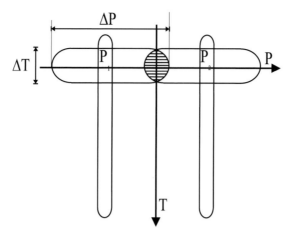

Fig. 4. Uncertainty areas of two perceptions P_1 and P_2 for two different durations of the spike trains. In the case of short ΔT, the overlap region is represented by a Wigner function with strong positive and negative oscillations which go as $\cos \frac{\Delta P}{C} T$ along the T axis; therefore with a frequency given by the ratio of the percept separation $\Delta P = P_2 - P_1$ to the perceptual 'Planck's constant' C

a comparison, is rather unrealistic. A finite available time ΔT places a crucial role in any decision, either if we are trying to identify an object within a fast sequence of different perceptions or if we are scanning through memorized patterns in order to decide about an action.

As a result the perceptual space P per se is meaningless. What is relevant for cognition is the joint (P, T)-space, since "in vivo" we have always to face a limited time ΔT which may truncate the whole spike sequence upon which a given perception has been coded. Only "in vitro" we allot to each perception all the time necessary to classify it.

A limited ΔT is not only due to the temporal crowding of sequential images, as e.g. reported clinically in behavioral disturbances of teenagers exposed to fast video games, but also to sequential conjectures that the semantic memory essays via different top-down signals. Thus in the metrical space (P, T), while the isolated localization of a percept P (however long is T) or of a time T (however spread is the perceptual interval ΔP) have a sense, a joint localization both in percept and time has an ultimate limitation when the corresponding domain is less than the quantum area C.

Let us consider the following thought experiment. Take two percepts P_1 and P_2 which for long processing times appear as the two stable states of a bistable optical illusion, e.g the Necker cube. If we let only a limited observation time ΔT, then the two uncertainty areas overlap. The contours drawn in Fig. 4 have only a qualitative meaning.

The situation looks like the noncommutative coordinate-momentum space of a single quantum particle. In this case, it is well known [24] that the quasiprobability Wigner function has strong non classical oscillations in the overlap region;

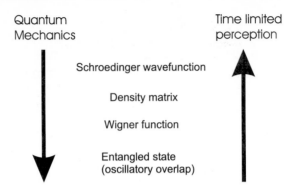

Fig. 5. Direction of the logical processes which lead from wavefunction to entangled states or viceversa.

we cannot split the coordinate-momentum space into two disjoint domains (sets) to which we can apply a Boolean logic or a classical Kolmogorov probability. This is the reason why the Bell inequalities are violated in an experiment dealing with such a situation [7].

The Wigner function formalism derives from a Schrödinger wavefunction treatment for a pure state, and corresponding density matrix for mixed states.

In the perceptual (P, T)-space no Schrödinger treatment is available, but we can apply a reverse logical path as follows.

The uncertainty relation $\Delta P \cdot \Delta T \geq C$ forbids a partition of the (P, T)-space into sets only if the (P, T)-space is noncommutative. Thus it must be susceptible of a Wigner function treatment and we can consider the contours of Fig. 4 as fully equivalent to isolevel cuts of a Wigner function. Hence we can introduce Schroedinger cat states and violations of Bell inequalities exactly as in quantum physics but with a reverse logical process, as illustrated in Fig. 5.

The equivalent of a superposition state should be a bistable situation observed for a time shorter than the whole decision time. An experimental test should discriminate such a situation. Precisely, suppose we adjust the duration ΔT such that the observation point falls in the overlap region of the (P, T)-space. If we repeat the exposure of the same subject to a sequence of uncorrelated tests of this kind, we expect two possibilities, namely,

i) *Fuzzy Classical Sets.* The subject provides in average an equal amount of P_1 and P_2 responses, even though each isolated response is either $P1$ or $P2$.

ii) *Quantum Entangled States.* In this case the Wigner function describing the overlap region is modulated by an oscillatory term proportional to

$$\cos \frac{\Delta P}{C} \Delta T \ .$$

For a fixed separation $P_2 - P_1 = \Delta P$ and a subject characterized by a given C, small adjustments of ΔT will bring the cosine function from $+1$ to -1. Since it is a weight function for P values, then the subject consistently reports the answer $+1$ or -1, depending on the setting of ΔT.

Such a test should provide an estimation of the C-value, which plausibly changes from individual to individual, and for a single one may be age and motivation dependent.

The proposed test is analogous to that suggested by Leggett [25] for a microscopic physical device. As in that case we must face decoherence problems [24]. Thus, the experiment requires a careful isolation of the subject from the environment, which we have seen to be an a-semantic requirement, that is, to avoid environmental disturbances consisting of the intrusion of top-down categories.

Let us assume for the moment that the test yields result ii).

This would mean that in neurophysics time occurs under two completely different meanings, that is, as the ordering parameter to classify the position of successive events and as the useful duration of a relevant spike sequence, that is the duration of a synchronized train. In the second meaning, time T is a variable conjugate to perception P.

The quantum character has emerged as a necessity from the analysis of an interrupted spike train in a perceptual process. It follows that the (P, T)-space cannot be partitioned into disjoint sets to which a Boolean yes/not relation is applicable and hence where ensembles obeying a classical probability can be considered. A set-theoretical partition is the condition to apply the Church-Turing thesis, which establishes the equivalence between recursive functions on a set and operations of a universal computing machine.

The evidence of quantum entanglement of overlapping perceptions should rule out in principle a finitistic character of the perceptual processes. This should be the negative answer to the Turing 1950 question whether the mental processes can be simulated by a universal computer [26].

Among other things, the characterization of the "concept" or "category" as the limit of a recursive operation on a sequence of individual related perceptions gets rather shaky, since recursive relations imply a set structure.

Quantum limitations were also put forward by Penrose [27], but on a completely different basis. In his proposal, the quantum character was attributed to the physical behavior of the "microtubules" which are microscopic components of the neurons playing a central role in the synaptic activity. However, speaking of quantum coherence at the $\hbar$-level in biological processes is not plausible, if one accounts for the extreme vulnerability of any quantum system due to "decoherence" processes, which make quantum superposition effects observable only in extremely controlled laboratory situations, and at sub-picosecond time ranges ,not relevant for synchronization purposes in the $10 - 100\ ms$ range. Our tenet is that the quantum C-level in a living being emerges from the limited time available in order to take vital decisions; it is logically based on a noncommutative set of relevant variables and, hence, it requires the logical machinery built for the $\hbar$ quantum description of the microscopic world, where noncommutativity emerges from use of variables coming from macroscopic experience, as coordinate and momenta, to account for new facts. The recent debate [7] has clearly excluded the need for hidden variables and hence the possibility of considering the quantum formalism as a projection of a more general classical one. Similarly, in our case, the fact that decisions are taken on, say, a 100 milliseconds sequence of neuronal spikes, excludes in principle a cognitive world where longer decision times (say, 300 ms) would disentangle overlapping states. In other words, the universe of perceptual 100 ms events would be correctly described by a C-quantum formalism, since the 300 ms perceptual facts would be considered not accessible in principle.

References

1. F.T. Arecchi: Cognitive Processing **1**, 23 (2000)
2. C. von der Malsburg: 'The correlation theory of brain function'. Reprinted in: *Models of neural networks II*, ed. by E. Domani, J.L. Van Hemmen, and K. Schulten (Springer-Verlag, Berlin 1981)
3. W. Singer and C.M. Gray: C.M., 1995, Annu. Rev. Neurosci. **18**, 555 (1995)
4. E.M. Izhikevich: Int. J. of Bifurcation and Chaos **10**, 1171 (2000)
5. S. Grossberg: The American Scientist, **83**, 439 (1995)
6. G.M. Edelman and G. Tononi: 'Neural Darwinism: The brain as a selectional system'. In: *Nature's Imagination: The frontiers of scientific vision*, ed. by J. Cornwell (Oxford University Press, New York 1995), p. 78
7. R. Omnès: *The interpretation of Quantum Mechanics* (Princeton University Press, Princeton, NJ, 1994)
8. E. Allaria, F.T. Arecchi, A. Di Garbo, and R. Meucci: Phys. Rev. Lett. **86**, 791 (2001)
9. R. Meucci, A. Di Garbo, E. Allaria, and F.T. Arecchi: Phys. Rev. Lett. **88**, 144101 (2002)
10. C.S. Zhou, E. Allaria, S. Boccaletti, R. Meucci, J. Kurths, and F.T. Arecchi: Phys. Rev. E **67**, 015205 (2003)
11. F.T. Arecchi, R. Meucci, E. Allaria, A. Di Garbo, and L.S. Tsimring: Phys. Rev. E **65**, 046237 (2002)
12. I. Leyva, E. Allaria, S. Boccaletti, and F.T. Arecchi: 'Competition of synchronization patterns in arrays of homoclinic chaotic systems', nlin.PS/0302008
13. F.T. Arecchi: 'Instabilities and chaos in single mode homogeneous line lasers'. In: *Instabilities and chaos in quantum optics*, ed. by F.T. Arecchi and R.G. Harrison, Springer Series Synergetics, Vol. 34 (Springer-Verlag, Berlin 1987), p. 9
14. F.T. Arecchi, R. Meucci, and W. Gadomski: Phys. Rev. Lett. **58**, 2205 (1987)
15. D.H. Hubel: *Eye, brain and vision*, Scientific American Library, n. 22, (W.H. Freeman, New York 1995)
16. E. Rodriguez, N. George, J.P. Lachaux, J. Martinerie, B. Renault, and F. Varela: Nature **397**, 340 (1999)
17. W. Softky: Current Opinions in Neurobiology **5**, 239 (1995)
18. W.J. Freeman: Sci. Am. **264** (2), 78 (1991)
19. M. Rabinovich, A. Volkovskii, P. Lecanda, R. Huerta, H. Abarbanel, and G. Laurent: Phys. Rev. Lett. **87**, 068102 (2001)
20. J.D. Victor and K.P. Purpura: Network: Comput. Neural Syst. **8**, 127 (1997)
21. M. van Rossum: Neural Computation **13**, 751 (2001)
22. K. MacLeod, A. Backer, and G. Laurent: Nature **395**, 693 (1998)
23. F. Rieke, D. Warland, R. de Ruyter van Steveninck, and W. Bialek: *Spikes: Exploring the neural code* (MIT Press, Cambridge, MA 1996)
24. W.H. Zurek: Phys. Today (October issue), 36 (1991)
25. A.J. Leggett and A. Garg: Phys. Rev. Lett. **54**, 857 (1985)
26. A. Turing: Mind **59**, 433 (1950)
27. R. Penrose: *Shadows of the Mind* (Oxford University Press, New York 1994)

Measurement and Information Extraction in Complex Dynamics Quantum Computation

Giulio Casati[1] and Simone Montangero[2]

[1] Center for Nonlinear and Complex Systems, Università dell'Insubria and
INFM, Unità di Como, Via Valleggio 11, 22100 Como, Italy
INFN, Sezione di Milano, Via Celoria 16, 20133 Milano, Italy
[2] Scuola Normale Superiore, NEST-INFM, Piazza dei Cavalieri 7, Pisa, Italy

Quantum Information processing has several different applications: some of them can be performed controlling only few qubits simultaneously (e.g. quantum teleportation or quantum cryptography) [1]. Usually, the transmission of large amount of information is performed repeating several times the scheme implemented for few qubits. However, to exploit the advantages of quantum computation, the simultaneous control of many qubits is unavoidable [2]. This situation increases the experimental difficulties of quantum computing: maintaining quantum coherence in a large quantum system is a difficult task. Indeed a quantum computer is a many-body complex system and decoherence, due to the interaction with the external world, will eventually corrupt any quantum computation. Moreover, internal static imperfections can lead to quantum chaos in the quantum register thus destroying computer operability [3]. Indeed, as it has been shown in [4], a critical imperfection strength exists above which the quantum register thermalizes and quantum computation becomes impossible. We showed such effects on a quantum computer performing an efficient algorithm to simulate complex quantum dynamics [5,6].

In this paper, we address a different and very general problem related to the extraction of the information. Indeed, the information is encoded in the wave function, and apart from very particular situations, it is hard to find a way to address and extract the useful information. Commonly, when trying to extract information, the efficiency of quantum information processing is lost. This is one of the main problem to solve while looking for new quantum algorithms. However, in some cases, these difficulties can be bypassed, as in Shor, Grover and other well known quantum algorithms [7].

The problem of extracting information is particularly difficult in one of the most general quantum computing applications: the simulation of many-body complex quantum systems. Indeed, the results of such simulations is, typically, the quantum state: the wave function as a whole. The problems is that, in order to measure all N wave function coefficients (coded in $n_q = log_2(N)$ qubits) by means of a projective measurement, one must repeat the calculus $O(N)$ times. This destroys the efficiency of any quantum algorithm even in the case in which such algorithm can compute the wave function with an exponential gain in the number n_q of elementary gates.

G. Casati and S. Montangero, Measurement and Information Extraction in Complex Dynamics
Quantum Computation, Lect. Notes Phys. **633**, 341–348 (2004)
http://www.springerlink.com/   © Springer-Verlag Berlin Heidelberg 2004

However, as it is the case for other quantum algorithms, there are some questions that can be answered in an efficient way. Here we present an interesting example where important information can be extracted efficiently by means of quantum simulations. We show how this methods work on a dynamical model, the so-called Sawtooth Map [5]. This map is characterized by very different dynamical regimes: from near integrable to fully developed chaos; it also exhibits quantum dynamical localization [8, 9]. We show how to extract efficiently the localization length and the mean square deviation. The results obtained here for the sawtooth map can shed some light for the study of different quantum systems. Indeed, as it is well known, any classical simulation of a quantum system will pretty soon be limited by lack of computational resources. In our work we show how some questions can be answered efficiently by means of a quantum computer, thus allowing the investigation of some general properties beyond the reach of classical supercomputers.

This paper is organized as follows: in Sect. 1 we introduce our model: the Sawtooth Map. In Sect. 2 the quantum algorithm to compute the quantum motion is presented in detail. In Sect. 3 we review the exponentially efficient calculation of dynamical localization length. Finally, in Sect. 4, we discuss the additional information that can be extracted. Our conclusions are summarized in Sect. 5.

1   Sawtooth Map

The classical sawtooth map is given by

$$\bar{n} = n + k(\theta - \pi), \quad \bar{\theta} = \theta + T\bar{n}, \tag{1}$$

where (n, θ) are conjugated action-angle variables ($0 \leq \theta < 2\pi$), and the bars denote the variables after one map iteration. Introducing the rescaled variable $p = Tn$, one can see that the classical dynamics depends only on the single parameter $K = kT$. The map (1) can be studied on the cylinder ($p \in (-\infty, +\infty)$), which can also be closed to form a torus of length $2\pi L$, where L is an integer. For any $K > 0$, the motion is completely chaotic and one has normal diffusion: $< (\Delta p)^2 > \approx D(K)t$, where t is the discrete time measured in units of map iterations and the average $< \cdots >$ is performed over an ensemble of particles with initial momentum p_0 and random phases $0 \leq \theta < 2\pi$. It is possible to distinguish two different dynamical regimes [8]: for $K > 1$, the diffusion coefficient is well approximated by the random phase approximation, $D(K) \approx (\pi^2/3)K^2$, while for $0 < K < 1$ diffusion is slowed down, $D(K) \approx 3.3K^{5/2}$, due to the sticking of trajectories close to broken tori (cantori). For $-4 < K < 0$ the motion is stable, the phase space has a complex structure of elliptic islands down to smaller and smaller scales, and we observed anomalous diffusion, $< (\Delta p)^2 > \propto t^\alpha$, (for example, $\alpha = 0.57$ when $K = -0.1$). Inside each island the motion can be approximated by an harmonic oscillator, as can be easily view by the fixed points stability analysis. In Fig. 1 we show a typical phase space of the classical sawtooth map for $K = -0.3$.

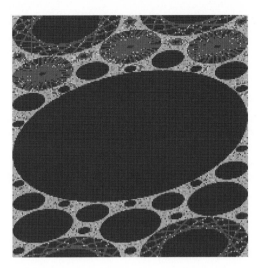

Fig. 1. The phase space of the classical Sawtooth map for $K = -0.3$. The phase space density plot is obtained form an ensemble of 10^4 trajectories with initial n_0 and random phases outside the main island (big black central area). $-\pi \leq p < \pi$ (vertical axis), $0 \leq \theta < 2\pi$ (horizontal axis). The probability density varies from zero (black) through intermediate (lightest) to maximal probability (intermediate grey shade)

The quantum evolution on one map iteration is described by a unitary operator $\hat{U}$ acting on the wave function ψ:

$$\overline{\psi} = \hat{U}\psi = e^{-iT\hat{n}^2/2}e^{ik(\hat{\theta}-\pi)^2/2}\psi, \tag{2}$$

where $\hat{n} = -i\partial/\partial\theta$ (we set $\hbar = 1$). Equation (2) is obtained by integrating over one period T the Schrödinger equation. As we set $\hbar = 1$, one has $[\theta, p] = T[\theta, n] = iT$ giving $\hbar_{eff} = T$. Thus, the classical limit corresponds to $T \to 0$, $k \to \infty$, and $K = kT = $ const. In this quantum model one can observe important physical phenomena like dynamical localization [9,10]. Indeed, similarly to other models of quantum chaos [14], the quantum interference in the sawtooth map leads to suppression of classical chaotic diffusion after a break time

$$t^\star \approx D_n \approx (\pi^2/3)k^2, \tag{3}$$

where D_n is the classical diffusion coefficient, measured in number of levels ($<(\Delta n)^2> \approx D_n t$). For $t > t^\star$ only $\Delta n \sim D_n$ levels are populated and the localization length $\ell \sim \Delta n$ for the average probability distribution is approximately equal [15]:

$$\ell \approx D_n \tag{4}$$

Thus the quantum localization can be detected if ℓ is smaller than the system size N.

In Sect. 3, we study the map (2) in the deep quantum regime of dynamical localization. For this purpose, we keep k, K constant. Thus the effective Planck constant is fixed and the number of cells L grows exponentially with the number of qubits ($L = TN/2\pi$). In this case, one studies the quantum sawtooth map on the cylinder ($n \in (-\infty, +\infty)$), which is cut-off to a finite number of cells due to the finite quantum (or classical) computer memory.

Notice that keeping K and L constant while increasing the number of qubits, allows the study of the quantum to classical transition. Moreover, in the stable case, $-4 < K < 0$, one can study the wave function evolution inside and outside the islands. In the first case, the motion follows the classical periodic motion with a given characteristic frequency ω, while outside the classical anomalous diffusion can be suppressed by quantum effects.

We stress again that, since in a quantum computer the memory capabilities grow exponentially with the number of qubits, already with less than 40 qubits one could make simulations inaccessible for today's supercomputers.

2   Quantum Algorithm

The quantum algorithm introduced in [5] simulates efficiently the quantum dynamics (2) using a register of n_q qubits. It is based on the forward/backward quantum Fourier transform [16] between the θ and n representations and has some elements of the quantum algorithm for kicked rotator [17]. Such an approach is rather convenient since the Floquet operator $\hat{U}$ is the product of two operators $\hat{U}_k = e^{ik(\hat{\theta}-\pi)^2/2}$ and $\hat{U}_T = e^{-iT\hat{n}^2/2}$: the first one is diagonal in the $\hat{\theta}$ representation, the latter in the $\hat{n}$ representation. Moreover both operators can be decomposed in a sequence of controlled phase shifts without using any temporary storing register.

The quantum algorithm for one map iteration requires the following steps:
I. The unitary operator $\hat{U}_k$ is decomposed in n_q^2 two-qubit gates

$$e^{ik(\theta-\pi)^2/2} = \prod_{i,j} e^{i2\pi^2 k(\alpha_i 2^{-i} - \frac{1}{2n_q})(\alpha_j 2^{-j} - \frac{1}{2n_q})}, \tag{5}$$

where $\theta = 2\pi \sum \alpha_i 2^{-i}$, with $\alpha_i \in \{0,1\}$. Each two-qubit gate can be written in the $\{|00\rangle, |01\rangle, |10\rangle, |11\rangle\}$ basis as $\exp(ik\pi^2 D)$, where D is a diagonal matrix with elements

$$\left\{ \frac{1}{2n_q^2}, -\frac{1}{n_q}\left(\frac{1}{2^j} - \frac{1}{2n_q}\right), -\frac{1}{n_q}\left(\frac{1}{2^i} - \frac{1}{2n_q}\right), \right.$$
$$\left. 2\left(\frac{1}{2^i} - \frac{1}{2n_q}\right)\left(\frac{1}{2^j} - \frac{1}{2n_q}\right) \right\}. \tag{6}$$

II. The change from the θ to the n representation is obtained by means of the quantum Fourier transform, which requires n_q Hadamard gates and $n_q(n_q-1)/2$ controlled-phase shift gates [16].
III. In the new representation the operator $\hat{U}_T$ has essentially the same form as $\hat{U}_k$ in step I and therefore it can be decomposed in n_q^2 gates similarly to (5).
IV. We go back to the initial θ representation via inverse quantum Fourier transform.

On the whole the algorithm requires $3n_q^2 + n_q$ gates per map iteration. Therefore it is exponentially efficient with respect to any known classical algorithm. Indeed the most efficient way to simulate the quantum dynamics (2) on a classical computer is based on forward/backward fast Fourier transform and requires $O(n_q 2^{n_q})$ operations. We stress that this quantum algorithm does not need any extra work space qubit. This is due to the fact that for the quantum sawtooth map the kick operator $\hat{U}_k$ has the same quadratic form as the free rotation operator $\hat{U}_T$.

3   Simulation of Dynamical Localization

In Fig. 2, we show that, using our quantum algorithm, exponential localization can be clearly seen already with $n_q = 6$ qubits. After the break time $t^\star$, the probability distribution over the momentum eigenbasis decays exponentially,

$$W_n = |\hat{\psi}(n)|^2 \approx \tfrac{1}{\ell} \exp\left(-\tfrac{2|n-n_0|}{\ell}\right), \tag{7}$$

with $n_0 = 0$ the initial momentum value. Here the localization length is $\ell \approx 12$, and classical diffusion is suppressed after a break time $t^\star \approx \ell$, in agreement with the estimates (3)-(4). This requires a number $N_g \approx 3n_q^2\ell \sim 10^3$ of one- or two-qubit quantum gates. The full curve of Fig. 2 shows that an exponentially localized distribution indeed appears at $t \approx t^\star$. Such distribution is frozen in time, apart from quantum fluctuations, which we partially smooth out by averaging over a few map steps. The localization can be seen by the comparison of the probability distributions taken immediately after $t^\star$ (full curve in Fig. 2) and at a much larger time $t = 300 \approx 25t^\star$ (dashed curve in the same figure).

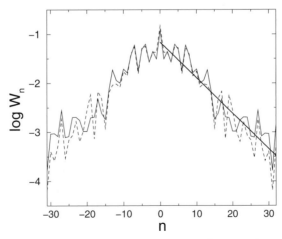

Fig. 2. Probability distribution over the momentum basis with $n_q = 6$ qubits for $k = \sqrt{3}$ and initial momentum $n_0 = 0$; the numerical data are averaged in the intervals $10 \leq t \leq 20$ (full curve) and $290 \leq t \leq 300$ (dashed curve). The straight line fit, $W_n \propto \exp(-2|n|/\ell)$, gives the localization length $\ell \approx 12$

We now discuss how it would be possible to extract information (the value of the localization length) from a quantum computer simulating the above described dynamics. The localization length can be measured by running the algorithm several times up to a time $t > t^*$. Each run is followed by a standard projective measurement on the computational (momentum) basis. The outcomes of the measurements can be stored in histogram bins of width $\delta n \propto \ell$, and then the localization length can be extracted from a fit of the exponential decay of this coarse-grained distribution over the momentum basis. In this way the localization length can be obtained with accuracy ν after the order of $1/\nu^2$ computer runs. It is important to note that it is sufficient to perform a coarse grained measurement to generate a coarse grained distribution. This means that it will be sufficient to measure the most significant qubits, and ignore those that would give a measurement accuracy below the coarse graining δn. Thus, the number of runs and measurements is independent of ℓ. However, it is necessary to make about $t^* \sim \ell$ map iterations to obtain the localized distribution (see equations (3,4)). This is true both for the present quantum algorithm and for classical computation. This implies that a classical computer needs $O(\ell^2 \log \ell)$ operations to extract the localization length, while a quantum computer would require $O(\ell(\log \ell)^2)$ elementary gates (classically one can use a basis size $N \sim \ell$ to detect localization). In this sense, for $\ell \sim N = 2^{n_q}$ the quantum computer gives a square root speed up if both classical and quantum computers perform $O(N)$ map iterations. However, for a fixed number of iterations t the quantum computation gives an exponential gain. For $\ell \ll N$ such a gain can be very important for more complex physical models, in order to check if the system is truly localized [18].

4   Efficient Measurements

We now discuss how some other important system's characteristic quantity can be extracted efficiently by means of a quantum simulation. We would like to stress that our procedure can be applied to any quantum system with a similar dynamics simulated on a quantum computer.

As for the localization length, also the mean squared moment can be efficiently extracted with a given precision. Indeed, for a fixed number of iterations t the quantum computation gives an *exponential gain* since one should compare $O(t(\log N)^2)$ gates (quantum computation) with $O(tN \log N)$ gates (classical computation). Starting from the wave function, by repeating ν projective measurements, it is possible to get the mean squared moment and therefore the diffusion coefficient $D_n \approx < (\Delta n(t))^2 > /t$. This is an important characteristic which determines the transport properties of the system.

In a similar way one can compute $< (\Delta n(t))^2 >$ in the classical limit which is obtained by keeping K, L constant and increasing N. Correspondingly, the number of qubits increases and therefore one can explore smaller and smaller scales of the phase space, that is, of the wave function. This can be fundamental in complex systems with fractal or self-similar structures in order to determine

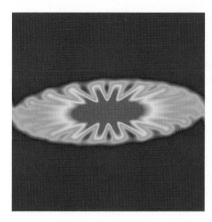

Fig. 3. Husimi function [19] of the Sawtooth Map for $K = -0.1, T = 2\pi/N, s = \Delta p \Delta \theta = 1, -\pi \leq p < \pi, 0 \leq \theta < 2\pi$. Probability varies from zero (black) through intermediate (lightest) to maximal probability (intermediate grey shade). Initial conditions: a momentum eigenstate (left), a coherent state (right). The Husimi functions are averaged over twenty map steps

the nature of the diffusion in the system, e.g. discriminating from anomalous or brownian diffusion. Notice that classically, investigating smaller scales needs exponential efforts. On the contrary, only a polynomial increasing of resources is needed when performing a quantum simulation.

Finally we would like to mention another important feature which can be efficiently extracted. Indeed if, as in our model for $-4 < K < 0$, an island is present in the system phase space, one can extract the island characteristic frequency ω. Two typical procedures are possible: the first one can be applied if the island size is comparable with the system size (at least in momentum or phase representation). This is the case of the principal island of our model (see Fig. 3, left). Indeed, starting from a momentum eigenstate inside the resonance, the wave function will expand and squeeze periodically. Thus, computing the squared moment deviation at different times one can recover ω. In the case that the island is small compared to system size, one can start with a localized wave function inside it, e.g. a coherent state. Although coherent states are more difficult to prepare with respect to momentum or phase eigenfunctions, in the case of fixed K, k, the number of operations needed to prepare such states is N-independent. Under the dynamical evolution a coherent state will periodically return back to its initial position (Fig. 3, right). As before, measuring the wave packet center of mass at different times, it is possible to estimate the frequency ω.

5   Conclusions

We discuss the problem of information extraction in quantum simulation of complex systems. We present some examples of how quantum computation can be useful to efficiently extract important information as localization length, mean square deviation and system's characteristic frequency. Notice that, in general, extracting the information embedded in the wave function is not an efficient process, however we show that in some cases quantum simulations can be exponentially efficient with respect to classical ones. We study in particular the quantum sawtooth map algorithm which can be an optimum test for such simulations and requires only a number of qubits less than ten.

Finally, we stress that the procedure discussed here is not restricted to the sawtooth map but can be applied to more general quantum systems.

This research was supported in part by the EC RTN contract HPRN-CT-2000-0156, the NSA and ARDA under ARO contracts No. DAAD19-02-1-0086, the project EDIQIP of the IST-FET programme of the EC and the PRIN-2002 "Fault tolerance, control and stability in quantum information processing".

References

1. A.Steane: Rep. Prog. Phys. **61**, 117 (1998).
2. See, e.g., M.A. Nielsen and I.L. Chuang: *Quantum Computation and Quantum Information* (Cambridge University Press, Cambridge 2000).
3. B. Georgeot and D.L. Shepelyansky: Phys. Rev. E **62**, 3504 (2000); **62**, 6366 (2000)
4. G. Benenti, G. Casati and D. L. Shepelyansky: Eur. Phys. J. D **17**, 265 (2001).
5. G. Benenti, G. Casati, S. Montangero and D.L. Shepelyansky, Phys. Rev. Lett. **87**, 227901 (2001).
6. G. Benenti, G. Casati, S. Montangero and D.L. Shepelyansky: Eur. Phys. J. D **20**, 293 (2002); Eur. Phys. J. D **22**, 285 (2003).
7. A. Ekert, P. Hayden and H. Inamori. In: *Les Houches Summer School on "Coherent Matter Waves"* (1999).
8. I. Dana, N.W. Murray and I.C. Percival: Phys. Rev. Lett. **62**, 233 (1989); Q. Chen, I. Dana, J.D. Meiss, N.W. Murray and I.C. Percival: Physica D **46**, 217 (1990).
9. G. Casati and T. Prosen: Phys. Rev. E **59**, R2516 (1999).
10. F. Borgonovi, G. Casati and B. Li: Phys. Rev. Lett. **77**, 4744 (1996); F. Borgonovi: Phys. Rev. Lett. **80**, 4653 (1998); G. Casati and T. Prosen: Phys. Rev. E **59**, R2516 (1999).
11. T. Geisel, G. Radons and J. Rubner: Phys. Rev. Lett. **57**, 2883 (1986).
12. R.S. MacKay and J.D. Meiss: Phys. Rev. A **37**, 4702 (1988).
13. R.E. Prange, R. Narevich and O. Zaitsev: Phys. Rev. E **59**, 1694 (1999).
14. B. Georgeot and D.L. Shepelyansky: Phys. Rev. Lett. **86**, 2890 (2001).
15. D.L. Shepelyansky: Physica D **28**, 103 (1987).
16. See, e.g., A. Ekert and R. Jozsa: Rev. Mod. Phys. **68**, 733 (1996).
17. B. Georgeot and D.L. Shepelyansky: Phys. Rev. Lett. **86**, 2890 (2001).
18. T. Prosen, I.I. Satija and N. Shah, Phys. Rev. Lett. **87**, 066601 (2001), and references therein.
19. The computation of Husimi functions is described in: S.-J. Chang and K.-J. Shi: Phys. Rev. A **34**, 7 (1986).

Spin Glasses:
Still Complex after All These Years?

Daniel L. Stein

Departments of Physics and Mathematics, University of Arizona, Tucson, AZ 85721, USA

Abstract. Spin glasses are magnetic systems exhibiting both quenched disorder and frustration, and have often been cited as examples of 'complex systems.' In this talk I review some of the basic notions of spin glass physics, and discuss how some of our recent progress in understanding their properties might lead to new viewpoints of how they manifest 'complexity'.

1   Introduction

This talk will probably be a change of pace for most of you; at least the topic is mostly orthogonal to those covered by the other talks. In particular, I'll be discussing a *classical* statistical mechanical problem. The origins of the interactions that define the spin glass are of course quantum mechanical; and quantum phenomena in many spin glass systems have become an active area of study over the past decade. Nevertheless, many of the important phenomena observed down to very low temperatures in a wide variety of spin glasses can be explained using classical statistical mechanics – or, more truthfully, *could* be explained if we could figure out how to treat the enormous complications arising from the quenched randomness inherent in these systems. Because of these complications, the most basic questions remain open, and the spin glass has often been touted as a model example of a complex system.

In the absence of a universally agreed definition of 'complex system', it is as difficult to argue with that claim as it is to justify it. Maybe spin glasses – as well as other systems discussed at this meeting – are merely 'complicated systems'. What I'll try to do in this talk is to convey some of the flavor of spin glass physics, and to show why its understanding requires the introduction of some new concepts and tools into statistical mechanics. You can then judge for yourself whether the classical spin glass fits your own understanding of 'complexity'.

The talk will be divided into four parts:

- What is a spin glass?
- Why are they interesting to physicists?
- What is the current level of our understanding?
- What – if anything – do they have to do with complexity?

D.L. Stein, Spin Glasses: Still Complex After All These Years?, Lect. Notes Phys. **633**, 349–361 (2004)
http://www.springerlink.com/   © Springer-Verlag Berlin Heidelberg 2004

2   Brief Review of Spin Glasses

Spin glasses are systems with localized electronic magnetic moments whose interactions are characterized by *quenched randomness*: a given pair of localized moments ('spins' for short) have a roughly equal *a priori* probability of having a ferromagnetic or an antiferromagnetic interaction. The prototype material is a dilute magnetic alloy, with a small amount of magnetic impurity randomly substituted into the lattice of a nonmagnetic metallic host; for example, CuMn or AuFe. However, insulators such as $Eu_xSr_{1-x}S$, with x roughly between .1 and .5, also display spin glass behavior. The underlying physics governing the random interactions differs for different classes of materials; for the dilute magneitc alloys, it arises from the conduction electron-mediated RKKY interactions between the localized moments.

Early experiments by Cannella and Mydosh [1] indicated that a phase transition occurred in AuFe alloys: the low-field ac magnetic susceptibility exhibited a cusp at a frequency-dependent temperature T_f. Similar behavior has since been seen in other spin glasses, and has become a signature feature of spin glass behavior. At the same time, specific heat curves show no singularities, but instead a smoothly rounded maximum at a temperature slightly above T_f (for a review, see [2]). Whether there exists a true thermodynamic phase transition to a low-temperature spin glass phase remains an open question.

Neutron magnetic scattering data and other probes of magnetic structure indicate that at low temperatures, the spins are frozen – at least on experimental timescales – in random orientations. Hence the name *spin glass*: the magnetic disorder is reminiscent of the translational disorder in the atomic arrangement of an ordinary glass.

3   Why Should We Care?

So why should we care?

• **Because It's There.** From the perspective of condensed matter physics, any new class of condensed matter systems is worth understanding. Of course, some classes of systems are more interesting than others. By more interesting, I mean that they may have great importance for technological application, and/or they give rise to powerful new ideas, and perhaps new physical or mathematical tools. Often these ideas and tools are applicable to different kinds of condensed matter systems, and perhaps to problems outside of condensed matter physics altogether. A well-known example is the broken gauge symmetry of superconductors providing a 'mass' to the photon, which was influential in the uncovering of the Higgs mechanism in particle physics.

Spin glasses are likely to belong to such a class of systems; indeed, as we'll describe below, they have already proved a fertile ground for uncovering new ideas and techniques with potentially wide applicability. But returning to the problem of spin glasses proper: unlike ordinary glasses, which must be cooled sufficiently rapidly to avoid the crystalline phase, the spin glass has no competing ordered phase. So if a thermodynamic phase transition does exist, then the low temperature phase would truly be an equilibrium condensed disordered phase – a new state of matter.

● **Statistical Mechanics of Disordered Systems.** Homogeneous systems, such as crystals, uniform ferromagnets, and superfluids, display spatial symmetries that greatly simplify their physical and mathematical analyses. The absence of such symmetries enormously complicates the understanding of disordered systems like spin glasses. This may lead to new types of broken symmetries, a breakdown of the thermodynamic limit for certain quantities, the emergence of new phenomena such as chaotic temperature dependence, the need for creation of new thermodynamic tools, and other unanticipated features to be described below. While it may not be necessary to completely revamp statistical mechanics in order to understand disordered systems, as has sometimes been suggested, it is at least necessary to carefully rethink some deeply held assumptions.

● **Applications to Other Areas.** Concepts that arose in the study of spin glasses have led to applications in areas as diverse as computer science [3–6], neural networks [7,8], prebiotic evolution [9–11], protein conformational dynamics [12], protein folding [13], and a variety of others. We will not have time to discuss these applications here, but extensive treatments can be found in [14–16].

4   Spin Glass Theory

The modern theory of spin glasses began with the work of Edwards and Anderson (EA) [17], who proposed that the essential physics of spin glasses lay not in the details of their microscopic interactions but rather in the *competition* between quenched ferromagnetic and antiferromagnetic interactions. It should therefore be sufficient to study the Hamiltonian

$$\mathcal{H}_{\mathcal{J}} = - \sum_{<x,y>} J_{xy}\sigma_x\sigma_y - h\sum_x \sigma_x \; , \tag{1}$$

where x is a site in a d-dimensional cubic lattice, $\sigma_x = \pm 1$ is the Ising spin at site x, h is an external magnetic field, and the first sum is over nearest neighbor sites only. To keep things simple, we take $h = 0$ and the spin couplings J_{xy} to be independent Gaussian random variables whose common distribution has mean zero and variance one. With these simplifications, the EA Hamiltonian (1) has global spin inversion symmetry. We denote by $\mathcal{J}$ a particular realization of the couplings, corresponding physically to a specific spin glass sample.

4.1   Frustration

The Hamiltonian (1) exhibits *frustration*: no spin configuration can simultaneously satisfy all couplings. If a closed circuit $\mathcal{C}$ in the edge lattice satisfies the

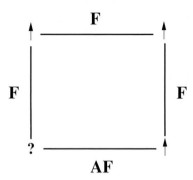

Fig. 1. A simple frustrated contour in a 2D lattice. Bonds marked "F" correspond to ferromagnetic couplings ($J_{xy} > 0$) and "AF" corresponds to an antiferromagnetic coupling ($J_{xy} < 0$). One possible arrangement of spins at the corner sites is shown

property

$$\prod_{<x,y>\in\mathcal{C}} J_{xy} < 0 \, . \tag{2}$$

then the spins along it cannot all be simultaneously satisfied [18] (Fig. 1).

Anderson [19] suggested a different formulation, namely that frustration manifests itself as free energy fluctuations scaling as the *square root* of the surface area of a typical sample. Either way, the spin glass is characterized by both *quenched disorder* and *frustration*. Their joint presence indicates the possibility that spin glasses might possess multiple pure thermodynamic states unrelated by any simple symmetry transformation. We will return to this question later.

4.2   Mean Field Theory

Within months of appearance of the EA model, an infinite-ranged version was proposed by Sherrington and Kirkpatrick (SK) [20]. For a system of N Ising spins, and in zero external field, their Hamiltonian is

$$\mathcal{H}_{J,N} = -\frac{1}{\sqrt{N}} \sum_{1\le i<j\le N} J_{ij}\sigma_i\sigma_j \tag{3}$$

where the independent, identically distributed couplings J_{ij} are again chosen from a Gaussian distribution with zero mean and variance one; the $1/\sqrt{N}$ rescaling ensures a sensible thermodynamic limit for free energy per spin and other thermodynamic quantities.

SK showed that their model had an equilibrium phase transition at $T_c = 1$. While the static susceptibility had a cusp there, so did the specific heat. This wasn't necessarily surprising given that infinite-ranged models aren't expected to correctly describe the behavior of low-dimensional systems at the critical point. More troubling was SK's observation that the low-temperature phase had an instability: in particular, the entropy became negative at very low temperature.

A mean field theory, employing the Onsager reaction field term, was proposed two years later by Thouless, Anderson, and Palmer [21]. Their approach indicated that there might be many low-temperature solutions, possibly corresponding to different spin glass 'phases'. (As a point of nomenclature, one should probably reserve use of the term 'mean field theory' for the TAP model. Nevertheless, to save space and time, I will follow general practice and use the term to refer to the infinite-ranged SK model also.) Other important early papers include the work of deAlmeida and Thouless [22], who considered the stability of the SK solution in the h-T plane, and the dynamical work of Sompolinsky and Zippelius [23–25].

We will not have time to discuss these papers here, and will focus instead on what is believed today to be the correct solution for the low-temperature phase of the SK model. This solution, due to Parisi [26], employed a novel *ansatz* and required several more years before a physical interpretation could be worked out [27–29]. The picture that finally arose was that of a system with an extraordinary new kind of symmetry breaking, known today as 'replica symmetry breaking', or RSB, after the mathematical procedures used to derive it. The essential idea is that the low-temperature phase consists not of a single spin-reversed pair of states, but rather of "infinitely many pure thermodynamic states" [27], not related by any simple symmetry transformations. In the next section, we describe the qualitative features of the Parisi solution in greater detail.

4.3   Broken Replica Symmetry

It had been pointed out by EA that a correct description of the spin glass phase needs to reflect the lack of orientational spin order with the spin 'frozenness', or long-range order in time. Denoting by Λ_L a cube of side L centered at the origin, and $\langle \cdot \rangle$ a thermal average, the magnetization per spin in a pure phase

$$M = \lim_{L \to \infty} \frac{1}{|\Lambda_L|} \sum_{x \in \Lambda_L} \langle \sigma_x \rangle \tag{4}$$

should vanish (for a.e. $\mathcal{J}$), while

$$q_{EA} = \lim_{L \to \infty} \frac{1}{|\Lambda_L|} \sum_{x \in \Lambda_L} \langle \sigma_x \rangle^2 \tag{5}$$

should not.

The quantity q_{EA} measures the breaking of time-reversal symmetry, and is now known as the 'EA order parameter', but by itself is not sufficient to describe the broken symmetry of the SK spin glass phase. The correct order parameter needs to describe the structure and relationships among the infinitely many states present at low temperature. (We ignore here the problems inherent in defining 'pure state' for the SK model; for more discussion on this, see [30–32].) To do this, we consider the *overlap* between two states α and β, at fixed $\mathcal{J}$ and T:

$$q_{\alpha\beta} \approx \frac{1}{N} \sum_{i=1}^{N} \langle \sigma_i \rangle_\alpha \langle \sigma_i \rangle_\beta , \tag{6}$$

where $\langle \cdot \rangle_\alpha$ is a thermal average in pure state α.

Given the infinity of states, quantities referring to individual pure states are of little use, even if such things could be defined. What is really of interest is the *distribution* of overlaps. Consider choosing two pure states randomly and independently from the Gibbs distribution at fixed N; this will be a mixture over many pure states α, with varying weights $W_{\mathcal{J}}^\alpha$ (dependence on N and T is suppressed for ease of notation). Let $P_{\mathcal{J}}(q)dq$ be the probability that the overlap of the two states lies between q and $q + dq$. $P_{\mathcal{J}}(q)$ is commonly referred to as the *Parisi overlap distribution*. It is equal to

$$P_{\mathcal{J}}(q) = \sum_\alpha \sum_\beta W_{\mathcal{J}}^\alpha W_{\mathcal{J}}^\beta \delta(q - q_{\alpha\beta}). \tag{7}$$

If there is a single pure state, such as the paramagnet at $T > T_c$, then $P_{\mathcal{J}}(q)$ is simply a δ-function at $q = 0$. For ferromagnets with free or periodic boundary conditions, there are only two pure states, namely the uniform positive and negative magnetization states, each appearing in any finite-volume Gibbs state with weight $1/2$. The overlap distribution function is now a pair of δ-functions, each with weight $1/2$, located at $\pm M^2(T)$.

What about in the SK model? According to the Parisi solution, for fixed $\mathcal{J}$ and (large) N, it has the form qualitatively sketched in Fig. 2. The nontrivial nature of the overlap structure reflects the presence of many states (although only a handful have weights of $O(1)$) that are not related to each other by a simple symmetry transformation.

Even more interesting is the *non-self-averaging* of the overlap distribution function. Suppose a new coupling realization $\mathcal{J}'$ is considered. Now, for any large N, the overlaps (except for the two at $\pm q_{EA}$, which are present for almost every $\mathcal{J}$) will appear at different values of q, and the set of corresponding weights will also differ. This is true no matter how large N becomes.

Let $P_N(q) = \overline{P_{\mathcal{J},N}(q)}$, where $\overline{[\cdot]}$ denotes an average over coupling realizations, and $P(q) = \lim_{N\to\infty} P_N(q)$. The resulting distribution $P(q)$ will be supported on all values of q in the interval $[-q_{EA}, q_{EA}]$; a sketch is shown in Fig. 3.

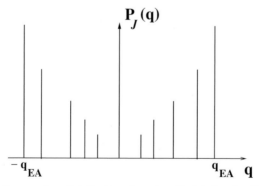

Fig. 2. Sketch of the overlap distribution function $P_{\mathcal{J}}(q)$ for the SK model below T_c

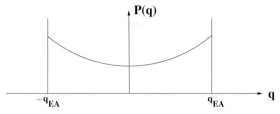

Fig. 3. Sketch of the averaged overlap distribution function $P(q)$ for the SK model below T_c. The spikes at $\pm q_{EA}$ are δ-functions

Together $P_{\mathcal{J}}(q)$ and $P(q)$ can be thought of as describing the broken symmetry of the SK model below T_c. There is another famous feature, namely the *ultrametricity* of the overlaps. Instead of considering pairs of states, one now considers triples. The Parisi solution predicts that, for almost every fixed $\mathcal{J}$, the distances $d_{\alpha\beta} = q_{EA} - q_{\alpha\beta}$ among the three pairs will form the sides of an equilateral or an acute isosceles triangle. These strong correlations among the overlaps correspond to a tree-like, or hierarchical, structure of the states.

So the infinite-ranged spin glass has a strikingly novel type of broken symmetry, much different from anything observed in homogeneous systems. Because mean-field theory has usually provided a reliable description of the *low-temperature* properties of finite-dimensional models, it was perhaps natural to expect that the RSB mean-field picture should similarly describe the nature of ordering in the EA and other short-ranged spin glass models. We now turn to these and related issues.

4.4   Some Open Questions

We saw from the preceding discussion that, already in mean field theory, spin glasses possess unusual and exotic properties. But what has been presented so far is only a piece of the story. Laboratory spin glasses display an array of puzzling thermodynamic and dynamical behavior. We confine ourselves here to a discussion of the nature of broken symmetry and ordering of the low-temperature spin glass phase – if there is one. At the same time, we emphasize that this represents only a part of the overall picture.

In this section we list some open – and fundamental – questions regarding the nature of broken symmetry in the short-ranged spin glass.

- Is there a phase transition to a low-temperature spin glass phase?

We have already seen that the answer is 'yes' for the infinite-ranged model. For the EA and other short-ranged models, the answer is not definitively known. There is some analytical [33, 34] and numerical [2, 35–37] work that supports existence of a phase transition in three – and even more likely, four – dimensional Ising spin glasses. But the issue remains unsettled [38]. It is usually assumed (though it doesn't necessarily follow) that a low-temperature phase will be non-unique and display broken spin-flip symmetry – that is, a nonzero q_{EA}.

If there is no broken-spin-flip symmetric spin glass phase in any finite dimension, then the study of spin glasses becomes one of dynamics. In what follows we will simply assume that such a phase does exist, above some lower critical dimension. Then an important question is:

- Are there infinitely many pure state pairs below T_c in the EA spin glass?

If RSB correctly describes the low-T phase, then the answer is yes. However, a competing picture [39–45] that arose in the early to mid-80's, based on domain-wall renormalization group ideas, leads to a very different picture of the low-T phase. This approach, known as droplet/scaling, leads to the conclusion that there is only a *single pair* of spin-flip-reversed pure states at low temperature in *any* finite dimension.

Which of these alternatives is the case is not known in any dimension greater than one (where of course there is only a single pair). In two dimension, it is believed that $T_c = 0$, so the issue becomes one of the number of *ground* state pairs only. Recent numerical experiments [46–48] support the possibility of only a single pair of ground states. Recent rigorous work [49, 50] also supports the notion that only a single pair of ground states occurs in two dimensions. In three dimensions numerical simulations give conflicting results [51, 52].

But let's suppose that there are infinitely many ground and/or pure states above some lower critical dimension. Even if that's the case, it does not follow that the mean-field picture of replica symmetry breaking holds in finite dimensions, because that picture requires a very intricate pattern of relationships among all of the states. So we come to our third question:

- If there do exist infinitely many equilibrium states in some finite dimensions, is their organization similar to that of the Parisi solution of the SK model?

Here the question *has* largely been answered, due to a series of rigorous and heuristic results.

4.5   Constraints on the Ordering of Short-Ranged Spin Glasses

In a series of papers [30, 53–57], Newman and Stein have shown that the complex structure of replica symmetry breaking cannot be supported by finite-dimensional spin glasses in any dimension and at any temperature. The arguments will not be presented here, but instead we will focus on alternative scenarios that remain viable.

One possibility is that $T_c = 0$ in any finite dimension, and all of the interesting features of spin glasses arise from dynamical processes. It could also happen that $T_c > 0$, but there remains a unique (non-paramagnetic) Gibbs state below T_c. While either of these possibilities may end up being the case, it is the feeling of most workers in the field that there is a low-temperature phase with broken spin-flip symmetry.

If this is so, then [53–57] leave open two main possibilities. (I emphasize here that these are not the only remaining *logical* possibilities, but rather the most plausible ones.) The first is a two-state picture like droplet/scaling, described in

Sect. 4.4. What about many-state pictures? There is one that would be consistent with the rigorous results of [53,54,57], called the *chaotic pairs* picture [54,56–58].

Consider a cube Λ_L with L large, and with periodic boundary conditions. In the chaotic pairs picture, the resulting finite-volume Gibbs state for any $0 < T < T_c$ would consist of a single pair of spin-flip-related pure states, just as in the droplet/scaling picture. But in the latter picture the *same* pair of pure states appears in every large volume, while in chaotic pairs the pure state pair appearing in Λ_L will vary chaotically with L.

So chaotic pairs resembles the droplet/scaling picture in any single finite volume, but differs over a collection of volumes and hence has a different thermodynamic structure. It is a many-state picture, but unlike the mean-field picture, it has a *trivial* overlap function. In any finite volume, $P_{\mathcal{J}}(q)$ would be a pair of δ-functions at $\pm q_{EA}$, just as in the droplet/scaling picture. If one instead constructed the overlap function of the infinite-volume pure states, then it would most likely be a single δ-function at the origin, for almost every $\mathcal{J}$. So, even though there are infinitely many states in this picture, there is no nontrivial replica symmetry breaking, no non-self-averaging, and no ultrametricity.

5 Are Spin Glasses Complex Systems?

The preceding overview enables us to return to the issue posed in the title of the talk. As I promised early on, I won't attempt to define 'complexity' or 'complex system', but will instead review some of the salient properties of spin glasses and leave it to you to decide whether they fit into your conception of a complex system. Whatever your answer, perhaps a more important characterization is whether you find them to be interesting and possibly relevant to problems that you work on. (And if the answer to all these is no, I hope that the talk at least was a pleasant diversion!)

I should begin by emphasizing that I did not cover, or even mention, many of the features of spin glasses that years ago helped to earn them the title of 'complex system'. These include some of the following: their property of displaying many metastable states, that is, states stable to flips of finite numbers of spins; their possessing a 'rugged energy landscape' (more or less equivalent to the preceding property, but sometimes also used to denote the presence of many pure or ground states); and their anomalous dynamical behaviors – slow relaxation, irreversibility, memory effects, hysteresis, and aging. I did briefly touch on their connections to problems in computer science, biology, and other areas, and of course that's important also.

But I'd like to emphasize here some of their more newly discovered properties that perhaps have not received as much attention. In what follows, I list several features – some recently uncovered, some not – in which spin glasses display unexpected behavior ('unexpected' meaning 'not familiar from our experience with the statistical mechanics of homogeneous systems'). A similar discussion appears in [31], and much of it represents joint work with Chuck Newman.

The appearance of broken replica symmetry in the infinite-ranged spin glass alone might merit the 'complexity' label, especially given its hierarchical structure in state space. A hierarchical organization of components has often been used to explain how a complex organization can be built out of simple components [59]. However, it was noted above that this type of broken symmetry is absent in short-ranged spin glass models.

But this in itself is an interesting piece of news, because mean field theory has almost always served as an invaluable guide to the low-temperature behavior of statistical mechanical systems, in particular the nature of the order parameter and broken symmetry. The failure of mean field theory to provide a correct description of the low-temperature phase in *any* finite dimension indicates that the $d \to \infty$ limit of the EA model is singular. This possibility was broached by Fisher and Huse [44], and our work confirms their conjecture.

The failure of mean-field theory to describe the broken symmetry of short-ranged models is perhaps just as interesting – and, if you're so inclined, just as good a candidate for the label 'complex' – as the exotic features of the Parisi solution. But *why* does mean field theory fail for realistic spin glasses? This is discussed at length in [31,32], and I'll refer the interested reader to those papers. For those who are moderately interested, but not enough to begin reading yet another paper on this subject, I'll just note here that one important reason lies in the combination of the physical couplings scaling to zero in the SK model, along with their statistical independence. This combination ensures that something like the following must happen: suppose that some fixed N_1 corresponds to a particular 'pure state' structure (roughly speaking). As N continues to increase, it will eventually reach an $N_2 \gg N_1$ in which the earlier pure state structure, corresponding to N_1, is completely washed out. Any fixed, finite set of couplings will eventually have no effect on the spin glass state for N large enough. In contrast, short-ranged models do not share this peculiar feature, or at least it should be considerably weaker there. For these models, the couplings outside of a particular fixed, finite region can act at most on its boundaries.

A second important feature is the possible nonexistence of a 'straightforward' thermodynamic limit for Gibbs states. By 'straightforward' I mean that a sequence of finite-volume Gibbs states, generated along an infinite sequence of volumes with boundary conditions chosen independently of the couplings, may not yield a limiting thermodynamic state. This should occur whenever there are many pure state pairs [58], although I haven't had time to discuss it here. But it's a reflection of the lack of any spatial symmetries that allow one to choose simple boundary conditions, like free or periodic, or an external symmetry-breaking field, that can lead to the existence of such a limit. (Two caveats: first, one can always choose subsequences that *do* lead to limiting thermodynamic states, but they would presumably have to be chosen in a coupling-dependent way, and at present there are no known ways of doing so. Second, this discussion is confined to Gibbs states, or equivalently correlation functions, which depend sensitively on the local details of the coupling realization. Global quantities, such as the free energy per spin, *do* have a limit for a given coupling-independent sequence of boundary conditions, for a.e. $\mathcal{J}$.)

This has an interesting consequence, related to our usual expectations for the behavior of large condensed matter systems. Statistical mechanical calculations typically rely on the assumption that the thermodynamic limit reveals the bulk properties of macroscopic systems. And in fact it almost certainly does, for disordered systems as well as homogeneous. But if, in the former case, there exist many competing pure states, then the connection between large (but finite) and infinite systems may be far more subtle than in homogeneous systems, where a straightforward extrapolation is generally sufficient. In order to connect the thermodynamic behavior of disordered systems with that of large but finite volumes, we have found it extremely useful to utilize a new thermodynamic tool which we call the *metastate* [54–57, 60].

Finally, it is possible that spin glasses display *chaotic temperature dependence* [41, 45], in which correlation functions on sufficiently large lengthscales change in an unpredictable (though deterministic) fashion under infinitesimal temperature changes.

So the study of spin glasses has provided us with a host of new phenomena and tools: analytical tools such as replica symmetry breaking for infinite-ranged systems, and metastates for both short-ranged and infinite-ranged systems; numerical tools such as simulated annealing; and others. The infinite-ranged spin glass displays a beautiful new type of order, and the short-ranged spin glass seems to break many of the old rules and assumptions. Perhaps most surprisingly, analytical – and even rigorous – progress has been achievable. As this progress continues, there is little doubt that more surprises remain in store.

Acknowledgments

I wish to thank Chuck Newman for a long and enjoyable collaboration on these questions, and for helpful comments on the manuscript. This work was supported in part by National Science Foundation Grant DMS-01-02541.

References

1. V. Cannella and J.A. Mydosh: Phys. Rev. B **6**, 4220 (1972)
2. K. Binder and A.P. Young: Rev. Mod. Phys. **58**, 801 (1986)
3. S. Kirkpatrick, C.D. Gelatt, Jr., and M.P. Vecchi: Science **220**, 671 (1983)
4. M. Mézard and G. Parisi: J. Phys. Lett. **46**, L771 (1985)
5. Y. Fu and P.W. Anderson: J. Phys. **A19**, 1605 (1986)
6. M. Mézard and G. Parisi: J. Phys. **47**, 1285 (1986)
7. J.J. Hopfield: Proc. Natl. Acad. Sci. USA **79**, 2554 (1982)
8. D.J. Amit, H. Gutfreund, and H. Sompolinsky: Phys. Rev. Lett. **55**, 1530 (1985)
9. P.W. Anderson: Proc. Natl. Acad. Sci. USA **80**, 3368 (1983)
10. P.W. Anderson and D.L. Stein: 'Broken Symmetries, Dissipative Structures, Emergent Properties, and Life'. In: *Self-Organizing Systems*, ed. by F. Yates (Plenum, NY, 1985)
11. D.S. Rokhsar, D.L. Stein, and P.W. Anderson: J. Mol. Evol. **23**, 119 (1986)
12. D.L. Stein: Proc. Natl. Acad. Sci. USA **82**, 3670 (1985)
13. J.D. Bryngelson and P.G. Wolynes: Biopolymers **30**, 177 (1990)
14. M. Mézard, G. Parisi, and M.A. Virasoro: *Spin Glass Theory and Beyond* (World Scientific, Singapore 1987)

15. D.L. Stein: *Spin Glasses and Biology* (World Scientific, Singapore 1992)
16. H. Nishimori: *Statistical Physics of Spin Glasses and Information Processing* (Oxford University Press, Oxford, 2001)
17. S.F. Edwards and P.W. Anderson: J. Phys. F **5**, 965 (1975)
18. G. Toulouse: Commun. Phys. **2**, 115 (1977)
19. P.W. Anderson: J. Less-Common Metals **62**, 291 (1978)
20. D. Sherrington and S. Kirkpatrick: Phys. Rev. Lett. **35**, 1792 (1975)
21. D.J. Thouless, P.W. Anderson, and R.G. Palmer: Phil. Mag. **35**, 593 (1977)
22. J.R.L. de Almeida and D.J. Thouless: J. Phys. A **11**, 983 (1978)
23. H. Sompolinsky and A. Zippelius: Phys. Rev. Lett. **47**, 359 (1981)
24. H. Sompolinsky, Phys. Rev. Lett. **47**, 935 (1981)
25. H. Sompolinsky and A. Zippelius, Phys. Rev. B **25**, 6860 (1982)
26. G. Parisi: Phys. Rev. Lett. **43**, 1754 (1979)
27. G. Parisi: Phys. Rev. Lett. **50**, 1946 (1983)
28. M. Mézard, G. Parisi, N. Sourlas, G. Toulouse, and M. Virasoro: Phys. Rev. Lett. **52**, 1156 (1984)
29. M. Mézard, G. Parisi, N. Sourlas, G. Toulouse, and M. Virasoro: J. Phys. (Paris) **45**, 843 (1984)
30. C.M. Newman and D.L. Stein: J. Stat. Phys. **106**, 213 (2002)
31. C.M. Newman and D.L. Stein: "Topical Review: Ordering and Broken Symmetry in Short-Ranged Spin Glasses", submitted to Journal of Physics: Condensed Matter
32. C.M. Newman and D.L. Stein: "Ultrademocracy of Ground States in the Infinite-Ranged Spin Glass", submitted to Phys. Rev. Lett.
33. M.E. Fisher and R.R.P. Singh: 'Critical Points, Large-Dimensionality Expansions, and the Ising Spin Glass'. In: *Disorder in Physical Systems*, ed. by G. Grimmett and D.J.A. Welsh (Clarendon Press, Oxford, 1990) pp. 87–111
34. M.J. Thill and H.J. Hilhorst: J. Phys. I **6**, 67 (1996)
35. A.T. Ogielski: Phys. Rev. B **32**, 7384 (1985)
36. A.T. Ogielski and I. Morgenstern: Phys. Rev. Lett. **54**, 928 (1985)
37. N. Kawashima and A.P. Young: Phys. Rev. B **53**, R484 (1996)
38. E. Marinari, G. Parisi, and F. Ritort: J. Phys. A **27**, 2687 (1994)
39. W.L. McMillan: J. Phys. C **17**, 3179 (1984)
40. A.J. Bray and M.A. Moore: Phys. Rev. B **31**, 631 (1985)
41. A.J. Bray and M.A. Moore: Phys. Rev. Lett. **58**, 57 (1987)
42. D.S. Fisher and D.A. Huse: Phys. Rev. Lett. **56**, 1601 (1986).
43. D.A. Huse and D.S. Fisher: J. Phys. A **20**, L997 (1987)
44. D.S. Fisher and D.A. Huse: J. Phys. A **20**, L1005 (1987)
45. D.S. Fisher and D.A. Huse: Phys. Rev. B **38**, 386 (1988)
46. A.A. Middleton: Phys. Rev. Lett. **83**, 1672 (1999)
47. M. Palassini and A.P. Young: Phys. Rev. B **60**, R9919 (1999)
48. A.K. Hartmann: Eur. Phys. J. B **13**, 539 (2000)
49. C.M. Newman and D.L. Stein: Phys. Rev. Lett. **84**, 3966 (2000)
50. C.M. Newman and D.L. Stein: Commun. Math. Phys. **224**, 205 (2001)
51. M. Palassini and A.P. Young: Phys. Rev. Lett. **83**, 5126 (1999)
52. E. Marinari and G. Parisi: Phys. Rev. B **62**, 11677 (2000)
53. C.M. Newman and D.L. Stein: Phys. Rev. Lett. **76**, 515 (1996)
54. C.M. Newman and D.L. Stein: Phys. Rev. Lett. **76**, 4821 (1996)
55. C.M. Newman and D.L. Stein: 'Thermodynamic Chaos and the Structure of Short-Ranged Spin Glasses'. In: *Mathematics of Spin Glasses and Neural Networks*, ed. by A. Bovier and P. Picco (Birkhäuser, Boston, 1997) pp. 243–287

56. C.M. Newman and D.L. Stein: Phys. Rev. E **55**, 5194 (1997)
57. C.M. Newman and D.L. Stein: Phys. Rev. E **57**, 1356 (1998)
58. C.M. Newman and D.L. Stein: Phys. Rev. B **46**, 973 (1992)
59. H. Simon: 'The Organization of Complex Systems'. In: *Hierarchy Theory – The Challenge of Complex Systems*, ed. by H.H. Pattee (George Braziller, NY 1973)
60. M. Aizenman and J. Wehr: Commun. Math. Phys. **130**, 489 (1990)

Information Theory and Generalized Statistics

Petr Jizba

Institute of Theoretical Physics, University of Tsukuba, Ibaraki 305-8571, Japan

Abstract. In this lecture we present a discussion of generalized statistics based on Rényi's, Fisher's and Tsallis's measures of information. The unifying conceptual framework which we employ here is provided by information theory. Important applications of generalized statistics to systems with (multi-)fractal structure are examined.

1 Introduction

One of the important approaches to statistical physics is provided by information theory erected by Claude Shannon in the late 1940s. Central tenet of this approach lies in a construction of a measure of the "amount of uncertainty" inherent in a probability distribution [1]. This measure of information (or Shannon's entropy) quantitatively equals to the number of binary (yes/no) questions which brings us from our present state of knowledge about the system in question to the one of certainty. The higher is the measure of information (more questions to be asked) the higher is the ignorance about the system and thus more information will be uncovered after an actual measurement. Usage of Shannon's entropy is particularly pertinent to the Bayesian statistical inference where one deals with the probability distribution assignment subject to prior data one possess about a system [2,3]. Here the prescription of maximal Shannon's entropy (i.e., maximal ignorance) – MaxEnt, subject to given constraints yields the least biased probability distribution which naturally protects against conclusions which are not warranted by the prior data. In classical MaxEnt the maximum-entropy distributions are always of an exponential form and hence the name generalized canonical distributions. Note that MaxEnt prescription, in a sense, resembles the principle of minimal action of classical physics as in both cases extremization of a certain functionals – the entropy or action functionals – yields physical predictions. In fact, the connection between information and the action functional was conjectured by E.T. Jaynes [3] and J.A. Wheeler [4], and most recently this line of reasonings has been formalized e.g., by B.R. Frieden in his "principle of extreme physical information – EPI" [5].

On a formal level the passage from information theory to statistical thermodynamics is remarkably simple. In this case a maximal-entropy probability distribution subject to constraints on average energy, or constant average energy and number of particles yields the usual canonical or grand-canonical distributions of Gibbs, respectively. Applicability in physics is, however, much wider.

Petr Jizba, Information Theory and Generalized Statistics, Lect. Notes Phys. **633**, 362–376 (2004)
http://www.springerlink.com/

Aside of statistical thermodynamics, MaxEnt has now become a powerful tool in non-equilibrium statistical physics [6] and is equally useful in such areas as astronomy, geophysics, biology, medical diagnosis and economics. For the latest developments in classical MaxEnt the interested reader may consult [7] and citations therein.

As successful as Shannon's information theory has been, it is clear by now that it is capable of dealing with only a limited class of systems one might hope to address in statistical physics. In fact, only recently it has become apparent that there are many situations of practical interest requiring more "exotic" statistics which does not conform with the generalized canonical prescription of the classical MaxEnt (often referred as Boltzmann–Gibbs statistics). Percolation, polymers, protein folding, critical phenomena or stock market returns provide examples. On the other hand, it cannot be denied that MaxEnt approach deals with statistical systems in a way that is methodically appealing, physically plausible and intrinsically nonspeculative (i.e., MaxEnt invokes no hypotheses beyond the sample space and the evidence that is in the available data). It might be therefore desirable to inspect the axiomatics of Shannon's information theory to find out whether some "plausible" generalization is possible. If so, such an extension could provide a new conceptual frame in which generalized measures of information (i.e., entropies) could find their theoretical justification. The additivity of independent mean information is the most natural axiom to attack. On this level three modes of reasoning can be formulated. One may either keep the additivity of independent information but utilize more general definition of means, or keep the the usual definition of linear means but generalize the additivity law or combine both these approaches together.

In the first case, the crucial observation is that the most general means compatible with Kolmogorov's axioms of probability are the so called quasi-linear means which are implemented via Kolmogorov–Nagumo functions [8]. This approach was pioneered by A. Rényi [9], J. Aczél and Z. Daróczy [10] in 60s and 70s. The corresponding measure of information is then called Rényi's entropy. Because the independent information are in this generalization still additive and because the quasi-linear means basically probe dimensionality of the sample space one may guess that this theory should play an important rôle in classical information-theoretical systems with a non-standard geometry, such as fractals, multi-fractals or systems with embedded self-similarity. These include phase transitions and critical phenomena, chaotic dynamical systems with strange attractors, fully developed turbulence, hadronic physics, cosmic strings, etc.

Second case amounts to a modification of the additivity law. Out of the infinity of possible generalizations the so called q-additivity prescription has found widespread utility. The q-calculus was introduced by F. Jackson [11] in 20's and more recently developed in the framework of Quantum Groups by V. Drinfeld [12] and M. Jimbo [13]. With the help of q-calculus one may formalize the entire approach in an unified manner by defining q-derivative (Jackson derivative), q-integration (Jackson integral), q-logarithms, q-exponentials, etc. The corresponding measure of information is called Tsallis or nonextensive entropy. The q-additivity is in a sense minimal generalization because the non-additive part is proportional to both respective information and is linearly parametrized by the only one "coupling" constant. The non-additivity prescription might be understood as a claim that despite given phenomena being statistically independent there still might be non-vanishing correlation between them and hence information might get "entangled". One may thus expect that Tsallis entropy should be important in systems with long-range correlations or long-time memories. One may even guess that quantum non-locality might become crucial playground for nonextensive statistics.

Third generalization is still not explored in the literature. It can be expected that it should become relevant in e.g., critical phenomena in a strong quantum regime. The latter can be found, for instance, in the early universe cosmological phase transitions or in currently much studied quantum phase transitions (frustrated spin systems or quantum liquids being examples).

The structure of this lecture is the following: In Sect. 2 we review the basic information-theoretic setup for Rényi's entropy. We show its relation to (multi-) fractal systems and illustrate how the Rényi parameter is related to multifractal singularity spectrum. The connection of Rényi's entropy with Fisher information and the metric structure of statistical manifolds (i.e., Fisher–Rao metric) are also discussed. In Sect. 3 the information theoretic rationale of Tsallis entropy are presented.

2 Rényi's Entropy

Rényi entropies (RE) were introduced into mathematics by A. Rényi [9] in mid 60's. The original motivation was strictly formal. Rényi wanted to find the most general class of information measures which preserved the additivity of statistically independent systems and were compatible with Kolmogorov's probability axioms.

Let us assume that one observes the outcome of two independent events with respective probabilities p and q. Additivity of information then requires that the corresponding information obeys Cauchy's functional equation

$$\mathcal{I}(pq) = \mathcal{I}(p) + \mathcal{I}(q). \tag{1}$$

Therefore, aside from a multiplicative factor, the amount of information received by learning that an event with probability p took place must be

$$\mathcal{I}(p) = -\log_2 p. \tag{2}$$

Here the normalization was chosen so that ignorant's probability (i.e. $p = 1/2$) sets the unit of information – bit. Formula (2) is known as Hartley's information measure [9]. In general, if the outcomes of some experiment are $A_1, \ldots, A_n$ with respective probabilities $p_1, \ldots, p_n$, and if A_k outcome delivers $\mathcal{I}_k$ bits of information then the mean received information reads

$$\mathcal{I} = g^{-1} \left(\sum_{k=1}^{n} p_k g(\mathcal{I}_k) \right). \tag{3}$$

Here g is an arbitrary invertible function - Kolmogorov–Nagumo function. The mean defined in (3) is the so-called quasi-linear mean and it constitutes the most general mean compatible with Kolmogorov's axiomatics [8, 14]. Rényi then proved that when the postulate of additivity for independent events is applied to (3) it dramatically restricts the class of possible g's. In fact, only two classes are possible; $g(x) = cx + d$ which implies the Shannon information measure

$$\mathcal{I}(\mathcal{P}) = -\sum_{k=1}^{n} p_k \log_2(p_k), \tag{4}$$

and $g(x) = c\, 2^{(1-q)x} + d$ which implies

$$\mathcal{I}_q(\mathcal{P}) = \frac{1}{(1-q)} \log_2 \left(\sum_{k=1}^{n} p_k^q \right), \tag{5}$$

with $q > 0$ (c and d are arbitrary constants). In both cases $\mathcal{P} = \{p_1, \ldots, p_n\}$. Note that for linear g's the quasi-linear mean turns out to be the ordinary linear mean and hence Shannon's information is the averaged information in the usual sense. Information measure defined by (5) is called Rényi's information measure (of order q) or Rényi's entropy. Term "entropy" is chosen in a close analogy with Shannon's theory because Rényi's entropy also represents the disclosed information (or removed ignorance) after performed experiment. On a deeper level it might be said that Rényi's entropy measures a diversity (or dissimilarity) within a given distribution [15]. In Sect. 2.5 we will see that in parametric statistics Fisher information plays a similar rôle. It will be shown that the latter measures a diversity between two statistical populations.

To find the most fundamental (and possibly irreducible) set of properties characterizing Rényi's information it is desirable to axiomatize it. Various axiomatizations can be proposed [9, 16]. For our purpose the most convenient set of axioms is the following [16]:

1. For a given integer n and given $\mathcal{P} = \{p_1, p_2, \ldots, p_n\}$ ($p_k \geq 0, \sum_k^n p_k = 1$), $\mathcal{I}(\mathcal{P})$ is a continuous with respect to all its arguments.
2. For a given integer n, $\mathcal{I}(p_1, p_2, \ldots, p_n)$ takes its largest value for $p_k = 1/n$ ($k = 1, 2, \ldots, n$) with the normalization $\mathcal{I}\left(\frac{1}{2}, \frac{1}{2}\right) = 1$.
3. For a given $q \in \mathrm{IR}$; $\mathcal{I}(A \cap B) = \mathcal{I}(A) + \mathcal{I}(B|A)$ with

$$\mathcal{I}(B|A) = g^{-1}\left(\sum_k \varrho_k(q) g(\mathcal{I}(B|A = A_k))\right),$$

   and $\varrho_k(q) = p_k^q / \sum_k p_k^q$ (distribution $\mathcal{P}$ corresponds to the experiment A).
4. g is invertible and positive in $[0, \infty)$.
5. $\mathcal{I}(p_1, p_2, \ldots, p_n, 0) = \mathcal{I}(p_1, p_2, \ldots, p_n)$, i.e., adding an event of probability zero (impossible event) we do not gain any new information.

Note particularly the appearance of distribution $\varrho(q)$ in axiom 3. This, so called, zooming (or escort) distribution will prove crucial is Sects- 2.4 and 3.

Further characteristics of expressions (5) were studied extensively in [9, 16]. We list here a few of the key ones.

(a) RE is symmetric: $\mathcal{I}_q(p_1, \ldots, p_n) = \mathcal{I}_q(p_{k(1)}, \ldots, p_{k(n)})$;
(b) RE is nonnegative: $\mathcal{I}_q(\mathcal{P}) \geq 0$;
(c) RE is decisive: $\mathcal{I}_q(0, 1) = \mathcal{I}_q(1, 0)$;
(d) For $q \leq 1$ Rényi's entropy is concave. For $q > 1$ Rényi's entropy in not pure convex nor pure concave ;
(e) RE is bounded, continuous and monotonous in q ;
(f) RE is analytic in $q \in \mathfrak{C}_{I \cup III} \Rightarrow$ for $q = 1$ it equals to Shannon's entropy, i.e., $\lim_{q \to 1} \mathcal{I}_q = \mathcal{I}$.

Despite its formal origin Rényi's entropy proved important in variety of practical applications. Coding theory [10], statistical inference [17], quantum mechanics [18], chaotic dynamical systems [19–22] and multifractals provide examples. The rest of Sect. 2 will be dedicated to applications in multifractal systems. For this purpose it is important to introduce the concept of renormalized information.

2.1   Continuous Probability Distributions – Renormalization

Let us assume that the outcome space (or sample space) is a continuous d-dimensional manifold. It is then heuristically clear that as we refine the measurement the information obtained tends to infinity. Yet, under certain circumstances a finite information can be extracted from the continuous measurement.

To show this we pave the outcome space[1] with boxes of the size $l = 1/n$. This divides the d-dimensional sample space into cells labelled by an index k which runs from 1 up to n^d. If $\mathcal{F}(\mathbf{x})$ is a continuous probability density function (PDF), the corresponding integrated probability distribution $\mathcal{P}_n = \{p_{nk}\}$ is generated via prescription

$$p_{nk} = \int_{\text{k-th box}} \mathcal{F}(\mathbf{x}) d^d \mathbf{x} . \tag{6}$$

Generic form of $\mathcal{I}_q(\mathcal{P}_n)$ it then represented as

$$\mathcal{I}_q(\mathcal{P}_n) = \text{divergent in } n + \text{finite} + o(1) , \tag{7}$$

where the symbol $o(1)$ means that the residual error tends to 0 for $n \to \infty$. The *finite* part ($\equiv \mathcal{I}_q(\mathcal{F})$) is fixed by the requirement (or by renormalization prescription) that it should fulfill the postulate of additivity in order to be identifiable with an information measure. Incidentally, the latter uniquely fixes the divergent part [16] as $d \log_2 n$. So we may write

$$\mathcal{I}_q(p_{nk}) \approx d \log_2 n + h + o(1) , \tag{8}$$

which implies that

$$\mathcal{I}_q(\mathcal{F}) \equiv h = \lim_{n \to \infty} (\mathcal{I}_q(\mathcal{P}_{nk}) - d \log_2 n) = \frac{1}{(1-q)} \log_2 \left(\int \mathcal{F}^q(\mathbf{x}) d^d \mathbf{x} \right). \tag{9}$$

---

[1] For simplicity's sake we consider that the outcome space has volume $V = 1$.

The latter might be generalized to piecewise-continuous $\mathcal{F}(\mathbf{x})$'s (Stiltjes integration) and to Lebesgue measurable sets [9]. Needless to say that Rényi's entropy $\mathcal{I}_q(\mathcal{F})$ exists iff. the integral on the RHS of (9) exists.

Note that (9) can be recast into a form

$$\mathcal{I}_q(\mathcal{F}) \equiv \lim_{n\to\infty} \left(\mathcal{I}_q(\mathcal{P}_n) - \mathcal{I}_q(\mathcal{E}_n)\right). \tag{10}$$

with $\mathcal{E}_n = \left\{\frac{1}{n^d}, \dots, \frac{1}{n^d}\right\}$ being the uniform distribution. Expression (10) represents nothing but Rényi's generalization of the Szilard-Brillouin negentropy.

2.2   Fractals, Multifractals, and Generalized Dimension

Aforementioned renormalization issue naturally extends beyond simple metric outcome spaces (like $\mathbb{R}^d$). Our aim in this and the next subsection is to discuss the renormalization of information in the cases when the outcome space is fractal or when the statistical system in question is multifractal. Conclusions of such a reneormalization will be applied in Sect. 2.4.

Fractals are sets with a generally non-integer dimension exhibiting property of self-similarity. The key characteristic of fractals is fractal dimension which is defined as follows: Consider a set M embedded in a d-dimensional space. Let us cover the set with a mesh of d-dimensional cubes of size l^d and let $N_l(M)$ is a number of the cubes needed for the covering. The fractal dimension of M is then defined as [23, 24]

$$D = -\lim_{l\to 0} \frac{\ln N_l(M)}{\ln l}. \tag{11}$$

In most cases of interest the fractal dimension (11) coincides with the Hausdorff–Besicovich dimension used by Mandelbrot [23].

Multifractals, on the other hand, are related to the study of a distribution of physical or other quantities on a generic support (be it or not fractal) and thus provide a move from the geometry of sets as such to geometric properties of distributions. Let a support is covered by a probability of some phenomenon. If we pave the support with a grid of spacing l and denote the integrated probability in the ith box as p_i, then the scaling exponent a_i is defined [23, 24]

$$p_i(l) \sim l^{a_i}. \tag{12}$$

The exponent a_i is called singularity or Lipshitz–Hölder exponent.

Counting boxes $N(a)$ where p_i has $a_i \in (a, a + da)$, the singularity spectrum $f(a)$ is defined as [23, 24]

$$N(a) \sim l^{-f(a)}. \tag{13}$$

Thus a multifractal is the ensemble of intertwined (uni)fractals each with its own fractal dimension $f(a_i)$. For further investigation it is convenient to define a "partition function" [23]

$$Z(q) = \sum_i p_i^q = \int da' \rho(a') l^{-f(a')} l^{qa'}. \tag{14}$$

In the small l limit the method of steepest descent yields the scaling [23]

$$Z(q) \sim l^\tau \,, \tag{15}$$

with

$$\tau(q) = \min_a(qa - f(a)), \ f'(a) = q \,. \tag{16}$$

This is precisely Legendre transform relation. So pairs $f(a), a$ and $\tau(q), q$, are conjugates with the same mathematical content.

Connection of Rényi entropies with multifractals is frequently introduced via generalized dimensions

$$D_q = \lim_{l \to 0} \left(\frac{1}{(q-1)} \frac{\log Z(q)}{\log l} \right) = - \lim_{l \to 0} \mathcal{I}_q(l) / \log_2 l \,. \tag{17}$$

These have direct applications in chaotic attractors [19–22] and they also characterize, for instance, intermittency of turbulence [17, 25] or diffusion-limited aggregates (DLA) like patterns [26]. In chaotic dynamical systems all D_q are necessary to describe uniquely e.g., strange attractors [22]. While the proof in [22] is based on a rather complicated self-similarity argumentation, by employing the information theory one can show that the assumption of a self-similarity is not really fundamental [16]. For instance, when the outcome space is discrete then all D_q with $q \in [1, \infty)$ are needed to reconstruct the underlying distribution, and when the outcome space is d-dimensional subset of IR^d then all D_q, $q \in (0, \infty)$, are required to pinpoint uniquely the underlying PDF. The latter examples are nothing but the information theoretic variants of Hausforff's moment problem of mathematical statistics.

2.3   Fractals, Multifractals, and Renormalization Issue

In a close analogy with Sect. 2.1 it can be shown [16] that for a fractal outcome space the following asymptotic expansion of Rényi's entropy holds

$$\mathcal{I}_q(p_{kn}) \approx D \log_2 n + h + o(1) \,, \tag{18}$$

where D corresponds to the Hausdorff dimension. The finite part h is, as before, chosen by the renormalization prescription - additivity of information for independent experiments. Then

$$\mathcal{I}_q(\mathcal{F}) \equiv h = \lim_{n \to \infty} (\mathcal{I}_q(\mathcal{P}_n) - D \log_2 n) = \lim_{n \to \infty} (\mathcal{I}_q(\mathcal{P}_n) - \mathcal{I}_q(\mathcal{E}_n))$$

$$= \frac{1}{(1-q)} \log_2 \left(\int_M d\mu \, \mathcal{F}^q(\mathbf{x}) \right) \,. \tag{19}$$

Measure μ in (19) is the Hausdorff measure

$$\mu(d; l) = \sum_{k\text{-th box}} l^d \xrightarrow{l \to 0} \begin{cases} 0 & \text{if } d < D \\ \infty & \text{if } d > D \end{cases} \,. \tag{20}$$

Technical issues connected with integration on fractal supports can be found, for instance, in [27, 28]. Again, renormalized entropy is defined as long as the integral on the RHS of (19) exists.

We may proceed analogously with multifractals. The corresponding asymptotic expansion now reads [16]

$$I_q(p_{nk}) \approx \frac{\tau(q)}{(1-q)} \log_2 n + h + o(1) \,. \tag{21}$$

This implies that

$$h \equiv I_q(\mu_{\mathcal{P}}) = \lim_{l \to 0} \left(I_q(\mathcal{P}_n) - \frac{\tau(q)}{(q-1)} \log_2 n \right) = \lim_{l \to 0} \left(I_q(\mathcal{P}_n) - I_q(\mathcal{E}_n) \right)$$

$$= \frac{1}{(1-q)} \log_2 \left(\int_a d\mu_{\mathcal{P}}^{(q)}(a) \right) \,. \tag{22}$$

Here the multifractal measure is defined as [24]

$$\mu_{\mathcal{P}}^{(q)}(d; l) = \sum_{k\text{-th box}} \frac{p_{nk}^q}{l^d} \xrightarrow{l \to 0} \begin{cases} 0 & \text{if } d < \tau(q) \\ \infty & \text{if } d > \tau(q) \,. \end{cases} \tag{23}$$

It should be stressed that integration on multifractals is rather delicate technical issue which is not yet well developed in the literature [28].

2.4   Canonical Formalism on Multifractals

We shall now present an important connection of Rényi's entropy with multifractal systems. The connection will be constructed in a close analogy with canonical formalism of statistical mechanics. As this approach is thoroughly discussed in [16] we will, for shortness's sake, mention only the salient points here.

Let us first consider a multifractal with a density distribution $p(x)$. If we use, as previously, the covering grid of spacing l then the coarse-grained Shannon's entropy of such a process will be

$$I(\mathcal{P}_n(l)) = -\sum p_k(l) \log_2 p_k(l) \,. \tag{24}$$

Important observation of the multifractal theory is that when $q = 1$ then

$$a(1) = \frac{d\tau(1)}{dq} = f(a(1)) = \lim_{l \to 0} \frac{\sum_k p_k(l) \log_2 p_k(l)}{\log_2 l} = -\lim_{l \to 0} \frac{I(\mathcal{P}_n(l))}{\log_2 l} \,, \tag{25}$$

describes the Hausdorff dimension of the set on which the probability is concentrated – measure theoretic support. In fact, the relative probability of the complement set approaches zero when $l \to 0$. This statement is known as Billingsley theorem [29] or curdling [23].

For the following considerations it is useful to introduce a one-parametric family of normalized measures $\varrho(q)$ (zooming or escort distributions)

$$\varrho_i(q, l) = \frac{[p_i(l)]^q}{\sum_j [p_j(l)]^q} \sim l^{f(a_i)} \,. \tag{26}$$

Because

$$df(a) \begin{cases} \leq da \text{ if } q \leq 1, \\ \geq da \text{ if } q \geq 1, \end{cases} \tag{27}$$

we obtain after integrating (27) from $a(q=1)$ to $a(q)$ that

$$f(a) \begin{cases} \leq a \text{ if } q \leq 1, \\ \geq a \text{ if } q \geq 1. \end{cases} \tag{28}$$

So for, $q > 1$, $\varrho(q)$ puts emphasis on the more singular regions of $\mathcal{P}_n$, while for $q < 1$ the accentuation is on the less singular regions. Parameter q thus provides a "zoom in" mechanism to probe various regions of a different singularity exponent.

As the distribution (26) alters the scaling of original $\mathcal{P}_n$, also the measure theoretic support changes. The fractal dimension of the new measure theoretic support $\mathcal{M}^{(q)}$ of $\varrho(q)$ is

$$d_h(\mathcal{M}^{(q)}) = \lim_{l \to 0} \frac{1}{\log_2 l} \sum_k \varrho_k(q, l) \log_2 \varrho_k(q, l). \tag{29}$$

Note that the curdling (29) mimics the situation occurring in equilibrium statistical physics. There, in the canonical formalism one works with (usually infinite) ensemble of identical systems with all possible energy configurations. But only the configurations with $E_i \approx \langle E(T) \rangle$ dominate at $n \to \infty$. Choice of temperature then prescribes the contributing energy configurations. In fact, we may define the "microcanonical" partition function as

$$Z_{mic} = \left(\sum_{a_k \in (a_i, a_i + da_i)} 1 \right) = dN(a_i). \tag{30}$$

Then the microcanonical (Boltzmann) entropy is

$$\mathcal{H}(\mathcal{E}(a_i)) = \log_2 dN(a_i) = \log_2 Z_{mic}, \tag{31}$$

and hence

$$\frac{\mathcal{H}(\mathcal{E}(a_i))}{\log_2 \varepsilon} \approx -\langle f(a) \rangle_{mic}. \tag{32}$$

Interpreting $E_i = -a_i \log_2 \varepsilon$ as "energy" we may define the "inverse temperature" $1/T = \beta / \ln 2$ (note that here $k_B = 1/\ln 2$) as

$$1/T = \left. \frac{\partial \mathcal{H}}{\partial E} \right|_{E=E_i} = -\frac{1}{\ln \varepsilon} \frac{1}{Z_{mic}} \frac{\partial Z_{mic}}{\partial a_i} = f'(a_i) = q. \tag{33}$$

On the other hand, with the "canonical" partition function

$$Z_{can} = \sum_i p_i(\varepsilon)^q = \sum_i e^{-\beta E_i}, \tag{34}$$

and $\beta = q \ln 2$ and $E_i = -\log_2(p_i(\varepsilon))$ the corresponding means read

$$a(q) \equiv \langle a \rangle_{can} = \sum_i \frac{a_i}{Z_{can}} e^{-\beta E_i} \approx \frac{\sum_i \varrho_i(q,\varepsilon) \log_2 p_i(\varepsilon)}{\log_2 \varepsilon}, \tag{35}$$

$$f(q) \equiv \langle f(a) \rangle_{can} = \sum_i \frac{f(a_i)}{Z_{can}} e^{-\beta E_i} \approx \frac{\sum_i \varrho_i(q,\varepsilon) \log_2 \varrho_i(q,\varepsilon)}{\log_2 \varepsilon}. \tag{36}$$

Let us note particularly that the fractal dimension of the measure theoretic support $d_h(\mathcal{M}^{(q)})$ is simply $f(q)$. By gathering the results together we have

micro-canonical ensemble - unifractals	canonical ensemble - multifractals
$Z_{mic}; \mathcal{H} = S_{mic} = \log_2 Z_{mic}$	$Z_{can}; S_{can} = \log_2 Z_{can} - q \langle a \rangle_{can} \log_2 \varepsilon$
$\langle a \rangle_{mic} = a_i = \sum_k a_k / Z_{mic}$	$\langle a \rangle_{can} = \sum_k a_k \, e^{-\beta E_k} / Z_{can}$
$\langle f(a) \rangle_{mic} = -S_{mic} / \log_2 \varepsilon$	$\langle f(a) \rangle_{can} = -S_{can} / \log_2 \varepsilon$
$q = \partial S_{mic} / \partial E \vert_{E=E_i}$	$q = \partial S_{can} / \partial \langle E \rangle_{can}$
$\beta = \ln 2 / T = q$	$\beta = \ln 2 / T = q$
$E_i = -\log_2 p_i = -a_i \log_2 \varepsilon$	$\langle E \rangle_{can} = -\langle a \rangle_{can} \log_2 \varepsilon$
$\langle f(a) \rangle_{mic} = q \langle a \rangle_{mic} - \tau$	$\langle f(a) \rangle_{can} = q \langle a \rangle_{can} - \tau$

Looking at fluctuations of a in the "canonical" ensemble we can establish an equivalence between unifractals and multifractals. Recalling (15) and realizing that

$$\partial^2 (\log_2 Z_{can}) / \partial q^2 = \langle E^2 \rangle_{can} - \langle E \rangle_{can}^2 \approx (\log_2 \varepsilon)^2, \tag{37}$$

$$\partial^2 (\tau \log_2 \varepsilon) / \partial q^2 = (\partial a / \partial q) \log_2 \varepsilon \approx \log_2 \varepsilon, \tag{38}$$

we obtain for the relative standard deviation of "energy"

$$\frac{\sqrt{\langle E^2 \rangle_{can} - \langle E \rangle_{can}^2}}{\log_2 \varepsilon} = \sqrt{\langle a^2 \rangle_{can} - \langle a \rangle_{can}^2} \approx \frac{1}{\sqrt{-\log_2 \varepsilon}} \to 0. \tag{39}$$

So for small ε (i.e., exact multifractal) the a-fluctuations become negligible and almost all a_i equal to $\langle a \rangle_{can}$. If q is a solution of the equation $a_i = \tau'(q)$ then in the "thermodynamic" limit ($\varepsilon \to 0$) $a_i \approx \langle a \rangle_{can}$ and the microcanonical and canonical entropies coincide. Hence

$$S_{mic} \approx -\sum_k \varrho_k(q,\varepsilon) \log_2 \varrho_k(q,\varepsilon) \equiv \mathcal{H}(\mathcal{P}_n)\vert_{f(q)}.$$

The subscript $f(q)$ emphasizes that the Shannon entropy $\mathcal{H}(\mathcal{P}_n)$ is basically the entropy of an unifractal specified by the fractal dimension $f(q)$. Legendre transform then implies that

$$\mathcal{H}(\mathcal{P}_n)|_{f(q)} \approx -qa(q)\log_2(\varepsilon) + (1-q)\mathcal{I}_q(\mathcal{P}). \tag{40}$$

Employing the renormalization prescriptions (19) and (22) we finally receive that

$$\mathcal{I}_q^r = \mathcal{H}^r|_{f(q)}. \tag{41}$$

So by changing the q parameter Rényie's entropy "skims over" all renormalized unifractal Shannon's entropies. Rényi's entropy thus provides a unified information measure which keeps track of all respective unifractal Shannon entropies.

The passage from multifractals to single-dimensional statistical systems is done by assuming that the a-interval gets infinitesimally narrow and that PDF is smooth. In such a case both a and $f(a)$ collapse to $a = f(a) \equiv D$ and $q = f'(a) = 1$. For instance, for a statistical system with a smooth measure and the support space $\mathbb{R}^d$ equation (41) constitutes a trivial identity. We believe that this is the primary reason why Shannon's entropy plays such a predominant role in physics of single-dimensional sets. Discussion of (41) can be found in [16].

2.5   Rényi's Entropy and Fisher's Information

Let us present here an interesting connection which exists between Riemaniann geometry on statistical parameter spaces and Rényi entropies.

Consider a family of PDF's characterized by a vector parameter θ

$$\mathcal{F}_\theta = \{p(x,\theta); x \in M; \theta \in \mathcal{M}, \text{a manifold in } \mathbb{R}^n\}. \tag{42}$$

We further assume that $p(x,\theta) \in \mathcal{C}^2$. The Gibbs PDF's (with θ_i being the inverse temperature β, the external field H, etc.) represent example of (42).

To construct a metric on $\mathcal{M}$ which reflects the statistical properties of the family (42) Rao and co-workers [30] proposed to adopt various measures of dissimilarity between two probability densities, and then use them to derive the metric. Important class of dissimilarity measures are measures based on information theory. Typically it is utilized the gain of information when a density $p(x,\phi)$ is replaced with a density $p(x,\theta)$. In the case of Rényi's entropy this is [9]

$$\mathcal{I}_q(\theta||\phi) = \frac{1}{q-1}\log_2 \int_M dx \, \frac{p(x,\theta)^q}{p(x,\phi)^{q-1}}. \tag{43}$$

The information metric on $\mathcal{M}$ is then defined via the leading order of dissimilarity between $p(x,\theta)$ and $p(x,(\theta+d\theta))$, namely

$$\mathcal{I}_q(\theta||\theta+d\theta) = \frac{1}{2!}\sum_{i,j} g_{ij}(\theta)\, d\theta_i d\theta_j + \dots. \tag{44}$$

Note that because $\mathcal{I}_q(\theta||\phi)$ is minimal at $\theta = \phi$, linear term in (44) vanishes. So we have

$$g_{ij}(\theta) = \left[\frac{\partial^2}{\partial\phi_i\partial\phi_j}\mathcal{I}_q(\theta||\phi)\right]_{\theta=\phi} = \frac{q}{2\ln 2}\left(\int_M dx\, p(x,\theta)\frac{\partial\ln p(x,\theta)}{\partial\theta_i}\frac{\partial\ln p(x,\theta)}{\partial\theta_j}\right)$$
$$= \frac{q}{2\ln 2}\, F_{ij}(p(x,\theta))\,. \tag{45}$$

Here F_{ij} is the Fisher information matrix (or Fisher–Rao metric) [5, 15]. Fisher matrix is the only Riemaniann metric which is invariant under transformation of variables as well as reparametrization [15]. In addition, the diagonal elements of Fisher's information matrix represent the amount of information on θ_i in an element of a sample chosen from a population of density functions $p(x,\theta)$. Due to its relation with Cramér–Rao inequality Fisher information matrix plays a crucial rôle in parametric estimation [5]. Let us stress that the latter is used in quantum mechanics to formulate information uncertainty relations [5, 18].

3   Tsallis' Entropy

Tsallis' entropy (or nonextensive entropy, or q-order entropy of Havrda and Charvát [15, 31]) has been recently much studied in connection with long-range correlated systems and with non-equilibrium phenomena. Although firstly introduced by Havrda and Charvát in the cybernetics theory context [31] it was Tsallis and co-workers [32] who exploited its nonextensive features and placed it in a physical setting. Applications of Tsallis' entropy are ranging from 3-dimensional fully developed hydrodynamic turbulence, 2-dimensional turbulence in pure electron plasma, Hamiltonian systems with long-range interactions to granular systems and systems with strange non-chaotic attractors. The explicit form of Tsallis' entropy reads

$$\mathcal{S}_q = \frac{1}{(1-q)}\left[\sum_{k=1}^n (p_k)^q - 1\right]\,, \quad q > 0\,. \tag{46}$$

This form indicates that $\mathcal{S}_q$ is a positive and concave function in $\mathcal{P}$. In the limiting case $q \to 1$ one has $\lim_{q\to 1}\mathcal{S}_q = \lim_{q\to 1}\mathcal{I}_q = \mathcal{H}$. In addition, (46) obeys a peculiar nonextensivity rule

$$\mathcal{S}_q(A\cap B) = \mathcal{S}_q(A) + \mathcal{S}_q(B|A) + (1-q)\mathcal{S}_q(A)\mathcal{S}_q(B|A)\,. \tag{47}$$

It can be proven that the following axioms uniquely specify Tsallis' entropy [33]:

1. For a given integer n and given $\mathcal{P} = \{p_1, p_2, \ldots, p_n\}$ ($p_k \geq 0, \sum_k^n p_k = 1$), $\mathcal{S}(\mathcal{P})$ is a continuous with respect to all its arguments.
2. For a given integer n, $\mathcal{S}(p_1, p_2, \ldots, p_n)$ takes its largest value for $p_k = 1/n$ ($k = 1, 2, \ldots, n$).

3. For a given $q \in \mathbb{R}$; $\mathcal{S}(A \cap B) = \mathcal{S}(A) + \mathcal{S}(B|A) + (1-q)\mathcal{S}(A)\mathcal{S}(B|A)$ with

$$\mathcal{S}(B|A) = \sum_k \varrho_k(q)\, \mathcal{S}(B|A = A_k),$$

and $\varrho_k(q) = p_k^q / \sum_k p_k^q$ (distribution $\mathcal{P}$ corresponds to the experiment A).

4. $\mathcal{S}(p_1, p_2, \ldots, p_n, 0) = \mathcal{S}(p_1, p_2, \ldots, p_n)$.

Note that these axioms bear remarkable similarity to the axioms of Rényi's entropy. Only the axiom of additivity is altered. We keep here the linear mean but generalize the additivity law. In fact, the additivity law in axiom 3 is the Jackson sum (or q-additivity) of q-calculus. The Jackson basic number $[X]_q$ of quantity X is defined as $[X]_q = (q^X - 1)/(q-1)$. This implies that for two quantities X and Y the Jackson basic number $[X+Y]_q = [X]_q + [Y]_q + (q-1)[X]_q[Y]_q$. The connection with axiom 3 is established when $q \to (2-q)$.

The former axiomatics might be viewed as the q-deformed extension of Shannon's information theory. Obviously, in the $q \to 1$ limit the Jackson sum reduces to ordinary $\mathcal{S}(A \cap B) = \mathcal{S}(A) + \mathcal{S}(B|A)$ and above axioms boil down to Shannon–Khinchin axioms of classical information theory [34].

Emergence of q-deformed structure allows to formalize many calculations. For instance, using the q-logarithm [35], i.e., $\ln_q x = (x^{1-q} - 1)/(1-q)$, Tsallis' entropy immediately equals to the q-deformed Shannon's entropy (again after $q \to (2-q)$), i.e.

$$\mathcal{S}_q(\mathcal{P}) = -\sum_{k=1}^n p_k \ln_q p_k . \tag{48}$$

The interested reader may find some further applications of q-calculus in nonextensive statistics, e.g., in [35]

Let us finally add a couple of comments. Firstly, it is possible to show [16] that Rényi's entropy prescribes in a natural way the renormalization for Tsallis' entropy in cases when the PDF is absolutely continuous. This might be achieved by analytically continuing the result for renormalized Rényi entropy from the complex neighborhood of $q = 1$ to the entire right half of the complex plane [16]. Thus, if $\mathcal{F}$ is the corresponding PDF then

$$\mathcal{S}_q(\mathcal{F}) \equiv \lim_{n \to \infty} \left(\frac{\mathcal{S}_q(\mathcal{P}_n)}{n^{D(1-q)}} - \frac{\mathcal{S}_q(\mathcal{E}_n)}{n^{D(1-q)}} \right) = \frac{1}{(1-q)} \int_M d\mu\, \mathcal{F}(\mathbf{x}) \left(\mathcal{F}^{q-1}(\mathbf{x}) - 1 \right) .$$

Extension to multifratals is more delicate as a possible non-analytic behavior of $f(a)$ and $\tau(q)$ invalidates the former argument of analytic continuation. These "phase transitions" are certainly an interesting topic for further investigation.

Secondly, note that Tsallis' and Rényi's entropies are monotonic functions of each other and thus both are maximized by the same $\mathcal{P}$. This particularly means that whenever one uses MaxEnt approach (e.g., in thermodynamics, image processing or pattern recognition) both entropies yield the same results.

4   Conclusions

In this lecture we have reviewed some information-theoretic aspects of generalized statics of Rényi and Tsallis.

Major part of the lecture – Sect. 2 – was dedicated to Rényi's entropy. We have discussed the information-theoretic foundations of Rényi's information measure and its applicability to systems with continuous PDF's. The latter include systems with both smooth (usually part of $\mathbb{R}^d$) and fractal sample spaces. Particular attention was also paid to currently much studied multifractal systems. We have shown how the Rényi parameter q is related to multifractal singularity spectrum and how Rényi's entropy provides a unified framework for all unifractal Shannon entropies.

In cases when the physical system is described by a parametric family of probability distributions one can construct a Riemannian metric over this probability space-information metric. Such a metric is then a natural measure of diversity between two populations. We have shown that when one employs Rényi's information measure then the information metric turns out to be Fisher–Rao metric (or Fisher's information matrix).

In Sect. 3 we have dealt with Tsallis entropy. Because detailed discussions of various aspects of Tsallis statistics are presented in other lectures of this series we have confined ourselves to only those characteristics of Tsallis' entropy which make it interesting from information-theory point of view. We have shown how the q-additive extension of original Shannon–Khinchin postulates of information theory gives rise to q-deformed Shannon's measure of information – Tsallis entropy.

Acknowledgements

We would like to thank the organizers of the Piombino workshop DICE2002 on "Decoherence, Information, Complexity and Entropy" for their kind invitation. It is a pleasure to thank Profs. C. Tsallis, T. Arimitsu and Dr A. Kobryn for reading a preliminary draft of this paper, and for their comments. The work was supported by the ESF network COSLAB and the JSPS fellowship.

References

1. C.E. Shannon and W. Weaver: *The Mathematical Theory of Communication* (University of Illinois Press, New York 1949)
2. E.T. Jaynes. In: *Information Theory and Statistical Mechanics*, Vol. 3, ed. by K.W. Ford, (W.A. Benjamin, Inc., New York 1963); *Papers on Probability and Statistics and Statistical Physics* (D. Reidel Publishing Company, Boston 1983)
3. E.T. Jaynes: *Probability Theory: The Logic of Science* (1993), http://omega.albany.edu:8008/JaynesBook
4. J.A. Wheeler: in *Physical Origins of Time Asymmetry*, ed. by J.J. Halliwell, J. Perez-Mercader, and W.H. Zurek (Cambridge University Press, Cambridge 1994), p. 1
5. B.R. Frieden: *Physics from Fisher Information* (Cambridge University Press, Cambridge 2000)
6. D.N. Zubarev: *Noequilibrium Statistical Thermodynamics* (Consultant Bureau, New York 1974)
7. http://astrosun.tn.cornell.edu/staff/loredo/bayes/

8. A. Kolmogorov: Atti della R. Accademia Nazionale dei Lincei **12**, 388 (1930); M. Nagumo: Japanese Jour. Math. **7**, 71 (1930)

9. A. Rényi: *Probability Theory* (North-Holland, Amsterdam 1970); *Selected Papers of Alfred Rényi*, Vol. 2 (Akadémia Kiado, Budapest 1976)

10. J. Aczel and Z. Daroczy: *On measures of information and their characterizations* (Academic Press, London 1975)

11. F.H. Jackson, Q. J. Pure Appl. Math. **41**, 193 (1910); Am. J. Math. **32**, 305 (1910)

12. V.G. Drinfeld: 'Quantum Groups': in *Proc. Intern. Congr. Math.*, ed. by A. Gleason (Berkeley, CA, 1986) p. 798

13. M. Jimbo: Lett. Math. Phys. **10**, 63 (1985)

14. G.H. Hardy, J.E. Littlewood and G. Pólya: *Inequalities* (Cambridge University Press, Cambridge 1952)

15. C.R. Rao. In: *Differential Geometry in Statistical Inference*, IMS-Lecture Notes, Vol. 10, 217 (1987)

16. P. Jizba and T. Arimitsu. In: AIP Conf. Proc. **597** (2001), p. 341; cond-mat/0108184

17. T. Arimitsu and N. Arimitsu: Physica A **295**, 177 (2001); J. Phys. A: Math. Gen. **33**, L235 (2000) [CORRIGENDUM: **34**, 673 (2001) 673]; Physica A **305**, 218 (2002); J. Phys.: Condens. Matter **14**, 2237 (2002)

18. H. Maassen and J.B.M. Uffink: Phys. Rev. Lett. **60**, 1103 (1988)

19. T.C. Halsey, M.H. Jensen, L.P. Kadanoff, I. Procaccia and B.I. Shraiman: Phys. Rev. A **33**, 1141 (1986)

20. M.H. Jensen, L.P. Kadanoff, A. Libchaber, I. Procaccia and J. Stavans: Phys. Rev. Lett. **55**, 2798 (1985)

21. K. Tomita, H. Hata, T. Horita, H. Mori, and T. Morita: Prog. Theor. Phys. **80**, 963 (1988)

22. H.G.E. Hentschel and I. Procaccia: Physica D **8**, 435 (1983)

23. B.B. Mandelbrot: *Fractal-Form, Chance and Dimension* (Freeman, London 1977)

24. J. Feder: *Fractals* (Plenum Press, New York 1988)

25. G. Paladin and A. Vulpiani: Phys. Rep. **156**, 148 (1987); G. Paladin, L. Peliti, and A. Vulpiani: J. Phys. A **18**, L991 (1986)

26. T.C. Halsey, P. Meakin and I. Procaccia: Phys. Rev. Lett. **56**, 854 (1986)

27. M.F. Barnsley: *Fractals everywhere* (Academic Press, Boston 1988)

28. G.A. Edgar: *Integral, Probability, and Fractal Measures* (Springer-Verlag, New York 1998)

29. P. Billingsley: *Ergodic Theory and Information* (Wiley, New York 1965)

30. J. Burbea and C. Rao: Probability Math. Statist. **3**, 115 (1982); J. Multivar. Anal. **12**, 575 (1982); J. Burbea: Expo. Math. **4**, 347 (1986)

31. M.E. Havrda and F.Charvát: Kybernetika **3**, 30 (1967)

32. C. Tsallis: J. Stat. Phys. **52**, 479 (1988); Braz. J. Phys. **29**, 1 (1999); E.M.F. Curado and C. Tsallis: J. Phys. A: Math. Gen. **24**, L69 (1991); C. Tsallis, R.S. Mandes and A.R. Plastino: Physica A **261**, 534 (1998); *Nonextensive Statistical Mechanics and Its Applications*, ed. by S. Abe and Y. Okamoto, (Springer Verlag, New York 2001); http://tsallis.cat.cbpf.br/biblio.htm

33. S. Abe: cond-mat/0005538

34. A.I. Khinchin: *Mathematical Foundations of Information Theory* (Dover Publications, Inc., New York 1957)

35. T. Yamano: Phys. Rev. E **63**, 046105 (2001); S. Abe and A.K. Rajagopal: Physica A **298**, 157 (2001); S. Abe: Phys. Lett. A **271**, 74 (2000)

Generalization of Boltzmann Equilibration Dynamics

Travis Sherman and Johann Rafelski

Department of Physics, University of Arizona, Tucson, Arizona 85721, USA

Abstract. We propose a novel approach in the study of transport phenomena in dense systems or systems with long range interactions where multiple particle interactions must be taken into consideration. Within Boltzmann's kinetic formalism, we study the influence of other interacting particles in terms of a random distortion of energy and momentum conservation occurring when multi-particle interactions are considered as binary collisions. Energy and momentum conservation still holds exactly but not in each model binary collision. We show how this new system differs from the Boltzmann system and we note that our approach naturally explains the emergence of Tsallis-like equilibrium statistics in physically relevant systems in terms of the long since neglected physics of interacting and dense systems.

1 Introduction

The Boltzmann equation [1] for a spatially independent system of particles describes the evolution of the one particle momentum $\boldsymbol{p}$ distribution $f(\boldsymbol{p}, t)$ of a "foreground" particle with mass m subject to interactions (here assumed to be number conserving) with N "background" particles $j = 1, \ldots, N$ with masses $m_1, m_2, \ldots, m_N$. In classical, non-relativistic regimes, the Boltzmann equation takes the form

$$\frac{\partial f(\boldsymbol{p}, t)}{\partial t} = \int \left[W(\boldsymbol{p}', \boldsymbol{p}) f(\boldsymbol{p}', t) - W(\boldsymbol{p}, \boldsymbol{p}') f(\boldsymbol{p}, t) \right] d^3 \boldsymbol{p}', \tag{1}$$

where we have suppressed a possible time dependence in the transition rate $W(\boldsymbol{p}, \boldsymbol{p}')$, the rate per unit time of the foreground particle making a momentum transition from $\boldsymbol{p}$ to $\boldsymbol{p}'$ due to interactions with the background particles.

In applications, the Boltzmann equation is restricted to sufficiently rarefied systems with short range interactions so that only two particle interactions (two body or binary collisions) are incorporated into the transition rate [2]. For these systems, the transition rate is given by [3]

$$W(\boldsymbol{p}, \boldsymbol{p}') = \sum_j \left[\int d^3 \boldsymbol{q_j} \int d^3 \boldsymbol{q_j}' \, \delta^3(\boldsymbol{p} + \boldsymbol{q_j} - \boldsymbol{p}' - \boldsymbol{q_j}') \right.$$

$$\left. \times \delta(\mathrm{E}_p + \mathrm{E}_{q_j} - \mathrm{E}_{p'} - \mathrm{E}_{q_j'})(\sigma v)_j b_j(\boldsymbol{q_j}, t) \right], \tag{2}$$

where the energies are given by $\mathrm{E}_p = |\boldsymbol{p}|^2/2m$ and $\mathrm{E}_{q_j} = |\boldsymbol{q_j}|^2/2m_j$, $b_j(\boldsymbol{q_j}, t)$ is the momentum $\boldsymbol{q_j}$ distribution for background particle j, and $(\sigma v)_j$ is the

T. Sherman and J. Rafelski, Generalization of Boltzmann Equilibration Dynamics, Lect. Notes Phys. **633**, 377–384 (2004)
http://www.springerlink.com/

differential cross section for the scattering of the foreground particle with a background particle j.

No satisfactory generalization of the transition rates has been given to describe transport phenomena in dense systems or systems with long range interactions, that is, in systems where multiple particle interactions must be taken into consideration [4]. The natural generalization of the transition rates is to assume that the transition rate is proportional to the sum over collections of background particles of

1) the expectation of a collision between the foreground particle and a collection of background particles and

2) the expectation of the conservation of energy and momentum constraint.

However, there is no clear way of systematically generalizing the Boltzmann equation by adding successive multi-particle-interaction correction terms (which are generally divergent). Attempts to generalize the Boltzmann equation for even "moderately" dense systems have largely been abandoned.

2   Energy-Momentum Conservation in Many-Body Reactions

In dense systems or systems with long range interactions, for the transition $p \to p'$ of the foreground particle to be physically possible in an interaction, we need the initial and final momenta and energies to satisfy

$$p + q_1 + \ldots + q_N \quad = p' + q'_1 + \ldots + q'_N \tag{3}$$
$$E_p + E_{q_1} + \ldots + E_{q_N} = E_{p'} + E_{q'_1} + \ldots + E_{q'_N} \tag{4}$$

by the conservation of momentum and energy. Singling out the dominant role that a single background particle j plays in making the transition physically possible, we write the above constraints as

$$p + q_j \quad = p' + {q_j}' + \epsilon \tag{5}$$
$$E_p + E_{q_j} = E_{p'}(2\gamma - 1) + E_{q'_j}. \tag{6}$$

The latter form of the constraints appears arbitrary. However, when combined with (3) and (4), it serves as a defining relation for the variables ϵ and γ. The above form is chosen here as such for mathematical convenience.

Thus, in dense systems or systems with long range interactions, for the transition to be "due" to an interaction with particle j, we need a foreground-background binary interaction as before, but with modified energy and momentum constraints. The transition rates for dense systems or systems with long range interactions are then given by simply replacing the δ-functions appearing in (2) according to:

$$\delta^3(p + q_j - p' - {q_j}') \quad \to \langle\langle \delta^3(p + q_j - p' - {q_j}' - \epsilon) \rangle\rangle_\epsilon \tag{7}$$
$$\delta(E_p + E_{q_j} - E_{p'} - E_{q'_j}) \to \langle\langle \delta[E_p + E_{q_j} - E_{p'}(2\gamma - 1) - E_{q'_j}] \rangle\rangle_\gamma. \tag{8}$$

So far, we have pushed all of the complexity of computing multiple scattering transition rates into computing the distributions of the random vector ϵ and the random variable γ. However, the form of the transition rate is now familiar and many aspects of the distributions are immediately clear from their definitions. As the density of a system and/or the range of interactions decrease,

$$\langle\langle\gamma\rangle\rangle_\gamma \to 1, \qquad \langle\langle\epsilon\rangle\rangle_\epsilon \to \mathbf{0}$$

and

$$\mathrm{Var}(\gamma) \to 0, \qquad \mathrm{Var}(\epsilon) \to \mathbf{0}$$

for only two particle collisions are increasingly present. Clearly, our transition rates simplify to the usual binary transition rates for rarefied systems with short range interactions and thus, incorporate the successes of the Boltzmann equation in describing transport phenomena in these systems.

However, as the density of a system and/or as the range of interactions increase, $\mathrm{Var}(\gamma)$ and $\mathrm{Var}(\epsilon)$ increase. For such physically relevant systems, the success of our proposed generalization to the Boltzmann equation will be made evident by indicating how our generalization determines generalized equilibration dynamics which differ from canonical equilibration dynamics and how our generalization very naturally reproduces experimentally measured equilibrium transport phenomena.

In an interaction involving a foreground particle and a background particle j we can interpret the presence of other interacting background particles in dense systems or systems with long range interactions as providing a mechanism for carrying off or providing the additional momentum and/or energy necessary to make transitions physically possible. We can also interpret the above transition rate as the usual transition rate in a system in which only binary collisions occur which do not necessarily conserve kinetic energy and momentum, as is the case for inelastic collisions.

Not only do these two interpretations aid us in understanding the nature of the γ and ϵ distributions, but we see that by extending our original definitions of ϵ and γ, the form of our transition rates is applicable to a very large class of systems: with elastic and/or inelastic interactions (here defined generally to incorporate all interactions which do not conserve energy and momentum), with low or high number densities, with short and/or long range interactions, and with any mechanisms which can carry off or provide additional momentum and energy to particles.

Indeed, for general systems, the calculation of the ϵ and γ distributions has become even more complicated. Before proceeding, it is helpful to note that the expectation of the parameters of our generalization appearing in the delta functions of Boltzmann's formalism has the effect of widening or broadening the delta functions. Thus, the existence of our parameters with some fluctuation is guaranteed in that no exact delta function exists in nature. As such, the importance of our proposed generalization to physically relevant systems depends on the extent to which systems exhibit fluctuations in the parameters γ and ϵ.

3   Fluctuation Distribution

It is natural to expect that systems with large number densities, long range interactions, inelastic interactions, and other such mechanisms (such as quantum mechanical interactions) exhibit sufficient fluctuations to, and indeed as they are known to, require a generalized formalism. Thus, to verify the necessity of generalizing Boltzmann's formalism to a system requires extensive theoretical calculations and/or precise experimental measurements of the γ and ϵ parameters of that system. One such theoretical calculation has already been given by H. Haug and C. Ell [5], whose derivation of a semiclassical Boltzmann equation for Coulomb quantum kinetics in a dense electron gas suggests, in the limit of completed collisions, an asymptotic and approximate distribution of gamma given by a peaked distribution about $\gamma = 1$:

$$f_\gamma(\gamma) = \frac{2\left(\Gamma/2E_p'\right)}{(\gamma - 1)^2 + \left(\Gamma/2E_p'\right)^2},\tag{9}$$

where Γ is the sum of the collision damping coefficients.

In the absence of other such calculations or measurements of the distributions of γ and ϵ in various physical systems, we may still proceed in the verification of our theory since the precise form of our proposed generalization can be verified by comparing the consequences of our proposal with well established and currently unexplained phenomena. For instance, we can study transport phenomena within our formalism which result from certain classes of ϵ and γ distributions which we expect to be present in many physically relevant systems.

In many systems, we expect ϵ and γ to be peaked distributions about $\epsilon = 0$ and $\gamma = 1$ and approximately independent of the incoming and outgoing momenta and energies (so that we interpret the influence of other interacting particles, unaccounted interaction processes and mechanisms as producing a random noise distortion of the conservation of energy and momentum constraints). In particular, we might expect ϵ to have a spherically symmetric distribution about $\epsilon = 0$. Moreover, since the family of Gamma distributions has a rich variety of shapes and can approximate many classes of distributions, in many systems we expect the peaked distribution of γ to be well approximated by a Gamma distribution with parameters α and λ, chosen so that the distribution is peaked about $\gamma = 1$:

$$f_\gamma(\gamma) = \lambda(\lambda\gamma)^{\alpha-1}e^{-\lambda\gamma}/\Gamma(\alpha),\tag{10}$$

where the average and variance of γ are

$$\text{Avg}(\gamma) \equiv \langle\langle\gamma\rangle\rangle_\gamma = \frac{\alpha}{\lambda}, \qquad \text{Var}(\gamma) \equiv \langle\langle\gamma^2\rangle\rangle_\gamma - \langle\langle\gamma\rangle\rangle_\gamma^2 = \frac{\alpha}{\lambda^2}.$$

The form of the distribution of γ as a Gamma distribution is chosen here for mathematical convenience and to analytically recover the precise form of Tsallis' proposed equilibrium distribution in the following example, but we emphasize that any peaked distribution of γ about $\gamma = 1$ will yield a Tsallis-like distribution, i.e., a deviation from the Boltzmann equilibrium distribution. Numerical calculations with the distribution of gamma suggested by the work of Haug and Ell resulted in Tsallis-like equilibrium distributions for a large range of Γ and the Boltzmann equation was recovered as the width $\Gamma \to 0$.

4   Example: Dense Classical Nonrelativistic System

To analytically illustrate some immediate results of our proposed generalization, we consider a system of N classical background particles denoted by $j = 1, ..., N$ with mass m_j. Suppose further that the interactions are completely elastic and that the number density of the system and the range of the interactions are sufficiently small to guarantee that the momentum distribution of the background particle $j = 1, ..., N$ approaches the Boltzmann equilibrium distribution with a common temperature T (measured in units of energy):

$$b_j^{eq}(q_j) = C_j \exp\left(-q_j^2/2m_j T\right)$$

where C_j is a normalization constant. After the N particles are sufficiently equilibrated, we inject a foreground particle with mass $m \gg m_j$ for all background particles $j = 1, ..., N$ and an initial distribution $f(\boldsymbol{p}, t = 0)$. Suppose that multiparticle collisions between the foreground and background particles occur (for instance, as the diameter of the foreground particle is much larger than the diameter of the background particles or if the range of interaction is larger for a foreground-background interaction than for a background-background interaction) and/or that the interactions are inelastic. Thus, the distributions of γ and ϵ exhibit variation. Further, suppose $(\sigma v)_j$ is constant as in the case of hard sphere interactions. We also suppose that the background particles are sufficient in extent (i.e., the background is an ideal heat bath) so that we may assume that the background approximately remains in equilibrium during the evolution of the foreground particle. Therefore, the time dependent distribution $f(\boldsymbol{p}, t)$ satisfies (1), where the transition rate is

$$W(\boldsymbol{p}, \boldsymbol{p}') = \sum_j \int d^3 q_j \int d^3 q_j{}' \langle\langle \delta^3(\boldsymbol{p} + \boldsymbol{q_j} - \boldsymbol{p}' - \boldsymbol{q}_j' - \boldsymbol{\epsilon}) \rangle\rangle_\epsilon$$

$$\times \langle\langle \delta[E_p + E_{q_j} - E_{p'}(2\gamma - 1) - E_{q_j'}] \rangle\rangle_\gamma (\sigma v)_j \, b_j^{eq}(q_j). \quad (11)$$

Interchanging the order of integration and using the δ^3-function to integrate out $\boldsymbol{q_j}'$ (so that $\boldsymbol{q_j}' = \boldsymbol{p} + \boldsymbol{q} - \boldsymbol{p}' - \boldsymbol{\epsilon}$), we obtain

$$W(\boldsymbol{p}, \boldsymbol{p}') = \left\langle\!\!\left\langle \sum_j \int q_j^2 \, dq_j \, d(\cos\theta_j) \, d\phi_j \right.\right.$$

$$\delta\left(\frac{p^2 + p'^2}{2m} - \gamma\frac{p'^2}{m} - \frac{|\boldsymbol{p} - \boldsymbol{p}' - \boldsymbol{\epsilon}|^2}{2m_j} - \frac{|\boldsymbol{p} - \boldsymbol{p}' - \boldsymbol{\epsilon}| q_j \cos\theta_j}{m_j}\right)$$

$$\left.\left. \times (\sigma v)_j \, C_j \, e^{-\frac{q_j^2}{2m_j T}} \right\rangle\!\!\right\rangle_{\epsilon, \gamma} \quad (12)$$

where we have introduced spherical coordinates (q_j, θ_j, ϕ_j) for the integration over $\boldsymbol{q}_j$. Note that we have chosen θ_j to measure the angle between $\boldsymbol{q}$ and $\boldsymbol{p} - \boldsymbol{p}' - \boldsymbol{\epsilon}$ so that the integration over ϕ_j is trivial and yields a factor of 2π. We can now integrate over $\cos\theta_j$ using the remaining delta function. A required transformation introduces a factor $(|\boldsymbol{p} - \boldsymbol{p}' - \boldsymbol{\epsilon}| \, q_j)^{-1} m_j$ and the single zero of the delta function uniquely determines $\cos\theta_j$:

$$\cos\theta_j = \left(\frac{p^2 + p'^2}{2m} - \gamma \frac{p'^2}{m} - \frac{|\boldsymbol{p} - \boldsymbol{p}' - \boldsymbol{\epsilon}|^2}{m_j} \right) \frac{m_j}{|\boldsymbol{p} - \boldsymbol{p}' - \boldsymbol{\epsilon}| \, q_j}. \tag{13}$$

The result is

$$W(\boldsymbol{p}, \boldsymbol{p}') = \left\langle\!\!\left\langle \sum_j \int dq_j \, \frac{2\pi q_j m_j (\sigma v)_j C_j}{|\boldsymbol{p} - \boldsymbol{p}' - \boldsymbol{\epsilon}|} e^{-\frac{q_j^2}{2m_j T}} \right\rangle\!\!\right\rangle_{\epsilon, \gamma} \tag{14}$$

where the integration over q_j is taken over all positive q_j which satisfy $-1 \leq \cos\theta_j \leq 1$. Solving this constraint implies that

$$q_j^2 \geq \frac{|\boldsymbol{p} - \boldsymbol{p}' - \boldsymbol{\epsilon}|^2}{4} + \left(p'^2 \gamma - \frac{p^2 + p'^2}{2} \right) \frac{m_j}{m} \tag{15}$$

to first order in m_j/m. The remaining integral over q_j is trivial and one obtains

$$W(\boldsymbol{p}, \boldsymbol{p}') = \sum_j 2\pi m_j^2 T \, (\sigma v)_j \, C_j \, e^{-\frac{E_p + E_{p'}}{2T}}$$

$$\times \left\langle\!\!\left\langle \frac{e^{-\frac{|\boldsymbol{p} - \boldsymbol{p}' - \boldsymbol{\epsilon}|^2}{8 m_j T}}}{|\boldsymbol{p} - \boldsymbol{p}' - \boldsymbol{\epsilon}|} \right\rangle\!\!\right\rangle_{\epsilon} \left\langle\!\!\left\langle e^{-\frac{\gamma E_{p'}}{T}} \right\rangle\!\!\right\rangle_{\gamma}. \tag{16}$$

The final form of transition rate indicates that the distribution of $\boldsymbol{\epsilon}$ determines the rate of equilibration, while the distribution of γ determines the shape of the resulting equilibrium. Now, if $\boldsymbol{\epsilon}$ is symmetrically distributed about $\boldsymbol{\epsilon} = \boldsymbol{0}$ (as we suggested above), we have that

$$\left\langle\!\!\left\langle \frac{e^{-\frac{|\boldsymbol{p} - \boldsymbol{p}' - \boldsymbol{\epsilon}|^2}{8 m_j T}}}{|\boldsymbol{p} - \boldsymbol{p}' - \boldsymbol{\epsilon}|} \right\rangle\!\!\right\rangle_{\epsilon} = \left\langle\!\!\left\langle \frac{e^{-\frac{|\boldsymbol{p} - \boldsymbol{p}' + \boldsymbol{\epsilon}|^2}{8 m_j T}}}{|\boldsymbol{p} - \boldsymbol{p}' + \boldsymbol{\epsilon}|} \right\rangle\!\!\right\rangle_{\epsilon}. \tag{17}$$

Multiplying (16) by $\langle\!\langle \exp(-\gamma E_p/T) \rangle\!\rangle_\gamma$, interchanging $\boldsymbol{p} \leftrightarrow \boldsymbol{p}'$ and making use of (17), we see that

$$W(\boldsymbol{p}', \boldsymbol{p}) \left\langle\!\!\left\langle e^{-\frac{\gamma E_{p'}}{T}} \right\rangle\!\!\right\rangle_{\gamma} = W(\boldsymbol{p}, \boldsymbol{p}') \left\langle\!\!\left\langle e^{-\frac{\gamma E_p}{T}} \right\rangle\!\!\right\rangle_{\gamma}. \tag{18}$$

Note that this is a detailed balance equation [3], implying

$$f^{\text{eq}}(\boldsymbol{p}) = C \left\langle\!\!\left\langle e^{-\frac{\gamma E_p}{T}} \right\rangle\!\!\right\rangle_{\gamma} \tag{19}$$

is a time-independent solution of (1), where C is a normalization constant. This is easily verified by substituting $f^{\mathrm{eq}}(\boldsymbol{p})$ in (19) for $f(\boldsymbol{p}, t)$ in (1) and by making use of (18). If the distribution of γ is well approximated by a Gamma distribution with parameters α and λ (as suggested above), we find that

$$f^{\mathrm{eq}}(\boldsymbol{p}) = C \left(1 + \frac{E_p}{T}\lambda\right)^{-\alpha}. \tag{20}$$

Thus, interpreting the Tsallis [6] nonextensivity parameter q_{T} as

$$q_{\mathrm{T}} = 1 + \frac{1}{\alpha} = \frac{\langle\langle\gamma^2\rangle\rangle_\gamma}{\langle\langle\gamma\rangle\rangle_\gamma^2} \tag{21}$$

and the inverse temperature of the foreground particle $\beta_{\mathrm{F}} = 1/T_{\mathrm{F}}$ as $\beta_{\mathrm{F}} = \frac{\alpha}{\lambda}\beta = \langle\langle\gamma\rangle\rangle_\gamma\beta$, we have that the equilibrium distribution of the foreground particle is a Tsallis equilibrium distribution with nonextensivity parameter q_{T}, inverse foreground temperature $\beta_F = 1/T_{\mathrm{F}}$, and normalization constant C:

$$f^{\mathrm{eq}}(\boldsymbol{p}) = C \left[1 - \beta_{\mathrm{F}}(1 - q_{\mathrm{T}})E_p\right]^{\frac{1}{1 - q_{\mathrm{T}}}}. \tag{22}$$

The Tsallis equilibrium distribution is often successfully used to model equilibrium distributions exhibiting power-tail behavior. Furthermore, Tsallis-like distributions are extensively measured in many physical systems, especially in systems with long range interactions and in turbulent flows [7].

We note that our approach differs from [8] in that our "natural" generalization does not up-front introduce exact Tsallis equilibration and, moreover, we retain the intuitive statistical form of Boltzmann's molecular chaos hypothesis, transport equation, and transition rates. Our proposal compliments the approach taken by C. Beck [9], which generates Tsallis equilibrium statistics by considering fluctuations in the parameters of the Langevin equation. We incorporate the approach taken by G. Wilk and Z. Wlodarczyk [10], which generates Tsallis equilibrium statistics from fluctuations in background temperatures in the Boltzmann exponential factor, by incorporating these fluctuations into our distribution of γ.

5   Conclusions

Concluding, we have proposed a natural generalization of Boltzmann equilibration dynamics to model transport phenomena in general, nonequilibrium systems. Our generalization resulted from our interpretation of the influence of other interacting particles as a random distortion of energy and momentum conservation occurring when multiple interactions are considered as binary collisions. As it turns out, this also explains the appearance of Tsallis distribution in Fokker-Planck dynamics [11], where local energy conservation was not maintained.

In outlook, we noted that our formalism is applicable to many different and large classes of systems: with elastic and/or inelastic interactions, with low or high number densities, with short- and/or long-range interactions, and with any mechanisms which can carry off or provide additional momentum and energy to particles. We also indicated how the result of our proposed generalization can be determined in systems by studying classes of reasonable distributions for the proposed parameters of our generalization. Although we have only considered a simple example where we could easily and explicitly determine some immediate results, our approach is easily extended to more general systems. We already have similar results for a relativistic, quantum mechanical foreground particle scattering in a thermal background of light relativistic particles, and it is clear that the general situation is similar.

Acknowledgments

Work supported in part by a grant from the U.S. Department of Energy, DE-FG03-95ER40937. Travis Sherman was in part supported by a UA/NASA Space Grant Undergraduate Research Internship Program.

References

1. L. Boltzmann: *Lectures on Gas Theory*, English Translation (University of California Press, Berkeley 1964)
2. G.E. Uhlenbeck: *Acta Physica Austriaca*, Suppl. X, 107 (1973)
3. N.G. van Kampen: *Stochastic Processes in Physics and Chemistry* (North-Holland, Amsterdam 1981)
4. E.G.D. Cohen: *Acta Physica Austriaca*, Suppl. X, 157 (1973)
5. H. Haug and C. Ell: Phys. Rev. B **46**, 2126 (1992)
6. C. Tsallis: J. Stat. Phys. **52**, 479 (1988)
7. C. Beck, G.S. Lewis, and H.L. Swinney: Phys. Rev. E **63**, 035303 (2001)
8. J.A.S. Lima, R. Silva and A.R. Plastino: Phys. Rev. Lett. **86**, 2938 (2001)
9. C. Beck: Phys. Rev. Lett. **87**, 180601 (2001)
10. G. Wilk and Z. Wlodarczyk: Phys. Rev. Lett. **84**, 2770 (2000)
11. D.B. Walton and J. Rafelski: Phys. Rev. Lett. **84**, 31 (2000)

On the Emergence of Nonextensivity
at the Edge of Quantum Chaos

Yaakov S. Weinstein[1], Constantino Tsallis[2], and Seth Lloyd[3]

[1] Massachusetts Institute of Technology, Department of Nuclear Engineering
Cambridge MA 02319, USA
[2] Centro Brasileiro de Pesquisas Fisicas, Xavier Sigaud 150, 22290-180,
Rio de Janeiro, RJ, Brazil
[3] d'Arbeloff Laboratory for Information Systems and Technology,
Massachusetts Institute of Technology, Department of Mechanical Engineering,
Cambridge, MA 02139, USA

Abstract. We explore the border between regular and chaotic quantum dynamics, characterized by a power law decrease in the overlap between a state evolved under chaotic dynamics and the same state evolved under a slightly perturbed dynamics. This region corresponds to the edge of chaos for the classical map from which the quantum chaotic dynamics is derived and can be characterized via nonextensive entropy concepts.

1   Introduction

Classical chaotic dynamics is identified by extreme sensitivity to initial conditions. Under chaotic dynamics, two arbitrarily close points in phase space diverge at an exponential rate, quantified by the Lyapunov exponent [1]. Non-chaotic dynamics does not show this extreme dependence on initial conditions, hence, the Lyapunov exponent is equal to zero. However, it has been conjectured that, though the Lyapunov exponent may vanish, a system may have a positive generalized Lyapunov coefficient [2] describing power-law, rather than exponential, divergence of classical trajectories. This is the case, at the border between chaotic and non-chaotic dynamics (the 'edge of chaos') where the Lyapunov exponent just goes to zero, but there remains a positive generalized Lyapunov coefficient.

This paper identifies a characteristic signature for the edge of quantum chaos. Quantum states maintain a constant overlap fidelity (heretofore referred to as the overlap or fidelity) , or distance, under all quantum dynamics, regular and chaotic. One way to characterize quantum chaos is to compare the evolution of an initially chosen state under the chaotic dynamics with the same state evolved under a perturbed dynamics [3–5]. When the initial state is in a regular region of a mixed phase space system, a system whose phase space has regular and chaotic regions, the overlap remains close to one. When the initial state is in a chaotic region, the overlap decay is exponential. This paper explores the edge of quantum chaos, a region of polynomial overlap decay [6].

Y.S. Weinstein, C. Tsallis, and S. Lloyd, On the Emergence of Nonextensivity at the Edge of Quantum Chaos, Lect. Notes Phys. **633**, 385–397 (2004)
http://www.springerlink.com/   © Springer-Verlag Berlin Heidelberg 2004

This paper is structured as follows, we first give a short review of the Lyapunov description of chaotic dynamics and the lack of correspondence between this description and quantum dynamics. We then discuss suggested characteristics of quantum chaos (known as signatures of quantum chaos), concentrating on the overlap decay first introduced by Peres [3,4] . Overlap decay proves to be a useful signature of quantum chaos from which to explore the border between regular and chaotic quantum dynamics. Next, we will review the nonextensive entropy form introduced in [7] and discuss its relevance to edge of chaos phenomenon. Finally, we locate the 'edge of quantum chaos' in a mixed phase space system and, using the nonextensive entropy formalism, show how the overlap decay at the edge of quantum chaos depends on perturbation strength and Hilbert space dimension.

2   Classical Chaos

The Lyapunov exponent description of classical chaos is as follows [1]. Let Δx be the distance between two points on phase space. We define $\xi = lim_{\Delta x(0) \to 0}(\frac{\Delta x(t)}{\Delta x(0)})$, to describe how far apart two initially arbitrarily close points on phase space become at some time t. Generally, $\xi(t)$ is the solution to the differential equation

$$\frac{d\xi(t)}{dt} = \lambda_1 \xi(t), \tag{1}$$

giving the solution

$$\xi(t) = e^{\lambda_1 t} \tag{2}$$

where λ_1 is the Lyapunov exponent (the use of the subscript will become clear later on). As seen from the above equation when the Lyapunov exponent is positive two arbitrarily close points on phase space diverge at an exponential rate. Thus, the dynamics described by $\xi(t)$ is strongly sensitive to initial conditions and we have chaotic dynamics.

3   Quantum Chaos

While the Lyapunov exponent description of chaos works well for points on a classical phase space it cannot hold true for quantum mechanical states or wavefunctions. A measure of distance between quantum wavefunctions is the overlap

$$O_i = \langle \Psi | \Phi \rangle. \tag{3}$$

However, the overlap between two quantum wavefunctions remains unchanged under unitary evolution governed by the linear Schrödinger equation. This is seen from

$$O(n) = \langle \Psi (U^n)^\dagger U^n | \Phi \rangle = O_i, \tag{4}$$

where U is the unitary system evolution. Hence, the distance between two arbitrarily close quantum mechanical wavefunctions, like the distance between two

Liouville probability densities, does not diverge and cannot be described by the Lyapunov exponent picture. This seeming lack of correspondence has led to the study of 'quantum chaos,' the search for characteristics of quantum dynamics that manifest themselves as chaos in the classical realm [8–11].

Many such characteristics, quantum signatures of chaos, have been detailed in the literature and tested for quantum analogs of classically chaotic systems. These signatures can be divided into two broad categories, static signatures and dynamic signatures. Static signatures look at characteristics of the Hamiltonian or unitary operator governing the system. The conjecture is that the evolution operator of quantum chaotic systems have statistical properties similar to those of random matrices. Hence, quantum analogs of classically chaotic systems show level repulsion, that is, if the energy eigenvalues of the system are ordered, the difference between nearest neighbors would result in a histogram with a Wigner-Dyson distribution [11] and not of a Poisonnian distribution expected for regular, integrable systems. In addition, the squared modulus of the elements of the eigenvectors of quantum chaotic operators follow χ^2_ν distributions [12] from appropriate random matrix ensemble.

Dynamic signatures of quantum chaos look at the evolution of a state under the quantum chaotic operator compared to the same evolution with some perturbed operator. An example of a dynamic signature of chaos is hypersensitivity to perturbation. For chaotic systems (both classical and quantum), the amount of information necessary to track the state of a system when the dynamics is interrupted by an unknown perturbation grows at an exponential rate with increasing time [5]. Other dynamic signatures of quantum chaos look at the entropy growth of chaotic systems versus regular systems [13].

4   Overlap Decay

Overlap decay as a signature of quantum chaos was first introduced by Peres [3, 4, 14] as a quantum analog of initial state sensitivity. Rather than look at two slightly different states and see how they evolve under a certain dynamics, Peres suggested looking at one state, $|\psi_i\rangle$, and see how it evolves under under two slightly different dynamics, an unperturbed Hamiltonian H, and the same Hamiltonian with a small perturbation $H + \delta V$, where δ is the perturbation strength. The overlap at time t is

$$O(t) = |\langle \psi_u(t)|\psi_p(t)\rangle|. \tag{5}$$

where $|\psi_u(t)\rangle = e^{-iHt}|\psi_i\rangle$ and $|\psi_p(t)\rangle = e^{-i(H+\delta V)t}|\psi_i\rangle$ are the unperturbed and perturbed states, respectively. The initial behavior of the overlap shows different behavior depending on whether or not H is chaotic. Recently, the study of overlap decay has seen a revival of interest which has served to identify several regimes of overlap decay behavior based on whether the system is chaotic or regular, the type and strength of perturbation, and the type of initial state. We note that many of the works cited use O^2 as the fidelity. Here, we follow [15] and simply use the overlap, O.

For quantum chaotic systems, quantum versions of classically chaotic maps or random matrix models, there are several regions of behavior based on the strength of the perturbation. The critical perturbation strength is a function of the size of a typical off diagonal element, σ, of the perturbation operator when the perturbation operator is written in the ordered eigenbasis of the unperturbed dynamics, H. If σ is less than the average level spacing of the unperturbed system, Δ, the perturbation is said to be weak. If $\sigma > \Delta$, the perturbation is said to be in the Fermi Golden Rule (FGR) regime [16, 17]. The average level spacing is equal to $2\pi/N$ where N is the dimension of the system Hilbert space. A typical off diagonal element of the perturbation operator in the ordered eigenbasis of the unperturbed system (where the unperturbed system is assumed to have eigenvector statistics of a random matrix see [18]) is equal to $\sqrt{\delta^2 \overline{V_{mn}^2}}$, where $\overline{V_{mn}^2}$ is the second moment of the matrix elements of the perturbation Hamiltonian.

For short enough time, the overlap decay for any perturbation strength is quadratic [3]. After this time, weak perturbations lead to a Gaussian overlap decay as expected from first order perturbation theory [14]. Perturbations in the FGR regime lead to an exponential decay of overlap whose rate, Γ, increases with perturbation strength. For many systems, Γ increases as the perturbation strength squared [16], however, some systems do not show this exact dependence [19]. For systems with a classical analog, initial coherent states, and classical perturbations, Γ will increase with perturbation strength until the decay rate reaches a value given by the Lyapunov exponent of the analog classical system [16, 20, 21]. This is a wonderful example of a dynamical property of the classical system emerging in quantum mechanics. The increase of Γ with perturbation strength may also saturate at the bandwidth of H [16].

For initial states that are coherent or random states, the Gaussian or exponential behavior continues until the fidelity reaches a saturation point at $\simeq 1/N$. For initial states that are system eigenstates, there is no decay in the weak perturbation regime, and in the FGR regime there is an exponential decay of overlap which saturates at $\simeq 1/\Gamma$ [19, 22].

Regular, non-chaotic, systems have a Gaussian decay in the FGR regime [15] that can be faster then the exponential decay of chaotic states. This result has been explained using correlation functions and may be understood as follows: a perturbation to a chaotic system is quickly spread out to the entire Hilbert space of the system. Repetition of the same perturbation does not lead to a compounded error (the correlation time is short). Regular systems, however, do not have this mechanism to spread out the perturbation. The same perturbation compounds the error (there is a long correlation time) leading to a faster fidelity decay. For some regular systems, a power-law decay has also been observed [23].

Here we study a mixed system, a system whose phase space has both chaotic and regular regimes. Coherent states within the regular regime are practically eigenstates of the system and the overlap of these states oscillates close to unity [4, 15], as shown in Fig. 1. Coherent states in the chaotic regime behave like other chaotic systems, they show exponential overlap decay in the FGR regime and

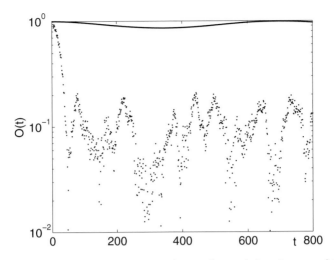

Fig. 1. Overlap decay of coherent states in the regular and chaotic zone of the $J = 480$ QKT with $k = 3$ and $\delta = .005$ (in the FGR regime). The overlap of the coherent state centered at the fixed point of the classical kicked top is practically an eigenstate of the QKT and oscillates close to unity. The overlap of the coherent state in the chaotic zone decays first quadratically and then exponentially

Gaussian overlap decay for weak perturbations. We explore the border between the regular and chaotic regions and show that, in both the FGR and weak perturbation regimes, coherent states near this border have a polynomial overlap decay [6].

5 Nonextensive Entropy

The Boltzmann-Gibbs formulation of statistical mechanics is one of the pillars of modern day physics. However, its applicability, indeed its formulation, is only for systems with short range interactions. Self-gravitating systems and other systems such as those with long-range forces, long-range memories (non-Markovian), or multifractal structures present significant difficulties for the application of the Boltzmann-Gibbs formulation. In 1988, in an attempt to understand the nature of some of these systems, one of us [7] proposed the generalization of Boltzmann-Gibbs statistical mechanics on the basis of a generalized nonextensive entropy, S_q:

$$S_q = k\frac{1-\sum_{i=1}^{W}p_i^q}{q-1} \qquad (6)$$

where k is a positive constant, p_i is the probability of finding the system in microscopic state i, and W is the number of possible microscopic states of the system; q is the entropic index which characterizes the degree of the system nonextensivity. In the limit $q \to 1$ we recover the usual Boltzmann entropy

$$S_1 = -k \sum_{i=1}^{W} p_i \, lnp_i. \tag{7}$$

To demonstrate how q characterizes the degree of nonextensivity of a system, we present the S_q entropy addition rule [24]. If A and B are two independent systems such that the probability $p(A+B) = p(A)p(B)$, the entropy of the total system $S_q(A+B)$ is given by the following:

$$\frac{S_q(A+B)}{k} = \frac{S_q(A)}{k} + \frac{S_q(B)}{k} + (1-q)\frac{S_q(A)S_q(B)}{k^2}. \tag{8}$$

From the above equation one can see that $q < 1$ corresponds to superextensivity, the combined system entropy is more then the sum of the two independent systems; $q > 1$ corresponds to subextensivity, the combined system entropy is less then the two independent systems. Using this entropy to generalize statistical mechanics and thermodynamics has helped to explain many natural phenomena in a wide range of fields.

One application of nonextensive entropy occurs in one dimensional dynamical maps. As explained above, when the Lyapunov exponent of a system is positive, the system dynamics is strongly sensitive to initial conditions and is called chaotic. When the Lyapunov exponent goes to zero it has been conjectured [2] (and proven for the logistic map [25]) that the distance between two initially arbitrarily close points can be described by $\frac{d\xi}{dt} = \lambda_{q_{sen}}\xi^{q_{sen}}$ leading to $\xi = [1 + (1 - q_{sen})\lambda_{q_{sen}}t]^{1/(1-q_{sen})}$ (sen stands for sensitivity). This requires the introduction of $\lambda_{q_{sen}}$ as a generalized Lyapunov coefficient. The Lyapunov coefficient scales inversely with time as a power law instead of the characteristic exponential of a Lyapunov exponent. Thus, there exists a regime, $q_{sen} < 1, \lambda_1 = 0, \lambda_{q_{sen}} > 0$, which is weakly sensitive to initial conditions and is characterized by having power law, instead of exponential, mixing. This regime is called the edge of chaos.

The polynomial overlap decay found for initial states of a mixed system near the chaotic border are at the 'edge of quantum chaos', the border between regular and chaotic quantum dynamics. This region is the quantum analog of the classical region characterized by the generalized Lyapunov coefficient.

6   The Quantum Kicked Top

The system studied in this work is the quantum kicked top (QKT) [26] defined by the operator:

$$U_{QKT} = e^{-i\pi J_y/2\hbar}e^{-i\alpha J_z^2/2J\hbar}. \tag{9}$$

J is the angular momentum of the top and α is the 'kick' strength. The representation is such that J_z is diagonal. The classical version of the kicked top has either regular, mixed, or fully chaotic dynamics depending upon the kick strength. We work with a QKT of $\alpha = 3$ whose classical analog has a mixed phase space, with clearly defined regions of chaotic and regular dynamics. The perturbed operator is simply a QKT with a stronger kick strength α'. Hence, the perturbation operator, $V = \delta \pi J_z^2/2J$ where $\delta \equiv \alpha' - \alpha$.

When J is even, the QKT has three invariant, dynamically independent subspaces [4, 26]. We can write basis functions for the invariant subspaces in terms of the eigenvectors of J_z which, following Peres [4], we will denote as $|m\rangle$. The ee subspace is even under a 180° rotation about x and even under a 180° rotation about y and has a Hilbert space dimension of $N = J/2 + 1$. The basis functions of the ee subspace are $|0\rangle$ and $(|2m\rangle + |-2m\rangle)/\sqrt{2}$, where m ranges from 1 to $J/2$. The oo subspace is even under a 180° rotation about x and odd under a 180° rotation about y and $N = J/2$. Its basis functions are $(|2m-1\rangle - |1-2m\rangle)/\sqrt{2}$. The oe subspace is odd under a 180° rotation about x, $N = J$, and has basis functions $(|2m\rangle - |-2m\rangle)/\sqrt{2}$ and $(|2m-1\rangle + |1-2m\rangle)/\sqrt{2}$. All of the numerical simulations in this work were done in the oo subspace of the QKT. This is to say, we construct the complete QKT and transform it into a basis such that it is block diagonal with three blocks corresponding to the three invariant subspaces. The dimensions of the three blocks are those mentioned above. The columns of the transformation matrix T to transform the QKT operator into block diagonal form are the states which form the bases of the invariant subspaces. We take only the block corresponding to the oo subspace and use that as our map. The initial angular momentum coherent states are also transformed into this basis and, again, only the part of the state corresponding to the oo subspace is used.

The phase space of the classical kicked top is the unit sphere, $x^2 + y^2 + z^2 = 1$ and the resulting action of the map is:

$$
\begin{aligned}
x' &= & z \\
y' &= & x\,sin(\alpha z) + y\,cos(\alpha z) \\
z' &= & -x\,cos(\alpha z) + y\,sin(\alpha z).
\end{aligned}
\tag{10}
$$

For $\alpha = 3$ there are two fixed points of order one at the center of the regular regions. They are located at

$$
x_f = z_f = \pm 0.6294126, \quad y_f = 0.4557187.
\tag{11}
$$

The regular and chaotic regions of the kicked top are clearly seen in the classical maps phase space shown in Fig. 2.

Hence, we expect quantum coherent states centered near the classical periodic point to exhibit significantly different behavior than coherent states centered in the classically chaotic region of the map. This is indeed the case as shown in Fig. 1.

7   Locating the Edge of Quantum Chaos

To locate the edge of quantum chaos we use initial angular momentum coherent states keeping y equal to y_f of the positive fixed point for the classical kicked

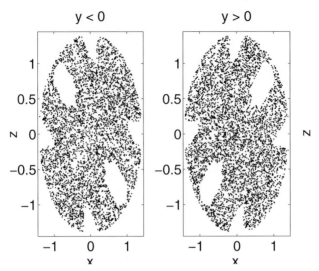

Fig. 2. 10000 points of a chaotic orbit on the phase space of the classical kicked top with kick strength $\alpha = 3$. The regular regions are clearly visible. The spherical phase space is projected onto the $x - z$ plane by multiplying the x and z coordinates of each point by R/r where $R = \sqrt{2(1 - |y|)}$ and $r = \sqrt{(1 - y^2)}$ [4]

top and changing z until a state which has a power-law overlap decay is found. The state which gives this behavior depends on the angular momentum of the QKT, J, but, given a fixed J, the power law emerges for perturbations in both the weak perturbation and FGR regimes. Examples of edge of quantum chaos behavior in both regimes is shown in Fig. 3 for a QKT of $J = 240$. As seen in the figure, the power-law overlap decay is transitory between the quadratic and exponential behavior of the overlap decay. This transitory region does not appear for chaotic states (as shown in Fig. 1) or states close to the fixed point of the map, it is a signature of the 'edge of quantum chaos.'

The overlap decay for the edge of quantum chaos state is very well fit by the solution of the differential equation

$$dO/d(t^2) = -O^{q_{rel}}/\tau^2_{q_{rel}}. \tag{12}$$

In the above equation, *rel* stands for relaxation which is an appropriate description of a $q > 1$ phenomenon. In classical systems, q_{rel} characterizes the relaxation of initial states towards an attractor. Although we do not know how to derive this differential equation from first principles, the numerical agreement is remarkable (see also [27]). A time-dependent q-exponential expression analogous to the one shown here has recently been proved for the edge of chaos and other critical points of the classical logistic map [25].

The values of q_{rel} and $\tau_{q_{rel}}$ depend on the perturbation strength, J, and on whether the perturbation strength is in the weak perturbation or in the FGR regime. For perturbation strengths below the FGR regime, δ is less than some δ_c, q_{rel} remains constant at a value of q^c_{rel}. The transition into the FGR regime

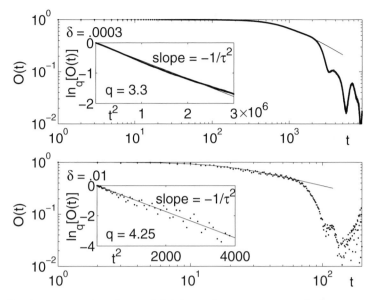

Fig. 3. Overlap versus time for an initial angular momentum coherent state located at the border between regular and chaotic zones of the QKT of spin 240 and $\alpha = 3$. This region, the edge of quantum chaos, shows the expected power law decrease in overlap. The top figure is for a perturbation strength in the weak perturbation regime, $\delta = .0003$ and the bottom figure is for a perturbation strength of $\delta = .01$, within the FGR regime. On the log-log plot the power law decay region, from about 600-2500 in the weak perturbation regime and 20-70 in the FGR regime, is linear. We can fit the decrease in overlap with the expression $[1 + (q_{rel} - 1)(t/\tau_{q_{rel}})^2]^{1/(1-q_{rel})}$ where, in the weak perturbation regime, the entropic index $q_{rel} = 3.3$ and $\tau_{q_{rel}} = 1300$ and in the FGR regime $q_{rel} = 4.25$ and $\tau_{q_{rel}} = 34$. The insets of both figures show $\ln_{q_{rel}} O \equiv (O^{1-q_{rel}} - 1)/(1 - q_{rel})$ versus t^2; since $\ln_q x$ is the inverse function of $e_q^x \equiv [1 + (1-q)\,x]^{\frac{1}{1-q}}$, this produces a straight line with a slope $-1/\tau^2$ (also plotted)

arises when the typical off diagonal elements of V are larger than Δ. We can approximate $\delta_c \simeq \sqrt{2\pi/N^3}$ [16] where, in the oo subspace of the kicked top $N = J/2$. Values for δ_c are shown in the table. In the FGR regime q_{rel} continues to increase with perturbation strength until a saturation perturbation, δ_s, after which $q_{rel} = q_{rel}^s$ remains constant. As the top becomes more classical, with increased J, there is a decrease in q_{rel}^c while in the weak perturbation regime, but an increase in the rate of increase of q_{rel} while in the FGR regime. Because of this larger rate of increased q_{rel} in the FGR regime, q_{rel}^s increases with increasing J. However, we see that this is not the case for the $J = 120$ case. For $J = 120$, q_{rel} increases beyond the expected saturation point. This may be because the $J = 120$ coherent state is so large that, at stronger perturbations, it leaks out of the regular region of the map in more than one place, due to the odd shape of the regular region (see Fig. 6). The value of $\tau_{q_{rel}}$ decreases with perturbation strength and is well fit by a line of slope ≈ -1 (log-log plot); $\tau_{q_{rel}}$ also decreases

Table 1. The edge of quantum chaos, the critical perturbation strength, δ_c, and q^c_{rel}, the q_{rel} for perturbations strengths below δ_c, for explored values of J. As J increases behavior characteristic of the edge of quantum chaos occurs further away from the fixed point of the classical map. The critical perturbation and q^c_{rel} decrease with increased J

J	Edge	δ_c	q^c_{rel}
120	$z_f - .124$	5.39×10^{-3}	3.8
150	$z_f - .139$	3.86×10^{-3}	3.7
180	$z_f - .151$	2.94×10^{-3}	3.6
210	$z_f - .160$	2.33×10^{-3}	3.4
240	$z_f - .176$	1.91×10^{-3}	3.3
280	$z_f - .183$	1.51×10^{-3}	3.1
360	$z_f - .190$	1.04×10^{-3}	2.8
480	$z_f - .194$	6.74×10^{-4}	2.6

with increasing J at fixed perturbation strength. The values of q_{rel} and $\tau_{q_{rel}}$ for a number of different perturbation strengths can be seen in Fig. 4.

The location of states exhibiting edge of quantum chaos behavior is not the same as the edge of chaos for the classical kicked top, due to the finite size of coherent states. Though the coherent state may be centered at a point that is classically regular, the state may 'leak out' into the chaotic region of the map. As J increases the coherent state gets smaller and the quantum edge approaches the classical one. The location of the edge and the size of the coherent state compared to regular regions of the map are shown in Fig. 6.

In conclusion, we have explored the region at the border between chaotic and non-chaotic quantum dynamics, the edge of quantum chaos. Coherent states located at this border exhibit a power-law decrease in overlap as opposed to practically no decay for coherent states near the periodic point of the classical map and exponential overlap decay exhibited by fully chaotic quantum dynamics. This region is the quantum parallel of the classical region at the border between regular and chaotic classical dynamics where the Lyapunov exponent goes to zero and the mixing is characterized by the generalized Lyapunov coefficient. Further studies of this rich system are certainly welcome.

Acknowledgement

One of us (C.T.) acknowledges warm hospitality by H.-T. Elze and the organizers during the interesting meeting.

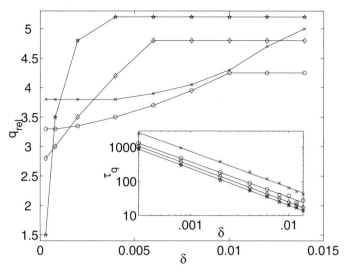

Fig. 4. Values of q_{rel} and τ_q (inset) for $J = 120$ (x), 240 (circles), 360 (diamonds) and 480 (stars). q_{rel} remains constant for perturbation strengths below the critical perturbation and above the saturation perturbation. In between q_{rel} increases with a rate dependent on J. The values of q_{rel}^c, q_{rel}^s, δ_c and δ_s can be seen in the figure. In addition the rate of growth of q_{rel} with increased perturbation strength can be seen. The inset shows a loglog plot of the value of τ_q versus δ for the above values of J. The data can be fit with a lines of slope -1.06, -1.03, -1.07, and -1.08 for $J = 120$, 240, 360 and 480 (top to bottom)

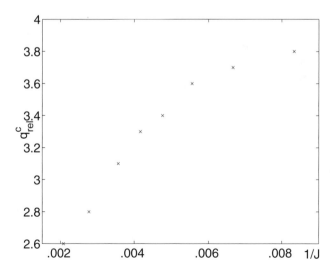

Fig. 5. Values of q_{rel}^c versus $1/J$. These are determined by exploring a number of perturbations much less than δ_c. We note that q_{rel}^c of the $J = 480$ QKT is larger than q_{rel} reported in Fig. 4. It is unclear why in this instance the value of q_{rel} decreases with increased perturbation strength

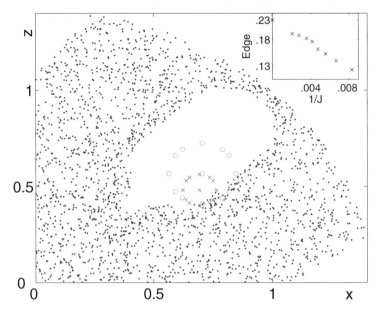

Fig. 6. Coherent states of $J = 120$ (circles) and $J = 480$ (x) at the edge of quantum chaos superimposed on the upper right quarter of the classical phase space (shown in Fig. 2). The spherical phase space and ellipsoidal coherent states are projected onto the $x - z$ plane as described above. The $J = 120$ coherent state is significantly larger then the $J = 480$ coherent state, causing edge of quantum chaos behavior to appear for the $J = 120$ wavefunction much closer to the classical periodic point than edge of chaos behavior for the $J = 480$ wavefunction. The inset is a semilog (y-axis) plot of $1/J$ versus distance from the fixed point of the classical map. Note that the point of appearance of edge of chaos behavior follows an inverse power law with the size of the coherent state

References

1. A.J. Lichtenberg and M.A. Lieberman: *Regular and Chaotic Dynamics* (Springer-Verlag, Berlin 1992)
2. C. Tsallis, A.R. Plastino, and W.M. Zheng: Chaos, Solitons and Fractals **8**, 885 (1997); M.L. Lyra and C. Tsallis: Phys. Rev. Lett. **80**, 53 (1998); V. Latora, M. Baranger, A. Rapisarda, and C. Tsallis: Phys. Lett. A, **273** 97 (2000)
3. A. Peres: in *Quantum Chaos, Quantum Measurement*, ed. by H.A. Cerdeira, R. Ramaswamy, M.C. Gutzwiller, and G. Casati (World Scientific, Singapore 1991)
4. A. Peres: *Quantum Theory: Concepts and Methods* (Kluwer Academic Publishers, 1995)
5. R. Schack and C.M. Caves: Phys. Rev. Lett. **71**, 525 (1993). R. Schack, G.M. D'Ariano, and C.M. Caves: Phys. Rev. E **50**, 972 (1994); R. Schack and C.M. Caves: Phys. Rev. E **53**, 3387 (1996); R. Schack and C.M. Caves, Phys. Rev. E **53**, 3257 (1996)
6. Y.S. Weinstein, S. Lloyd, and C. Tsallis: Phys. Rev. Lett. **89**, 214101 (2002)

7. C. Tsallis: J. Stat. Phys. **52**, 479 (1988). E.M.F. Curado and C. Tsallis: J. Phys. A **24**, L69 (1991) [Corrigenda: **24**, 3187 (1991) and **25**, 1019 (1992)]. C. Tsallis, R.S. Mendes, and A.R. Plastino: Physica A **261**, 534 (1998)
8. M.V. Berry: Proc. R. Soc. London Ser. A **413**, 183 (1987)
9. M.V. Berry: in *New Trends in Nuclear Collective Dynamics*, ed. by Y. Abe, H. Horiuchi and K. Matsuyanagi (Springer-Verlag, Berlin 1992), p. 183
10. F. Haake: *Quantum Signatures of Chaos* (Springer, New York 1991)
11. O. Bohigas, M.J. Giannoni, and C. Schmit: Phys. Rev. Lett. **52**, 1 (1984)
12. F. Haake and K. Zyczkowski: Phys. Rev. A **42**, 1013 (1990)
13. W.H. Zurek and J.P. Paz: Physica D, **83**, 300 (1995)
14. A. Peres: Phys. Rev. A **30**, 1610 (1984)
15. T. Prosen and M. Znidaric: J. Phys. A **35** 1455 (2002)
16. Ph. Jacquod, P.G. Silvestrov, and C.W.J. Beenakker: Phys. Rev. E **64**, 055203 (2001)
17. N.R. Cerruti and S. Tomsovic: Phys. Rev. Lett. **88**, (2002)
18. J. Emerson, Y.S. Weinstein, S. Lloyd, and D.G. Cory: Phys. Rev. Lett. **89**, 284102 (2002)
19. D. Wisniacki: nlin.CD/0208044
20. R.A. Jalabert and H.M. Pastawski: Phys. Rev. Lett. **86**, 2490 (2001); F.M. Cucchietti, C.H. Lewenkopf, E.R. Mucciolo, H.M. Pastawski, and R.O. Vallejos: Phys. Rev. E **65**, 046209 (2002); F. Cucchietti, H.M. Pastawski, and D.A. Wisniacki, Phys. Rev. E **65**, 045206 (2002)
21. G. Benenti and G. Casati: Phys. Rev. E **65**, 066205 (2002)
22. Y. S. Weinstein, J. Emerson, S. Lloyd, and D.G. Cory, Quat. Inf. Proc., to be published
23. Ph. Jacquod, I. Adagideli and C.W.J.Beenakker: Europhys. Lett. **61**, 729 (2003)
24. C. Tsallis: Brazilian Journ. Phys. **29**, 1 (1999). Further reviews can be found in: *Nonextensive Statistical Mechanics and Its Applications*, ed. by S. Abe and Y. Okamoto, *Lecture Notes in Physics* (Springer-Verlag, Berlin 2001); *Non Extensive Thermodynamics and Physical Applications*, ed. by G. Kaniadakis, M. Lissia, and A. Rapisarda, Physica A **305** (Elsevier, Amsterdam 2002); *Nonextensive Entropy – Interdisciplinary Applications*, ed. by M. Gell-Mann and C. Tsallis (Oxford University Press, Oxford 2002), in press. A complete bibliography on the subject can be found at http://tsallis.cat.cbpf.br/biblio.htm
25. F. Baldovin and A. Robledo: Europhys. Lett. **60**, 518 (2002); Phys. Rev. E **66**, R045104 (2002); cond-mat/0304410
26. F. Haake, M. Kus, and R. Scharf: Z. Phys. B, **65**, 381 (1987)
27. E.P. Borges, C. Tsallis, G.F.J. Ananos, and P.M.C. Oliveira: Phys. Rev. Lett. **89**, 254103 (2002)

Index

Lecture Notes in Physics

For information about Vols. 1–592
please contact your bookseller or Springer-Verlag
LNP Online archive: springerlink.com

Druck: Strauss Offsetdruck, Mörlenbach
Verarbeitung: Schäffer, Grünstadt